Light Scattering Reviews 6

Light Scattering and Remote Sensing of Atmosphere and Surface

Alexander A. Kokhanovsky (Editor)

Light Scattering Reviews 6

Light Scattering and Remote Sensing of Atmosphere and Surface

Editor
Dr. Alexander A. Kokhanovsky
Institute of Environmental Physics
University of Bremen
Bremen
Germany

ISBN 978-3-642-15530-7 e-ISBN 978-3-642-15531-4
DOI 10.1007/978-3-642-15531-4
Springer Heidelberg Dordrecht London New York

Cover design: Jim Wilkie
Project copy editor: Mike Shardlow
Author-generated LaTex, processed by EDV-Beratung Herweg, Germany

Printed on acid-free paper

Springer is part of Springer Science+Business Media (www.springer.com)

Contents

List of Contributors

Saswatee Banerjee
IT-related Chemicals Research Laboratory
Sumitomo Chemical Co. Ltd
1-1 Ohe-cho, Niihama City
Ehime 792-0015
Japan
banerjees@sc.sumitomo-chem.co.jp

Brian Barkey
Dept. of Atmospheric and Oceanic Sciences
University of California, Los Angeles
7127 Math Sciences Bldg
405 Hilgard Avenue
Los Angeles, CA 90095
USA
brian_barkey@juno.com

James B. Cole
Department of Computer Science
Graduate School of Systems and
Information Engineering
University of Tsukuba
1-1-1 Tennodai, Ibaraki
Tsukuba 305-8573
Japan
cole@cs.tsukuba.ac.jp

Qiang Fu
Dept of Atmospheric Sciences,
University of Washington, Box 351640
Seattle, WA 98195
USA
qfu@atmos.washington.edu

Yongxiang Hu
NASA Langley Research Center
Mail Stop 420
Hampton, Virginia 23681
USA
yongxiang.hu-1@nasa.gov

Jianping Huang
College of Atmospheric Sciences
Lanzhou University
Lanzhou, Gansu
China 730000
hjp@lzu.edu.cn

Alexander A. Kokhanovsky
Institute of Environmental Physics
University of Bremen,
O. Hahn Allee 1
D-28334 Bremen
Germany
alexk@iup.physik.uni-bremen.de

Bing Lin
NASA Langley Research Center
Mail Stop 420
Hampton, Virginia 23681
USA
bing.lin-1@nasa.gov

Kuo Nan Liou
University of California, Los Angeles
Los Angeles, CA 90095-1565
USA
knliou@atmos.ucla.edu

Zhaoyan Liu
NASA Langley Research Center
Mail Stop 420
Hampton, Virginia 23681
USA
zhaoyan.liu-1@nasa.gov

Naoki Okada
Department of Computer Science
Graduate School of Systems and
Information Engineering,
University of Tsukuba

Room 3F936, Building F, Third area,
1-1-1 Tennodai, Ibaraki
Tsukuba 305-8573
Japan
okada.com@hotmail.co.jp

Suzanne Paulson
Department of Atmospheric and
Oceanic Sciences
University of California, Los Angeles
Los Angeles, CA 90095-1565
USA
paulson@atmos.ucla.edu

Peter Pilewskie
University of Colorado, Laboratory
for Atmospheric and Space Physics
Campus Box 392, Boulder, Colorado,
80309
USA
peter.pilewskie@lasp.colorado.edu

Vladimir V. Rozanov
Institute of Environmental Physics
University of Bremen,
O. Hahn Allee 1
D-28334 Bremen
Germany
rozanov@iup.physik.uni-bremen.de

Sebastian Schmidt
University of Colorado, Laboratory
for Atmospheric and Space Physics
Campus Box 392
Boulder, Colorado, 80309
USA
sebastian.schmidt@lasp.colorado.edu

Jakob J. Stamnes
Department of Physics and Technology
University of Bergen
AllËgt. 55, N-5007 Bergen
Norway
jakobj.stamnes@ift.uib.no

Knut Stamnes
Department of Physics and
Engineering Physics
Stevens Institute of Technology Hoboken
New Jersey 07030
USA
Knut.Stamnes@stevens.edu

Wenbo Sun
Mail Stop 420
NASA Langley Research Center
Hampton, VA 23681
USA
wenbo.sun-1@nasa.gov

Stoyan Tanev
Integrative Innovation Management Group
Institute of Technology and Innovation
University of Southern Denmark
Niels Bohrs Alle 1
DK-5230 Odense M
Denmark
tan@iti.sdu.dk

Gorden Videen
US Army Research Laboratory
2800 Powder Mill Road, Adelphi, MD
20783
USA
gorden.videen@us.army.mil

Evgenij Zubko
Department of Physics
Division of Geophysics and Astronomy
P.O.Box 64
Gustaf Hällströmin katu 2
FI-00014 University of Helsinki
Finland
evgenij.zubko@helsinki.fi

Notes on the contributors

Saswatee Banerjee is currently working as a researcher with Sumitomo Chemical Co. Ltd, Japan. She received her PhD degrees both from Calcutta University, India, and Tsukuba University, Japan, in 2001 and 2005 respectively. Her research interests are in design and simulation of engineering optics. She is the author of over 13 peer-reviewed papers and has co-authored three book chapters.

Brian Barkey holds the position of assistant researcher at Department of Atmospheric and Oceanic Sciences, University of California, Los Angeles. He received his PhD in 1996 at the University of Utah under Prof. K. N. Liou. His thesis was on the measurement of light scattering from ice particle analogs. He has authored and co-authored several papers on experimental measurements of single and multiple scattering from water drops, ice crystals and aerosols and has developed several instruments and a laboratory cloud chamber to aid in his work. At present, he is studying the multiple scattering properties of laboratory-developed snow crystals and the absorption properties of coated back carbon aerosols.

James B. Cole graduated from the University of Maryland (USA) with a PhD in particle physics. During post-doctorate research at the NASA Goddard Space Flight Center be began his career in numerical simulations (cosmic ray antiproton flux). Later he worked on stochastic simulations at the Army Research Laboratory, and visited the NTT Basic Research Laboratory (Japan) for one year. As a research physicist at the Naval Research Laboratory, working on the Connection Machine, he developed the earliest nonstandard finite difference (NS-FD) models for acoustic simulations. After joining the faculty of the University of Tsukuba (Japan), where is a professor, he extended NS-FD models to computational electromagnetics and optics.

Qiang Fu received his BSc in 1983 and MSc in 1985 from Peking University. He received his PhD in 1991 from University of Utah. Dr Qiang Fu has been a full professor at the Department of Atmospheric Sciences, University of Washington since 2006. He has authored/co-authored more than 100 refereed journal papers in the areas of the light scattering and radiative transfer, parameterization of atmospheric radiation and cloud processes, cloud/aerosol/radiation/climate interactions, atmospheric remote sensing, and climate change detection. Dr Qiang Fu is an elected fellow of the American Meteorological Society.

Yongxiang Hu is a senior research scientist in the Science Directorate, NASA Langley Research Center (LaRC), Hampton, Virginia, USA. He graduated with a BSc degree in Meteorology from Peking University in 1985, with a MSc degree from the National Satellite Meteorological Center of China in 1988, and with a PhD degree in atmospheric sciences from University of Alaska in 1994. His expertise is in radiative transfer, lidar remote sensing and climate modeling.

Jianping Huang is Professor and Dean of College of Atmospheric Sciences, Lanzhou University. He got his PhD from Lanzhou University in 1988. Dr Huang's research is focused on arid/semi-arid climate change, especially on the impact of dust aerosol and atmosphere–land surface interaction on semi-arid climate change. He has published 104 journal papers.

Alexander A. Kokhanovsky graduated from the Physical Department of the Belarussian State University, Minsk, Belarus, in 1983. He received his PhD degree in optical sciences from the B. I. Stepanov Institute of Physics, National Academy of Sciences of Belarus, Minsk, Belarus, in 1991. The PhD work was devoted to modeling the light scattering properties of aerosol media and foams. Alexander Kokhanovsky is currently a member of the SCIAMACHY/ENVISAT algorithm development team (Institute of Environmental Physics, University of Bremen). His research interests are directed toward modeling light propagation and scattering in the terrestrial atmosphere. Dr Kokhanovsky is the author of the books *Light Scattering Media Optics: Problems and Solutions* (Chichester: Springer-Praxis, 1999, 2001, 2004), *Polarization Optics of Random Media* (Berlin: Springer-Praxis, 2003), *Cloud Optics* (Dordrecht: Springer, 2006) and *Aerosol Optics* (Berlin: Springer-Praxis, 2008). He has published more than 150 papers in the field of environmental optics, radiative transfer, and light scattering. Dr Kokhanovsky is a member of the American and European Geophysical Unions.

Bing Lin is a senior research scientist in the Science Directorate, NASA Langley Research Center (LaRC), Hampton, Virginia, USA. He received his MSc and PhD in atmospheric sciences from the Columbia University in the city of New York in 1992 and 1995, respectively. He also earned his BEng and MEng in electrical engineering from the University of

Sci. and Tech. of China, Hefei, China and the Institute of Electronics, Chinese Academy of Sciences, Beijing, China ,in 1983 and 1986, respectively. Dr. Lin?s research areas include atmospheric radiation, cloud and climate feedbacks, passive and active remote sensing of atmosphere, land and oceans, radiative transfer theory, the Earth?s water and energy cycle, and the interaction among major climate components.

Kuo-Nan Liou is a Distinguished Professor of Atmospheric Sciences and Director of the Joint Institute for Regional Earth System Science and Engineering at the University of California, Los Angeles (UCLA). He received his BS from the National Taiwan University and his PhD from New York University. Professor Liou has authored and co-authored more than 200 peer-reviewed papers, invited book chapters, and review articles. He is best known for his two monographs, *An Introduction to Atmospheric Radiation* and *Radiation and Cloud Processes in the Atmosphere: Theory, Observation and Modeling*, which provide students and researchers with contemporary knowledge of radiative transfer, its means of acquisition, and the most salient uncertainties and challenges still faced by the field. Dr Liou recently edited a monograph, *Recent Progress in Atmospheric Sciences: Applications to the Asia-Pacific Region* (2008).

Zhaoyan Liu received his BEng degree in laser physics and MEng degree in optics in 1984 and 1987, respectively, from Harbin Institute of Technology (HIT) in China, and his DEng degree in system design in 1996 from Fukui University in Japan. Dr Liu is currently with the National Institute of Aerospace, Hampton, Virginia, USA. His research interests include nonlinear optics, laser and lidar system development, and lidar remote sensing. In recent years, Dr Liu has been primarily involved in two space lidar projects. He participated in the Japanese satellite-borne lidar project ELISE where he was in charge of developing the lidar data processing algorithms, while he developed a ground-based, high-spectral-resolution lidar (HSRL) for validation and algorithm testing. He currently works on the satellite-borne lidar project CALIPSO and studies the generation and long-range transport of airborne mineral dust using the unique vertical-resolved global data measured by the CALIPSO lidar.

Naoki Okada is a PhD student in the Graduate School of Systems and Information Engineering, University of Tsukuba, Japan. His main research interests are in computational optics, electromagnetic simulations, and algorithm development. In particular, he focuses on the use of high-accuracy nonstandard finite-difference time-domain (FDTD) algorithms for light scattering simulation and optical waveguide device design. He is also interested in parallel computation using the graphics processing unit (GPU), and photorealistic rendering of structural colors such as Morpho butterfly. Please visit his web page at http://nsfdtd.org/.

Suzanne Paulson is a professor and vice-chair of the Department of Atmospheric & Oceanic Sciences and professor of the Institute of the Environment and Sustainability at UCLA. Her current research studies the impact of tiny naturally occurring and human-made particles on human health and the Earth's climate. She is the recipient of a National Science Foundation CAREER award. Currently Dr Paulson serves on the Research Screening Committee for the California Air Resources Board. Before joining the UCLA faculty in 1994, she was an Advanced Studies Program Fellow at the National Center for Atmospheric Research. She holds a PhD in Environmental Engineering Science from the California Institute of Technology.

Peter Pilewskie joined the University of Colorado in fall 2004 with a joint appointment in the Department of Atmospheric and Oceanic Sciences and in the Laboratory for Atmospheric and Space Physics. He teaches courses in radiative transfer, remote sensing, and environmental instrumentation. Peter is a co-Investigator for the NASA Solar Radiation and Climate Experiment (SORCE) and the current NASA mission measuring the total and spectral solar irradiance from space, and he is Principal Investigator for the Total and Spectral Solar Irradiance Sensor (TSIS), the future mission to insure the continuity of those same climate data records. His research interests include solar spectral variability and its effects on terrestrial climate; quantifying the Earth-atmosphere radiative energy budget; surface, airborne, and satellite remote sensing of clouds and aerosols; and theoretical atmospheric radiative transfer. Prior to his arrival at the University of Colorado, Peter spent 15 years at the NASA Ames Research Center where his research centered on airborne measurements of atmospheric radiation, cloud and aerosol remote sensing, and analysis of the atmospheric radiative energy budget.

Vladimir V. Rozanov graduated from the University of St Petersburg, Russia, in 1973. He received his Ph.D. degree in physics and mathematics from the University of St Petersburg, Russia in 1977. From 1973 until 1991 he was a research scientist at the Department of Atmospheric Physics of the University of St Petersburg. In 1990–1991 he worked at the Max-Planck Institute of Chemistry, Mainz, Germany. In July 1992 he joined the Institute of Remote Sensing at the University of Bremen, Germany. The main directions of his research are atmospheric radiative transfer and remote sensing of atmospheric parameters (including aerosols, clouds, and trace gases) from space-borne spectrometers and radiometers. He is author and co-author of about 100 papers in peer-reviewed journals.

K. Sebastian Schmidt studied physics in Leipzig, Germany, and Edinburgh, Scotland, and received his PhD in 2005 from the Leibniz Institute for Tropospheric Research at the University of Leipzig, Germany. He works as a research scientist at the Laboratory for Atmospheric and Space Physics at the University of Colorado in Boulder. The foundation of his research lies in airborne measurements of solar radiation collected during numerous cloud and aerosol experiments as well as in 1D and 3D radiative transfer modeling. He is particularly interested in the relationship between cloud-aerosol spatial structure and their spectral signature. He has developed hybrid measurement-model approaches for studying the radiative properties and energy budget of cloud-aerosol layers. Currently, he is working on new spectral retrieval algorithms and on new hardware for airborne irradiance measurements.

Jakob J. Stamnes is professor of physics in the Department of Physics and Technology at the University of Bergen, Norway. He has published a total of more than 160 papers on wave propagation, diffraction, and scattering as well as on remote sensing of the atmosphere and ocean and of biological tissue. Stamnes is author of *Waves in Focal Regions* (Adam Hilger, 1986) and editor of the SPIE Milestone Series on *Electromagnetic Fields in the Focal Region* (SPIE Optical Engineering Press, Bellingham, USA, 2001). He is fellow of the OSA, founding member and fellow of the EOS (European Optical Society), and a member of the SPIE, and was elected member of the Norwegian Academy of Technological Sciences in 2009. Stamnes obtained his MSc degree in applied physics from the Norwegian Technical University, Trondheim, Norway, in 1969 and his PhD degree in optics from the University of Rochester, USA, in 1975.

Knut Stamnes is a professor in the Department of Physics and Engineering Physics, and Director of the Department of Physics and Engineering Physics as well as the Light and Life Laboratory at Stevens Institute of Technology in Hoboken, New Jersey. Stamnes began his career in upper atmospheric physics, and has since specialized in atmospheric radiation, remote sensing, and climate-related studies. He has published more than 175 research papers, and is the co-author of *Radiative Transfer in the Atmosphere and Ocean* (Cambridge University Press, 1999). He is a fellow of OSA, and a member of SPIE, and was elected member of the Norwegian Academy of Technological Sciences in 2009. Stamnes earned his BSc, and MSc in physics from the University of Oslo, Norway, and his PhD in astro-geophysics from the University of Colorado, USA.

Wenbo Sun is a senior research scientist of Science Systems and Applications Inc. (SSAI) in Virginia, United States. He works in NASA Langley Research Center for climate-related satellite remote sensing missions. He gained his BSc degree in atmospheric physics from Peking University in 1988; he got his MSc degree in ocean remote sensing from the State Ocean Administration (SOA) of China in 1991. He obtained his PhD degree in atmospheric sciences from Dalhousie University, Canada, in 2000. Wenbo Sun's research interests involve light scattering by arbitrarily shaped atmospheric particles, radiation transfer in clouds and aerosol layers, and satellite remote sensing of the Earth–atmosphere system.

Stoyan Tanev is an associate professor in the Department of Industrial and Civil Engineering and member of the Integrative Innovation Management (I^2M) Research Unit at the University of Southern Denmark (SDU), Odense, Denmark. Before joining the I^2M unit at SDU in August 2009, Dr Tanev was a faculty member in the Department of Systems and Computer Engineering, Carleton University, Canada. Stoyan Tanev got his MSc and PhD in physics in 1995 from the University of Sofia, Bulgaria and the University Pierre and Marie Curie, France, jointly. He gained his MEng in technology management in 2005 from Carleton University, Canada, and his MA in 2009 from University of Sherbrooke, Canada. His main research interests are in the field of nanobiophotonics modeling, technology innovation management, and value co-creation in technology-driven businesses.

Gorden Videen is a physicist with the Army Research Laboratory and the Space Science Institute. His primary focus recently has been on characterizing irregularly shaped aerosols using light-scattering techniques. Other research topics include analyzing surface structures and remote sensing. He got his BSc degree from the Physics Department of University of Arizona in 1986, his MSc degree from the Optical Sciences Center of University of Arizona in 1991, and his PhD degree from the Optical Sciences Center of University of Arizona in 1992. Publications include over 130 peer-reviewed journal articles, 8 patents, and several book chapters; he has edited 7 books and 6 special journal issues.

Evgenij S. Zubko holds positions of a postdoctoral researcher at Physics Department, University of Helsinki (Finland), and a senior research fellow at Astronomical Institute of Kharkov National University (Ukraine). He defended a PhD thesis in 2003 at the Physics Department of Kharkov National University under the supervision of Prof. Yu. Shkuratov. The thesis was focused on numerical simulation of light scattering by single irregularly shaped particles comparable with wavelength as well as powder-like surfaces consisting of such particles. After the thesis defense, he worked at University of Helsinki (Finland), Hokkaido University (Japan), and Tohoku University (Japan). At present, he studies light scattering by single irregularly shaped particles and performs interpretation of photo-polarimetric observations of small solar system bodies. Evgenij Zubko is co-author of about forty papers published in scientific peer-reviewed journals on light scattering by irregularly shaped particles or random discrete media consisting of such particles.

Preface

Light scattering and radiative transfer is central to a number of diverse scientific disciplines ranging from atmospheric to medical optics. This volume is composed of two parts. In the first part, current problems and methods used in modern studies of single light scattering are considered. The chapter by Barkey et al. is aimed at experimental studies of single light scattering by ice crystals. The work is of great importance for ice cloud remote sensing and also for climate studies because it enhances our knowledge of the single scattering patterns of ice crystals. There are a lot of theoretical results in the area of light scattering by nonspherical particles, but ice clouds also contain irregularly shaped particles and the characterization of corresponding shape and size distributions for realistic cloud scenarios is not a trivial one. Light scattering by small nonspherical particles as studied in the framework of discrete dipole approximation (DDA) is considered by Zubko. Unfortunately, the technique cannot be used for scatterers much larger than the wavelength of light, such as those occurring in ice and dust clouds. On the other hand, the technique is very powerful with respect to modeling of particles of complex shapes, chains, and aggregates. The papers of Sun et al. and Cole et al. address recent advances in the finite-difference time-domain (FDTD) methods. Sun et al. reviews the FDTD technique for modeling of light scattering by arbitrarily shaped dielectric particles and surfaces. The emphasis is on the fundamentals of the FDTD algorithms for particle and surface scattering calculations and the uniaxial perfectly matched layer and the novel scattered-field uniaxial perfectly matched layer absorbing boundary conditions for truncation of the FDTD grid. Both DDA and FDTD are based on the direct solution of Maxwell equations without reference to the wave equation, which is usually used in the treatment of light scattering by some simple shapes such as spheres and spheroids. These approaches are capable of considering the particles of arbitrary shapes. However, the computation speed is low and particles much larger than the wavelength cannot be considered. The chapter of Cole et al. introduces some recent developments of the finite-difference time-domain method and some new applications. Using what is called a nonstandard finite-difference model, the accuracy of the FDTD algorithm can be greatly enhanced without using higher-order finite difference approximations on a coarse numerical grid. This algorithm was checked by computing whispering gallery modes in the Mie regime for infinite dielectric cylinders. To compute light propagation in dispersive materials the FDTD algorithm must be modified. The recursive convolution algorithm is one such modification, but it is computationally intensive and sometimes unstable. The authors introduce stability criteria, and show how to improve the accuracy while decreasing computational cost.

The second part of the book is aimed at the application of radiative transfer theory and respective optical measurements for the characterization of various turbid media. Stamnes et al. considers radiative transfer in coupled systems such as, e.g., the ocean–atmosphere interface. Airborne spectral measurements of shortwave radiation are reviewed by Schmidt and Pilewskie. It is demonstrated how spectral information can be used for cloud and aerosol remote sensing. The final chapter of the book, prepared by Kokhanovsky and Rozanov, is aimed at the determination of snow grain sizes using satellite observations. The asymptotic radiative transfer theory is used for the satellite snow grain sizing. Nowadays, this technique has become standard for the solution of inverse problems of snow optics.

The editor thanks F. Herweg, C. Horwood, and M. Shardlow for the help in production of this volume, and J. Sterritt and C. Witschel for inclusion of Light Scattering Reviews in the general collection of e-books produced by Springer.

This book is dedicated to D. Tanre, a pioneer in the area of aerosol remote sensing from space, on the occasion of his 60th birthday.

Bremen, Germany
January, 2011

Alexander A. Kokhanovsky

Part I

Single Light Scattering

1 Polar nephelometers for light scattering by ice crystals and aerosols: design and measurements

Brian Barkey, Suzanne Paulson and Kuo-Nan Liou

1.1 Introduction

The angular distribution of light scattered from a particle is dependent on its size, shape, composition and on the wavelength and polarization state of the incident light. This information is of tremendous interest to many researchers due to implications in the fields of remote sensing, climatic effects of radiative transfer, and in industrial and scientific laboratories in applications such as aerosol monitoring (Hansen and Lacis 1990; Liou et al. 1999; Mishchenko et al. 1995). The measured angular scattering patterns can lend insights into the chemical and physical properties of the particles (i.e., Pluchino 1987; Swanson et al. 1999; Shaw 1979). The interest in these data is shown by the over 25 polar nephelometers (PN) described in this chapter, that have been built since the 1960s to experimentally examine the angular scattering properties of various small particles. The goals of these measurements are either to verify the increasingly complex theoretical methods of calculating the scattering properties of non-spherical particles, or to use the angular intensities to infer optical, physical or chemical properties of the particles.

The scattering angle (θ) is referenced from the direction of the incident light as shown in Fig. 1.1 with $\theta = 0°$ coincident with the incident light vector and $\theta = 180°$ is the direction towards the light source. The azimuthal angle (ϕ) is defined from an arbitrary axis oriented in the plane perpendicular to the incident light direction and ranges from $0°$ to $360°$. The angular distribution of light intensities from a randomly oriented particle is generally a function of both θ and ϕ.

Theoretically, there are many ways in which to calculate the light scattering properties of particles, or phase functions, including homogeneous spheres (Mie 1908), coated spheres (Aden and Kerker 1951), and other non-spherical shapes (Mishchenko 1991; Takano and Liou 1989; Yang and Liou 1996). Although these models have a strong theoretical basis and are able to capture the scattering properties of very complex morphologies, experimental measurements are necessary because the physical, chemical and optical properties of the scattering particles are largely unknown. For instance, measured and inferred refractive indices for several types of black carbon based aerosols range from 1.1 to 2.75 in the real part and from 0.01 to 1.46 in the imaginary part (Bond et al. 2006; Fuller et al. 1999; Horvath 1993; Seinfeld and Pandis 1998), although the refractive index of pure carbon

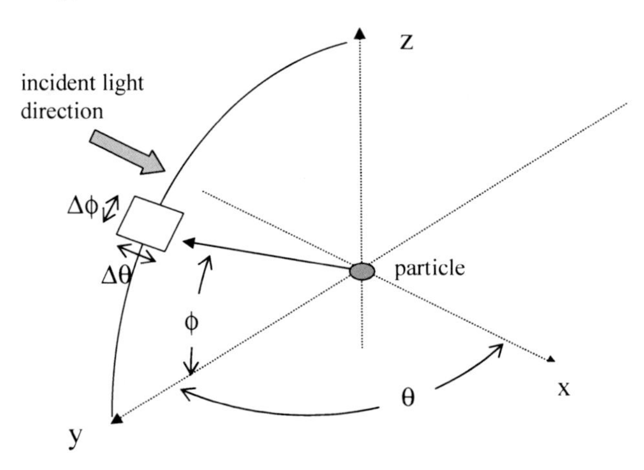

Fig. 1.1. The scattering angle (θ) is referenced from the direction of the incident light. The particle is located at the intersection of the axis. In this figure $\theta = \pi/2$. The azimuthal angle (ϕ) is defined from an arbitrarily defined axis, but is usually referenced to the polarization plane of the incident light if it is linearly polarized.

(2.67 – i1.34) (Borghesi and Guizzetti 1991) is well known. These unknowns can have significant effects on the radiative properties of aerosols. For example, the asymmetry parameter, which is the cosine-weighted integral of the light scattered from a particle, is a key parameter in radiative transfer models. The asymmetry parameter varies by a factor of at least 10 as the number of agglomerated spherules changes from 1 to 200 (Liu and Mishchenko 2005). Cirrus clouds are composed of ice particles with highly complex shapes that vary significantly in space and time which affects their single scattering and bulk radiative properties. The theoretically determined asymmetry parameter can vary from 0.77 to 0.84 at a wavelength of 0.55 μm for simple hexagonal ice particle morphologies and distributions seen in nature (Takano and Liou 1989). However measured values of the asymmetry parameter for low-latitude cirrus clouds range from 0.74 to 0.77 (Garrett et al. 2003). An increase in g of 0.06 was shown to cause a decrease in radiative forcing by at least 12% for non-absorbing particles (Marshall et al. 1995). Accurate and reliable morphologically based scattering information for these particles is necessary as cirrus clouds contribute significantly to the earth's radiative balance (Liou 1986).

In concept, polar nephelometers are simple instruments. A particle is illuminated with a highly collimated beam of light, most often from a laser, and a detector (or detectors) measure the light intensities at the desired angles. But, complexities arise in the detection geometry, sample presentation and in the analysis of the measurements. This chapter presents a review of some of the designs of polar nephelometers, and presents a short summary of the measurements made, including our previous ice particle and aerosol studies. Finally, we describe the current dual polarization polar nephelometer (UCLA PN) developed at UCLA, non-absorbing and absorbing aerosol particle scattering measurements, and a detailed analysis of the refractive index retrieval accuracy.

1.2 Measuring the intensity of scattered light

Measuring the angular dependence of light scattered by a particle is done by placing a detector at the proper angular location with respect to the incident light and determining the intensity of the light that falls onto the detector's sensing area. In practice, however, there are many complications to the experimental setup, i.e.: How many particles are scattering light? What volume do the particles occupy? What is the orientation of the particles? Detectors have finite sensing areas, how wide is it? And so on. In this section, a few of these questions are addressed; however, it should be realized that due to the many possible instrument designs, a complete analysis of all polar nephelometer design considerations cannot be included.

1.2.1 Geometry

The direction of light scattered from a particle is defined by the scattering (or polar) angle (θ) and the azimuthal angle (ϕ) as shown in Fig. 1.1. The azimuthal reference angle or $\phi = 0°$ is arbitrarily defined, however, if the incident light has a preferred polarization orientation, this direction often defines the $\phi = 0°$ direction. θ and ϕ thus defined represent all the directions in the space about the particle. It is impossible to measure the scattering at 0° and 180° as the light scattered at $\theta = 0°$ cannot be separated from the incident light, and any detector positioned at the reverse angle of $\theta = 180°$ would obscure the incident beam. The instruments reviewed in this paper measure the light scattered from particles with relevance to atmospheric radiative transfer, such as aerosols, cloud water drops and cirrus cloud particles. These particles usually have maximum dimensions smaller than 1 to 100 μm and thus are usually randomly oriented when suspended in air. Perrin (1942) used symmetry relationships to show that there is no azimuthal dependence to the light scattering from a small volume of randomly oriented particles when the incident light is not polarized. There are situations in which these symmetry relations do not hold, such as for larger ice particles which have preferred orientations when falling (Klett 1995).

Detectors have a finite field of view defined in Fig. 1.1 as the area inside the box with width $\Delta\theta$ and $\Delta\phi$. Here, the detector's field of view is rectangular, but detection apertures can be circular, slits or other shapes. Unless the scattering is from a single particle, light is scattered from a volume containing several particles which is defined by the either the field of view of the detector (which is often defined by collection optics on the detector) and/or the volume of particles illuminated by the incident light (i.e., a stream of particles that intersect the laser beam). In general, for linearly polarized incident light, the measured scattered intensity, $I(\theta_i)$, is

$$I(\theta_i) = \frac{\lambda^2}{4\pi^2} \int_{\theta_1}^{\theta_2} \int_{\phi_1}^{\phi_2} (I_1(\theta,\phi) + I_2(\theta,\phi))\, k(\theta,\phi)\, \sin\theta \, d\theta \, d\phi \tag{1}$$

where λ is the wavelength of the incident light, $k(\theta,\phi)$, defines the detector response to light scattered into the angle θ and ϕ, and is a function of the sensing geometry, the scattering volume and the detector response properties. $I_1(\theta,\phi)$ and $I_2(\theta,\phi)$ are the light intensities scattered parallel and perpendicular to the reference plane and are dependent on the polarization direction of the incident light, the refractive index (m) of the scattering particles and the number, shape and size of particles in the scattering volume. For randomly oriented or symmetric particles with non-polarized incident light, the azimuthal dependence can be ignored (Van de Hulst 1957; Perrin 1942). For detectors with a narrow angular field of view, detector surface response characteristics introduce very small errors (Jones et al. 1994).

Determining the response characteristics of a polar nephelometer depends on the geometry of each design and usually requires three-dimensional numerical integration across the various scattering and sensing angles. An example is shown in Fig. 1.2(a) which is a top view of the UCLA polar nephelometer showing the sensing geometry of two detector positions. Variables are described in Table 1.1.

Table 1.1. Parameters/descriptions for selected figures.

Figure	Variable	Definition
2(a)	$D_L, -D_L$	Extent of particle volume which is defined by the sample guide tubes.
	a_l	Position of a particle on the line defined by the laser beam.
	R_d	Distance from the center of the scattering plane to the detector.
	r_d	Detector radius
	θ_i	Angular position of the detector at channel i.
	θ	Scattering angle of the photon that origines from the particle at position a_l and that lands on the detector at a position a_d.
Figure	**Variable**	**Description**
4(a)	♦	Experiment: Irregular ice particles, $T = -41°$C, mean maximum dimension = 7.5 μm. Theory: Randomly oriented bullet rosettes and plate crystals calculated using the unified theory of light scattering by ice crystals (Liou et al. 1999)
	□	Experiment: water droplets, mean diameter = 7.5 μm. Theory: Mie–Lorentz, mean diameter = 7.5 μm, standard deviation = 0.2 μm, refractive index = 1.33 – i0.0.
4(b)	▲	Experiment: Growth $T \cong -6°$C. Mean maximum dimension = $36 \pm 0.5\,\mu$m. Theory: based on the observed hollow column and short column habits with rough surfaces on 80% of the particles, calculated using geometric ray-tracing (Takano and Liou 1989).
4(c)	♦	Experiment: Growth $T < -10°$C. Mean maximum dimension = $17 \pm 0.5\,\mu$m. Theory: based on the observed short column and plate habits with rough surfaces on 80% of the particles, calculated using geometric ray-tracing (Takano and Liou 1989).
7	♦, •	Experiment: Ammonium sulfate $(NH_4)_2SO_4$-H_20 droplets. Incident light polarized parallel (♦) and perpendicular (•) to the scattering plane. Theory: Mie–Lorentz, mean diameter = 60 nm. GA determined m = 1.414 – i0.0.

More information on this instrument is given in section 1.4 and Fig. 1.6. In this instrument the particles are confined to the center of the scattering volume via a sample guide tube with a rectangular cross-section in a swath about 19 mm long and 4 mm wide, parallel and coincident with the laser beam. To simplify this anal-

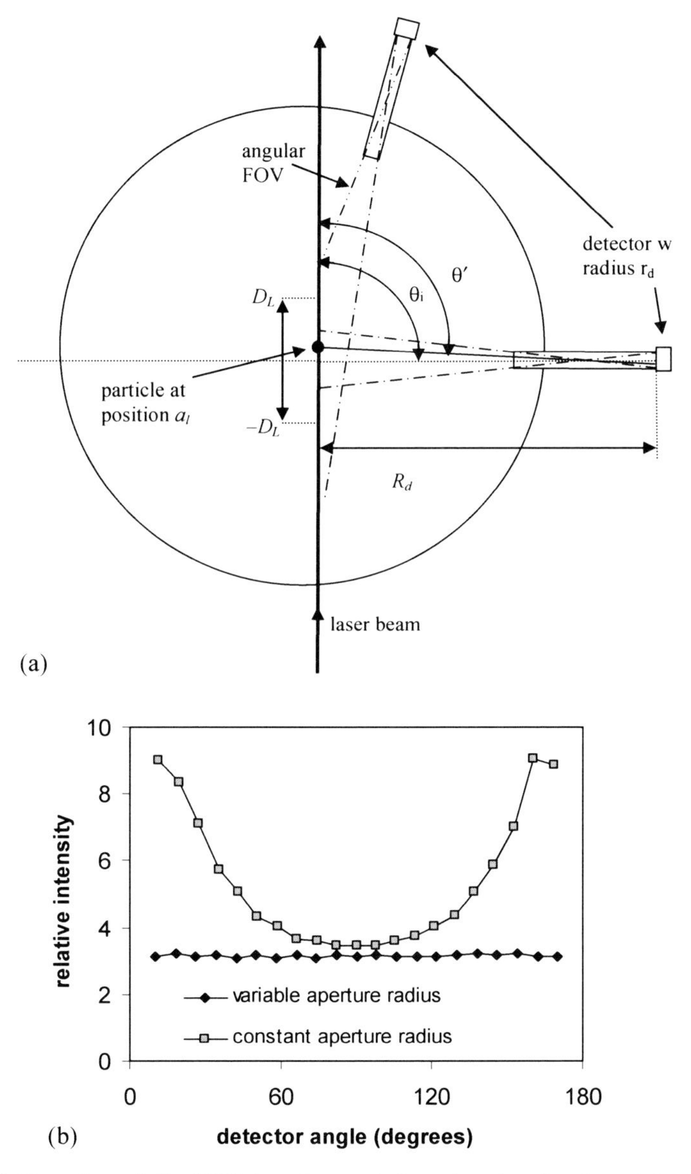

Fig. 1.2. (a) Top view of the UCLA polar nephelometer showing two detector positions to detail dimensions relevant to calculating the scattering response. Particles intersect the incident laser beam in the rectangle at the center of the scattering array. Definitions of the variables are in the text and in Table 1.1. (b) The UCLA polar nephelometer relative response to particles that scatter isotropically with detector apertures that are equal in size and with apertures sizes adjusted to remove the non-uniform response.

ysis, particles are assumed to lie on the laser beam line which is reasonable as the laser beam width is about 1 mm and the azimuthal scattering dependence is ignored as the maximum azimuthal sensor extent is only 3° above and below the scattering plane. Each detector has a finite area, defined by its radius r_d. The scattering angle, θ', for a particle located between $-D_L$ and D_L that scatters a photon that falls at a location between $-r_d$ and $+r_d$ on the detector is not usually the same as the angular position of the detector θ_i as shown in Fig. 1.2. The relative intensity, $I_r(\theta_i)$, of light scattered into each channel i is determined via;

$$I_r(\theta_i) = C_i \int_{a_l=-D_L}^{a_l=D_L} \int_{a_d=-r_d}^{a_d=r_d} I_p(\theta'(a_l, a_d)) \left(\frac{R_d}{R'(a_d, a_l, \theta_i)} \right)^2 D(\theta', a_d, a_l)\, da_d\, da_l \quad (2)$$

where the integral is for all the particle positions along the laser beam from $a_l = +D_L$ to $-D_L$ and from all the scattered photon landing positions on the detector between $a_d = -r_d$ to $+r_d$. Due to inherent differences in detector gain, a calibration constant C_i is applied and discussed in more detail in section 1.4.3.1 and eq. (7). The intensity of the light scattered into the angle θ' for the types and distribution of particles is denoted by $I_p(\theta'(a_l, a_d))$ where the subscript 'p' indicates that the incident light is polarized parallel or perpendicular to the measurement plane. From simple geometry θ' is a function of the position of the particle (a_l) and where its photon lands on the detector (a_d) via;

$$\theta'(a_l, a_d) = \tan^{-1}\left[\frac{a_l \cos\theta_i + R_d}{a_l \sin\theta_i - a_d}\right] \quad (3)$$

where R_d is the detector's distance from the center of the scattering plane. Because the distance between the particle location a_l and where its photon lands on the detector at a_d is not the same as R_d, the scattered intensity is adjusted by the second term in the integral of eq. (2) where;

$$R'(a_d, a_l, \theta_i) = \left[(a_l \sin\theta_i - a_d)^2 + (a_l \cos\theta_i + R_d)^2\right]^{\frac{1}{2}}. \quad (4)$$

$D(\theta',\ a_d,\ a_l)$, which is not explicitly defined here, accounts for the circular shape of the detector aperture as the amount of light getting to the detector is dependent on the height of the hole where the photon enters the aperture.

In this design, the scattering volume is rectangular in order to maximize the number of particles scattering light into the angles near 90° as the intensities at this angle are usually much lower than those in the forward directions. However, because of this, much more light is detected in the forward and reverse directions as shown in Fig. 1.2(b) as the plot marked 'constant aperture radius' in which the relative response for a particle which scatters isotropically is calculated as discussed above with a constant aperture diameter of 4.75 mm. The larger forward response limits the instrument in that the dynamic range of the instrument is reduced, i.e., the intensities at the forward and reverse angles cause the detectors to saturate at relatively low particle concentrations, and also post-processing of the measured signal is required in order to compare measured response to theoretical results. To reduce this problem, the aperture diameters in the newest UCLA nephelometer have diameters which range from about 3 mm in the forward and reverse directions and increase to about 4.7 mm at the side scattering angles. Thus, the response to an isotropic source is more constant as shown by the plot labeled 'variable aperture radius' in Fig. 1.2(b).

1.2.2 Beam considerations

The intensity of the light across a laser beam is not constant, and for Gaussian TEM_{00} (single mode lasers) the radial intensity profile of the laser beam is described by $I(r) = I_{c0} \exp(-2r^2/w^4)$, where $I(r)$ is the intensity of the laser light at the beam radius r, I_{c0} is the laser beam intensity at the center of the laser beam, and w is the laser beam waist, or the laser beam radius at which the laser beam intensity falls to a value $1/e^2$ of the axial value. Colak et al. (1979) has shown that for a sphere if the beam waist is 5 times greater than the particle maximum dimension, then differences between the light scattered by the Gaussian beam and that by a plane wave differ by less than 5%. Small deviations in the collimation of the light source do not affect the scattered intensities greatly unless the beam size is comparable to the size of the particle (Barkey et al. 1999).

The overall sensitivity of the instrument is proportional to the intensity of the incident beam. However, it is possible to 'burn' the particles, which can cause erroneous measurements. The maximum amount of power is a function of the type of particle studied, i.e., particles which absorb very little can tolerate much more incident light, the wavelength of the light, the beam profile and size and the amount of time the particle spends in the beam (Lushnikov and Negin 1993). A convenient upper limit is 210 W/cm^2 as this intensity will cause 150 μm diameter carbon particles, which are the most absorptive particles likely to be encountered, to ignite in 7 milliseconds in a 1 μm wavelength laser beam (Bukatyi et al. 1983).

1.2.3 Stray light

Detectors in polar nephelometers monitor the light scattered into a small fraction of the full 4π solid angle about the sample. Particles however scatter light into the full 4π solid angle and this light needs to be handled. The intensity of light scattered into the near forward angles is orders of magnitude higher than that scattered into the side and reverse directions. If only a small percentage of this light undergoes multiple internal reflections within the working volume of the instrument it can 'leak' to the detectors and effect the measurement. Common methods to reduce stray light includes blackening the interior surfaces of the instrument and designing optics or apertures (discussed above in section 1.2.1), to restrict the detectors' field of view. A beam dump to collect the light exiting the scattering volume is necessary to prevent the incident light from scattering back into the measurement volume. We have found that a properly designed 'Rayleigh' horn beam dump is much more effective than the stacked razor-blade style.

In the multi-detector design of the UCLA polar nephelometer, all surfaces in direct view of the sample volume and the detectors are polished and angled such that scattered light is directed away from the detection plane. The large area photodiode detectors in this instrument are also effective reflective surfaces, hence the detectors are positioned such that each detector is not in the field of view of another detector. For example, for the detector at 10°, a detector is not also placed directly across the scattering plane at the position of 100°.

1.3 Polar nephelometer designs and some measurements

There are many different designs of instruments that measure the angular scattering properties of particles. The discussion here is limited to those instruments that have small angular resolutions, i.e. their detector(s) span only about 1° to 2° of the scattering angles. Instruments that have wider angular detection limits, i.e., greater than 45°, and which approximate the measurement of integrating nephelometers are not considered (Szymanski et al. 2002). There are also instruments that measure the scattering properties in the microwave region (Chýlek et al. 1988; Zerull et al. 1977; Zerull et al. 1980) but these are not considered as this review is limited to measurements in the visible and infrared wavelengths. Although the optical equivalence theorem allows us to apply results from this measurement to smaller wavelengths, microwave experiments face the problem of constructing equivalently sized particles that are truly analogous to the aerosol, dust or ice particles of interest in the visible and infrared regions.

A few instruments isolate a single particle in an electromagnetic field (Bacon and Swanson 2000; Bacon et al. 1998; Pluchino 1987) but most polar nephelometer designs measure the light scattered from a narrow stream of particles intersecting a collimated light beam. As such, they are not strictly single scattering measurements as they measure the light scattered from many particles in a small volume. However, if the average distance between each particle is large compared to the average particle size, and the wavelength and the scattering volume is small compared to the detector-to-sample distance, then single scattering can be assumed (Mishchenko et al. 2004). A few instruments (Castagner and Bigio, 2006, 2007; Schnaiter and Wurm, 2002) isolate the sample in a glass tube; however, interaction of the scattered light with the container material can potentially introduce errors.

Polar nephelometers can be categorized broadly according to their sensing geometry, illustrated in Fig. 1.3. Designs include goniometer instruments (Fig. 1.3(a)), multi-detector devices (Fig. 1.3(b)) and elliptical mirror (Fig. 1.3(c)) devices. This last category describes instruments that use elliptical mirrors to redirect the scattered light to intensity detectors in a manner such that the scattering angle information is preserved.

1.3.1 Goniometer-type polar nephelometers

Shown in Table 1.2 is a compilation of several polar nephelometers that rotate a detector (or detectors) to the desired angular position. These instruments have the advantage of high angular resolution as the detector can be stepped at any desired angular increment. This process does have the drawback of requiring more time to measure all of the desired angles. Thus the sample stream needs to be maintained at a constant concentration for a relatively long period of time. Methods are available to correct for variations in the particle concentration (i.e., Sassen and Liou

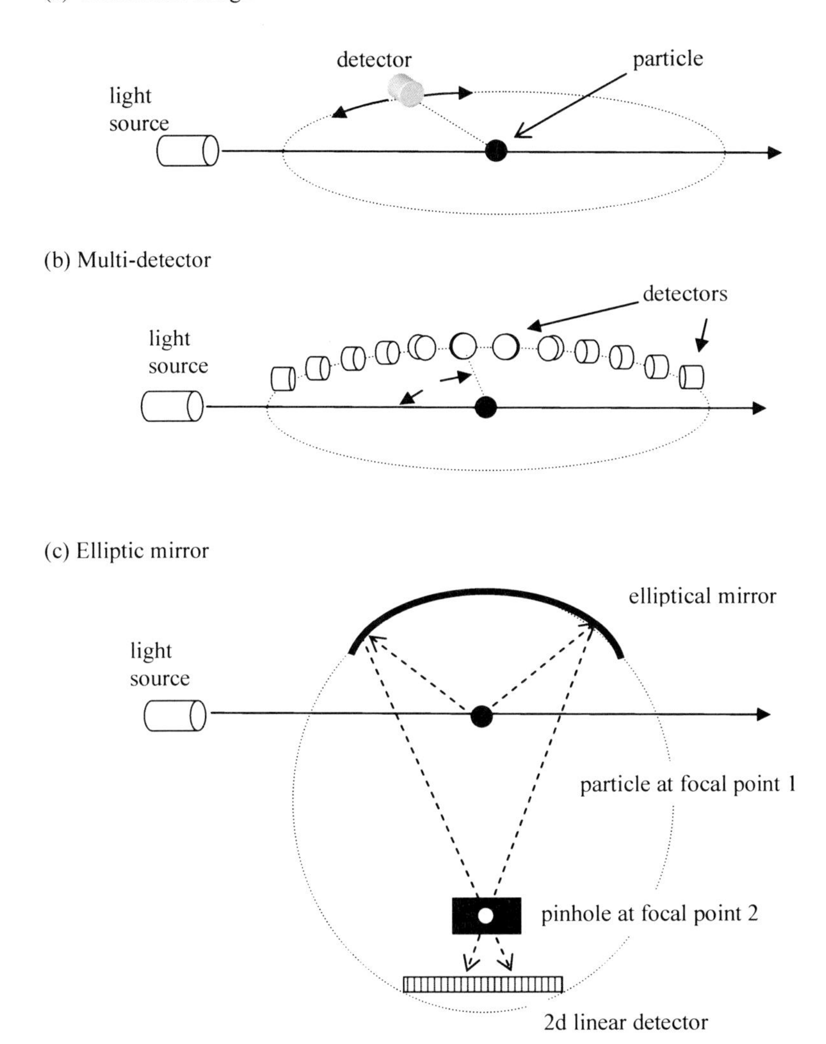

Fig. 1.3. (a) Goniometer type nephelometer geometry. A few detectors (1–3 usually) are mounted on a gimbal device that allows the detector to move to the selected scattering angle. (b) Multi-detector geometry in which several detectors are fixed at discrete angular positions. (c) In the elliptical mirror configuration, an elliptical mirror is used to focus the scattered light onto a linear detection array (or a 3D array in which 2D information is derived). The pinhole prevents light scattered from undesirable angles from getting to the sensor.

1979a) by using a separate fixed-angle detector to monitor sample concentration consistency.

Because there is only one detector, several goniometer instruments have the capability to measure the Mueller matrix. The Mueller matrix describes completely how the polarization properties of the scattered light are affected by the particles and has implications in remote sensing (Ou et al. 2005) and in the inversion of par-

Table 1.2. Compilation of polar nephelometers based on the goniometer design.

Light source	Measured polarization quantity	Angular extent	Particles	Reference
496 nm 552 nm 630 nm	Mueller matrix via rotating compensator	10°–170°	Fogs, light rain	Gorchakov 1966
Various visible	Unpolarized intensity	∼ 5°–175°	Talc powder	Holland and Draper 1967
486 nm 546 nm	Mueller matrix via difference method	18°–166°	Crystalline silica (sand)	Holland and Gagne 1970
475 nm 515 nm 745 nm	I^{**}_{parallel} $I^{**}_{\text{perpendicular}}$	∼5°–175°	Ice crystals	Huffman 1970
632.8 nm 325 nm	Mueller matrix via electro-optical polarization modulator	∼30°–160°	NaCl crystals, Ammonium sulfate spheres	Perry et al. 1978
632.8 nm	Intensity, DLP*	10°–165°	Water, ice crystals, mixed phase	Sassen and Liou 1979a; 1979b
408 nm 450 nm 570 nm 546 nm 578 nm	Mueller matrix elements via difference method	∼5°–180°	Water, ice clouds	Dugin et al. 1971; Dugin and Mirumyants 1976; Dugin et al. 1977
514.5 nm	Mueller matrix via difference method	2°–178°	N_2, ambient aerosols	Hansen and Evans 1980
514.5 nm	Intensity, DLP*	7°–170°	Ambient aerosols	Tanaka et al. 1983
532 nm	Mueller matrix via electro-optical polarization modulator	∼11°–170°	Marine boundary area like aerosols	Quinby-Hunt et al. 1997
632.8 nm	Selected Mueller matrix elements	4°–170°	N2 gas, ambient aerosols	Zhao 1999
632.8 nm 441.6 nm	Mueller matrix via electro-optical polarization modulator	3°–174°	Various mineral dusts	Volten et al. 2001
680 nm	Intensity, DLP*	∼ 30°–155°	PSL with various agglomerations	Schnaiter and Wurm 2002
441.6 nm, 632.8 nm	Mueller matrix via electro-optical polarization modulator	3°–174°	Water, quartz dust	Kuik et al. 1991 Hovenier et al. 2003
532 nm	$I^{**}_{\text{perpendicular}}$	2°–178°	Ambient aerosols	Lienert et al. 2003

* Degree of linear polarization.

** Incident or measured light polarized parallel or perpendicular to the scattering plane.

ticle properties (Zhao 1999; Kuik et al. 1991; Quinby-Hunt et al. 1997). Gorchakov (1966) rotated a mica compensator placed at the exit of the incident light and used tuned amplifiers at the photomultiplier detector to select various harmonics from the scattered light to determine the Mueller matrix elements. The matrix elements can also be measured via the phase-sensitive detection of various components of the scattered light in which the polarization of the incident light is time modulated as described by Hunt and Huffman (1973). These polarization-sensitive instruments have been used for measurements of various mineral aerosols that can be aerosolized by mechanical means (Hovenier et al. 2003) or nebulized in an aqueous solution and then dried (Perry et al. 1978). Mueller matrix elements have been measured for ice crystals (Dugin et al. 1971; Dugin and Mirumyants 1976; Dugin et al. 1977), but using the less accurate and more time-consuming difference method (Liou 1975), in which various polarization elements are placed in front of the source and detectors. Measurement errors inherent in this approach are significant as derivation of the Mueller matrix elements requires the determination of small differences between large values. The goniometer instrument developed by (Zhao 1999) was specifically designed to derive only a few of the Mueller matrix elements in order to determine the refractive index of aerosols via inversion of the Mie–Lorenz solution for scattering from a homogeneous sphere.

1.3.2 Multi-detector polar nephelometers

Table 1.3 lists instruments in which several detectors are placed at discrete and fixed angular locations as shown in Fig. 1.3(b). The design by West et al. (1997) uses 6 linear detector arrays that measure the intensity of light scattered into 6 angular swathes defined by focusing optics. Pluchino (1987) isolates a single particle in an electric field and uses fiber-optic light guides to direct scattered light to photodiode detectors. These instruments have simpler designs and the advantage that they can make fast (real-time) measurements. This is particularly desirable for measuring the time evolution of scattering properties, which can provide insights into particle growth or decay properties. The multi-detector designs are also relatively easy to ruggedize and thus suitable for measurements in the field and in more demanding laboratory environmental conditions. Because the response characteristics between each detector are different, calibration is required. Due to the complexity and expense of providing analyzing optics at each channel none of these instruments measure the polarization state of the scattered light. In a nephelometer design by Dick et al. (2007), along with scattering measurements between $\theta = 40°$ to $140°$ there are eight detectors at various azimuthal angles at $\theta = 55°$. These azimuthal detectors ensure that only scattering from spherical particles is recorded as the scattering from spherical particles with unpolarized incident light is not azimuthally dependent.

1.3.3 Elliptical mirror polar nephelometers

Table 1.4 lists several instruments that are based on the measurement of scattered light that is redirected by an ellipsoidal mirror to a detector as shown in Fig. 1.3(c). Elliptic mirrors have two focal points, thus light scattered at one focal point will be

Table 1.3. Compilation of multi-channel polar nephelometers.

Light source	Measured polarization quantity	Angular extent	Particles	Reference
855 nm	Intensity	23.1°–128.3°	Gases, PSL spheres, Ambient aerosols	Leong et al. 1995
1064 nm	Intensity	5°–175°	Evaporating water drops and ice crystals	Pluchino 1987
804 nm	Intensity	3°–169°	Ice crystals	Gayet et al. 1998
488 nm	Intensity	40°–140°	Aerosols	Dick et al. 1998; 2007
670 nm	$I^{**}_{\text{perpendicular}}$	5°–175°	Ice crystals	Barkey and Liou 2001; Barkey et al. 2002
670 nm	I^{**}_{parallel} $I^{**}_{\text{perpendicular}}$	5°–175°	PSL spheres, ammonium sulfate	Barkey et al. 2007
840 nm	Intensity	23°–129°	Freon-12, PSL spheres for refractive index inversion	Jones et al. 1994
470 nm 652 nm 937 nm	Intensity, DLP*	15°–170°	Mineral dusts	West et al. 1997

* Degree of linear polarization.
** Incident or measured light polarized parallel or perpendicular to the scattering plane.

focused onto the other focal point. A linear array detector placed just beyond the second focal point can thus derive the angular scattering information as a direct correlation exists between the scattered angle and the position the scattered light falls on the detector. In the instrument devised by Kaye et al. (1992) and Hirst et al. (1994) the light scattered between about 30° and 141° and all of the azimuthal angles are measured simultaneously by focusing them onto a two-dimensional detection array. The degree of linear polarization is determined by selecting the measured intensities parallel or perpendicular to the polarization plane of the incident light. Castagner and Bigio (2006; 2007) use a clever arrangement of two parabolic mirrors and a rotating mirror to scan across the angular scattering directions. To ensure only light focused by the elliptical mirror falls onto the detector, this design requires a slit (or pinhole) at the focal point of the elliptical mirror. The size of this aperture defines the angular resolution of the measurement. The slit also reduces the amount of scattered light reaching the detector. Depending on the response characteristics of the detector, measured voltages must be integrated for a significant period of time (minutes) in order to obtain a clear signal (Curtis et al. 2007). None of the elliptical mirror instruments developed to date derive the Mueller scat-

Table 1.4. Compilation of polar nephelometers that use elliptical mirrors.

Incident wave-length	Measured polar-ization quantity	Angular extent	Particles	Reference
632.8 nm	$I^{**}_{\text{perpendicular}}$	70°–125°	PSL	Castagner and Bigio 2006; 2007
685 nm	Intensity	10°–160°	Ambient aerosols	Kaller 2004
632.8 nm	Intensity, DLP*	30°–141° all azimuthal angles	PSL, various non-spherical dusts	Kaye et al. 1992; Hirst et al. 1994
550 nm	Intensity, DLP*	19°–175°	PSL, ammonium sulfate, 1uartz dusts	Curtis et al. 2007

* Degree of linear polarization.
** Incident or measured light polarized parallel or perpendicular to the scattering plane.

tering matrix elements. However, it is conceivable that a single analyzing optic can be placed at the secondary focal point to achieve this task.

1.3.4 Calibration

Calibration is necessary to ensure the validity of the measurements of all polar nephelometer designs. In multi-detector polar nephelometer designs there are differences in the response characteristics of each detector resulting from variations inherent in the manufacturing processes. At least four methods have been developed to calibrate scattering responses. Barkey and Liou (2001) and Barkey et al. (2002) used an isotropic light point-source placed at the scattering center to ensure equitable response at all channels. More commonly, spherical scattering particles with known refractive indices and size distributions are used. Non-absorbing polystyrene latex (PSL) microspheres which are available in several mono-disburse sizes have a well characterized refractive index as a function of wavelength. Exact scattering expectations can be developed from the Mie–Lorenz solution as these particles are known to be spherical and homogeneous. A few nephelometers are calibrated using molecular scattering from a gas (Jones et al. 1994; Zhao 1999). Ammonium sulfate and water mixtures can also be aerosolized to form spherical particles and have a well-defined refractive index, which varies somewhat with relative humidity (Tang and Munkelwitz 1991; 1994) and are also used for calibration and verification (Barkey et al. 2007; Curtis et al. 2007).

There is no accepted calibration standard for absorbing particles. Aerosolized and dried nigrosin (or more commonly 'India' ink) is used as a calibration standard by many instruments that measure aerosol absorption (Abo Riziq et al. 2007); however, there is significant variation between the reported refractive indices (Spindler et al. 2007). The uncertainty in these values probably arises from the fact that nigrosin is not manufactured for use as a standard, thus there are slight changes in the formulation from batch to batch. Also there are differences in how the nigrosin is aerosolized and dried.

PSL particles have been infused with various colored dyes, but these are not designed as absorption standards and there are differences in the refractive indices measured by various researchers. Inverting the measured angular scattering properties of black dyed PSL spheres suspended in water, Chae and Lee (1993) found a complex refractive index of $1.569 - i0.0$. Lack et al. (2009) found a refractive index value of $1.60(\pm 0.03) - i0.045(\pm 0.004)$ for aerosolized black dyed spheres using a cavity ring down transmissometer. We have found experimentally (from an unpublished experiment) that the refractive index of 1003 nm in diameter black dyed PSL spheres has a refractive index of $1.73 - i0.11$. The source of the particles used by Chae and Lee (1993) was different than that used by Lack et al. (2009) and by our group. The non-absorbing behavior of the Chae and Lee (1993) particles was attributed to the insensitivity of their scattering apparatus to their particles and the small amount of dye relative to the volume of PSL. We have found that treating the particles as a concentric sphere produced a core refractive index of $1.61 - i0.0054$ and a shell refractive index $= 1.64 - i0.03$. It is believed that the PSL spheres are not homogeneously infused with the dye, which is supported by the similarity between our shell refractive index and that of Lack et al. (2009). Also, Lack et al. (2009) found that the coated-sphere model better explained their results before they requested a special batch of PSL particles that were 'cooked' longer in order to more completely infuse the dye (via personal communication with D. Lack).

1.3.5 Applications

Polar nephelometer measurements are used for a variety of goals. These include experimental verification of existing theoretical methods of calculating single scattering characteristics using particles with well-known properties. They have also been applied to quantify and identify differences between the measured and expected scattering when either the particle morphology or scattering theory is less established. The validity of the Mie–Lorenz scattering solution has been demonstrated repeatedly for spherical and homogeneous particles and provides a method for calibration, as discussed above. In contrast, several studies of non-spherical particles, including mineral dusts, readily confirm that there are differences between the Mie–Lorenz assumption and scattering from non-spherical particles (Curtis et al. 2007; Kuik et al. 1991; Perry et al. 1978; Volten et al. 2001). In this section, we review some of the ice particle and aerosol scattering measurements made by our group.

1.3.5.1 Ice particle measurements

Ice crystals have highly irregular shapes, but the refractive index as a function of wavelength is well known (Warren and Brandt 2008). Predicting the angular scattering properties of hexagonally shaped ice crystals requires computer-intensive algorithms such as ray-tracing (Takano and Liou 1989), finite-difference time-domain (Yang and Liou 1996) or T-matrix methods (Baran et al. 2001). The light scattered from ice particles is very different from that scattered from water droplets. For instance, shown in Fig. 1.4(a) (the plot labeled 'ice measurement') is the 670 nm unpolarized light scattered from irregular ice particles as measured by a 33-channel polar nephelometer (Barkey and Liou 2001). This polar nephelometer used fiber-optic light guides to couple light from the two-dimensional scattering plane to silicon photodiode detectors. Due to the small diameter of the light guides (2 mm) it was possible to concentrate a few detectors near 22° to study the expected halo features from ice crystals. The ice particles are made from water drops generated in an ultrasonic humidifier that are injected into a dry-ice-cooled cold box with a temperature of about −30°C and has a volume of about 0.5 m^3. The resulting particles have an average maximum dimension of about 7.5 μm and non-spherical shapes, but they do not have the clearly defined hexagonal features normally expected for ice particles, as shown in the photomicrograph of the particles replicated on a slide using the vapor method (Takahashi and Fukuta 1988). It is believed that these smaller, ill-defined particle habits arise because all of the water drops produced by an ultrasonic humidifier are homogeneously nucleated immediately by the stainless steel walls of the cloud chamber, which are in direct contact with the dry ice. The rainbow peak, an intensity feature seen in the reverse direction at about 140° is seen for spherical water drop particles as shown in Fig. 1.4(a), the plot labeled 'water measurement'. The water drops are produced from the same ultrasonic humidifier as the ice particle experiment and also have an average diameter of about 7.5 μm. Not only is the rainbow peak absent from the ice particle measurement, but the ice particle scattering is more isotropic. Although the theoretical expectation for the water drop matches the measurement within the 10% measurement error, the level of agreement for the ice particle comparison is much lower. This is most likely due the difficulty in describing the irregular ice particle shape using randomly oriented bullet rosettes with rough surfaces calculated with the unified theory of light scattering by ice crystals (Liou et al. 1999).

The Desert Research Institute (University of Nevada, Reno) ice particle growth column can produce larger ice crystals with more selectable habits. Cooled water droplets from ultrasonic humidifiers are injected at the top of the column and are homogeneously nucleated by a liquid-nitrogen-cooled wire near the middle of the 4 m tall column. These nucleated particles grow with water vapor supplied by the remaining water drops as the saturation vapor pressure of water over ice is lower than that of water vapor over liquid water. The habit of the particles is largely controlled by the temperature in which they grow (Nakaya 1954). Fig. 1.4(b) shows the measured angular scattering intensities for predominately columnar ice particles, while Fig. 1.4(c) shows the same for particles that are mostly plates (Barkey et al. 2002). The columns are produced when the growth temperature is kept at about −6°C and plates are produced when the temperature is less than −10°C. The theoretical expectations shown in Fig. 1.4(b) and 1.4(c) are calculated with the ray-tracing method (Takano and Liou 1989) using particle habit statistics based on coincidently acquired images of the particles taken by a video microscope. The same 33-channel polar nephelometer used for the measurements of Fig. 1.4(a) was used for Figs. 1.4(b) and 1.4(c), but the 0.95 milliwatt unpolarized laser was replaced with a 35 milliwatt 670 nm unit with the polarization plane oriented parallel to the measurement plane. Both the theoretical and experimental results show the 22° intensity peak, which occurs from the refraction of light through the hexagonal faces that have a 60° angle between them, and the 46° peak which arises from re-

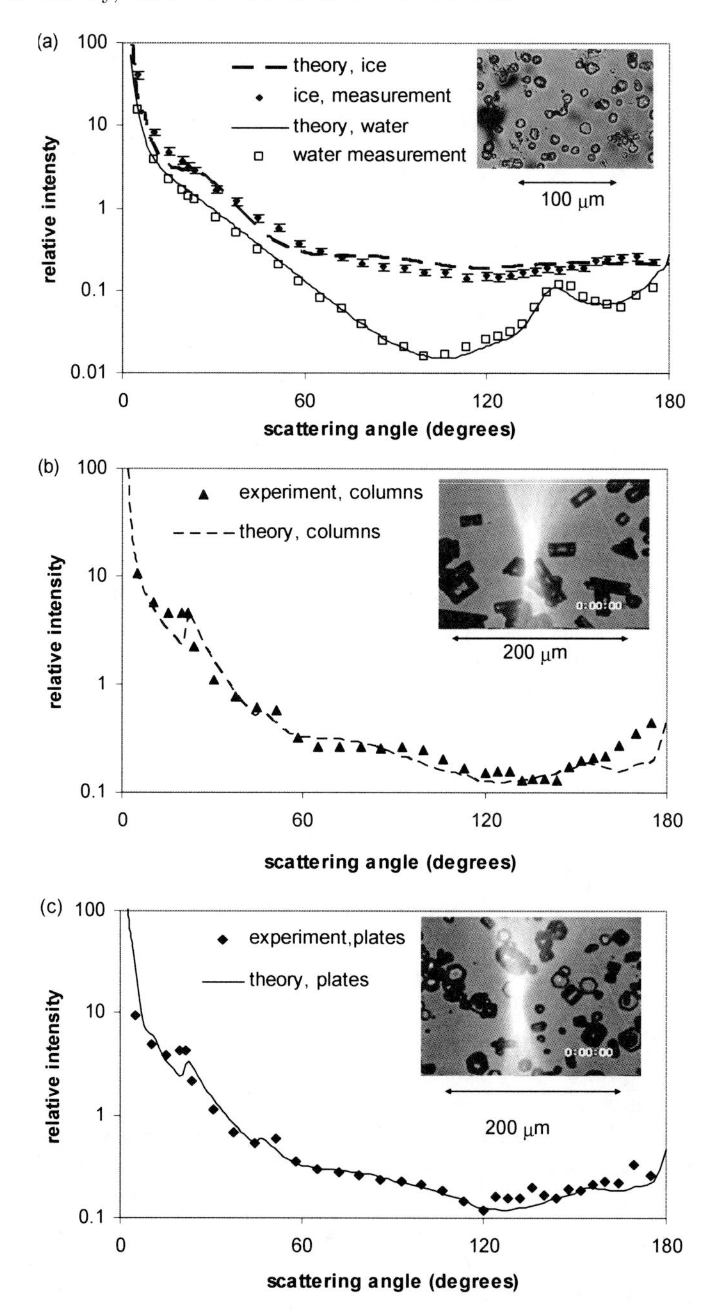

Fig. 1.4. (a) Measured and theoretical angular scattering properties of irregular ice crystals (♦) and water drops (♦) both with an average maximum dimension of about 7.5 μm. The incident 670 nm light is unpolarized and the theoretical expectations are fitted to the measurements using the method of least squares (Barkey and Liou 2001). (b) Measured and theoretical angular intensities with incident light polarized parallel to the measurement plane for ice particles that are predominately columnar in shape as seen in the image (Barkey et al. 2002). (c) is the same as (b) except for ice particles that are more plate like. Experimental and theoretical parameters are listed in Table 1.1.

fraction from end faces of the crystals that have a 90° angle between them. These refractions cause the halo or parhelia (or, more commonly, 'sun dogs') sometimes seen around the sun. Because the 60° faces occur more often than the 90° faces, the 22° intensity features in both Fig. 1.4(b) and 1.4(c) are more prominent than the 46° peak. The 46° peak for the column case (Fig. 1.4(c)) is less intense than that for the plate case because the occurrence of the 90° angle prisms are more prevalent for plates. Additionally, many of the columns have hollow ends as seen in the video image. Without these prisms, the intensity peaks are non-existent as seen for the irregular particle case seen in Fig. 1.4(a). The theoretical expectations, which are based on the observed particle habits, show similar relationships between the relative intensities of the halo features, thus providing direct verification of the complex ray-tracing algorithms on which they are based. Other polar nephelometer studies have also seen these halo intensity features in the laboratory (Sassen and Liou 1979a) and in the field (Gayet et al. 1998).

1.3.5.2 Aerosol scattering and inversion of PN data

Angular scattering information can be used to determine the optical and physical properties of particles. It has been shown theoretically that the determination of the real refractive index and size distribution parameters is possible from angular scattering measurements (Hodgson 2000; Shaw 1979) provided the particles are spherical and homogeneous. Spherical homogeneity allows the use of the Mie–Lorenz scattering solution. Researchers have successfully used several approaches to invert scattering data. These include direct inversion (Zhao 1999; Jones et al. 1994), manual trial and error (Kuik et al. 1991; Quinby-Hunt et al. 1997), table lookup (Verhaege et al. 2008) and optimization methods (Barkey et al. 2007; Lienert et al. 2003). The intensity of light as a function of the scattering angle is very sensitive to the size and composition of a single particle thus providing a means to study the particle's composition, optical properties and growth and evaporation rates (Pluchino 1987; Swanson et al. 1999).

An important application of PN measurements is the determination of the real refractive index of secondary organic aerosols (SOA). SOAs are ubiquitous in nature (Hallquist et al. 2009). While estimates of optical properties are available from organic materials with similarities to some of the components that make up complex SOA (Kanakidou et al. 2005) measurement of real SOA are few and until recently very rough. Shown in Fig. 1.5 are UCLA PN angular scattering measurements from SOA particles generated from ozonolysis of α-pinene in an outdoor solar reaction chamber. The particles grew from about 200 nm at 12:58 pm to about 450 nm in diameter at 14:55 pm. The experiment is discussed in detail in Kim et al. (2010). The retrieved real refractive index and fitness values for each plot are derived as described in section 1.4.2 below. The UCLA PN monitors scattering with incident light polarized both perpendicular (lower plot of each pair in Fig. 1.5) and parallel (upper plots) to the scattering plane. The earlier measurements are at the lower limit of the UCLA PN response. As the particle sizes and concentrations increase, the signal-to-noise ratio in the measurement increases and the angular scattering measurements become smoother. The 'shape' of the angular intensities change as the particles grow. For example the minimum intensity for the measured

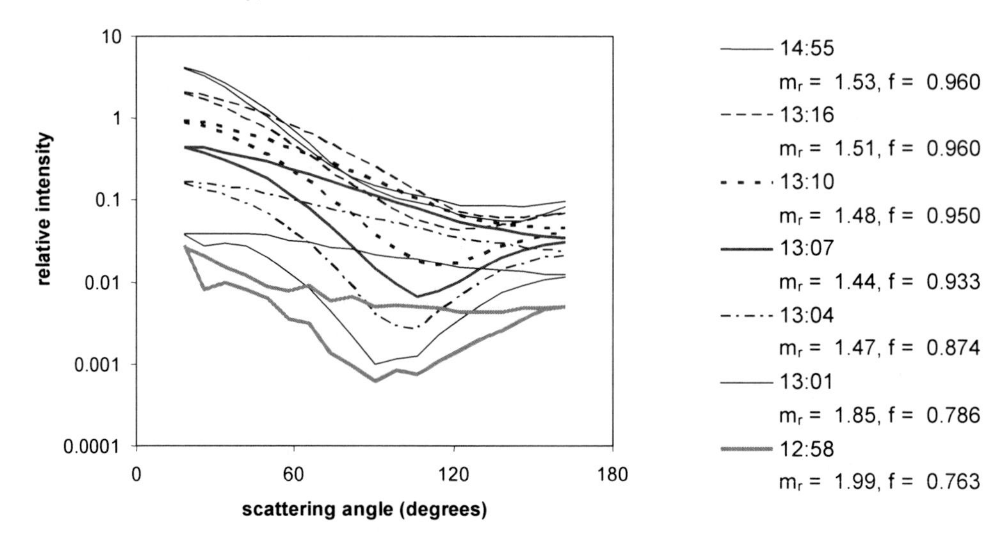

Fig. 1.5. UCLA PN scattering measurements of α-pinene-based SOA in which the particles grow from a diameter of about 200 nm at 12:58 pm to 450 nm at 14:55 pm. The two lines per time are with incident light polarized parallel (upper line) and perpendicular (lower line) to the measurement plane (Kim et al. 2010).

intensity with parallel incident light moves from about 90° to near 120° as the particle diameter increases. As the signal-to-noise ratio increases, the fitness value increases to near 0.96, indicating more confidence in the retrieved refractive index. A fitness value of '1' is optimal as described in section 1.4.2. This PN measurement is invaluable in not only providing values of the refractive indices for these volatile particles, but because of the fast response time of the measurement provides clues to the growth process of SOA.

1.4 UCLA polar nephelometer

This section describes the polar nephelometer (Fig. 1.6) developed at UCLA for the purpose of measuring the scattering properties of ice particles and aerosols. The first version was used had 33 fiber-optic light guides arranged to collect light scattered from 5° to 175° and was used to measure light scattering properties of ice particles as discussed above in section 1.3.5.1. Later, the instrument was modified to allow for the measurement of aerosol scattering. Enhancements included a sheath flow to better control the placement of the aerosols, installation of a higher power laser (670 nm, 350 milliwatts) and the incorporation of a half-wave plate and mechanical actuator to allow the incident light to be polarized parallel or perpendicular to the scattering plane. This prototype also replaced the fiber-optic light guides with detector aperture holes that more precisely controlled the angular resolution and greatly reduced the amount of stray light resulting from incomplete absorption of multiply reflected light in the scattering volume.

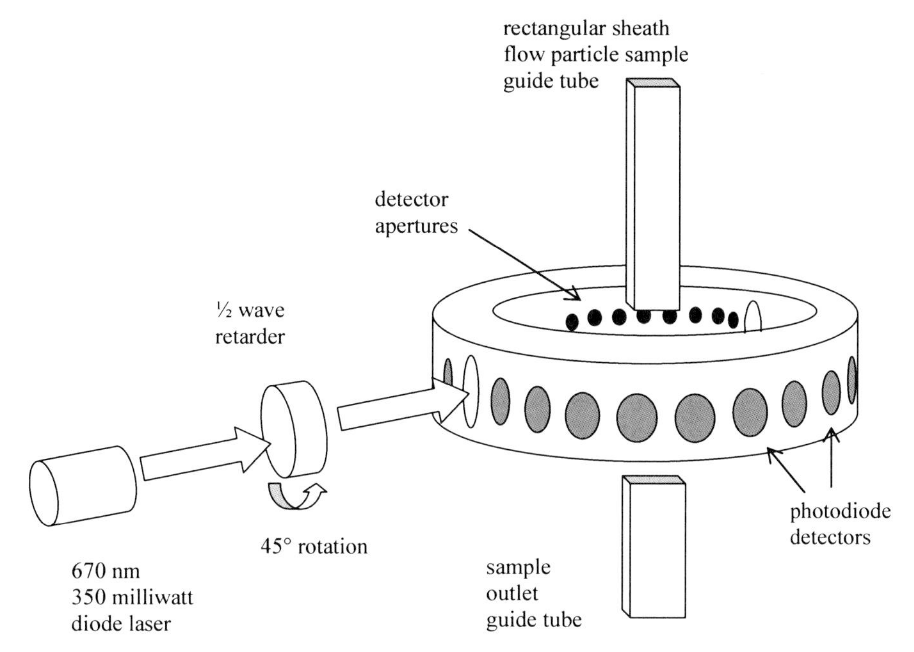

Fig. 1.6. Three-dimensional view of the UCLA polar nephelometer sensing geometry. The 21 photodetectors view the volume where the laser beam intersects the aerosol stream through 40 mm long holes that define the optical field of view. The aerosol stream is confined to the center of the detection array in a sheath flow. A half-wave plate rotates the polarization plane of the incident light to be either parallel or perpendicular to the detection plane.

1.4.1 Instrument description

The current UCLA polar nephelometer is an upgraded version of the unit described in (Barkey et al. 2007). Geometrically, the new polar nephelometer is similar to that unit but with significant improvements to the geometry, sample volume sealing and to the electromagnetic shielding to improve the sensitivity of the instrument. The original prototype was fabricated from plywood, while the new unit is machined to precise specifications from aluminum. The instrument consists of 21 large-area silicon photodiode detectors arranged in a two-dimensional plane (Figs. 1.6 and Fig. 1.2(b)) to measure the light scattered from a stream of particles that intersect a collimated laser beam directed across the plane. The stream of particles which is directed downwards perpendicular to the scattering plane is confined to the center of the array by a sheath flow of filtered air. The aerosol stream has a rectangular cross-section about 3 by 12 mm, oriented so that its long axis is parallel to the laser beam direction. The laser is easily changed. Data shown here was collected with a 350 milliwatt diode laser that directs a 670 nm wavelength beam of light with a profile height of about 3 mm and a width of 1 mm across the scattering volume. An electromechanically operated half-wave plate periodically changes the polarization plane of the incident light to be either parallel or perpendicular to the scattering measurement plane. The detectors are placed behind 40 mm long holes that act as apertures to define the field of view of each detector. The sensing geometry is described above in section 1.2.1. Detector signals are amplified and an embedded computer and data acquisition system sends the measurements to an external PC for storage. The detectors are positioned to sense light scattered from 10° to 170° in 8° increments and have improved electromagnetic shielding such that the detector noise level is at least 10 times lower than the previous version. Single particle scattering as defined by Mishchenko et al. (2004) can be assumed because the detector distance to the sample volume (R_d) is about 75 mm, which is larger than the maximum dimension (12 mm) of the scattering volume. The average distance between each particle for a high particle concentration of $2.5 \times 10^6\,\mathrm{cm}^3$ is about 40 μm, which is much larger than the largest average particle size of about 0.5 μm. Finally $R_d/\lambda \gg 1$, where λ is the wavelength of the laser light.

1.4.2 Method of GA refractive index retrieval

The light scattered from a particle is dependent on the particle size, shape and refractive index. Real refractive indices (m_r) can be retrieved effectively using the genetic algorithm (GA) approach (Goldberg 1989). This optimization method has been shown to be effective despite the complex solution space of this problem and is effective in the presence of significant noise levels (Hodgson 2000). The GA method mimics the way in which biological systems find the 'best' individual. For instance, a 'population' of solutions consisting of real refractive indices and size distribution parameters are randomly selected from within a predefined search space. Then theoretical scattering expectations are determined for each 'member' and adjusted to match the UCLA PN sensing geometry as discussed in section 1.2.1. To date, we have primarily used the Mie–Lorenz method of determining the scattering properties for the spherical homogeneous particle, although any theoretical solution can be used as the GA method is an optimization scheme, and not a direct inversion. The particle size distribution for N particles with diameter d is assumed to be single mode lognormal via;

$$N(d, \mu, \sigma) = \frac{N}{x\sigma\sqrt{2\pi}} \exp\left[-\frac{1}{2}\left(\frac{\ln(d) - \mu}{\sigma}\right)^2\right] \tag{5}$$

in which $\mu = \sum_1^N N_i \ln(d_i)/N$ and, $\sigma = \left[\sum_1^N N_i(\mu - \ln(d_i))^2/N\right]^{1/2}$ are the mean and standard deviation of the log transformed size distribution data. Experimentally, size distribution data is usually obtained in discrete bins in which N_i is the number of particles in the bin centered about d_i. The parameters of each set are then compared to the measured scattering intensities with a fitness parameter defined via:

$$F_p = 1 - \frac{1}{N_c} \sum_{i=1}^{N_c} \left|\log\left(I_{p,thy}(\theta_i, m_r, \mu, \sigma)\right) - \log\left(I_{p,meas}(\theta_i)\right)\right| \tag{6}$$

for each scattering channel for $i = 1$ to N_c and the subscript 'p' indicates the polarization state of the incident light. $I_{p,thy}(\theta_i, m_r, \mu, \sigma)$ is the expected intensity value at the angular position θ_i as calculated via eq. (2) using the GA parameters m_r, μ and σ, and the Mie–Lorentz solution for scattering from a sphere. $I_{p,meas}(\theta_j)$

is fitted to the theory using the method of least squares. The log factor ensures that the fitness is not biased toward higher intensities (Lienert et al. 2003). The total fitness value is $F_t = (F_l + F_r)/2$, where r corresponds to the incident light polarized perpendicular to the scattering plane and l is for light polarized parallel. The population members with the best fitness values are digitally 'mated' to produce 'offspring' that should fit the solution better. The number of optimal members in the GA search population is determined operationally. For the case in which the sample is assured to be homogeneous and spherical and the size distribution is known, a population level of 40 to 50 is sufficient for the solution of synthetic results (discussed below) converging to the correct result in 3 generations. In practice, doubling the population level for experimental results assures convergence and successful retrieval has been demonstrated using both PSL particles and particles developed from a solution of ammonium sulfate and water. The GA results used in this paper are calculated using the C++ GA library developed by Wall (1996). Fig. 1.7 shows the measured scattering properties of ammonium sulfate water droplets ($(NH_4)_2SO_4$-H_2O). The droplets were generated from a mixture of 0.25% by weight ammonium sulfate in deionized water with a Colison (BGI Inc.) spray nebulizer and then partially dried by passage through a dessicant drier. Also shown are the GA determined angular scattering intensities determined using a population of 100 that ran for 3 generations. The retrieved $m_r = 1.414 \pm 0.03$, is in excellent agreement with the expected value of 1.414–1.413 for the measured relative humidity of about 60% (Tang and Munkelwitz 1991; 1994).

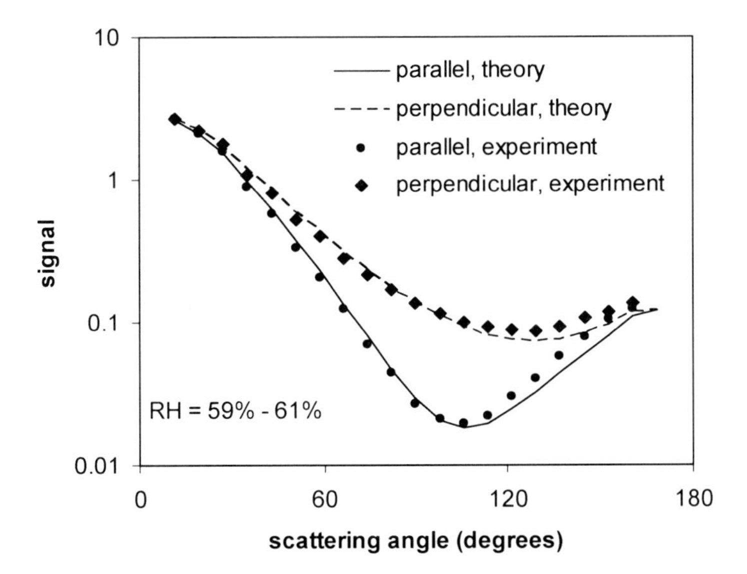

Fig. 1.7. The measured angular relative intensities from droplets consisting of a mixture of ammonium sulfate and water with the incident light polarized parallel (•) and perpendicular (♦) to the measurement scattering plane. The lines indicate the expectation developed from a GA search of these experimental results in which a real refractive index of 1.414 ± 0.03 was found to fit the measurement. The expected value, based on the measured humidity, is 1.413 to 1.414 (Barkey et al. 2007).

1.4.3 Noise/accuracy analysis

Accuracy of the GA retrieved refractive index depends on errors from several sources including electronic noise, calibration errors, due to saturation in one or more of the detector channels or from the GA method itself. Measurements made at the lower limit of the instrument response have a low signal-to-noise ratio which for this instrument occurs when the signal voltage level is less than about 0.0001 Vdc. Also the zero light signal levels for each detector are not the same, which introduces an error that becomes more prominent at lower concentrations. At these low levels the measured intensities can be very different than the expectation despite extensive signal averaging. It has been shown that the GA retrieval of the real refractive index is insensitive to these low-intensity differences up to a maximum of 15% between the expected value and that measured (Barkey et al. 2007). Errors due to the GA retrieval method are also possible. The search space of a particular problem can be insufficiently searched, defined improperly, i.e., too wide or narrow, or the problem may be ill-constrained. Because the GA method is not an inversion scheme, it is possible for the problem to be ill-posed and for the GA program to still return an 'optimal' solution. The viability of the solution is checked by examining the convergence behavior of the GA retrieved values by performing the search several times and examining the %CV (100 $*$ standard deviation/average) of the retrieved values of any parameter of interest. If the solution is tenable, the GA search algorithm will converge to a single solution. For searches of theoretically derived data developed from the Mie–Lorentz scattering solution, the %CV of the m_r of 6 GA searches are less than 1%. These same theoretical GA results with noise synthetically applied can have %CV values from 2–5%. GA searches of scattering measurements of particles that are known to be spherical and homogeneous at the lower limit of the polar nephelometer sensitivity also have %CV values from 2–5%. Operationally it has been determined that %CV values greater than this means either the noise level is too high or the problem is ill-constrained.

Key to the accuracy analysis of the PN GA analysis scheme is the fact that the GA retrieval scheme is highly insensitive to instrument noise (Hodgson 2000). The retrieved result is based on the best fit of all of the data and is thus not unduly affected by erroneous data values. It has been shown that for this instrument the GA retrieval of m_r with detector noise levels of over 15% are accurate to within ± 0.03 (Barkey et al. 2007) as long as the particles are spherical and homogeneous. Thus, in the following analysis, as long as error levels are less than 15%, the m_r retrieval is accurate to within 0.03.

1.4.3.1 Errors due to calibration

Because the response of each detector is slightly different, deriving channel specific calibrations is necessary. Polystyrene latex (PSL) microspheres with well characterized size distributions and a manufacturer specified m_r of 1.5854 at 670 nm are used for this purpose. Shown in Fig. 1.8(a) is a plot of the light scattered by 800 nm in diameter PSL spheres (Duke Scientific Inc., 3800A) as measured by the PN before calibration and the Mie–Lorenz-determined scattering properties based on the manufacturer's specified size distribution parameters and refractive index. The

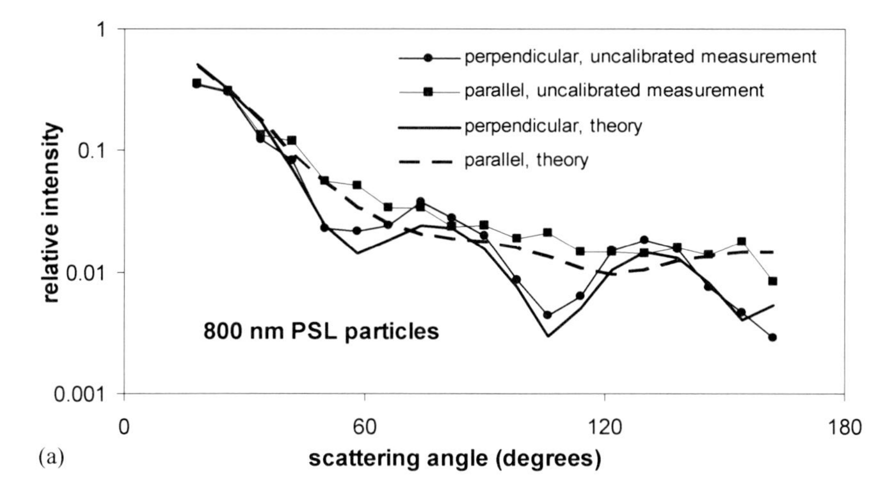

(a)

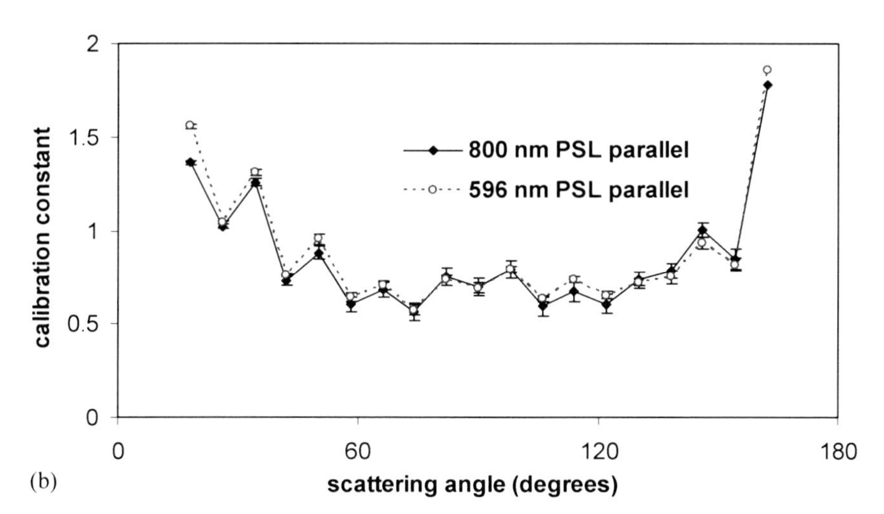

(b)

Fig. 1.8. (a) The measured uncalibrated angular scattering intensities from an almost monodisperse distribution of 800 nm in diameter PSL particles with the incident light polarized parallel and perpendicular to the scattering plane along with expectations using the Mie–Lorenz scattering solution based on the particle distribution and a refractive index of 1.5854 – i0.0. (b) Multiplicative calibration constants developed separately from measurements of the 800 nm PSL particles and similar monodisperse PSL particles with a diameter of 596 nm. The calibration constants are similar showing that the detector response is not a function of the scattering particle size.

PSL particles were aerosolized in a Colison spray nebulizer (BGI Inc.) and then directed through sufficient desiccant tubes to dry them completely. Multiplicative calibration constants are developed from these measurements via:

$$C_i = \frac{I_{thy}(\theta_i, m_r, \mu, \sigma)}{I_{meas}(\theta_i)} \tag{7}$$

where μ and σ are from the manufacturers specifications, $I_{thy}(\theta_i,\ m_r,\ \mu,\ \sigma)$ are as defined in eq. (6) (as discussed in section 1.4.2) and $I_{meas}(\theta_i)$, is the intensity at detector i, which is linearly proportional to the measured voltage. The theoretical results are fitted to the measurement using the method of least squares. The values for the far forward (10°) and far reverse angles (170°) cannot be calibrated at this time due to their highly variable signals levels caused by laser beam drift. Work is under way to correct this problem. Shown in Fig. 1.8(b) are the calibration constants, ranging from about 0.5 to 1.8, developed using measurements of the 800 nm PSL particles (part a of Fig. 1.8), as well as for those developed using similar measurements of PSL particles with a mean diameter of 596 nm (Duke Scientific Inc., 5060A). These constants, C_i, correct the measured voltages to match voltages expected if each detector had identical response characteristics. These calibration constants should be similar to one another as the detector response should not be dependent on the particle size. Any differences in C_i are due to instrument noise or differences in the amount of unwanted signal from stray reflections within the scattering volume. Calibration constants are derived from scattering measurements that are relatively isotropic, i.e., the parallel incident light results of the 800 nm PSL particles shown in Fig. 1.8(a) rather than the highly variable scattering pattern for the perpendicular incident light for the same particle. Signal response at angles with low intensities (i.e., the dips near 60° and 100° in Fig. 1.8(a)) are more susceptible to noise and multiple internal reflections within the scattering volume and thus produce calibration constants that are very different from those derived from more isotropic scattering patterns. The average percent difference between the calibration constants for 596 nm and 800 nm PSL size particles is about 2.5%, with the maximum difference of 8% at 74°. These results indicate that spurious reflections within the scattering volume are at an acceptable level. Also, these error levels are well below 15% thus calibration errors do not contribute to m_r retrieval uncertainties greater than 0.03 as discussed above. PSL particles smaller than about 500 nm, are not used for calibration as they have a tendency to clump into 2 particle dimers which cannot be modeled with the single particle Mie–Lorenz solution.

1.4.3.2 The number of available channels

Although the laser power in the polar nephelometer can be adjusted over a range of about a factor of 10 and the instrument detectors have a dynamic range of about 5 decades, light scattered in the instrument can easily saturates the detectors in the forward and reverse directions. When this occurs, the saturated signal values are nonlinear and cannot be used in the refractive index retrieval. Therefore here we discuss a test of the number of channels required to accurately determine the real refractive index. The goal of this analysis is to determine the validity of the retrieved refractive index under the specific saturation conditions which are sometimes seen in our experiments.

Synthetic scattering intensities were developed using the Mie–Lorentz scattering theory for particles with a refractive index of 1.4 − i0.0, and with a lognormal distribution average radius of 0.13 μm, and a standard deviation of 0.04 μm. These are similar to the particles seen in some of our aerosol experiments. Random noise was applied to the synthetic scattering intensities using the following expression:

$$I'(\theta_i) = I(\theta_i)\left[1 + N_{\max}(1 - 2n_r)\right] \tag{8}$$

where $I(\theta_\iota)$ is the original non-noisy synthetic signal at scattering angle θ_i, $I(\theta_i)$ is the noisy result, $N_{\max}$ is the maximum noise level and n_r is a random number between 0 and 1. Fig. 1.9(a) shows the GA determined m_r vs. the number of missing channels. The channels are eliminated from the test in the manner in which they would saturate, i.e., the far forward channel at 10° is removed first, and then the far reverse channel at 170°, then the next forward channel and so on. As the retrieved value is within 0.015 of the expected m_r, it is seen that for synthetic data, the problem is over constrained and the fitness level during these test remain constant at about 0.99. A fitness value (F_t in eq. (9)) of 1.0 indicates a perfect fit. The high fitness value of 0.99 is expected for synthetic data that do not have calibration errors combined with the insensitivity of the GA method to noise as discussed above.

In Fig. 1.9(b) is a similar analysis of the number of available channels but using experimental data from aerosols formed from the ozonolysis of β-pinene without scavenger (Kim et al. 2010). The aerosols are generated in a $24\,\mathrm{m}^3$ Teflon bag in which gaseous precursors are injected. When ozone was well mixed and its concentration became stable in the chamber, the hydrocarbon (β-pinene) was injected and the chamber mixed manually to minimize inhomogeneities. The GA retrieval of the refractive index for two PN measurements made once the particles had grown to 255 nm in diameter (labeled '255 nm' in Fig. 1.9(b)) at a particle concentration of about $9 \times 10^5\,\mathrm{cm}^{-3}$. Also shown is a similar result for particles formed earlier in the experiment when the particle mean diameter was about 145 nm at a lower concentration of $2 \times 10^5\,\mathrm{cm}^{-3}$. None of the detectors in any of these UCLA PN measurements were saturated. Because of the laser beam drift (above) the forward and reverse channels are not included hence the number of missing channels starts at 2 in Fig. 1.9(b). As in the synthetic case, the retrieved refractive index is relatively insensitive to the number of omitted channels although the retrieved m_r values varied about 0.042 for the 255 nm particles and 0.075 for the 145 nm particles. The range of the retrieved m_r for the synthetic results is 0.015. The retrieved m_r varies more with the number of missing channels for the 145 nm case than either the 255 nm cases or the synthetic cases which we believe is because there is more noise in the retrieved signal for the 145 nm case which had an overall smaller scattering signal. The fitness values for all the experimental cases range from 0.96 to 0.97.

1.4.4 GA retrieval of the imaginary index

In this section GA retrieval of synthetic data is used to determine how well the GA retrieval scheme can determine the real and imaginary (m_i) refractive indices of absorbing particles using data from the UCLA PN. There are many ways to measure the absorption (or imaginary component) of bulk materials (Toon et al, 1976) but aerosols do not come in the pure state of the measured components and are often mixed inhomogeneously with various, often unknown materials. Therefore there is a need for measuring the absorption component of materials in their aerosol state.

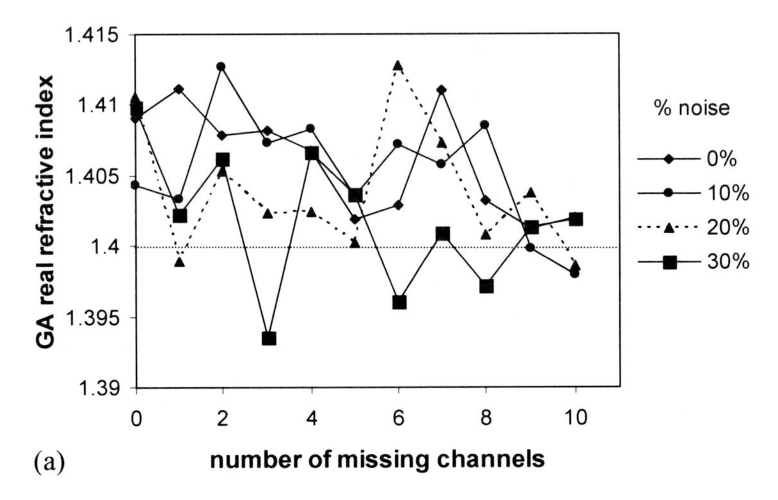

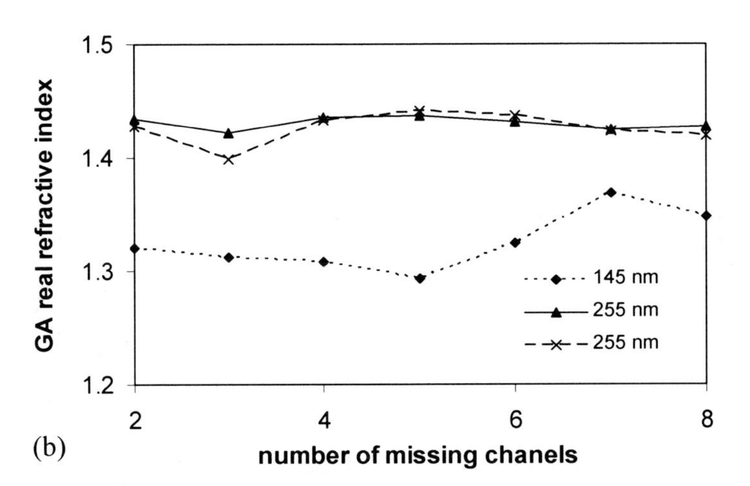

Fig. 1.9. (a) The GA determined real refractive index for Mie–Lorenz synthetic results developed for particles with a refractive index of 1.4 – i0.0. It is seen that the inversion problem is overdetermined as the GA retrieved m_r are within 0.015 of the expected value even when the number of missing channels or data points is increased to 10. (b) The same as (a) but for experimental measurements of α-pinene based secondary organic aerosol at 3 points in its growth cycle where the measured particle mean diameter is indicated.

Cavity ring down technology offers the capability to measure the extinction properties of ambient level aerosols in reasonable path lengths and are well suited for this task (Abo Riziq et al. 2007; Lang-Yona 2010). A previous attempt at retrieval of the imaginary component by (Jones et al. 1994) was unsuccessful, but the instrument in that study had only one incident polarization. (Zhao 1999) however achieved success by the including several Stokes parameters in his measurement; however, the accuracy of the imaginary component was only 50%. A sensitivity analysis of the retrieval capabilities of the imaginary index using another goniometer type dual polarization nephelometer design was shown to be possible for particles with moderate absorption, $10^{-4} \leq m_i \leq 0.5$ (Verhaege et al. 2008). Thus, the possibility of determining the imaginary component of the refractive index is dependent on the instrument configuration and which scattering properties are measured. In order to determine the feasibility of m_i determination with the UCLA PN, a sensitivity test is performed on synthetic data. Finally the refractive index developed from the GA inversion of aerosolized nigrosin is presented.

1.4.4.1 Theoretical examination of the search space

Synthetic data for aerosol particles with a refractive index of 1.7−i0.3 was generated for lognormal size distributions with mean and standard deviation values between 20 and 140 nm in 10 nm steps. The size distribution parameters are based on the sizes common to our experiments. The refractive index was chosen to be similar to that of nigrosin, which is a short-chain polyaniline compound consisting of 8 nitrogen-linked aromatic carbon rings ($C_{48}N_9H_{51}$) (Bond et al. 1999). Nigrosin forms spheres when aerosolized by the Colison nebulizer and dried completely when passed through a dessicant drier. The measured refractive index varies from 1.65 to 1.7 in the real part and from 0.24 to 0.31 in the imaginary component (Abo Riziq et al. 2007; Garvey and Pinnick 1983; Lack et al. 2006). The GA search was performed with a population of 50 and ran for 3 generations. A noise level of 15% is applied to the synthetic data as described by eq. (8). Shown in Fig. 1.10(a) is a plot of the retrieved average m_i of 6 separate GA searches of these synthetic data at each distribution size and standard deviation. The results show that the imaginary component is returned reasonably accurately for most combinations of the mean and standard deviation (within 7% for standard deviations greater than 50 nm), however for narrow distributions, the retrieval becomes consistently inaccurate and can deviate from the expected value by over 200%.

In Fig. 1.10(b) is shown the %CV of the 6 searches, indicating the convergence properties of the GA retrieval as discussed above in section 1.4.3. As in the m_i retrieval, the uncertainty increases as the distribution width narrows. For standard deviations values greater than 50 nm, the average %CV is about 15%, however, the maximum %CV is over 50% at the larger mean sizes in this region. The real refractive index (not shown) is much better retrieved with an average $m_r = 1.70 \pm 0.02$ when the standard deviation is greater than 60 nm. The %CV for the m_r retrievals is about 2.5%. As in the imaginary case, the GA m_r retrievals are slightly more uncertain for smaller standard deviations.

These results are somewhat contrary to the results of Verhaege et al. (2008) who have shown that the retrieval of the real and imaginary component of the refractive index is possible with moderate absorption ($10^{-4} \leq m_i \leq 0.5$) while our results show that the retrieved imaginary component is uncertain by at least 10%, but often by much more. There are several possible reasons for this, our retrieval method and experimental setup is different and we have not analyzed a large space of possible mean sizes, refractive indices and distributions widths, only those that we have seen in our experiments. Also, the resolution of this experimental study (i.e., the 10-nm step size in the mean and standard deviation) is much larger than that in the Verhaege study. There may be a localized minimum in the %CV of GA retrieval that we have missed.

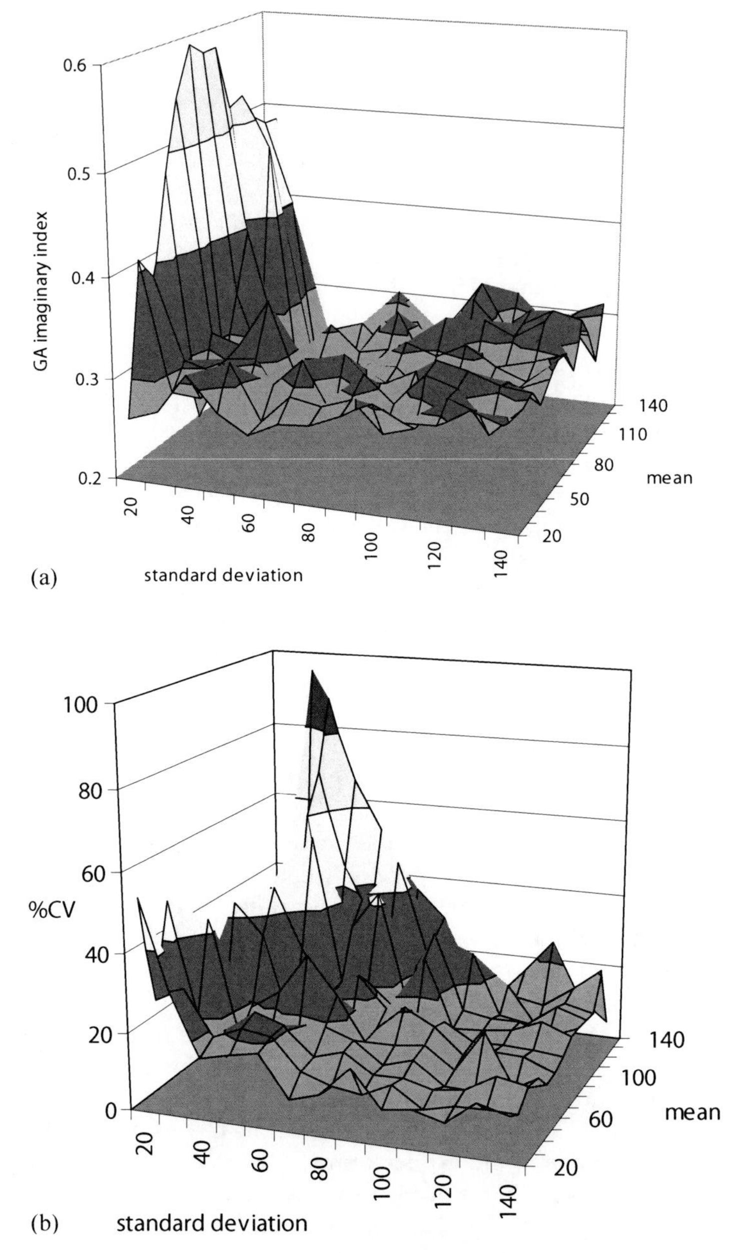

Fig. 1.10. (a) The GA determined imaginary refractive index for Mie–Lorenz-based synthetic results for particles with a refractive index similar to that of nigrosin $(1.7 - i0.3)$ for distribution parameters chosen to be similar to those seen in our experiments. Except for the distributions with standard deviations less than 60 nm, the maximum retrieved m_i is 0.33 and the minimum is 0.23. (b) The %CV of 6 separate searches for the m_i in part (a). As in part (a), the %CV, or uncertainty in the retrieved m_i due to the inversion being ill-posed is much higher for the narrower distributions.

1.4.4.2 Experimental results

Droplets of nigrosin and water from a Colison spray aerosolizer are sent through a dessicant drier and then separately sized by a scanning mobility particle sizer (SMPS, TSI Inc. Model 3080) or sent to the UCLA PN. Various particle distribution mean sizes and widths were generated by varying the concentration of the nigrosin in the solution. GA searches for the real and imaginary refractive indices using a population of 200 for 3 generations was performed for the measured PN angular scattering data. The size distribution search space was set at ±5% of the SMPS mean size and standard deviation as listed in Table 1.5. As expected, in Table 1.5, the real refractive indices were determined more accurately than the imaginary component. In almost all cases the levels of error are much higher than the errors seen in the theoretical study. This may arise from a handful of sources, the synthetic results do not have any noise or effects from multiple internal reflections, the actual refractive index of the nigrosin may be somewhat different from the accepted value due to batch-to-batch variations in the nigrosin, and the fact that it is difficult to be certain the particles are completely dry, and the particle size distributions may be distorted from lognormal. The retrieved refractive index values for the particles with the largest sizes differ more than the smaller sizes from the expected value, and their fitness values are lowest in these tests, which is probably because at these larger sizes, the measured distributions exhibit a dual mode behavior. A study of how distortions in the distribution, i.e., departure from lognormal, have been done for non-absorbing particles (Barkey et al. 2010). It was shown that widening the GA search space on both the measured mean and standard deviation beyond the SMPS accepted error of ±5% for UCLA PN measurements made from aerosols with distorted distributions (within a quantifiable level) can be done to retrieve the real refractive index to within 0.014. However, this analysis has yet to be performed for absorbing particles.

Table 1.5. GA determined nigrosin real and imagined refractive indicies from experimental measurements.

Mean diameter (nm)	Standard deviation	GA m_r	%difference*	GA m_i	%difference**	Fitness
71.0	38.5	1.54	8%	0.45	73%	0.94
47.7	21.9	1.56	7%	0.40	54%	0.96
70.7	38.1	1.63	3%	0.48	86%	0.92
72.9	39.4	1.59	5%	0.52	100%	0.92
134.0	96.0	1.69	1%	0.46	75%	0.96
136.8	126.2	1.43	14%	0.07	74%	0.88
112.0	94.0	1.40	16%	0.15	41%	0.83

* Percent difference between GA determined m_r and the accepted value of 1.67.

** Percent difference between the GA determined m_i and the accepted value of 0.26.

1.4.4.3 Discussion of the retrieval of the imaginary index

The UCLA PN and the GA retrieval method of determining the real refractive index has been shown to be accurate to within ±0.03 for particles which are known to be spherical and homogeneous. Uncertainties arising from errors in the calibration, from instrument noise and from missing, i.e., saturated, channels affect the retrieved refractive indices only slightly because of the insensitivity of the GA retrieval method to noise and because this retrieval problem is over-constrained. Preliminary studies of the effectiveness of the retrieval of the imaginary component of the refractive index is at most accurate to within 10% under optimal, i.e., no noise, conditions. Experimentally, however, errors up to 100% in the retrieved value of the index are seen although additional investigations may yield more accurate retrievals in the future. Within the size limits seen, in our laboratory, the retrieval accuracy of the imaginary refractive index is more dependent on the mean size and standard deviation of the measured particles than is the real part.

Acknowledgments

We would like to thank Hwajin Kim for her extensive help with the secondary organic aerosol experiments. This review included many projects funded by various agencies: US Department of Energy's Atmospheric Science Program (Office of Science, BER, Grant No. DE-FG02-05ER64011); National Science Foundation grant ATM-9907924; US Air Force Office of Scientific Research grant F499620-98-1-0232.

References

Abo Riziq, A., C. Erlick, E. Dinar, and Y. Rudich, 2007: Optical properties of absorbing and non-absorbing aerosols retrieved by cavity ring down (CRD) spectroscopy. *Atmos. Chem. Phys.*, **7**, 1523–1536.

Aden, A. L., and M. Kerker, 1951: Scattering of electromagnetic waves from two concentric spheres. *J. Appl. Phys.*, **22**, 1242–1246.

Bacon, N. J., and B. D. Swanson, 2000: Laboratory measurements of light scattering by single levitated ice crystals, *J. Atmos. Sci.*, **57**, 2094–2104.

Bacon, N. J., B. D. Swanson, M. B. Baker, and E. J. Davis. 1998: Laboratory measurements of light scattering by single ice particles. *J. Aerosol Sci., Proceedings of the 1998 International Aerosol Conference Part 2,* 29, S1317-S1318.

Baran, A. J., P. Yang, and S. Havemann, 2001: Calculation of the single-scattering properties of randomly oriented hexagonal ice columns: A comparison of the T-matrix and the finite-difference time-domain methods. *Appl. Opt.*, **40**, 4376–4386. doi:10.1364/AO.40.004376.

Barkey, B., H. Kim, and S. E. Paulson, 2010: Genetic algorithm retrieval of real refractive Index from aerosol distributions that are not lognormal. *Aerosol Sci. Technol.*, **44**, 1089–1095.

Barkey, B., S. E. Paulson, and A. Chung, 2007: Genetic algorithm inversion of dual polarization polar nephelometer data to determine aerosol refractive index. *Aerosol Sci. Technol.*, **41**, 751–760.

Barkey, B., M. Bailey, K. N. Liou, and J. Hallett, 2002: Light scattering properties of plate and column ice crystals generated in a laboratory cold chamber. *Appl. Opt.*, **41**, 5792–5796.

Barkey, B., and K. N. Liou, 2001: Polar nephelometer for light-scattering measurements of ice crystals. *Opt. Lett.*, **26**, 232–234.

Barkey, B., K. N. Liou, W. Gellerman, and P. Sokolsky, 1999: An analog light scattering experiment of hexagonal icelike particles. Part I: Experimental apparatus and test measurements. *J. Atmos. Sci.*, **56**, 605–612.

Bond, T. C., T. L. Anderson and D. Campbell, 1999: Calibration and intercomparison of filter-based measurements of visible light absorption by aerosols. *Aerosol Sci. Technol.*, **30**, 582–600.

Bond, T. C., G. Habib, and R. W. Bergstrom, 2006: Limitations in the enhancement of visible light absorption due to mixing state. *J. Geophys. Res.*, **111**, D20211. doi:10.1029/2006JD007315.

Borghesi, A., and G. Guizzetti, 1991: *Graphite in Handbook of Optical Constants of Solids II*. San Diego: Academic Press.

Bukatyi, V. I., Y. D. Kopytin and V. A Pogodaev, 1983: Combustion of carbon particles initiated by laser radiation. *Russ. Phy. J.*, **26**, 113–120.

Castagner, J. and I. J Bigio, 2006: Polar nephelometer based on a rotational confocal imaging setup. *Appl. Opt.*, **45**, 2232–2239.

Castagner, J. and I. J Bigio, 2007: Particle sizing with a fast polar nephelometer. *Appl. Opt.*, **46**, 527–532.

Chae, S., and H. S. Lee, 1993: Determination of refractive indices of dyed polymer onospheres. *Aerosol Sci. Technol.* **18**, 403–417. doi:10.1080/02786829308959613.

Chýlek, P., V. Srivastava, R. G. Pinnick, and R. T. Wang, 1988: Scattering of electromagnetic waves by composite spherical particles: experiment and effective medium approximations. *Appl. Opt.*, **27**, 2396–2404.

Colak, S., C. Yeh, and L. W. Casperson, 1979: Scattering of focused beams by tenuous particles. *Appl. Opt.* **18**, 294–302. doi:10.1364/AO.18.000294.

Curtis, D. B., M. Aycibin, M. A. Young, V. H. Grassian, and P. D. Kleiber, 2007: Simultaneous measurement of light-scattering properties and particle size distribution for aerosols: Application to ammonium sulfate and quartz aerosol particles. *Atmos. Environ.* **41**, 4748–4758.

Dick, W. D., P. J. Ziemann, and P. H. McMurry, 1998: Shape and refractive index of submicron atmospheric aerosols from multiangle light scattering measurements. *J. Aerosol Sci.*, **29**, S103–S104. doi:10.1016/S0021-8502(98)00148-7.

Dick, W. D., P. J. Ziemann, and P. H. McMurry, 2007: Multiangle light-scattering measurements of refractive index of submicron atmospheric particles. *Aerosol Sci. Technol.*, **41**, 549. doi:10.1080/02786820701272012.

Dugin, V. P., B. M. Golubitskiy, S. O. Mirumyants, P. I. Paramonov and M. V. Tantashev, 1971: Optical properties of artificial ice clouds, *Bull. (Izv.) Acad. Sci. USSR, Atmospheric and Oceanic Physics,* **7**, 871–877.

Dugin, V. P., and S. O. Mirumyants, 1976: The light scattering matrices of artificial crystalline clouds. *Bull. (Izv.) Acad. Sci. USSR, Atmospheric and Oceanic Physics,* **9**, 988–991.

Dugin, V. P., O. A. Volkovitskiy, S. O. Mirumyants and N. K. Nikiforova, 1977: Anisotropy of light scattering by artificial crystalline clouds. *Bull. (Izv.) Acad. Sci. USSR, Atmospheric and Oceanic Physics*, **13**, 22–25.

Fuller, K. A., W. C. Malm and S. M. Kreidenweiss, 1999: Effects of mixing on extinction by carbonaceous particles. *J. Geophys. Res.*, **104**, 15941–15954.

Garrett, T. J., H. Gerber, D. G. Baumgardner, C. H. Twohy and E. M. Weinstock, 2003: Small, highly reflective ice crystals in low-latitude cirrus, *Geophys. Res. Lett.*, **30**, 2132, doi:10.1029/2003GL018153.

Garvey, D. M., and R. G. Pinnick, 1983: Response characteristics of the particle measuring systems active scattering aerosol spectrometer probe (ASASP–X). *Aerosol Sci. Technol.*, **2**, 477. doi:10.1080/02786828308958651.

Gayet, J.-F., F. Auriol, S. Oshchepkov, F. Schröder, C. Duroure, G. Febvre, J.-F. Fournol, O. Crèpel, P. Personne, and D. Daugereon, 1998: In situ measurements of the scattering phase function of stratocumulus, contrails and cirrus. *Geophys. Res. Lett.*, **25**, 971–974.

Goldberg, D. E., 1989: *Genetic Algorithms in Search, Optimization, and Machine Learning*, 1st edn. Reading, MA: Addison-Wesley.

Gorchakov, G. I., 1966: Light scattering matrices in the atmospheric surface layer. *Bull. (Izv.) Acad. Sci. USSR, Atmospheric and Oceanic Physics,* **2**, 595–605.

Hallquist, M., J. C. Wenger, U. Baltensperger, Y. Rudich, D. Simpson, M. Claeys, J. Dommen, et al., 2009: The formation, properties and impact of secondary organic aerosol: current and emerging issues. *Atmos. Chem. Phys.*, **9**, 5155–5236.

Hansen, J. E. and A. A. Lacis, 1990: Sun and dust versus greenhouse gases: an assessment of their relative roles in global climate change. *Nature,* **346**, 713–719. doi:10.1038/346713a0.

Hansen, M. Z. and W. H. Evans, 1980: Polar nephelometer for atmospheric particulate studies. *Appl. Opt.*, **19**, 3389–3395. doi:10.1364/AO.19.003389.

Hirst, E., P. H Kaye and J. R Guppy, 1994: Light scattering from nonspherical airborne particles: Experimental and theoretical comparisons. *Appl. Opt.*, **33**, 7180.

Hodgson, R. J. W., 2000: Genetic algorithm approach to particle identification by light scattering, *J. Colloid Interface Sci.*, **229**, 399–406.

Holland, A. C., and G. Gagne, 1970: The scattering of polarized light by polydisperse systems of irregular particles. *Appl. Opt.* **9**, 1113–1121. doi:10.1364/AO.9.001113.

Holland, A. C., and J. S. Draper, 1967: Analytical and experimental investigation of light scattering from polydispersions of mie particles. *Appl. Opt.* **6**, 511–518. doi:10.1364/AO.6.000511.

Horvath, H., 1993: Atmospheric light absorption-A review. *Atmos. Environ.*, **27a**, 293–317.

Hovenier, J. W., H. Volten, O. MuÚoz, W. J. van der Zande, and L. B. Waters, 2003: Laboratory studies of scattering matrices for randomly oriented particles: Potentials, problems, and perspectives. *J. Quant. Spectrosc. Radiat. Transfer*, **79–80**. Electromagnetic and Light Scattering by Non-Spherical Particles, 741–755.

Huffman, P., 1970: Polarization of light scattered by ice crystals. *J. Atmos. Sci.*, **27**, 1207–1208.

Hunt, A. J. and D. R. Huffman, 1973. A new polarization-modulated light scattering instrument. *Rev. Sci. Instrum.*, **44**, 1753–1762.

Jones, M. R., K. H. Leong, M. Q. Brewster and B. P. Curry, 1994: Inversion of light-scattering measurements for particle size and optical constants: Experimental study. *Appl. Opt.*, **33**, 4035–4041.

Kaller, W., 2004: A new polar nephelometer for measurement of atmospheric aerosols. *J. Quant. Spectrosc. Radiat. Transfer,* **87**, 107–117.

Kanakidou, M., J. H. Seinfeld, S. N. Pandis, I. Barnes, F. J. Dentener, M. C. Facchini, R. Van Dingenen, et al., 2005: Organic aerosol and global climate modelling: A review. *Atmos. Chem. Phys.* **5**, 1053–1123.

Kaye, P. H., E. Hirst, J. M. Clark, and F. Micheli, 1992: Airborne particle shape and size classification from spatial light scattering profiles. *J. Aerosol Sci.*, **23**, 597–611.

Kim, H., B. Barkey, and S. E. Paulson, 2010. Real refractive indices of α- and β-pinene and toluene secondary organic aerosols generated from ozonolysis and photo-oxidation (in press).

Klett, J. D., 1995: Orientation model for particles in turbulence. *J. Atmos. Sci.*, **52**, 2276–2285.

Kuik, F., P. Stammes and J. W Hovenier, 1991: Experimental determination of scattering matrices of water droplets and quartz particles. *Appl. Opt.*, **30**, 4872–4881.

Lack, D. A., E. R. Lovejoy, T. Baynard, A. Pettersson and A. R. Ravishankara, 2006: Aerosol absorption measurement using photoacoustic spectroscopy: Sensitivity, calibration and uncertainty developments. *Aerosol Sci. Technol.*, **40**, 697. doi:10.1080/02786820600803917.

Lack, D. A., C. D. Cappa, D. Christopher, E. B. Cross, P. Massoli, A. T. Ahern, P. Davidovits and T. B. Onasch, 2009: Absorption enhancement of coated absorbing aerosols: Validation of the photo-acoustic technique for measuring the enhancement, *Aerosol Sci. Technol.*, **43**, 1006–1012.

Lang-Yona, N., Y. Rudich, Th. F. Mentel, A., Bohne, A. Buchholz, A. Kiendler-Scharr, E. Kleist, C. Spindler, R. Tillmann, and J. Wildt, 2010: The chemical and microphysical properties of secondary organic aerosols from Holm Oak emissions, *Atmos. Chem. Phys.*, **10**, 7253–7265, doi:10.5194/acp-10-7253-2010.

Leong, K. H., M. R. Jones, D. J. Holdridge and M. Ivey, 1995: Design and test of a polar nephelometer. *Aerosol Sci. Technol.*, **23**, 341–356.

Lienert, B. R, J. N Porter and S. K Sharma, 2003: Aerosol size distributions from genetic inversion of polar nephelometer data. *J. Atmos. Oceanic Technol.*, **20**, 1403–1410.

Liou, K. N., 1975: Theory of the scattering-phase-matrix determination for ice crystals. *J. Opt. Soc. Am.*, **65**, 159–162.

Liou, K. N., 1986: Influence of cirrus clouds on weather and climate processes: A global perspective. *Mon. Weather Rev.*, **114**, 1167–1199.

Liou, K. N., Y. Takano and P. Yang, 1999: Light scattering and radiative transfer in ice crystal clouds: Applications to climate research. In *Light Scattering by Nonspherical Particles: Theory, Measurements and Applications*, ed. M. I. Mishchenko, J. W. Hovenier and L. D. Travis, 417–449. New York: Academic Press.

Liu, L. and M. Mishchenko, 2005: Effects of aggregation on scattering and radiative properties of soot aerosols. *J. Geophys. Res.*, **110**, D11211. doi:10.1029/2004JD005649.

Lushnikov, A.A. and A.E. Negin, 1993: Aerosols in strong laser beams. *J. Aerosol Sci.*, **24**, 707–735. doi: 10.1016/0021-8502(93)90042-8.

Marshall, S. F., D. S. Covert and R. J. Charlson, 1995: Relationship between asymmetry parameter and hemispheric backscatter ratio: implications for climate forcing by aerosols. *Appl. Opt.*, **34**, 6306–6311. doi:10.1364/AO.34.006306.

Mie, Gustav, 1908. Beiträge zur optik trüber medien, speziell kolloidaler metallösungen. *Annalen der Physik*, **330**, 377–445.

Mishchenko, M. I., 1991: Light scattering by randomly oriented axially symmetric particles. *J. Opt. Soc. Am. A,* **8**, 871–882. doi:10.1364/JOSAA.8.000871.

Mishchenko, M. I., A. A. Lacis, B. E. Carlson and L. D. Travis, 1995: Nonsphericity of dust-like tropospheric aerosols: Implications for aerosol remote sensing and climate modeling. *Geophys. Res. Lett.*, **22**, 1077-1080.

Mishchenko, M. I., J. W. Hovenier and D. W. Mackowski, 2004: Single scattering by a small volume element. *J. Opt. Soc. Am. A,* **21**, 71–87. doi:10.1364/JOSAA.21.000071.

Nakaya, U., 1954. *Snow Crystals.* Harvard Univerity Press.

Ou, S. C., K. N. Liou, Y. Takano and R. L. Slonaker, 2005: Remote sensing of cirrus cloud particle size and optical depth using polarimetric sensor measurements. *J. Atmos. Sci.*, **62**, 4371–4383.

Perrin, F., 1942: Polarization of light scattered by isotropic opalescent media. *J. Chem. Phys.*, **10**, 415. doi:10.1063/1.1723743.

Perry, R. J., A. J. Hunt and D. R. Huffman, 1978. Experimental determinations of Mueller scattering matrices for nonspherical particles. *Appl. Opt.*, **17**, 2700–2710.

Pluchino, A. B., 1987. Scattering photometer for measuring single ice crystals and evaporation and condensation rates of liquid droplets. *J. Opt. Soc. Am. A,* **4**, 614.

Quinby-Hunt, M. S., L. L. Erskine and A. J. Hunt, 1997. Polarized light scattering by aerosols in the marine atmospheric boundary layer. *Appl. Opt.*, **36**, 5168–5184.

Sassen, K. and K. N. Liou, 1979a. Scattering of polarized laser light by water droplet, mixed-phase and ice crystal clouds. Part I: Angular scattering patterns. *J. Atmos. Sci.*, **36**, 838–851.

Sassen, K. and K. N. Liou, 1979b. Scattering of polarized laser light by water droplet, mixed-phase and ice crystal clouds. Part II: Angular depolarizing and multiple-scattering behavior. *J. Atmos. Sci.*, **36**, 852–861.

Schnaiter, M. and G. Wurm, 2002: Experiments on light scattering and extinction by small, micrometer-sized aggregates of spheres. *Appl. Opt.* **41**, 1175–1180.

Seinfeld, J. H. and S. N. Pandis, 1998: *Atmospheric Chemistry and Physics: From Air Pollution to Climate Change.* New York: John Wiley.

Shaw, G. E., 1979. Inversion of optical scattering and spectral extinction measurements to recover aerosol size spectra. *Appl. Opt.*, **18**, 988–993. doi:10.1364/AO.18.000988.

Spindler, C., A. Abo Riziq and Y. Rudich, 2007: Retrieval of aerosol complex refractive index by combining cavity ring down aerosol spectrometer measurements with full size distribution information. *Aerosol Sci. Technol.* **41**, 1011–1017.

Swanson, B. D., N. J. Bacon, E. J. Davis and M. B. Baker, 1999: Electrodynamic trapping and manipulation of ice crystals. *Q. J. R. Meteorolog. Soc.*, **125**, 1039–1058.

Szymanski, W. W., A. Nagy, A. Czitrovszky and P. Pani, 2002: A new method for the simultaneous measurement of aerosol particle size, complex refractive index and particle density. *Meas. Sci. Technol.* **13**, 303–307.

Takahashi, T., and N. Fukuta, 1988: Ice crystal replication with common plastic solutions. *J. Atmos. Oceanic Technol.*, **5**, 129–135.

Takano, Y., and K. N. Liou, 1989: Solar radiative transfer in cirrus clouds. Part I: Single-scattering and optical properties of hexagonal ice crystals. *J. Atmos. Sci.*, **46**, 3–19.

Tanaka, M., T. Takamura and T. Nakajima, 1983: Refractive index and size distribution of aerosols as estimated from light scattering measurements. *J. Climate Appl. Meteor.*, **22**, 1253–1261.

Tang, I. N., and H. R. Munkelwitz, 1991: Simultaneous determination of refractive index and density of an evaporating aqueous solution droplet. *Aerosol Sci. Technol.*, **15**, 201. doi:10.1080/02786829108959527.

Tang, I. N. and H. R. Munkelwitz, 1994: Water activities, densities, and refractive indices of aqueous sulfates and sodium nitrate droplets of atmospheric importance. *J. Geophys. Res.*, **99**, 18801–18808.

Toon, O., J. Pollack and B Khare. 1976: The optical constants of several atmospheric aerosol species: Ammonium sulfate, aluminum oxide, and sodium chloride. *J. Geophys. Res.*, **81**, 5733–5748.

Van de Hulst, H. C., 1957: *Light Scattering by Small Particles.* New York: John Wiley.

Verhaege, C., V. Shcherbakov and P. Personne, 2008: Limitations on retrieval of complex refractive index of spherical particles from scattering measurements. *J. Quant. Spectrosc. Radiat. Transfer,* **109**, 2338–2348. doi:10.1016/j.jqsrt.2008.05.009.

Volten, H., O. MuÚoz, E. Rol, J. F. de Haan, W. Vassen, J. W. Hovenier, K. Muinonen and T. Nousiainen, 2001: Scattering matrices of mineral aerosol particles at 441.6 nm and 632.8 nm. *J. Geophys. Res.*, **106**, 17,375–17,401.

Wall, M., 1996: *Galib: A C++ Library of Genetic Algorithm Components,* Mechanical Engineering Department, Massachusetts Institute of Technology. http://lancet.mit.edu/galib-2.4/

Warren, S. G., and R. E. Brandt, 2008: Optical constants of ice from the ultraviolet to the microwave: A revised compilation. *J. Geophys. Res.*, **113**, D14220, doi:10.1029/2007JD009744.

West, R. L, L. Doose, A. Eibl, M. Tomasko and M. Mishchenko, 1997: Laboratory measurements of mineral dust scattering phase function and linear polarization. *J. Geophys. Res.*, **102**, 16871–16881.

Yang, P., and K. N. Liou, 1996: Finite-difference time domain method for light scattering by small ice crystals in three-dimensional space. *J. Opt. Soc. Am. A,* **13**, 2072–2085. doi:10.1364/JOSAA.13.002072.

Zerull, R., R. Giese and K. Weiss, 1977: Scattering measurements of irregular particles vs. Mie theory. *Proc. Soc. Photo-Optical Instrum. Engrs.*, **12**. Instrumentation and Applications: 191–199.

Zerull, R. H., R. H. Giese, S. Schwill and K. Weis, 1980: Scattering by particles of nonspherical shape. In *Light Scattering by Irregularly Shaped Particles*, ed. D. W. Schuerman, 273–282. New York: Plenum.

Zhao, F., 1999: Determination of the complex index of refraction and size distribution of aerosols from polar nephelometer measurement. *Appl. Opt.*, **38**, 2331–2336.

2 Light scattering by irregularly shaped particles with sizes comparable to the wavelength

Evgenij S. Zubko

2.1 Introduction

Light scattering by single irregularly shaped particles whose sizes are comparable with wavelength plays an important role in numerous remote-sensing applications. It especially concerns applications dealing with both terrestrial and cosmic dust particles having truly irregular and random structure. The knowledge of the scattering by single irregular particles is absolutely necessary for a realistic modeling and successful interpretation of measurements of light scattering by a powder-like surface, such as, soils, sand-drift or planetary regolith. Note that, though the multiple scattering between constituent particles could significantly dominate over single scattering, it is quite evident that the one is a function of other.

The study light scattering by cosmic dust particles is a difficult problem, in particular, because there is an obvious lack of information on the structure of such particles. Indeed, unlike terrestrial dust, a sampling of cosmic dust within the Solar system is an extremely delicate and expensive undertaking; whereas, the sampling of dust particles beyond the Solar system is practically impossible. So far, there is only one successful attempt at such sampling, namely, the *Stardust* mission to comet 81P/Wild 2 (see, e.g., Brownlee *et al.*, 2006). This space probe collected dust particles in the environment of a cometary nucleus (the closest distance was 236 km) onto extremely porous silica foam, called aerogel. Though the density of aerogel varies in the range 0.01–0.05 g/cm^3, this substance is quite tenacious. For instance, a piece of aerogel weighing only 2 grams can support a brick having a weight of 2.5 kg (see picture in http://stardust.jpl.nasa.gov/photo/aerogel.html). When capturing dust particles, *Stardust* approached comet 81P/Wild 2 at a relative velocity of 6 km/s. Therefore, cometary particles experienced a highly energetic collision with the aerogel collector. Interestingly, the typical length of tracks of dust particles captured in aerogel is about one centimeter (Burnett, 2006), which implies a very fast deceleration of particles. As a consequence, micron-sized dust particles did not preserve their original structure; they were broken down into many small compact constituents and heated above $\sim$2000 K (Brownlee *et al.*, 2006).

A significantly less-destructive method to sample cosmic dust particles is to collect them in the Earth stratosphere using high-altitude airplanes (e.g., Brownlee *et al.*, 1995; Jessberger *et al.*, 2001; Busemann *et al.*, 2009). For example, Fig. 2.1

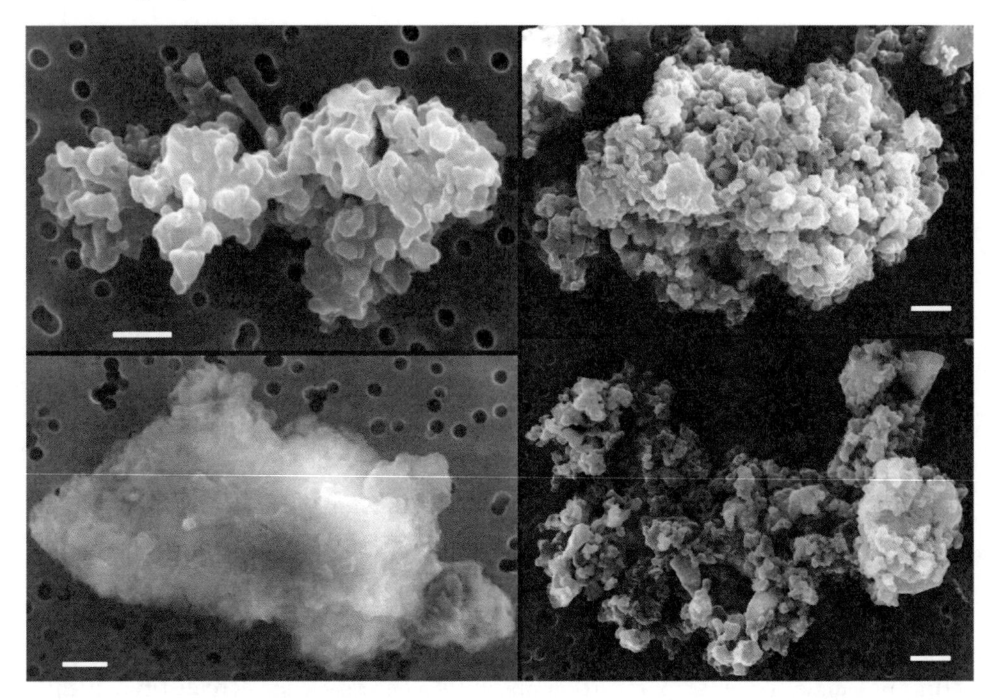

Fig. 2.1. Images of four interplanetary dust particles (IDPs) collected in the stratosphere, having probable cometary origin (adapted from Dai and Bradley, 2001; Hanner and Bradley, 2001; Jessberger *et al.*, 2001). The white line in each panel corresponds to 1 μm size.

presents images of four interplanetary dust particles (IDPs) collected in this way (adapted from Dai and Bradley, 2001; Jessberger *et al.*, 2001; Hanner and Bradley, 2004) and probably having a cometary origin. In general, IDPs originated either from asteroids or comets (e.g., Brownlee *et al.*, 1995). Though IDPs enter the atmosphere at velocities typically higher than 12 km/s, the air does not affect their structure as strongly as aerogel. Therefore, micron-sized particles may preserve their original structure. Indeed, as one can see in Fig. 2.1, IDPs with fluffy structure can be captured in the stratosphere. On the other hand, the specific parent bodies of IDPs captured in the stratosphere are not as well known as in the case of sampling directly near a comet or asteroid and, thus, additional efforts are required to determine the origin of collected particles (Brownlee *et al.*, 1995). Nevertheless, when sampling dust particles shortly after Earth passes through the dust tail of some comet, the origin of the collected particles can be established with some confidence. Using such a dedicated sampling method in the stratosphere, there were successful campaigns to catch dust particles from comet 26P/Grigg–Skjellerup (Busemann *et al.*, 2009). Fig. 2.2 reproduces images of some samples obtained by Busemann *et al.* (2009).

As one can see in Figs. 2.1 and 2.2, all IDPs are quite irregular in appearance. On the other hand, there is also a dramatic difference between particles; evidently, they do not belong to one morphologic type. Particles shown in Figs. 2.1 and 2.2 could be classified into, at least, two distinctive groups, namely, relatively compact

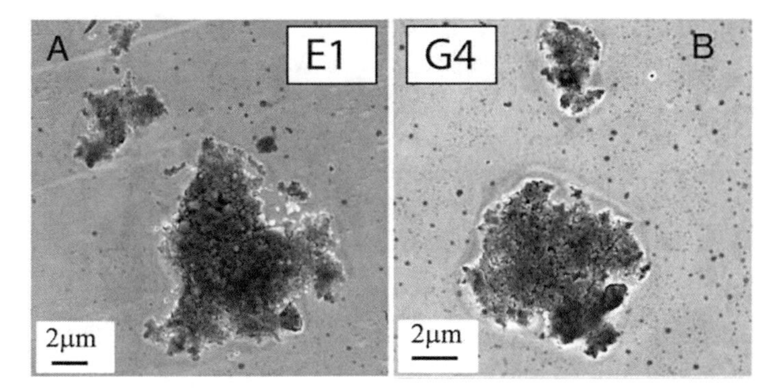

Fig. 2.2. Images of dust particles originating from comet 26P/Grigg-Skjellerup (adapted from Busemann *et al.*, 2009).

and fluffy. However, the difference in morphology of dust particles has an influence on their light scattering and understanding that interrelationship is necessary for the successful development of remote-sensing techniques.

The primary goal of this review is to show how different parameters of light scattering, such as angular profiles of intensity and degree of linear polarization, geometric and single-scattering albedo, asymmetry parameter, cross-sections of extinction and absorption, vary for different types of irregularly shaped particles. Note that, due to random variations in the structure of different samples belonging to one morphologic type, their light-scattering properties can vary dramatically. Therefore, in order to discriminate the impact of a particular characteristic morphology type from that caused by peculiarities of a given sample particle, it is necessary to perform a statistically reliable averaging of light-scattering properties over many samples for each type of particles.

Using the discrete dipole approximation (DDA), we simulate light scattering by six substantially different types of irregularly shaped particles. In most of cases, light-scattering properties of each type of irregular particles have been averaged over a few hundreds of samples. Simultaneously, each type of irregularly shaped particle is studied at three different refractive indices m, representing the abundant species of cosmic dust, such as, water ice, magnesium-rich silicates, and organic refractory in the visible band. Note that these refractive indices also can be associated with some terrestrial aerosols. Furthermore, each case of m is considered over a wide range of particle sizes comparable with wavelength λ of the incident radiation. The large generated data set allows us to make a systematic comparison of light-scattering properties of irregularly shaped particles with different morphology.

2.2 Modeling light scattering by irregularly shaped particles

2.2.1 Models of irregularly shaped particles

We study six types of irregularly shaped particles that could be labeled as follows: agglomerated debris particles, pocked spheres, rough-surface spheres, strongly damaged spheres, debris of spheres, and Gaussian random particles. Example images

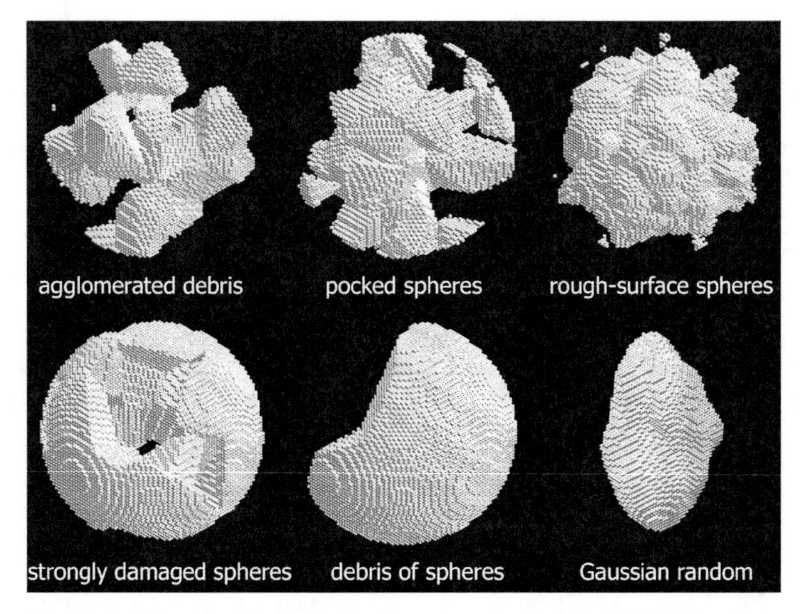

Fig. 2.3. Samples of irregularly shaped particles of six different types.

of these particles are shown in Fig. 2.3. One can see that all particles appear to be essentially non-spherical and reveal a wide variety of particle structures. Indeed, there are highly compact structures, such as rough-surface spheres and Gaussian random particles. Fluffy structure is represented in agglomerated debris particles and pocked spheres, while heavily damaged spheres and debris of spheres correspond to moderate cases.

Except for the Gaussian random particles, irregularly shaped particles have been generated using one algorithm, which is as follows. In computer memory, a spherical volume is filled with a regular cubic lattice that is considered as the initial matrix of the irregular particles. We consider the initial matrix consisting of 137,376 cells. In the process, the elements of the cubic lattice inside the initial matrix are assigned material properties corresponding to the refractive indices of the particles. All cubic cells forming this initial matrix are divided into two groups: cells belonging to the surface layer and cells internal to the surface layer. The depth of the surface layer is a parameter of our model. For instance, in the case of agglomerated debris particles and rough-surface spheres, the depth takes a value of only 0.5% of the radius of the initial matrix; i.e., the surface layer is formed only by dipoles having direct contact with the surrounding empty space. In the case of pocked spheres, the depth is 12.5% of the radius of the initial matrix. Strongly damaged spheres and debris of spheres are generated having no surface layer; i.e., all cells of the initial matrix are treated as internal.

Once cells forming the initial matrix are divided into two sub-groups, we choose seed cells for empty space and material at random. In general, we distinguish two types of seed cells for empty space: those belonging to the surface layer and those belonging to the internal volume; whereas, seed cells for the material are allocated only among internal cells. For instance, agglomerated debris particles are generated with 100 seed cells of empty space randomly chosen in the surface layer, 20 seed cells of empty space, and 21 seed cells of material randomly allocated throughout the internal volume. For rough-surface spheres, the numbers of seed cells are 1,200 for empty space in the surface layer, 150 for material and 0 for empty space within the interior. In the case of pocked spheres they are 100, 50, and 0, respectively. Because heavily damaged spheres and debris of spheres are generated with zero depth of the surface layer, it implies that the number of seed cells for empty space allocated in the surface layer is 0. For strongly damaged spheres, there are 20 seed cells of empty space and 21 seed cells of material; whereas, for debris of spheres the numbers of seed cells of empty space and material are both equal to 4. The final stage of generating a target particle is to evaluate the rest of the cells forming the initial matrix: step-by-step, each cell distinct from the seed cells is assigned with the same optical properties as that of the nearest seed cell. Agglomerated debris particles, pocked spheres, rough-surface spheres, and heavily damaged spheres have been previously studied in Zubko *et al.* (2006; 2009a); in these papers, further images of particles can be also found.

Using the algorithm described in Muinonen *et al.* (1996), we have generated a set of 100 samples of random Gaussian particles. This type of particle is parameterized by the relative radius standard deviation σ and power law index in the covariance function of the logarithmic radius ν, which take values of 0.245 and 4, respectively. Unlike other types of irregular particles involved in this study, random Gaussian particles have a quite smooth surface; nevertheless, their shape is significantly non-spherical. Note that light scattering by exactly the same ensemble of shapes of random Gaussian particles have been studied in Zubko *et al.* (2007), and additional images of sample particles can be found therein.

Two important parameters characterizing irregularly shaped particles are the radius of the circumscribing sphere r_{cs} and the packing density ρ of particle material. Note that these parameters do not describe completely the properties of a particle. The more complicated the structure, the more parameters are needed to describe it. Nevertheless, even considering the effects of varying these two parameters may significantly improve our understanding of scattering peculiarities of irregularly shaped particles. In cases of non-Gaussian irregular particles, the radius of the circumscribing sphere is close to the radius of the initial matrix; the difference does not exceed 1%. We approximate the circumscribing sphere with the largest sample from the set of 100 particles to preserve the size distribution of random Gaussian particles obtained with original algorithm described by Muinonen *et al.* (1996). We define the packing density ρ of a particle as the ratio of volume occupied by the particle material to volume of the circumscribing sphere. We would like to stress that, in the general case, different sample particles of given morphology do not have the same volume of material, so ρ varies randomly from one sample to another. Fig. 2.4 shows the probability distribution of packing density ρ for all six types of irregular particles. Here, curve (1) corresponds to agglomerated debris particles, (2) to pocked spheres, (3) to rough-surface spheres, (4) to heavily damaged spheres, (5) to debris of spheres, and (6) to random Gaussian particles. As one can see, the compact morphologies, such as rough-surface spheres and random Gaussian particles reveal quite narrow distributions of ρ. Pocked spheres also demonstrate a rather narrow distribution of ρ, though not as much as particles

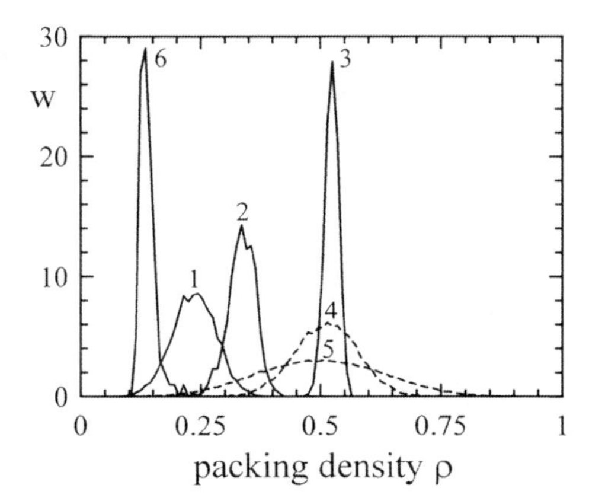

Fig. 2.4. Probability distribution of packing density ρ for six types of irregularly shaped particles.

with compact structure. Simultaneously, agglomerated debris particles have a significantly relaxed distribution of packing density, which, to some extent, is similar to that of heavily damaged spheres. Finally, debris of spheres reveal a quite wide distribution of ρ. Note that three types of irregular particles, namely, rough-surface spheres, heavily damaged spheres, and debris of spheres have almost the same average values of packing density $\langle\rho\rangle$: 0.523, 0.512, and 0.500, respectively. In the case of pocked spheres, $\langle\rho\rangle$ is 0.336; whereas, the average packing density of agglomerated debris particles is 0.236. Note that, according to our definition of packing density, its average is an anomalously low value of 0.139 in the case of random Gaussian particles. However, as one can see in Fig. 2.3, these particles are quite compact in appearance. Obviously, such an underestimation of the average packing density results from the elongated shape of the Gaussian particles. Therefore, in what follows, we refer random Gaussian particles as those having quite compact structure. Since non-Gaussian irregular particles are more or less equally elongated in all directions, in these cases, our definition provides accurate estimations of the average packing density.

2.2.2 The DDA method and parameters

We use the DDA to perform our light-scattering calculations (e.g., Draine, 1988; Draine and Flatau, 1994; Yurkin and Hoekstra, 2007; Zubko *et al.*, 2010). This approach allows us to consider particles with arbitrary shape and internal composition. In the DDA, a target particle is modeled with an array of small constituent volumes that together reproduce the shape and internal optical properties of the original particles. These constituent volumes must be significantly smaller than the wavelength of the incident electromagnetic wave, and their scattering properties take the form of a simple analytic expression (i.e., Rayleigh approximation). Due to such a replacement, one can reduce the light-scattering problem to a system of linear algebraic equations. One optional restriction on constituent volumes is that they be located in a regular cubic lattice to allow use of the fast Fourier transformation (FFT) and, thus, dramatically accelerate computations (Goodman *et al.*, 1991). We use our well-tested implementation of the DDA to perform the calculations of this manuscript (Zubko *et al.*, 2003, 2010; Penttilä *et al.*, 2007).

An important parameter specifying the DDA applicability is the size of the cubic lattice d. In application to irregular particles, the DDA provides accurate numerical result under the condition $kd|m| \leq 1$, where wavenumber $k = 2\pi/\lambda$, λ is the wavelength of the incident electromagnetic wave, and m is the refractive index of the particle (Zubko *et al.*, 2010). Note that throughout this study, the parameter $kd|m|$ remains less than 0.85.

In general, light scattering by a small particle is determined primarily by the complex refractive index m of the particle material and the ratio of the particle radius expressed as r_{cs} to the wavelength λ: $x = 2\pi r_{cs}/\lambda$ (e.g., Bohren and Huffman, 1983); note, this ratio is referred to as the *size parameter*. While we use the radius of the circumscribing sphere r_{cs} to express particle size, another parameter used is the radius of an equal-volume sphere r_{eq}. We denote the size parameter of the equal-volume sphere as x_{eq} and note that x_{eq} can be derived from x and the average packing density $\langle \rho \rangle$ as follows: $x_{eq} = \langle \rho \rangle^{1/3} x$.

We consider irregularly shaped particles having 3 different refractive indices m, which are representative of materials in cosmic and terrestrial dust: $m = 1.313 + 0i$, $1.6 + 0.0005i$, and $1.5 + 0.1i$. In all these cases, x is varied from 2 to 14 with a fixed step of 2.

2.2.3 Averaging of light scattering characteristics

Light-scattering properties of irregularly shaped particles are averaged over sample shape and orientation at each set of m and x. In all cases, except those of random Gaussian particles, we consider a minimum of 500 sample particle shapes. Light scattering by each sample particle has been computed for one random orientation of the incident electromagnetic wave and averaged over 100 scattering planes evenly distributed around the propagation direction of the incident light. This averaging over scattering planes does not require significant computational efforts; however, it improves significantly the statistical reliability of numerical results.

To test for convergence, we perform averaging in two sub-processes, each based on half of the scattering planes used in each sample (i.e., on 50 scattering planes). We compare results of both sub-processes to the results obtained when averaging over 100 scattering planes. As a quantitative indicator of the averaging, we use the standard deviation of the linear degree of polarization obtained with 100 and 50 scattering planes. This standard deviation depends stochastically on the geometry of light scattering, which could be described, for instance, with phase angle α. Note, α is the angle between two lines, one connecting the source of light to the target particle and the other connecting the target particle to the observer. Obviously, α is equal to 0° in the backscattering direction and 180° in the forward-scattering direction. It also needs to be mentioned that phase angle α is supplementary to the scattering angle θ. We continue averaging over particle shape while fluctuations of the standard deviation of the degree of linear polarization over the entire range of phase angle α exceed 1%; therefore, the actual number of sample particles considered very often exceeds 500.

For the random Gaussian particles, we consider a set of 100 sample particles. Therefore, in order to achieve a desirable accuracy, we consider each sample particle in more than one random orientation of the incident electromagnetic wave. The minimum number of orientations is 5 per sample particle. Although the criterion for termination of the averaging process is the same as for non-Gaussian irregular particles, there is a significant difference: when a given number of orientations of random Gaussian particles does not provide a desirable accuracy of the averaging, we add one additional random orientation for each sample particle from the ensemble and only then check the accuracy again. Therefore, at the termination of the averaging process, light-scattering properties of each sample particle are averaged over the same number of random orientations.

It should be emphasized that while many of the irregularly shaped particles considered in the current review have been studied in Zubko *et al.* (2006; 2007), we have significantly improved the quality of the averaging, so the present results are more statistically reliable. For instance, the minimum number of sample particles used in Zubko *et al.* (2006) was 200; whereas, here it is 500.

2.3 Comparative study of light scattering by irregular particles with different morphology

In general, the parameters describing light scattering by a particle can be classified into two groups, sometimes, referred as integral and differential parameters. For instance, cross-sections for absorption $C_{\rm abs}$ and extinction $C_{\rm ext}$ belong to integral parameters. Simultaneously, intensity I and degree of linear polarization P of the scattered light are examples of differential parameters. An essential feature of integral parameters is that they are independent of the conditions of observation; whereas, differential parameters are functions of two angles specifying the direction of scattered light to a detector. However, in the case of azimuthally symmetric targets, the angular dependence of differential parameters takes a significantly simpler form depending upon only phase angle α or, equivalently, the scattering angle θ. Note that statistically reliable averaging of light-scattering properties over sample particles and/or their orientations obviously makes differential parameters azimuthally averaged; therefore, in this case, the differential parameters also depend upon only one angle.

In the first part of this section, we analyze integral parameters of light scattering by irregularly shaped particles, such as the cross-sections for absorption $C_{\rm abs}$ and extinction $C_{\rm ext}$, single-scattering albedo ω, asymmetry parameter g, and radiation pressure efficiency $Q_{\rm pr}$. Most of these parameters play a key role in the simulation of the transparency of the interstellar medium and radiation transfer in planetary atmospheres including Earth; whereas, the radiation pressure efficiency determines the motion of cosmic dust particles. In the second part of the section, we focus on differential parameters, such as, geometric albedo A, linear and circular polarization ratios $\mu_{\rm L}$ and $\mu_{\rm C}$ at backscattering $\alpha = 0°$, and the dependencies of intensity I and degree of linear polarization P on phase angle α. These parameters are widely exploited in remote sensing applications of atmospheric aerosols, comets and other Solar system bodies.

2.3.1 Comparison of integral parameters of light scattering by irregularly shaped particles with different morphology

2.3.1.1 Cross-sections for absorption and extinction C_{abs} and C_{ext}

Interaction of electromagnetic radiation with particles decreases the energy flux of the incident wave. The total loss of the energy flux can be quantified in terms of area, which is normal to the incident beam and intercepts the lost flux of energy. Such an area is referred to as the cross-section for extinction C_{ext}. In the general case, the interaction of electromagnetic radiation with a particle results in absorption and scattering. The part of the total area that corresponds to loss due to absorption is referred to as the cross-section for absorption C_{abs}; whereas, the rest corresponds to the cross-section for scattering C_{sca}. These three values are obviously related as follows: $C_{\text{ext}} = C_{\text{abs}} + C_{\text{sca}}$.

In Fig. 2.5, we present cross-sections for absorption C_{abs} (left) and extinction C_{ext} (right) as functions of size parameter x for irregularly shaped particles of six different types of morphology. Here, the top panels correspond to the case of $m = 1.313 + 0i$, the middle to $m = 1.6 + 0.0005i$, and the bottom to $m = 1.5 + 0.1i$. Note that, though light scattering by particles comparable with wavelength depends on the dimensionless ratio of particle size to wavelength, C_{abs} and C_{ext} are not dimensionless; they are measured in units of area. For instance, the quantities in Fig. 2.5 are given for wavelength $\lambda = 0.628\,\mu\text{m}$ (wavenumber $k = 2\pi/\lambda \approx 10$). However, if necessary, they can be easily recalibrated for another λ.

As one can see in Fig. 2.5, except the case of C_{abs} at $m = 1.313 + 0i$, cross-sections for absorption and extinction grow rapidly with size parameter x; whereas, in the case of non-absorbing material, the cross-section for absorption remains predictably equal to zero through all x. Interestingly, curves in Fig. 2.5 reveal qualitatively similar behavior for all types of irregular particles, though there are visible quantitative distinctions. The latter are caused by differences in morphology and material volume between various types of irregularly shaped particles. On the other hand, for equidimensional target particles at small x, one could expect that the impact of particle morphology on light scattering has to be rather small because, in this case, light-scattering properties can be described quite well with only an isotropic electric dipole (e.g., Bohren and Huffman, 1983). However, the upper limit for the range of x where C_{abs} and C_{ext} are insensitive to morphology is unknown and, thus, its determination is of high practical interest.

In order to eliminate the impact of different volumes of material, one can consider ratios $C_{\text{abs}}/\langle\rho\rangle$ and $C_{\text{ext}}/\langle\rho\rangle$ instead of pure cross-sections for absorption and extinction. These normalized quantities are shown in Fig. 2.6. As one can see, agreement between different types of irregularly shaped particles becomes much better, namely, all curves nearly coincide, at least for $x \leq 4$. Moreover, in some cases, the coincidence is extended to much larger x. Particularly, the extinction cross-section of icy particles (i.e., $m = 1.313 + 0i$) does not depend on their morphology for $x \leq 8$. Interestingly, the differential parameters of light scattering, such as the angular profiles of intensity I and degree of linear polarization P, reveal significant deviations depending on morphology at $m = 1.313 + 0i$ and $x = 6$ (see top panels in Fig. 2.10). Surprisingly, in the case of weak absorption, the absorption cross-section

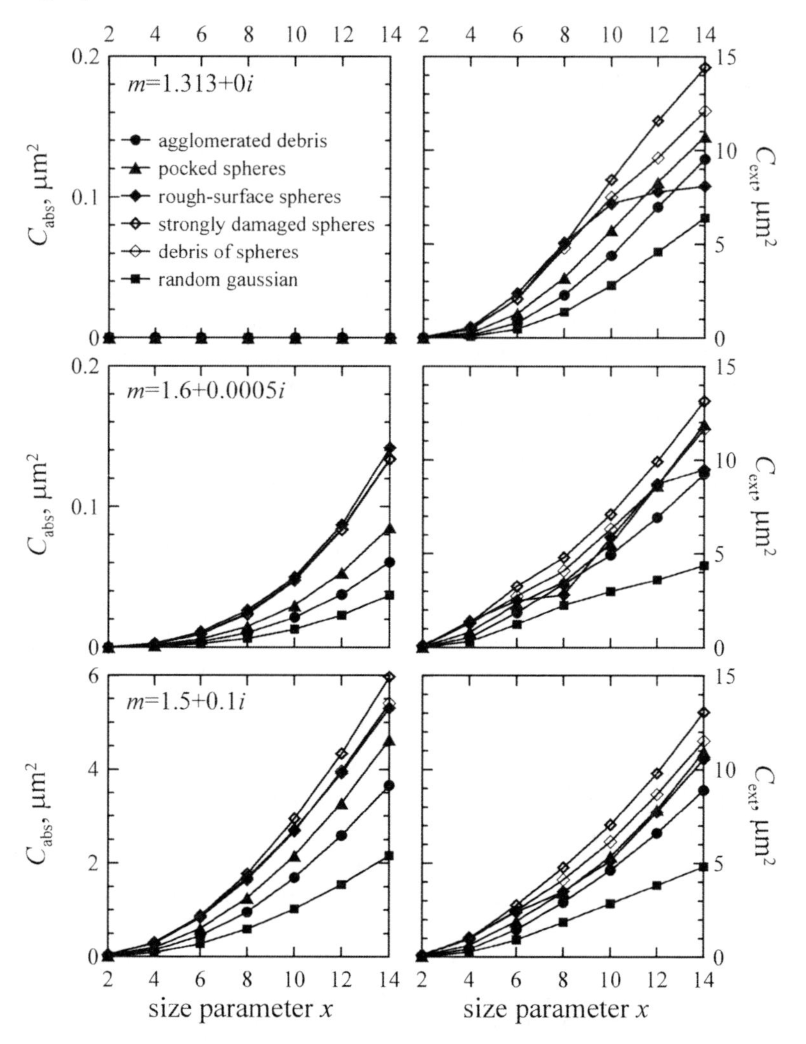

Fig. 2.5. Cross-sections for absorption $C_{\rm abs}$ (left) and extinction $C_{\rm ext}$ (right) computed at $\lambda = 0.628\,\mu$m, as functions of size parameter x for six types of irregularly shaped particles. The top panels show the case of $m = 1.313 + 0i$, the middle panels show the case of $m = 1.6 + 0.0005i$, and the bottom panels show the case of $m = 1.5 + 0.1i$.

remains insensitive to particles morphology over the entire range of x investigated in the current work (see left middle panel in Fig. 2.6).

Finally, we would like to note that within each morphology class of irregular particles, the cross-section for extinction $C_{\rm ext}$ attains similar absolute values at m $=1.5 + 0.1i$ and $m = 1.6 + 0.0005i$ (see Figure 5). One can conclude that the imaginary part of refractive index Im(m) does not significantly affect $C_{\rm ext}$. Therefore, considerable extinction of radiation does not necessarily mean the presence of highly absorbing particles. However, Im(m) has a significant impact on absorption cross-section $C_{\rm abs}$, and therefore also on the scattering cross-section $C_{\rm sca}$.

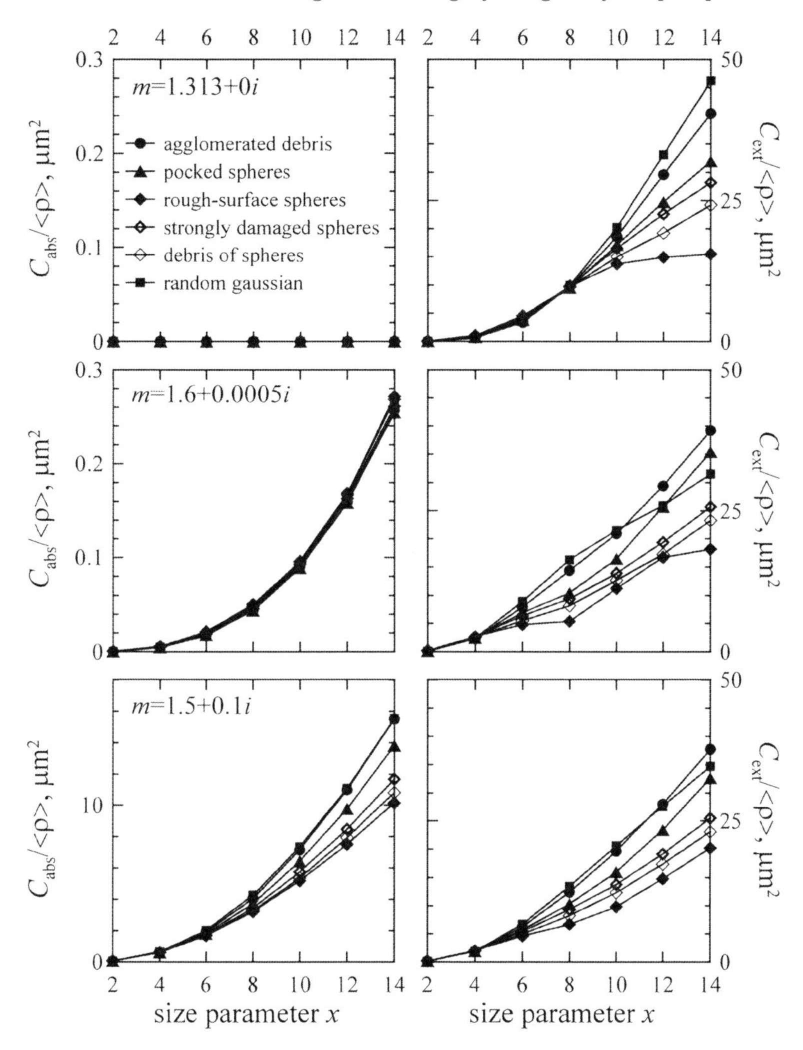

Fig. 2.6. The same as Fig. 2.5 but, $C_{\rm abs}$ and $C_{\rm ext}$ are normalized to an average packing density $\langle\rho\rangle$ of irregularly shaped particles.

2.3.1.2 Single-scattering albedo ω and asymmetry parameter g

Fig. 2.7 shows another pair of integral parameters of light scattering, namely, single-scattering albedo ω and asymmetry parameter g. The former is defined as the ratio $\omega = C_{\rm sca}/C_{\rm ext} = (C_{\rm ext} - C_{\rm abs})/C_{\rm ext}$; it varies from 0 and to 1 and presents a sort of efficiency of light scattering. The parameter g indicates the distribution of the scattered electromagnetic energy between forward and backward hemispheres with respect to the direction of the incident beam propagation:

$$g = \frac{\int\limits_{2\pi}\int\limits_{\pi} I(\theta,\varphi)\,\cos\theta\,\sin\theta\,d\theta\,d\varphi}{\int\limits_{2\pi}\int\limits_{\pi} I(\theta,\varphi)\,\sin\theta\,d\theta\,d\varphi}\,. \tag{1}$$

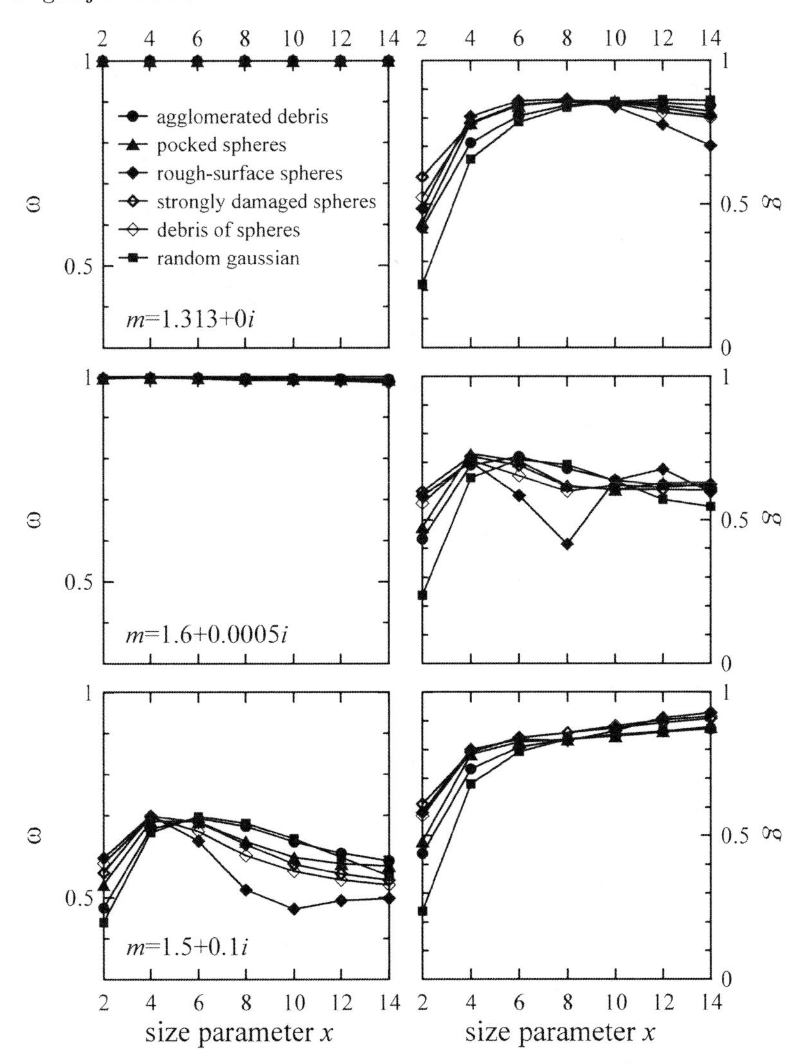

Fig. 2.7. Single-scattering albedo ω (left) and asymmetry parameter g (right) as a function of size parameter x for six types of irregularly shaped particles. The top panels show the case of $m = 1.313 + 0i$, the middle panels show the case of $m = 1.6 + 0.0005i$, and the bottom panels show the case of $m = 1.5 + 0.1i$.

Here, θ and φ are the scattering and azimuthal angles, $I(\theta, \varphi)$ is the intensity of scattering of unpolarized light. The denominator is equal to the scattering cross-section C_{sca}. Note that scattering intensity is a function of two angles in the general case; whereas, averaging over sample particles and/or their orientation removes the dependence on azimuthal angle φ. Asymmetry parameter g varies in the range from -1 to 1. It takes a positive value if the scattering into the forward hemisphere dominates over that into backward hemisphere, and $g = 0$ if the energy is distributed equally between hemispheres and is negative otherwise. Note that both single-scattering albedo ω and asymmetry parameter g are important parameters in computations of radiative transfer in discrete media (e.g., Bowell *et al.*, 1989; Bauer *et al.*, 2007).

As one could expect, single-scattering albedo ω of non-absorbing particles is exactly equal to unity. This means that all incident electromagnetic energy intercepted by a target particle is being scattered into surrounding space (see Fig. 2.7). However, a small imaginary part of refractive index $\mathrm{Im}(m) = 0.0005$ does not significantly affect ω either; for instance, the albedo remains higher than 98.5% for all types of particles and sizes. In case of highly absorbing material with $\mathrm{Im}(m) = 0.1$, there is a significant dependence of ω on x. Nevertheless, for the case of $m = 1.5 + 0.1i$ shown in Fig. 2.7, all profiles ω vs. x are quite similar to each other. Namely, in the range of x from 2 to 4–6, single-scattering albedo grows rapidly with size parameter x; then, it slowly decreases. Note that ω remains higher than 0.4 through all cases considered.

Interestingly, except in the case of rough-surface spheres, there is quite good quantitative agreement between albedo ω of particles with different morphologies. On the other hand, as was shown in Zubko *et al.* (2006), light scattering by the rough-surface spheres is very close to that of a perfect sphere of equivalent volume. Therefore, one may conclude, once irregular particles substantially differ from a perfect sphere, their single-scattering albedo ω does not depend significantly on their morphology. This statement is consistent with findings in Nousiainen and Muinonen (2007), where random Gaussian spheres having appreciably different shapes were studied. Nevertheless, in the case of irregular particles with some specific morphologies, such as extremely sparse ballistic cluster–cluster aggregates (BCCAs) and ballistic particle–cluster aggregates (BPCAs), the size dependence of single-scattering albedo ω may be substantially different from the profiles shown in Fig. 2.7. Notice, there is quite good quantitative agreement between single-scattering albedo of BCCAs and BPCAs (Kimura and Mann, 1998).

As one can see in the right panels of Fig. 2.7, the asymmetry parameter g shows more or less the same dependence on the size parameter x for all types of irregularly shaped particles. Note that, as in the case of single-scattering albedo ω, the most notable deviations from the general behavior are caused by rough-surface spheres. These deviations become apparent in cases of non- and weakly absorbing materials; whereas, in the case of $m = 1.5 + 0.1i$, rough-surface spheres do not reveal visible distinctions from the behavior of irregular particles with other morphologies.

Interestingly, irregular particles with different degrees of absorption reveal visibly different trends of curves g vs. x. Indeed, as one can see in Fig. 2.7, curves corresponding to weakly absorbing particles show a maximum at $x \approx 10$–12 ($m = 1.313+0i$) and $x \approx 4$–6 ($m = 1.6+0.0005i$). Note that similar non-monotonic behavior was found for particles with regular structure (Mishchenko, 1994; Bauer *et al.*, 2007), although in the case of irregular particles the maximum is located at systematically higher x. Simultaneously, profiles for highly absorbing particles approach the level $g = 1$ asymptotically, having no extremum. This behavior is also consistent with findings for spheres with large material absorption (Mishchenko, 1994). However, in this case, there is also rather good quantitative agreement between asymmetry parameters g for spheres and irregular particles.

We would like to note that the sign of the asymmetry parameter g for all six types of irregular particles is positive throughout the entire range of size parameter

x considered in Fig. 2.7. In practice this means that the amount of electromagnetic energy scattered in the forward hemisphere is greater than that scattered in the backward hemisphere. This result is consistent with findings for other types of particles, such as single spheres, randomly oriented oblate spheroids and Chebyshev particles (Mishchenko, 1994), particles with multiple internal inclusions (Macke and Mishchenko, 1996), BCCA and BPCA (Kimura and Mann, 1998), and layered spheres (Bauer *et al.*, 2007).

Putting a target particle into some electromagnetic field disturbs its initial structure. In order to express analytically the resulting (i.e., disturbed) electromagnetic field in space with respect to the particle, it is assumed that the total electromagnetic field is the sum of the initial electromagnetic field (i.e., the field existing in the absence of the obstacle) and the scattered electromagnetic field (i.e., all the rest). As it turns out, such a decomposition of the electromagnetic field in space is quite convenient for a subsequent analysis of electromagnetic scattering (Bohren and Huffman, 1983). However, in what follows, only the scattered electromagnetic field is studied; whereas, the incident electromagnetic field is ignored. Note that the definition of g eq. (1) presents exactly this case. Evidently, omitting the incident field corresponds to the case when the detector of radiation is isolated from the direct incident electromagnetic field. While this occurs in the vast majority of experiments and observations of light scattering, it does not hold true in the case of multiple scattering between closely packed particles. Indeed, in this case, the detector is a neighboring particle, which is irradiated by not only the electromagnetic wave scattered from the target particle but by the incident wave as well. Therefore, when considering the asymmetry parameter g for constituent particles forming a discrete medium, the intensity of scattering in relationship (1) has to be replaced with the intensity of the full electromagnetic field, which is based on the amplitude of the full electromagnetic field. Therefore, one can summarize that data for asymmetry parameter g shown in Fig. 2.7 can be used only in applications to single-scattering particles or constituents of clusters with very sparse and random structure; whereas, in the case of closely packed clusters, asymmetry parameter g needs to be computed in a way different from eq. (1).

2.3.1.3 Efficiency for radiation pressure Q_{pr}

The radiation-pressure efficiency Q_{pr} determines the motion of cosmic dust particles. It is defined as follows:

$$Q_{\mathrm{pr}} = C_{\mathrm{pr}}/G = (C_{\mathrm{ext}} - gC_{\mathrm{sca}})/G. \qquad (2)$$

Here, G is the geometric cross-section of the particle and C_{pr} is the cross-section for radiation pressure (e.g., van de Hulst, 1981); whereas, C_{ext}, C_{sca}, and g are, as given previously, the cross-sections for extinction and scattering, and asymmetry parameter, respectively. Note that the motion of cosmic dust particles near a star depends on the ratio of the radiation-pressure force to the star's gravitational force, which is designated as β (e.g., Burns et al., 1979; Artymowicz, 1988; Fulle, 2004). Some details on the difference between the orbit of the parent body and an ejected dust particle caused by radiation pressure acting on the particle can be found, e.g., in Augereau and Beust (2006). By definition, the ratio β is in direct proportion to radiation pressure: $\beta \propto Q_{pr}$ (e.g., Fulle, 2004). However, there is an obvious lack of Q_{pr} data for irregularly shaped particles. For most applications, computations of β are based on the radiation-pressure efficiency of a single sphere obtained using Mie theory (Burns et al., 1979; Artymowicz, 1988; Fulle, 2004; Augereau and Beust, 2006). Nevertheless, in some cases, particles with more complicated structure have been considered, such as, BCCA and BPCA clusters (Kimura and Mann, 1998; Kimura et al., 2002) or layered spheres (Wickramasinghe and Wickramasinghe, 2003).

In Fig. 2.8, we present radiation-pressure efficiencies Q_{pr} vs. size parameter x of irregularly shaped particles having six different types of morphology. One can immediately conclude that particles with $m = 1.313 + 0i$ behave peculiarly. In

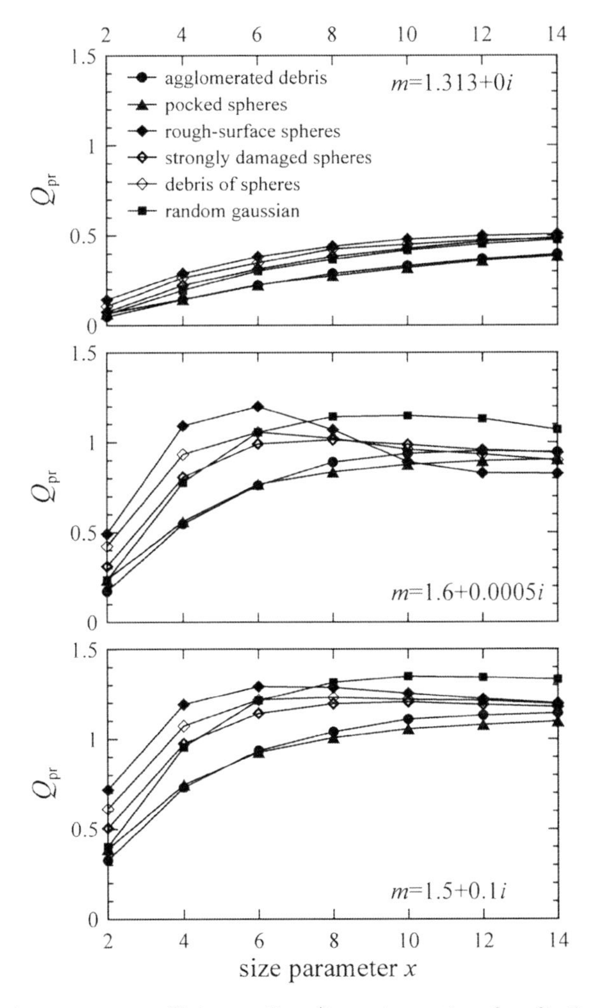

Fig. 2.8. Radiation-pressure efficiency Q_{pr} (i.e., the ratio of radiation pressure cross-section C_{pr} to geometric cross-section G) as a function of size parameter x for six types of irregularly shaped particles. The top panels show the case of $m = 1.313 + 0i$, the middle panels show the case of $m = 1.6 + 0.0005i$, and the bottom panels show the case of $m = 1.5 + 0.1i$.

comparison to this case, values of $Q_{\rm pr}$ for particles with $m = 1.6 + 0.0005i$ and $m = 1.5 + 0.1i$ are surprisingly close to each other; although radiation-pressure efficiency for highly absorbing particles systematically exceeds that for weakly absorbing particles. Interestingly, through all the studied refractive indices m and size parameters x, $Q_{\rm pr}$ for agglomerated debris particles and pocked spheres nearly coincide with each other. Both types of particles have fluffy irregular structure but with different packing densities. Quantitative resemblance also can be seen between curves for strongly damaged spheres and debris of spheres; whereas, rough-surface spheres and Gaussian random particles show individual dependences of $Q_{\rm pr}$ on x.

Note that in numerous practical applications, it is assumed that $Q_{\rm pr} \approx 1$ for very large particles with sizes significantly exceeding 1 μm (e.g., Burns et al., 1979; Fulle, 2004). However, as one can see in Fig. 2.8, while increasing size parameter x, the radiation-pressure efficiency $Q_{\rm pr}$ of non-optically soft irregular particles clearly approaches 1. For instance, at the largest size parameter $x = 14$ achieved in this study, the relative deviations of $Q_{\rm pr}$ from 1 do not exceed 20–25%; whereas, in the case of particles with fluffy structure (i.e., agglomerated debris particles and pocked spheres), these deviations are even smaller. Therefore, in the case of non-optically soft irregular particles, the efficiency for radiation pressure $Q_{\rm pr}$ can be assumed approximately equal to 1 when the particle size is larger than only a few micrometers in visible.

2.3.2 Comparison of differential parameters of light scattering by irregularly shaped particles with different morphology

There are a few different ways to describe properties of the scattered electromagnetic radiation. One of them is the formalism of the four-dimensional *Stokes vector*:

$$\boldsymbol{S} = \begin{pmatrix} I \\ Q \\ U \\ V \end{pmatrix}. \tag{3}$$

In eq. (3), the Stokes parameter I represents the total intensity of the electromagnetic wave, the Stokes parameters Q and U describe its linear polarization with respect to a given scattering plane, e.g., the plane containing the source of incident light, target particle, and observer; finally, the Stokes parameter V characterizes circular polarization of the electromagnetic wave. The most important advantage of such an approach is the simplicity in representation of the natural unpolarized light. Indeed, a monochromatic electromagnetic wave is polarized always 100%; whereas, unpolarized electromagnetic radiation appears as a superposition of numerous completely but stochastically polarized waves. However, for instance, within formalism based on amplitude and phase of electromagnetic wave, it is practically impossible to represent unpolarized light; whereas, in terms of the Stokes vector it can be quite easily expressed as follows:

$$\boldsymbol{S} = \begin{pmatrix} I \\ 0 \\ 0 \\ 0 \end{pmatrix}, \tag{4}$$

i.e., in the case of unpolarized light, the Stokes parameters Q, U, and V are assumed to be equal to zero. For convenience, the intensity of the incident light is assumed very often to be equal to 1.

Within the formalism of the Stokes vectors, the act of light scattering by a particle can be described with the help of a matrix–vector product $\mathbf{S}^{sc} = \mathbf{M}\mathbf{S}^{inc}$. Here, $\mathbf{S}^{inc}$ and $\mathbf{S}^{sc}$ are Stokes vectors for the incident and scattered light, respectively; whereas, $\mathbf{M}$ denotes a special 4×4 matrix, which describes the light-scattering properties of the target particle. Note that this matrix contains all possible differential parameters of light scattering. It is called the *Mueller matrix* (e.g., van de Hulst, 1981; Bohren and Huffman, 1983) and sometimes the *scattering matrix* (Bohren and Huffman, 1983). In general, all sixteen elements of the Mueller matrix are nonzero and depend on phase angle α.

It should be emphasized that the Stokes vector and the Mueller matrix are both specified within a given scattering plane. In order to obtain the actual values of the Mueller matrix elements for a given target particle and a given scattering plane, it is necessary to compute the electromagnetic scattering by that particle for two incident waves having orthogonal polarization states. The complete Mueller matrix can be derived from the resulting scattered fields with the help of simple formulae presented, e.g., in Bohren and Huffman (1983). Note that a change of the scattering plane, for instance, caused by rotation of the target particle around the direction of incident wave propagation, requires a complete recalculation of the Mueller matrix. In other words, the Mueller matrix of a single particle corresponding to one scattering plane cannot be easily transformed to the Mueller matrix associated with another scattering plane. Nevertheless, this procedure can be substantially accelerated owing to the linearity of Maxwell equations.

Averaging of light-scattering properties over sample particles and their orientations substantially simplifies the resulting Mueller matrix. In particular, half of the elements in the average Mueller matrix are equal to zero; whereas, some of the nonzero elements are not independent. In general, the average Mueller matrix takes on the following form (e.g., van de Hulst, 1981):

$$\mathbf{M} = \frac{1}{(kR)^2} \begin{pmatrix} M_{11} & M_{12} & 0 & 0 \\ M_{12} & M_{22} & 0 & 0 \\ 0 & 0 & M_{33} & M_{34} \\ 0 & 0 & -M_{34} & M_{44} \end{pmatrix}. \tag{5}$$

As one can see in (5), there are only six truly independent elements, although this number can be even less in the case of some special target particles. For instance, the Mueller matrix for a sphere consists of four independent elements; whereas $M_{11} = M_{22}$ and $M_{33} = M_{44}$ (Bohren and Huffman, 1983). When defining the Mueller matrix, it is common to distinguish the factor $(kR)^{-2}$, which describes the energy attenuation of the spherical wave with distance R and is common for all elements of the matrix (Bohren and Huffman, 1983).

In practice, when measuring light scattering by micron-sized particles, a cumulative signal coming simultaneously from enormous numbers of particles, is registered. Except in a few quite specific cases (e.g., Rosenbush et al., 2007), these particles are randomly oriented. Therefore, the form of the matrix retrieved in practice usually takes the form of eq. (5) (e.g., Muñoz et al., 2000; Hovenier et al., 2003; Volten et al., 2007). In astronomical applications, the incident light typically is emitted by some star, and such electromagnetic radiation is substantially unpolarized. As a consequence, the light-scattering parameters which are most commonly available in astronomical observations are defined by product of the Mueller matrix (eq. (5)) and the Stokes vector (eq. (4)), i.e.:

$$\mathbf{S}^{sc} = \frac{1}{(kR)^2} \begin{pmatrix} M_{11} \\ M_{12} \\ 0 \\ 0 \end{pmatrix}. \tag{6}$$

Obviously, the Stokes vector (6) remains valid for passive measurements of light scattering by atmospheric aerosols, i.e., with the Sun as the source of light (e.g., Kokhanovsky, 2008).

As one can see in (6), there are only two nonzero Stokes parameters containing information. However, the commonly measured values are the intensity of the scattered light $I = (kR)^{-2}M_{11}$ and its degree of linear polarization $P = -M_{12}/M_{11}$. Typically polarization is expressed in percent. Note that, taking into account the definitions for the Mueller matrix elements M_{11} and M_{12} (Bohren and Huffman, 1983), one can reformulate definitions for the intensity and degree of linear polarization alternatively as follows: $I = I_\perp + I_{||}$ and $P = (I_\perp - I_{||})/(I_\perp + I_{||})$. Here, $I_\perp$ denotes the intensity of the component of scattered light that is polarized perpendicular to the scattering plane; whereas, $I_{||}$ denotes the intensity of the component polarized within the scattering plane. While intensity I always takes positive and nonzero values, the degree of linear polarization P can take positive and negative values and also be equal to zero. Note also that sometimes the dependence of intensity I on phase angle α, which is normalized to the scattering cross-section C_{sca}, is called the *phase function* (e.g., van de Hulst, 1981; Bohren and Huffman, 1983; Kokhanovsky, 2008).

2.3.2.1 Full phase dependencies of the intensity and degree of linear polarization

Figs. 2.9–2.12 show the phase dependencies of intensity I normalized to its value at the backscattering (left) and degree of linear polarization P (right) for particles with size parameter $x = 2$, 6, 10, and 14, correspondingly. The upper panels show the results for particles with $m = 1.313 + 0i$, results with $m = 1.6 + 0.0005i$ are shown in the middle, and results with $m = 1.5 + 0.1i$ are shown in the bottom. For all the refractive indices presented in Fig. 2.9, the phase dependencies of intensity are not symmetric, having a wide peak of forward scattering. The intensity of forward scattering is up to about two orders of magnitude higher than that of bsackscattering; however, in the case of optically soft particles with $m = 1.313 + 0i$, this difference is less pronounced. As one can see in Fig. 2.9, even the smallest particles with $x = 2$ reveal noticeable differences in the intensity depending upon the particle morphology. On the other hand, curves for particles with similar morphology are generally in good quantitative agreement. For instance, the intensity profiles for agglomerated debris particles and pocked spheres are quite similar to

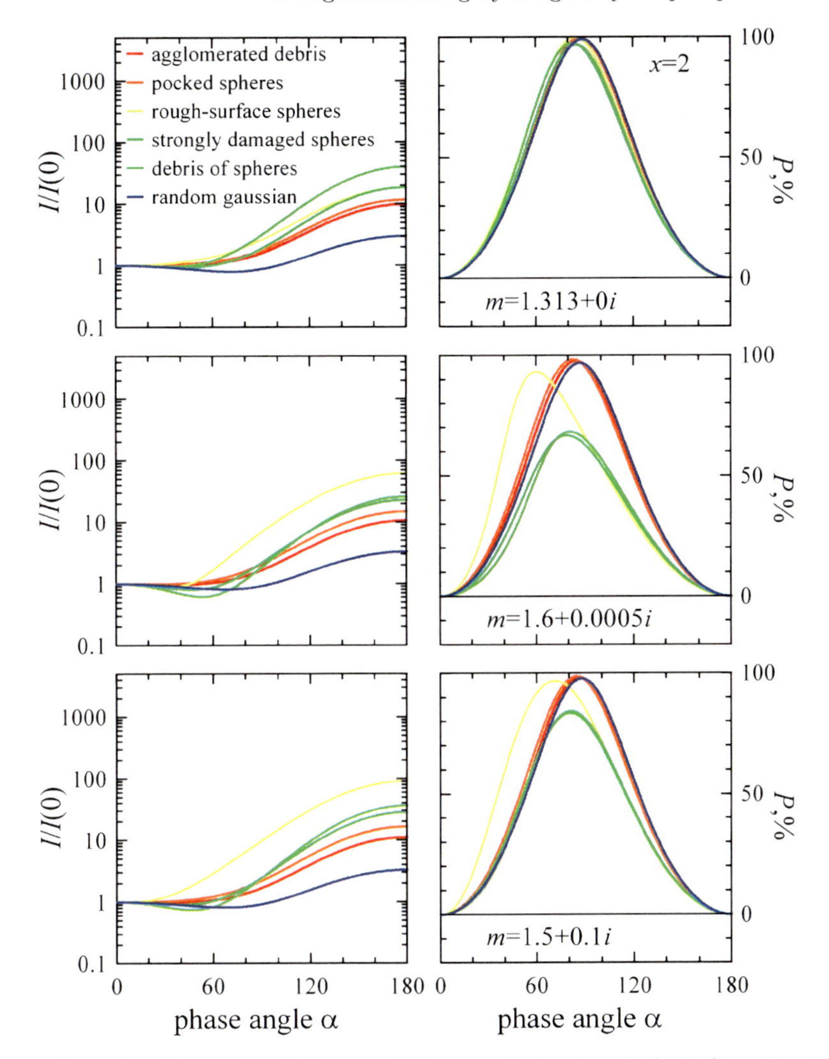

Fig. 2.9. Intensity I (left) and degree of linear polarization P (right) as functions of phase angle α for six types of irregularly shaped particles with $x = 2$. The top panels show the case of $m = 1.313 + 0i$, the middle panels show the case of $m = 1.6 + 0.0005i$, and the bottom panels show the case of $m = 1.5 + 0.1i$.

each other for all refractive indices studied. The same is observed for strongly damaged spheres and debris of spheres, except the case of $m = 1.313 + 0i$. We would like to emphasize that such similarities remain in the angular profiles of the degree of linear polarization (see right panels in Fig. 2.9) for agglomerated debris particles and pocked spheres, and for strongly damaged spheres and debris of spheres.

As one can see in Fig. 2.9, the degree of linear polarization, for all types of irregular particles with $x = 2$, remains substantially positive through all phase angles and refractive indices. The phase curves of linear polarization have a bell-like shape. In most cases, the maximum of positive polarization is located near $\alpha = 90°$; and only in the case of the rough-surface spheres, the maximum of linear polarization is noticeably shifted toward smaller phase angles. Through all the cases considered in Fig. 2.9, the maximum amplitude of linear polarization exceeds 60%; however, in many cases, the amplitude is even higher, reaching almost 100%. For instance, at $m = 1.313 + 0i$, all irregularly shaped particles have a very high maximum linear polarization. Such angular profiles of linear polarization nearly coincide with the result of the *Rayleigh approximation* for an extremely small particle, i.e., with $x \ll 1$ (e.g., van de Hulst, 1981; Bohren and Huffman, 1983). On the other hand, the Rayleigh approximation also predicts a symmetric phase function of small particles, which is not observed for the irregular particles with $x = 2$. Therefore, one can conclude that, in the case of small size parameter x, the angular dependence of the intensity is considerably more sensitive to particles' morphology than the degree of linear polarization. Interestingly, while the size of the particles grows, becoming comparable with the wavelength, the degree of linear polarization becomes more sensitive to properties of particles than the intensity (e.g., Zubko *et al.*, 2007; Nousiainen, 2009).

Fig. 2.10 shows phase dependencies of intensity I and degree of linear polarization P for irregular particles of $x = 6$. There is a dramatic difference from the curves for $x = 2$ presented in Fig. 2.9. The forward-scattering peak is substantially narrower in the case of $x = 6$. For optically soft and highly absorbing particles, the difference between intensities in the forward and backward scattering is an order of magnitude higher than for $x = 2$; however, in the case of $m = 1.6 + 0.0005i$, this difference is not so great. Unlike for the case of $x = 2$, angular profiles of the intensity and degree of linear polarization are accompanied with oscillations, which are most apparent for rough-surface spheres. The locations of these oscillations are close with those of the corresponding equal volume sphere (Zubko *et al.*, 2006). Among morphologies considered through this study, particles with agglomerated structure (i.e., agglomerated debris particles and pocked spheres) show the smoothest overall profiles of the intensity and degree of linear polarization. Moreover, as for $x = 2$, one can see quite good quantitative agreement between the curves corresponding to these two types of particles. Interestingly the similarities between strongly damaged spheres and debris of sphere, which are found for small particles with $x = 2$, do not seem to be present for $x = 6$.

Weakly absorbing particles reveal enhancement of the intensity near backscattering $\alpha = 0°$; whereas, in case of highly absorbing particles, the intensity curve is flattened around $\alpha = 0°$. In general, the enhancement of intensity near backscattering correlates qualitatively with the *negative polarization branch* (NPB), which also appears at small phase angles (see right panels in Fig. 2.10). Nevertheless, a quantitative interrelation between the enhancement of intensity near backscattering and the NPB does not take a simple form. For instance, at $m = 1.6 + 0.0005i$, the enhancement of intensity produced by rough-surface spheres does not significantly differ from that for strongly damaged spheres. However, NPBs of these particles diverge dramatically. One can see also that the width of the intensity surge near backscattering does not correlate unambiguously with the width of the NPB. Finally, there exists a dramatic decrease of the NPB due to the high absorption. Indeed, particles of all morphologies with $m = 1.5 + 0.1i$ almost do not show NPB.

Fig. 2.11 shows the phase curves of intensity I and degree of linear polarization P for particles with $x = 10$. As one can see, the further increase of particles size

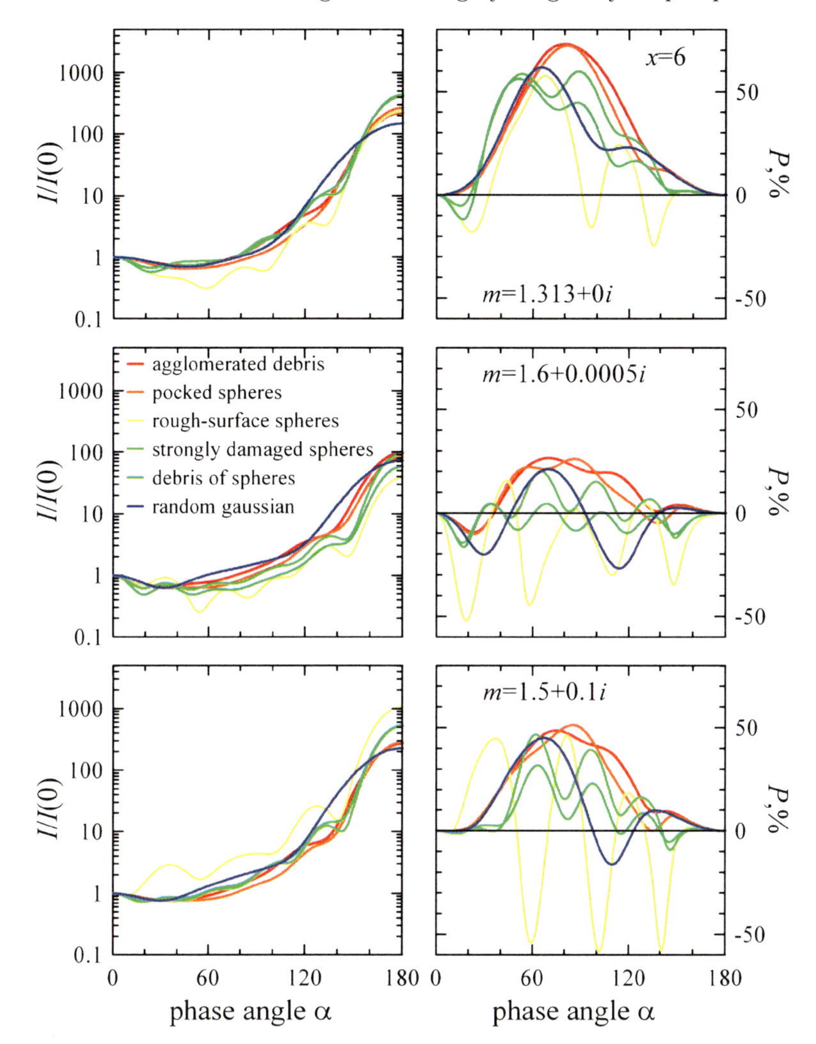

Fig. 2.10. The same as Fig. 2.9 but, for $x = 6$.

causes changes in the curves of intensity and linear polarization, which generally are consistent with those previously found between $x = 2$ and 6; however, they are much less dramatic. The smaller impact of size increase can be explained as follows. The growth of x from 2 to 6 equates to a threefold increase of particle size; whereas, the change from $x = 6$ to 10 equates to less than a twofold increase. The most noticeable change in light-scattering properties concerns the oscillations on phase curves of intensity and polarization. Indeed, in the case of $x = 10$, their number is significantly higher than at $x = 6$; simultaneously, the amplitude of oscillations is substantially damped, though, for rough-surface spheres, the oscillations remain quite dramatic in appearance.

It is interesting to compare the phase dependencies of intensity and polarization for rough-surface spheres and random Gaussian particles presented in Figs. 2.10–2.12. As was previously mentioned, these are two types of particles having compact

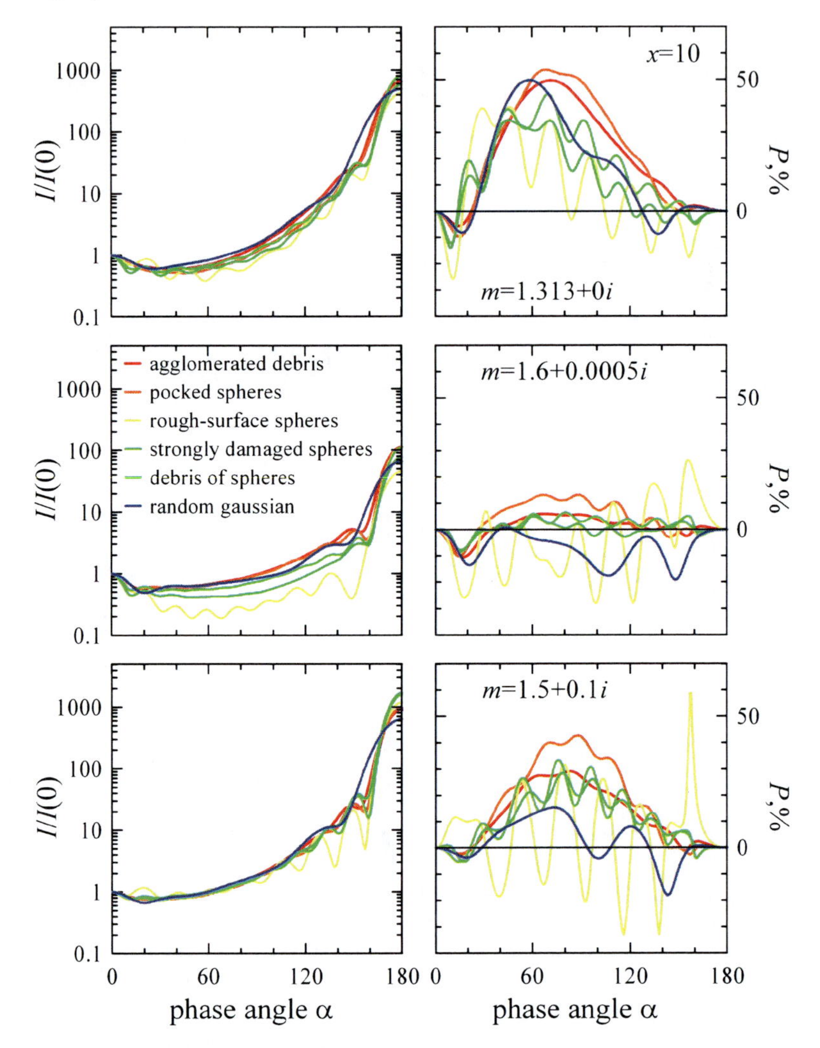

Fig. 2.11. The same as Fig. 2.9 but, for $x = 10$.

morphology. However, rough-surface spheres have an overall shape that is more or less spherical, but with significant surface roughness. The overall shape of a random Gaussian particle differs significantly from that of a sphere, yet its surface remains quite smooth (see in Fig. 2.3). As one can see, the phase curves of intensity and degree of linear polarization for rough-surface spheres reveal visibly more oscillatory behavior than random Gaussian particles. One can conclude that overall nonsphericity in particle shape is more efficient at eliminating light-scattering resonances than the surface roughness.

As one can see in Fig. 2.11, there is not a systematic resemblance between curves for agglomerated debris particles and pocked spheres with $x = 10$, like that seen for $x = 2$ and 6. Except for the angular profile of intensity for $m = 1.5{+}0.1i$, irregularly shaped particles with different morphology produce completely distinctive phase

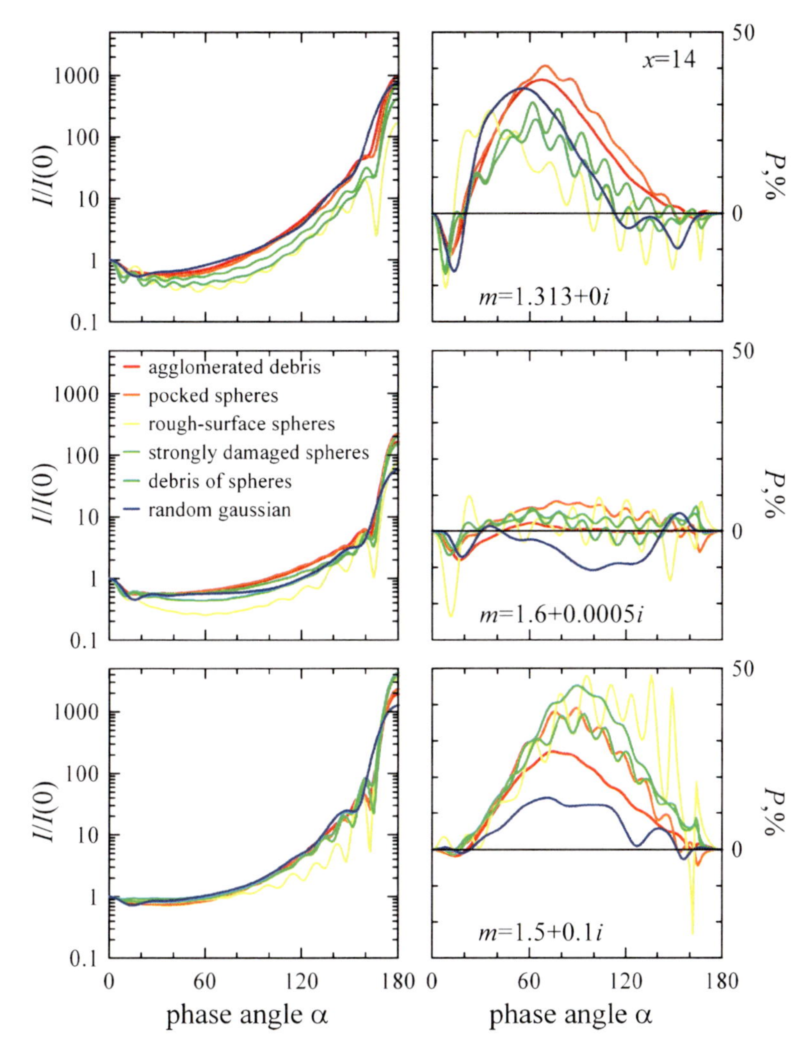

Fig. 2.12. The same as Fig. 2.9 but, for $x = 14$.

dependencies of intensity and degree of linear polarization at $x = 10$. Note that in Fig. 2.12, the same conclusions are valid for the case of $x = 14$.

Except for the case of rough-surface spheres with $m = 1.5 + 0.1i$, all irregular particles with $x = 10$ show a noticeable enhancement of intensity near backscattering and a NPB. Manifestation of both effects is visibly damped in highly absorbing particles. Note that, in the case of $x = 10$, as well as $x = 6$, the shape of the NPB is visibly non-symmetric, as the minimum of polarization is shifted toward the inversion point of polarization sign, i.e., $P(\alpha \neq 0°) = 0\%$, rather than toward $\alpha = 0°$. This kind of non-symmetry in shape of NPB has been found for other non-spherical particles with $x < 10$ (e.g., Lumme *et al.*, 1997; Zubko *et al.*, 2001; 2003; 2004; 2006; 2007; Kimura and Mann, 2004; Vilaplana *et al.*, 2006; Muinonen *et al.*, 2007; Nousiainen and Muinonen, 2007; Lindqvist *et al.*, 2009; Shen *et al.*,

2009). For weakly absorbing particles, while size parameter x increases, the NPB tends to become more symmetric (e.g., Zubko et al., 2004; 2006). This effect can be observed in Fig. 2.12. Note that further growth of particle size can make the shape of the NPB again non-symmetric but inversely to that of $x < 10$; for instance, in Fig. 2.12, consider the curve for agglomerated debris particles with $x = 14$ and $m = 1.6 + 0.0005i$. For highly absorbing particles, an increase of x does not significantly affect the symmetry of the NPB. However, it may cause a small additional branch of positive polarization near $\alpha = 0°$. Examples of such angular profiles of the NPB can be found for debris of spheres at $x = 10$ and $m = 1.5+0.1i$ (Fig. 2.11) or for a few types of particles with $x = 14$ and $m = 1.5+0.1i$ (Fig. 2.12). Note that, this additional positive polarization branch near backscattering also was found in other numerical simulations of light scattering by finite targets (see also Videen, 2002; Zubko *et al.*, 2009a); it seems that the effect appears predominately in the case of weak NPB.

Finally, we consider the rather deep negative polarization (up to -20%) produced by random Gaussian particles with $x = 10$–14 and $m = 1.6 + 0.0005i$, which is located at intermediate phase angles (see, Figs. 2.11 and 2.12). Interestingly, the presence of the negative polarization under given phase angles was also found in laboratory measurements of rutile (Muñoz *et al.*, 2006); whereas, an exhaustive explanation of this effect has been recently proposed by Tyynelä *et al.* (2010).

2.3.2.2 Parameters P_{min}, α_{min}, P_{max}, and α_{max}, describing the angular profile of degree of linear polarization

The angular dependence of the degree of linear polarization observed for the Moon (e.g., Dollfus and Bowell, 1971; Shkuratov *et al.*, 1992; Shkuratov and Opanasenko, 1992), asteroids (e.g., Zellner and Gradie, 1976), comets (e.g., Kiselev and Chernova, 1981; Chernova *et al.*, 1993), and zodiacal light (e.g., Dumont and Sanchez, 1975; Levasseur-Regourd *et al.*, 1990) consists of two prominent features: the NPB at small phase angles and a positive polarization branch (PPB) in the remaining range of phase angles. Transition of NPB to PPB happens at the *inversion angle* α_{inv}, which varies from 15°, in the cases of F-type asteroids (Belskaya *et al.*, 2005) and nucleus of comet 2P/Enke (Boehnhardt *et al.*, 2008), up to 30° in cases of circumnuclear haloes of various comets (Hadamcik and Levasseur-Regourd, 2003) and asteroid 234 Barbara (Cellino *et al.*, 2006). Simultaneously, within the Moon, α_{inv} varies within a relatively narrow range of values 19–24° (Shkuratov *et al.*, 1992).

Numerous laboratory measurements of light scattering by single particles having various properties show that their average angular profile of degree of linear polarization is qualitatively similar to that found for astronomical targets (Muñoz *et al.*, 2000; 2001; Hovenier *et al.*, 2003). The same is found in measurements of light scattering by powder-like surfaces (e.g., Woessner and Hapke, 1987; Shkuratov and Opanasenko, 1992; Hadamcik *et al.*, 2002; Shkuratov *et al.*, 2006).

The principal parameters characterizing the overall profile of angular dependence of linear polarization degree are the amplitudes of the NPB and PPB, P_{min} and P_{max}, and their locations, α_{min} and α_{max} (e.g., Dollfus and Bowell, 1971; Shkuratov *et al.*, 1992). Note that the inversion angle α_{inv} and slope of the polarization curve h near the inversion angle are also considered as valuable characteristics of the linear polarization profile (e.g., Zellner and Gradie, 1976; Chernova *et al.*, 1993). Nevertheless, it is obvious that the latter two parameters are not independent; whereas, to a large extent, they are defined through peculiarities of the NPB and PPB. As a consequence, the slope h has no simple connection with dust-particle properties (Zubko *et al.*, 2011).

In the top and middle panels of Figs. 2.13–2.15, we present the principal characteristics of angular dependence of linear polarization $P_{\min}$, $\alpha_{\min}$, $P_{\max}$, and $\alpha_{\max}$ for irregular particles with different morphology and refractive indices m. Note that the minimum of negative polarization was sought only within a small phase angle range $\alpha = 0$–$45°$. Such a limitation on the range of phase angles is introduced in order to prevent misidentification of the NPB, which can be provoked by very deep negative polarization produced by some particles at intermediate phase angles, as, for instance, in the case of random Gaussian particles with $x = 14$ and $m = 1.6 + 0.0005i$ (see Fig. 2.12). Similarly, we search for the maximum of positive polarization in a range of phase angles $\alpha = 0$–$135°$, in order to exclude from our consideration the resonant-like spikes in the polarization profile occurring near forward scattering angles, which can be observed, for example, for rough-surface spheres with $x = 10$ and $m = 1.5+0.1i$ (see Fig. 2.11). Obviously, such features are not related to the angular dependence of the degree of linear polarization observed for cosmic dust particles and most samples measured under laboratory conditions.

As one can see in Fig. 2.13, irregularly shaped particles with various morphologies have surprisingly similar behavior in the parameters $P_{\min}$, $\alpha_{\min}$, $P_{\max}$, and $\alpha_{\max}$ with size parameter x; whereas, for fixed x, their angular profiles of polarization are completely different (see Figs. 2.10–2.12). For instance, all types of particles do not produce the negative polarization at $x = 2$; whereas, the effect appears within the range of $x = 4$–8 and, then, grows almost monotonically up to largest size parameter that we considered $x = 14$. The profiles for rough-surface spheres and debris of spheres show slightly oscillatory behavior around the average profile. We have already mentioned the resonance behavior of rough-surface spheres, and this could be another manifestation. While the amplitude of negative polarization substantially varies for different types of irregular particles, the locations of the polarization minima reveal good quantitative agreement between some particles types with $x \geq 8$. In general, one can distinguish two groups of particles and one specific case of rough-surface spheres. The first group consists of strongly damaged spheres and debris of spheres. Simultaneously, the second group includes agglomerated debris particles, pocked spheres, and random Gaussian particles. Interestingly, the particles within these groups produce quantitatively similar profiles for $P_{\min}$ even though these particles have completely different structure and values for packing density $\langle \rho \rangle$ (see sub-section 2.2.1).

Fig. 2.14 shows data for irregular particles with $m = 1.6 + 0.0005i$. The size dependence of $P_{\min}$ has more structure than for the previous case of $m = 1.313+0i$. The size dependence of $P_{\min}$ has the largest negative polarization around $x = 4$–8; whereas, the negative polarization appears in tighter range within $x = 4$–6. The NPBs for all the morphologies appear at smaller size parameter x with $m = 1.6 + 0.0005i$ than for $m = 1.313 + 0i$. Similar conclusions can be reached for the highly absorbing particles with $m = 1.5 + 0.1i$. As one can see in

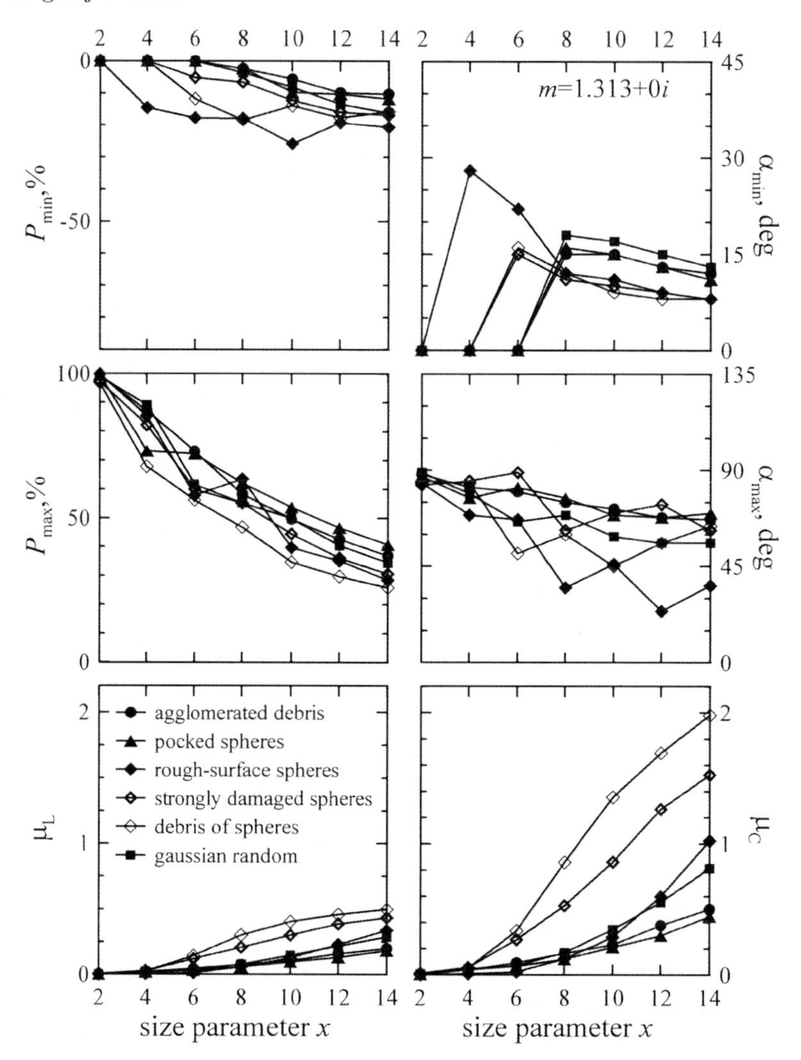

Fig. 2.13. Parameters describing the NPB at small phase angles (P_{min} and α_{min}) and PPB at intermediate phase angles (P_{max} and α_{max}), linear and circular polarization ratios μ_{L} and μ_{C} as functions of size parameter x for six types of irregularly shaped particles with $m = 1.313 + 0i$.

Fig. 2.15, except for the case of rough-surface spheres, the NPB of highly absorbing particles appears in the range $x = 6$–8, and the largest values of the negative polarization occurs around $x = 8$–10. Owing to such rapid growth, the spectral dependence of the negative polarization may exhibit a *blue polarimetric color*, i.e., when the decrease in wavelength λ invokes the increase of negative polarization. Indeed, according to the definition of size parameter x, if some irregular particles have $x = 4$ at $\lambda = 0.7\,\mu\text{m}$, then the same particles have $x = 8$ at $\lambda = 0.35\,\mu\text{m}$. However, as one can see in Figs. 2.13–2.15, the particles with $x = 4$–5 are generally too small in order to produce NPB; whereas, at $x = 8$–10, almost all of them show

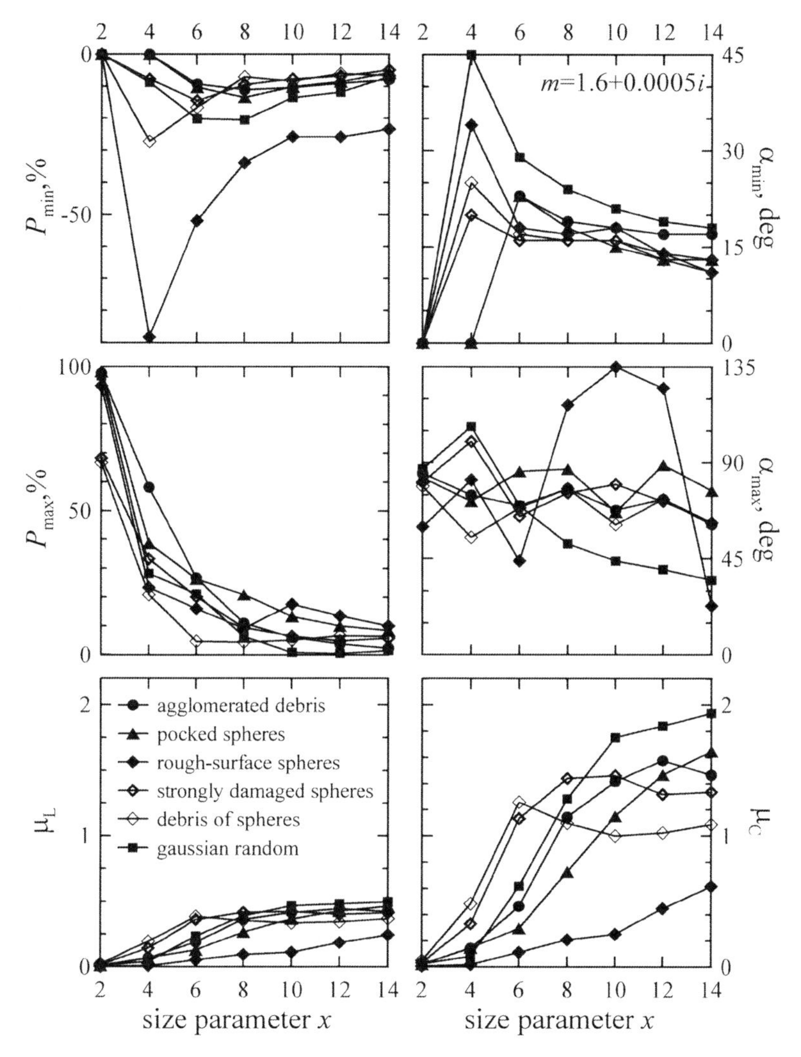

Fig. 2.14. The same as Fig. 2.13 but, for $m = 1.6 + 0.0005i$.

quite noticeable NPBs. As a consequence, the decrease of λ may cause a significant increase of the NPB. This very simple explanation has been successfully used in the quantitative interpretation of blue polarimetric color observed for comet 17P/Holmes during its mega-outburst in October, 2007 (Zubko *et al.*, 2009b).

Again, we consider the distinctive behavior of rough-surface particles with $m = 1.6 + 0.0005i$ and $1.5 + 0.1i$. At $x = 4$, both give rise to an extremely pronounced negative polarization with $|P_{\mathrm{min}}| = 80$–90%. In the case of highly absorbing particles, a similar but less dramatic spike of negative polarization also happens at $x = 8$. Obviously, such a pronounced negative polarization results from the significant sphericity preserved in the shape of the rough-surface particles.

The size dependencies of the phase angle of the polarization minimum α_{min} for irregular particles with $m = 1.6+0.0005i$ (see Fig. 2.14) are qualitatively consistent

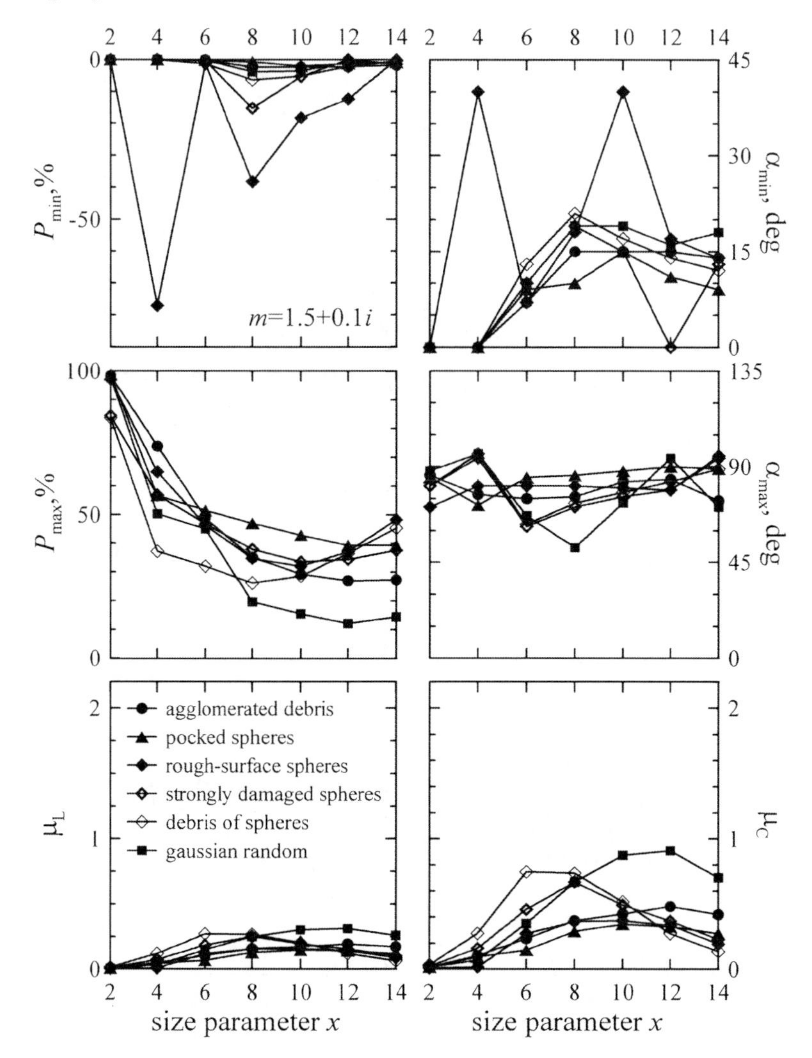

Fig. 2.15. The same as Fig. 2.13 but, for $m = 1.5 + 0.1i$.

with those for $m = 1.313+0i$ (compare with Fig. 2.13). That is to say, once the NPB becomes well-developed, the further increase of x decreases α_{min} monotonically, though nonlinearly. However, particles with larger real part of refractive index produce, in general, higher values of α_{min}. The profiles of α_{min} as a function of x are different for highly absorbing particles (Fig. 2.15) and weakly absorbing particles. Again, the most irregular behavior corresponds to rough-surface spheres. The other cases have qualitatively similar size-dependences of α_{min}. For instance, in the size range $x < 8$–10, the increase of size parameter x increases the phase angle of polarization minimum α_{min}; whereas, the further increase of size parameter results in a decrease of α_{min}. Note that such a profile of α_{min} vs. x is qualitatively consistent with what was found for weakly absorbing particles, though, in the case of highly absorbing particles, the growth of α_{min} to its maximal value is not that rapid. Finally, for particles with $m = 1.5 + 0.1i$, the profiles $\alpha_{\min}$ vs. x can be distorted by a small surge of positive polarization, appearing near backscattering $\alpha = 0°$ (see Figs. 2.11 and 2.12).

As one can see in Fig. 2.13, while the size parameter x grows, the maximal value of linear polarization $P_{\max}$ almost systematically decreases with only one exception for rough-surface spheres near $x = 8$. Although differences between profiles $P_{\max}$ vs. x for particles with different morphology are quite visible, they are not dramatic. Qualitatively the same conclusions can be made for particles with $m = 1.6 + 0.0005i$ (see Fig. 2.14). However, the particles with higher refractive index have substantially greater nonlinear profiles $P_{\max}$ vs. x. Note also that, for the case of $m = 1.313 + 0i$, $P_{\max}$ decreases about three times through the studied range x; whereas, for $m = 1.6 + 0.0005i$, that decrease is about an order of magnitude. Unlike the cases of weak absorption, irregular particles with $m = 1.5 + 0.1i$ show non-monotonic behavior through the range of $x = 2$–14 (see Fig. 2.15). Except for the case of pocked spheres, all size-dependencies of $P_{\max}$ show a minimum in the $x = 8$–12 range. For instance, in the case of agglomerated debris particles, the minimum occurs at $x = 12$; whereas, growth of $P_{\max}$ between $x = 12$ and 14 is only slight. Therefore, one can suppose that the case of pocked spheres is not really exceptional; instead, the profile $P_{\max}$ vs. x a has minimum at $x > 14$ and is beyond our calculated range.

Among four parameters describing the angular profile of linear polarization presented in Figs. 2.13–2.15, phase angle of the maximum of positive polarization $\alpha_{\max}$ has the most ambiguous dependence on size parameter x. In the case of optically soft particles (Fig. 2.13), $\alpha_{\max}$ tends to decrease while x increases; whereas, the common profiles are accompanied with significant oscillations. Nevertheless, for particles with $m = 1.6 + 0.0005i$, it is very difficult to identify a systematic behavior (see Fig. 2.14). Note that the significant oscillations are presented also on the generally flat profiles $\alpha_{\max}$ vs. x, corresponding to highly absorbing particles (Fig. 2.15).

2.3.2.3 Linear and circular polarization ratios μ_L and μ_C, and geometric albedo A

In radar applications, two important parameters quantifying electromagnetic scattering are the *linear* and *circular polarization ratios* μ_L and μ_C (e.g., Ostro, 1993; Mishchenko and Liu, 2007; Zubko *et al.*, 2008). In the case of a fully polarized incident light, the scattered light can be expressed as consisting of two fully polarized components. One component has the same state of polarization as the incident light and is often referred to as the co-polarized component. The cross-polarized component is orthogonally polarized to the incident wave. In the case of linearly polarized incident light, the linear polarization ratio is defined as the ratio of cross- to co-polarized parts of the scattered light; whereas, for circularly polarized incident light, it is more convenient do define the circular polarization ratio as the ratio of co- to cross-polarized parts of the scattered light (Ostro, 1993). In terms of the Mueller matrix elements the ratios are expressed as follows (e.g., Mishchenko and Liu, 2007):

$$\mu_L = \frac{M_{11} - M_{22}}{M_{11} + 2M_{12} + M_{22}}, \quad \mu_C = \frac{M_{11} + M_{44}}{M_{11} - M_{44}}. \tag{7}$$

Obviously, in the general case, both ratios depend on phase angle α. However, in radar applications, the most frequently used case corresponds to backscattering $\alpha = 0°$. Note that, for such specific geometry of electromagnetic scattering, averaging over an ensemble of sample particles and/or their orientations provides $M_{12} = 0$. Therefore, the definition for μ_L in eq. (7) can be simplified. According to radar measurements of all planetary targets, $\mu_L < 1$ and $\mu_L < \mu_C$ (Ostro, 1993). Moreover, as shown in previous studies of agglomerates of perfect spheres (e.g., Mishchenko and Liu, 2007; Mishchenko *et al.*, 2009) and irregularly shaped compact and fluffy particles (Zubko et al., 2008), the same relations hold true for targets comparable with wavelength. It is interesting that at exact backscattering, the ensemble of independent randomly oriented dipoles result in $\mu_L = 1/3$ and $\mu_C = 1$ (Long, 1965; Ostro, 1993).

In the bottom panels of Figs. 2.13–2.15, we show data for the linear (left) and circular (right) polarization ratios of irregular particles with different morphology. The relationships $\mu_L < 1$ and $\mu_L < \mu_C$ found for planetary targets, agglomerates of spherical and non-spherical grains, and independently scattering dipoles remain valid for all morphologies, refractive indices m, and size parameters x, presented in the current review. For instance, the ratio of linear polarization μ_L remains less than 0.5. In the case of optically soft particles (see Fig. 2.13) μ_L grows almost monotonically with x. To a large extent, such behavior can be found for $m = 1.6 + 0.0005i$ (Fig. 2.14). However, in the latter case, there are some exceptions; the most visible of them being for the debris of spheres. In the case of highly absorbing particles (Fig. 2.15), the size-dependencies of μ_L for all particle morphologies are not monotonic, having a maximum in the range $x = 6$–12. Note that in the case of high material absorption, the amplitude of curves μ_L vs. x is systematically less than in the case of weak material absorption.

As mentioned previously, the circular polarization ratio μ_C systematically exceeds the linear polarization ratio μ_L; however, it remains less than 2 for the case of weak absorption, and 1 for highly absorbing particles. The profiles μ_C vs. x are consistent with those for μ_L. In particular, from Figs. 2.13–2.15, one could conclude that for a given morphology of irregular particles and refractive index m, curves for μ_C and μ_L are different only by a scaling factor.

Another parameter of light scattering by single particles defined at $\alpha = 0°$ is the *geometric albedo A*. This parameter describes the ratio of the intensity backscattered by the particle to that scattered by a white disk of the same geometric cross-section G in accordance with Lambert's law (e.g., Hanner *et al.*, 1981; Hanner, 2003):

$$A = \frac{M_{11}(0°)\pi}{k^2 G}. \tag{8}$$

Here, $M_{11}(0°)$ is the corresponding element of the Mueller matrix at $\alpha = 0°$ and k is the wavenumber. In other words, the geometric albedo A equates to the backscattering efficiency of target particles.

As one can see in Fig. 2.16, the geometric albedo A is substantially varied through the range of size parameter $x = 2$–14. In general, the profile A vs. x has one of two distinctive features. The geometric albedo A may continuously grow with size parameter x, as in the case of optically soft particles and random Gaussian particles and both types of fluffy particles with $m = 1.6 + 0.0005i$. Alternatively, the size-

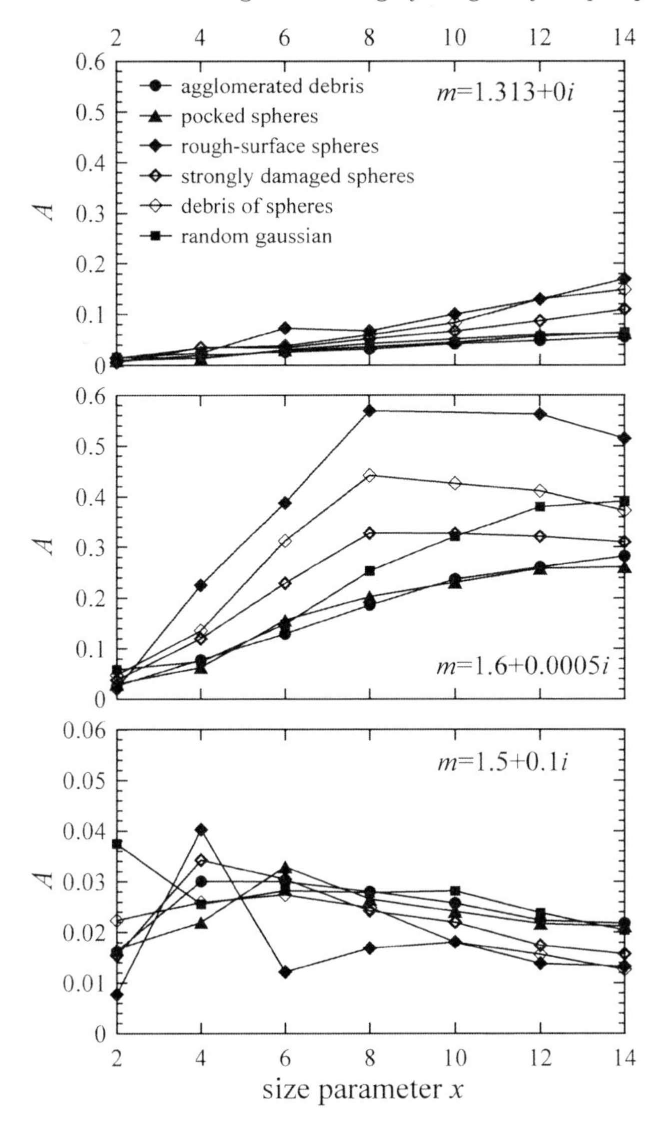

Fig. 2.16. Geometric albedo A as a function of size parameter x for six types of irregularly shaped particles. The top panels show the case of $m = 1.313 + 0i$, the middle panels show the case of $m = 1.6 + 0.0005i$, and the bottom panels show the case of $m = 1.5 + 0.1i$.

dependence of A may be non-monotonic, having a maximum in the range $x = 2$–8. Note that all the highly absorbing particles reveal a maximum in their geometric albedo A in the $x = 2$–4 range; whereas, rough-surface spheres, strongly damaged spheres, and debris of spheres produce maximum A at larger size parameters $x \sim 8$.

We note that the differences in the overall amplitude of albedo for particles with different refractive indices m are especially great. For instance, the highest geometric albedo A (up to 0.6) is produced by irregular particles with $m = 1.6 + 0.0005i$. Simultaneously, optically soft particles present a moderate amplitude in curves A vs. x (up to 0.18); whereas, the lowest albedo (up to only 0.04) can be found for highly absorbing particles. Interestingly, the lowest geometric albedo for particles with $m = 1.6 + 0.0005i$ observed at $x = 2$, is nearly coincident with the highest albedo for particles with $m = 1.5 + 0.1i$. Note, the geometric albedo of cometary dust particles averaged over the entire coma is estimated as $A = 0.05$ (Hanner, 2003). As one can see in Fig. 2.16, such a low geometric albedo can be obtained with either very small ($x \leq 2$) or highly absorbing particles. However, the average coma produces a NPB at small phase angles (e.g., Kiselev and Chernova, 1981; Chernova *et al.*, 1993; Hadamcik and Levasseur-Regourd, 2003), which is inconsistent with small particles (see Figs. 2.13–2.15 for size-dependences of $P_{\min}$). Therefore, low geometric albedo A of cometary dust particles suggests high absorption material.

2.4 Conclusion

In the present review, we have presented and discussed numerous aspects of light scattering by irregularly shaped particles having different morphologies and sizes comparable with the wavelength of incident radiation. Almost all characteristics of light scattering substantially depend on the morphology, constituent material, and size of particles. One principle goal is to make some principal characterization of the target particles (such as, the packing density, degree of absorption, real part of refractive index and size parameter x) from their light scattering. Of course, some *a priori* information on target particles will significantly simplify the remote-sensing inversion.

In general, the differential light-scattering parameters are more sensitive to the physical properties of particles than are the integral parameters. One of the most informative characteristic of light scattering is the angular dependence of the degree of linear polarization. In particular, the presence of the NPB at small phase angles indicates the particles are comparable with wavelength. Moreover, the shape of NPB and location of the polarization minimum $\alpha_{\min}$ strongly and, more or less predictably, depend on the size parameter of the particles; whereas, a simultaneous analysis of characteristics of NPB and positive polarization branch (PPB) at intermediate phase angles can be used to retrieve material absorption.

Acknowledgments

I thank Prof. Yuriy Shkuratov (Kharkov National University, Ukraine), Dr Gorden Videen (Army Research Laboratory, USA), and Prof. Karri Muinonen (University of Helsinki, Finland) for their permanent support of my research and, in particular, for the valuable comments on this review. This work is partially supported by the Academy of Finland (contract 1127461). This work was partially supported by NASA program for Outer Planets Research (grant NNX10AP93G).

References

Artymowicz, P., 1988: Radiation pressure forces on particles in the Beta Pictoris system, *Astrophys. J.*, **335**, L79–L82.

Augereau, J.-C., Beust, H., 2006: On the AU Microscopii debris disk. Density profiles, grain properties, and dust dynamics, *Astron. Astrophys.*, **455**, 987–999.

Bauer, S. E., Mishchenko, M. I., Lacis, A. A., Zhang, S., Perlwitz, J., Metzger, S. M., 2007: Do sulfate and nitrate coatings on mineral dust have important effects on radiative properties and climate modeling?, *J. Geophys. Res.*, **112**, D06307.

Belskaya, I. N., Shkuratov, Yu. G., Efimov, Yu. S., Shakhovskoy, N. M., Gil-Hutton, R., Cellino, A., Zubko, E. S., Ovcharenko, A. A., Bondarenko, S. Yu., Shevchenko, V. G., Fornasier, S., Barbieri, C., 2005: The F-type asteroids with small inversion angles of polarization, *Icarus*, **178**, 213–221.

Boehnhardt, H., Tozzi, G.P., Bagnulo, S., Muinonen, K., Nathues A., Kolokolova, L., 2008: Photometry and polarimetry of the nucleus of comet 2P/Encke, *Astron. Astrophys.*, **489**, 1337–1343.

Bohren, C.F., Huffman, D.R., 1983: *Absorption and Scattering of Light by Small Particles*, New York: John Wiley.

Bowell, E., Hapke, B., Domingue, D., Lumme, K., Peltoniemi, J., Harris, A. W., 1989: Application of photometric models to asteroids, in *Asteroids II*, R. Binzel *et al.* (eds), University of Arizona Press, Tucson, 524–556.

Brownlee, D. E., Joswiak, D. J., Schlutter, D. J., Pepin, R. O., Bradley, J. P., Love, S. G., 1995: Identification of individual cometary IDP's by thermally stepped He release, *Lunar Planet. Sci. Conf. XXVI*, 183–184.

Brownlee, D., Tsou, P., Aléon, J., and 180 coauthors, 2006: Comet 81P/Wild 2 Under a Microscope, *Science*, **314**, 1711–1716.

Burnett, D. S., 2006: NASA returns rocks from a comet, *Science*, **314**, 1709–1710.

Burns, J. A., Lamy, P. L., Soter, S., 1979: Radiation forces on small particles in the solar system, *Icarus*, **40**, 1–48.

Busemann, H., Nguyen, A. N., Cody, G. D., Hoppe, P., Kilcoyne, A. L. D., Stroud, R. M., Zega, T. J., Nittler, L. R., 2009: Ultra-primitive interplanetary dust particles from the comet 26P/Grigg-Skjellerup dust stream collection, *Earth and Planetary Science Letters*, **288**, 44–57.

Cellino, A., Belskaya, I. N., Bendjoya, Ph., Di Martino, M., Gil-Hutton, R., Muinonen, K., Tedesco, E. F., 2006: The strange polarimetric behavior of Asteroid (234) Barbara, *Icarus*, **180**, 565–567.

Dai, Z. R., Bradley, J. P., 2001: Iron-nickel sulfides in anhydrous interplanetary dust particles, *Geochimica et Cosmochimica Acta*, **65**, 3601–3612.

Dollfus, A., Bowell, E., 1971: Polarimetric Properties of the Lunar Surface and its Interpretation. Part I. Telescopic Observations, *Astron. Astrophys.*, **10**, 29–53.

Draine, B., 1988: The discrete-dipole approximation and its application to the interstellar graphite grains, *Astrophys. J.*, **333**, 848–872.

Draine, B. T., Flatau, P. J., 1994: The discrete dipole approximation for scattering calculations, *J. Opt. Soc. Am. A*, **11**, 1491–1499.

Dumont, R., Sanchez, F., 1975: Zodiacal light photopolarimetry. II. Gradients along the ecliptic and the phase functions of interplanetary matter, *Astron. Astrophys.*, **38**, 405–412.

Fulle, M., 2004: Motion of cometary dust, in *Comets II*, M. C. Festou *et al.* (eds), University of Arizona Press, Tucson, 565–575.

Chernova, G. P., Kiselev, N. N., Jockers, K., 1993: Polarimetric characteristics of dust particles as observed in 13 comets – Comparisons with asteroids, *Icarus*, **103**, 144–158.

Goodman, J. J., Draine, B. T., Flatau, P. J., 1991: Application of fast-Fourier-transform techniques to the discrete-dipole approximation, *Opt. Lett*, **16**, 1198–1200.

Hadamcik, E., Levasseur-Regourd, A.C., 2003: Imaging polarimetry of cometary dust: different comets and phase angles, *J. Quant. Spectrosc. Radiat. Transfer*, **79–80**, 661–678.

Hadamcik, E., Renard, J. B., Worms, J. C., Levasseur-Regourd, A. C., Masson, M., 2002: Polarization of light scattered by fluffy particles (PROGRA^2 experiment), *Icarus*, **155**, 497–508.

Hanner, M. S., 2003: The scattering properties of cometary dust, *J. Quant. Spectrosc. Radiat. Transfer*, **79–80**, 695–705.

Hanner, M. S., Bradley, J. P., 2004: Composition and mineralogy of cometary dust, in *Comets II*, M. C. Festou *et al.* (eds), University of Arizona Press, Tucson, 555–564.

Hanner, M. S., Giese, R. H., Weiss, K., Zerull, R., 1981: On the definition of albedo and application to irregular particles, *Astron. Astrophys.*, **104**, 42–46.

Hovenier, J. W., Volten, H., Muñoz, O., van der Zande, W. J., Waters, L. B. F. M., 2003: Laboratory studies of scattering matrices for randomly oriented particles: potentials, problems, and perspectives, *J. Quant. Spectrosc. Radiat. Transfer*, **79–80**, 741–755.

van de Hulst, H.C., 1981: *Light scattering by small particles*, New York: Dover.

Jessberger, E. K., Stephan, T., Rost, D., Arndt, P., Maetz, M., Stadermann, F. J., Brownlee, D. E., Bradley, J. P., Kurat, G., 2001: Properties of Interplanetary Dust: Information from Collected Samples, in *Interplanetary Dust*, E. Grün *et al.* (eds), Springer-Verlag, Berlin, 253–294.

Kimura, H., Mann, I., 1998: Radiation pressure cross-section for fluffy aggregates, *J. Quant. Spectrosc. Radiat. Transfer*, **60**, 425–438.

Kimura, H., Mann, I., 2004: Light scattering by large clusters of dipoles as an analog for cometary dust aggregates, *J. Quant. Spectrosc. Radiat. Transfer*, **89**, 155–164.

Kimura, H., Okamoto, H., Mukai, T., 2002: Radiation pressure and the Poynting-Robertson effect for fluffy dust particles, *Icarus*, **157**, 349–361.

Kiselev, N. N., Chernova, G. P., 1981: Phase functions of polarization and brightness and the nature of cometary atmosphere particles, *Icarus*, **48**, 473–481.

Kokhanovsky, A. A., 2008: *Aerosol Optics: Light absorption and scattering by particles in the atmosphere*, New York: Springer–Praxis.

Levasseur-Regourd, A. C., Dumont, R., Renard, J. B., 1990: A comparison between polarimetric properties of cometary dust and interplanetary dust particles, *Icarus*, **86**, 264–272.

Lindqvist, H., Muinonen, K., Nousiainen, T., 2009: Light scattering by coated Gaussian and aggregate particles, *J. Quant. Spectrosc. Radiat. Transfer*, **110**, 1398–1410.

Long, M., 1965: On the polarization and the wavelength dependence of sea echo, *IEEE Transactions on Antennas and Propagation*, **13**, 749–754.

Lumme, K., Rahola, J., Hovenier J. W., 1997: Light scattering by dense clusters of spheres, *Icarus*, **126**, 455–469.

Macke, A., Mishchenko, M. I., 1996: Asymmetry parameters of the phase function for large particles with multiple internal inclusions, *Bull. Am. Astron. Soc.*, **28**, 1122.

Mazets, E. P., Aptekar, R. L., Golenetskii, S. V., Guryan, Yu. A., Dyachkov, A. V., Ilyinskii, V. N., Panov, V. N., Petrov, G. G., Savvin, A. V., Sagdeev, R. Z., Sokolov, I. A., Khavenson, N. G., Shapiro, V. D., Shevchenko, V. I., 1986: Comet Halley dust environment from SP-2 detector measurements, *Nature*, **321**, 276–278.

Mishchenko, M. I., 1994: Asymmetry parameters of the phase function for densely packed scattering grains, *J. Quant. Spectrosc. Radiat. Transfer*, **52**, 95–110.

Mishchenko, M. I., Liu, L., 2007: Weak localization of electromagnetic waves by densely packed many-particle groups: Exact 3D results, *J. Quant. Spectrosc. Radiat. Transfer*, **106**, 616–621.

Mishchenko, M. I., Dlugach, J. M., Liu, L., Rosenbush, V. K., Kiselev, N. N., Shkuratov, Yu. G., 2009: Direct solutions of the Maxwell equations explain opposition phenomena observed for high-albedo Solar system objects, *Astrophys. J.*, **705**, L118–L122.

Muinonen, K., Nousiainen, T., Fast, P., Lumme, K., Peltoniemi, J.I., 1996: Light scattering by Gaussian random particles: ray optics approximation, *J. Quant. Spectrosc. Radiat. Transfer*, **55**, 577–601.

Muinonen, K., Zubko, E., Tyynelä, J., Shkuratov, Yu. G., Videen G., 2007: Light scattering by Gaussian random particles with discrete-dipole approximation, *J. Quant. Spectrosc. Radiat. Transfer*, **106**, 360–377.

Muñoz, O., Volten, H., de Haan, J. F., Vassen, W., Hovenier, J. W., 2000: Experimental determination of scattering matrices of olivine and Allende meteorite particles, *Astron. Astrophys.*, **360**, 777–788.

Muñoz, O., Volten, H., de Haan, J. F., Vassen, W., Hovenier, J. W., 2001: Experimental determination of scattering matrices of randomly oriented fly ash and clay particles at 442 and 633 nm, *J. Geophys. Res.*, **106**, 22833–22844.

Muñoz, O., Volten, H., Hovenier, J. W., Min, M., Shkuratov, Yu. G., Jalava, J. P., van der Zande, W. J., Waters, L. B. F. M., 2006: Experimental and computational study of light scattering by irregular particles with extreme refractive indices: hematite and rutile. *Astron. Astrophys.*, **446**, 525–535.

Nousiainen, T., 2009: Optical modeling of mineral dust particles: A review, *J. Quant. Spectrosc. Radiat. Transfer*, **110**, 1261–1279.

Nousiainen, T., Muinonen, K., 2007: Surface-roughness effects on single-scattering properties of wavelength-scale particles, *J. Quant. Spectrosc. Radiat. Transfer*, **106**, 389–397.

Ostro, S. J., 1993: Planetary radar astronomy, *Rev. Mod. Phys.*, **65**, 1235–1279.

Penttilä, A., Zubko, E., Lumme, K., Muinonen, K., Yurkin, M. A., Draine, B., Rahola, J., Hoekstra, A. G., Shkuratov, Yu., 2007: Comparison between discrete dipole implementations and exact techniques, *J. Quant. Spectrosc. Radiat. Transfer*, **106**, 417–436.

Rosenbush, V., Kolokolova, L., Lazarian, A., Shakhovskoy, N., Kiselev N., 2007: Circular polarization in comets: Observations of Comet C/1999 S4 (LINEAR) and tentative interpretation, *Icarus*, **186**, 317–330.

Shen, Y., Draine, B. T., Johnson, E. T., 2009: Modeling porous dust grains with ballistic aggregates. II. Light scattering properties, *Astrophys. J.*, **696**, 2126–2137.

Shkuratov, Yu.G., Opanasenko, N.V., 1992: Polarimetric and photometric properties of the Moon: Telescope observation and laboratory simulation. 2. The positive polarization, *Icarus*, **99**, 468–484.

Shkuratov, Yu.G., Opanasenko, N.V., Kreslavsky, M.A., 1992: Polarimetric and photometric properties of the Moon: Telescope observation and laboratory simulation. 1. The negative polarization, *Icarus*, **95**, 283–299.

Shkuratov, Yu., Bondarenko, S., Ovcharenko, A., Pieters, C., Hiroi, T., Volten, H., Munos, O., Videen, G., 2006: Comparative studies of the reflectance and degree of linear polarization of particulate surfaces and independently scattering particles, *J. Quant. Spectrosc. Radiat. Transfer*, **100**, 340–358.

Tyynelä, J., Zubko, E., Muinonen, K., Videen G., 2010: Interpretation of single-particle negative polarization at intermediate scattering angles, *Appl. Opt.*, submitted.

Videen, G., 2002: Polarization opposition effect and second-order ray tracing, *Appl. Opt.*, **41**, 5115–5121.

Vilaplana, R., Moreno, F., Molina, A., 2006: Study of the sensitivity of size-averaged scattering matrix elements of nonspherical particles to changes in shape, porosity and refractive index, *J. Quant. Spectrosc. Radiat. Transfer*, **100**, 415–428.

Volten, H., Muñoz, O., Hovenier, J. W., Rietmeijer, F. J. M., Nuth, J. A., Waters, L. B. F. M., van der Zande, W. J., 2007: Experimental light scattering by fluffy aggregates of magnesiosilica, ferrosilica, and alumina cosmic dust analogs, *Astron. Astrophys.*, **470**, 377–386.

Wickramasinghe, N. C., Wickramasinghe, J. T., 2003: Radiation pressure on bacterial clumps in the solar vicinity and their survival between interstellar transits, *Astrophysics and Space Science*, **286**, 453–459.

Woessner, P., Hapke, B., 1987: Polarization of light scattered by clover, *Remote Sensing of Environ*, **21**, 243–261.

Yurkin, M.A., Hoekstra, A.G., 2007L The discrete dipole approximation: An overview and recent developments, *J. Quant. Spectrosc. Radiat. Transfer*, **106**, 558–589.

Zellner, B., Gradie, J., 1976: Minor planets and related objects. XX - Polarimetric evidence for the albedos and compositions of 94 asteroids, *Astron. J.*, **81**, 262–280.

Zubko, E., Shkuratov, Yu., Muinonen, K., 2001: Light scattering by composite particles comparable with wavelength and their approximation by systems of spheres, *Optics and Spectroscopy*, **91**, 273–277.

Zubko, E. S., Shkuratov, Yu. G., Hart, M., Eversole, J., Videen, G., 2003: Backscattering and negative polarization of agglomerate particles, *Opt. Lett.*, **28**, 1504–1506.

Zubko, E. S., Shkuratov, Yu. G., Hart, M., Eversole, J., Videen, G., 2004: Backscatter of agglomerate particles, *J. Quant. Spectrosc. Radiat. Transfer.*, **28**, 163–171.

Zubko, E., Shkuratov, Yu., Kiselev, N., Videen, G., 2006: DDA simulations of light scattering by small irregular particles with various structure, *J. Quant. Spectrosc. Radiat. Transfer*, **101**, 416–434.

Zubko, E., Muinonen, K., Shkuratov, Yu., Videen, G., Nousiainen, T., 2007: Scattering of light by roughened Gaussian random particles, *J. Quant. Spectrosc. Radiat. Transfer*, **106**, 604–615.

Zubko, E., Kimura, H., Shkuratov, Yu., Muinonen, K., Yamamoto, T., Okamoto, H., Videen, G., 2009a: Effect of absorption on light scattering by agglomerated debris particles, *J. Quant. Spectrosc. Radiat. Transfer*, **110**, 1741–1749.

Zubko, E., Furusho, R., Yamamoto, T., Videen, G., Muinonen, K., 2009b: Interpretation of photo-polarimetric observations of comet 17P/Holmes during outburst in 2007, *Bull. Am. Astron. Soc.*, **41**, 1035.

Zubko, E., Petrov, D., Grynko, Y., Shkuratov, Yu., Okamoto, H., Muinonen, K., Nousiainen, T., Kimura, H., Yamamoto, T., Videen, G., 2010: Validity criteria of the discrete dipole approximation, *Appl. Opt.*, **49**, 1267–1279.

Zubko, E., Videen, G., Shkuratov, Yu., Muinonen, Yamamoto, T., 2011: The Umov effect for single irregularly shaped particles with sizes comparable with wavelength, *Icarus*, **212**, 403–415.

3 Finite-difference time-domain solution of light scattering by arbitrarily shaped particles and surfaces

Wenbo Sun, Gorden Videen, Qiang Fu, Stoyan Tanev, Bing Lin, Yongxiang Hu, Zhaoyan Liu, and Jianping Huang

3.1 Introduction

The scattering and absorption of electromagnetic waves by irregularly shaped particles and arbitrary surfaces occur in the atmosphere, ocean, and optical devices. In this chapter, we present the finite-difference time-domain (FDTD) method [1–6] that can be used to calculate light scattering by arbitrary particles and surfaces. The FDTD technique is a numerical solution to Maxwell's equations and is formulated by replacing temporal and spatial derivatives in Maxwell's equations with their finite-difference equivalences. This method can be accurately applied to general electromagnetic structures including arbitrary particles and surfaces. The FDTD technique has been successfully applied to calculate light scattering and absorption by particles of different shapes in free space [5] and in absorbing medium [6]. Recently, an advanced FDTD model to calculate the interaction of electromagnetic radiation with arbitrary dielectric surfaces has been developed [7]. In the following sections, these FDTD light-scattering models are reviewed.

3.2 Finite-difference time-domain method for light scattering by particles

Remote-sensing studies require precise knowledge of scattering and absorption by non-spherical particles. To date, however, except for some simple particle shapes, such as spheres [8], double-sphere systems [9, 10], spheroids [11, 12], circular cylinders [13, 14], Chebyshev particles [15, 16], finite circular cylinders [17–19], and cube-like particles [20, 21], analytic solutions are not available for light scattering by irregularly shaped particles. In the small particle limit, the Rayleigh theory [22] and the quasi-static approximation [23] can be applied, and when the size parameter is larger than ∼40, the geometric optics method [24] can be used for non-spherical particles. However, in the resonant region [25], these methods completely break down. To obtain accurate solutions for light scattering by particles of arbitrary shapes, numerical approaches such as the discrete dipole approximation [26] (DDA) and the FDTD algorithms [5, 6] have been developed.

Pioneered by the research of Yee [1] and many other electrical engineers, the FDTD solutions of Maxwell's equations have been applied extensively to electromagnetic problems such as antenna design, radar cross-section computation, waveguide analysis, and some other open-structure problems. This method can be applied accurately to general electromagnetic structures including particles of arbitrary shapes [5,6].

3.2.1 Scattered/total-field finite-difference time-domain method

The FDTD technique can be used to calculate the electromagnetic scattering and absorption in the time domain by directly solving the finite-differenced Maxwell's equations [1–6]. The spatial and temporal derivatives of the electric and magnetic fields are approximated using a finite-difference scheme with spatial and temporal discretizations selected to limit the numerical dispersion errors and ensure numerical stability of the algorithm [3]. The scatterer is embedded in a finite computational domain bounded by a truncation boundary [27–36]. The electromagnetic properties of the scatterer and the host medium are specified by assigning the permittivity, permeability, and conductivity at each grid point. A time-stepping iteration is used to simulate the field variation with time. The time series at each grid point is transformed into the fields in frequency domain using the discrete Fourier transform (DFT) [5, 6]. The scattering and absorption quantities are calculated with the fields in the frequency domain.

The FDTD scheme with the split-field perfectly matched layer (PML) absorbing boundary condition (ABC) [33, 34] was applied successfully to calculate the light scattering and absorption by optically thin particles with size parameters as large as 40 in a free space [5], which was the first accurate FDTD result for light scattering by particles with such size parameter. The same scheme also has been used to calculate light scattering and absorption by particles with refractive indices as large as $7.1499 + 2.914i$ [37]. In this section, we will review the FDTD technique, which is suitable to calculate light scattering and absorption by particles embedded in a more general host medium, one that can be absorbing (with non-absorbing as its special case) [6].

In the finite-difference time-domain-algorithm, the field updating equations are simply the finite-difference correspondences of Maxwell's equations in the time domain:

$$\nabla \times \boldsymbol{E} = -\mu \frac{\partial \boldsymbol{H}}{\partial t}, \tag{2.1a}$$

$$\nabla \times \boldsymbol{H} = \varepsilon \frac{\partial \boldsymbol{E}}{\partial t}, \tag{2.1b}$$

where μ and ε denote the absolute permeability and absolute permittivity, respectively. Assuming the absolute permeability and absolute permittivity in vacuum are μ_0 and ε_0, respectively, the absolute permeability and absolute permittivity of a medium can be expressed as $\mu = \mu' \mu_0$ and $\varepsilon = \varepsilon' \varepsilon_0$, where μ' and ε' denote the relative permeability and relative permittivity, respectively. Since the (relative) refractive index $m = (\varepsilon' \mu')^{1/2}$ and for non-ferromagnetic medium $\mu' = 1$, the real and imaginary parts of ε' can be expressed by the real and imaginary parts of m in the form $\varepsilon'_r = m_r^2 - m_i^2$ and $\varepsilon'_i = 2m_r m_i$, respectively.

Following Sun *et al.* [5] the explicit finite-difference approximation of Maxwell's equations can be derived by assuming that the time-dependent part of the electromagnetic field is $\exp(-i\omega t)$, so the electric and the magnetic fields can be written in the form

$$\mathbf{E}(x, y, z, t) = \mathbf{E}(x, y, z) \exp(-i\omega t), \tag{2.2a}$$

$$\mathbf{H}(x, y, z, t) = \mathbf{H}(x, y, z) \exp(-i\omega t), \tag{2.2b}$$

where ω is the angular frequency of the light. Inserting Eqs. (2.2) into Eq. (2.1b) we have

$$\nabla \times \mathbf{H}(x, y, z) = -i\omega(\varepsilon_r + \varepsilon_i i)\mathbf{E}(x, y, z), \tag{2.3}$$

where ε_i and ε_r are the absolute imaginary and real permittivity, respectively. Multiplying Eq. (2.3) with $\exp(-i\omega t)$ and using Eqs. (2.2), we obtain

$$\nabla \times \mathbf{H}(x, y, z, t) = \omega\varepsilon_i\mathbf{E}(x, y, z, t) + \varepsilon_r\frac{\partial\mathbf{E}(x, y, z, t)}{\partial t}. \tag{2.4}$$

Equation (2.4) can be rewritten in a form

$$\frac{\exp(\tau t)}{\varepsilon_r}\nabla \times \mathbf{H}(x, y, z, t) = \frac{\partial[\exp(\tau t)\mathbf{E}(x, y, z, t)]}{\partial t}, \tag{2.5}$$

where $\tau = \omega\varepsilon_i/\varepsilon_r$. Using the central finite-difference approximation for the temporal derivative in Eq. (2.5) over the time interval $[n\Delta t,\ (n+1)\Delta t]$, we have

$$\begin{aligned}\mathbf{E}^{n+1}(x, y, z) =& \exp(-\tau\Delta t)\mathbf{E}^n(x, y, z) + \exp(-\tau\Delta t/2)\frac{\Delta t}{\varepsilon_r}\nabla \\ &\times \mathbf{H}^{n+1/2}(x, y, z),\end{aligned} \tag{2.6}$$

where Δt is the time increment and n is an integer denoting time step. Eq. (2.6) shows that the electric and magnetic fields are evaluated at alternating half time steps.

By discretizing Eq. (2.1a) over the time interval of $[(n-1/2)\Delta t,\ (n+1/2)\Delta t]$, which is a half time step earlier than the time step when the electric field is evaluated, we have

$$\mathbf{H}^{n+1/2}(x, y, z) = \mathbf{H}^{n-1/2}(x, y, z) - \frac{\Delta t}{\mu}\nabla \times \mathbf{E}^n(x, y, z). \tag{2.7}$$

Furthermore, the spatial derivatives in Eqs. (2.6) and (2.7) can be approximated by the ratio of fields' differences at discrete spatial points and the spatial increment (Δs). For example, in a Cartesian grid system the x components of magnetic and electric fields, e.g., are in the forms

$$\begin{aligned}&H_x^{n+1/2}(i, j+1/2, k+1/2)\\ &= H_x^{n-1/2}(i, j+1/2, k+1/2) + \frac{\Delta t}{\mu(i, j+1/2, k+1/2)\Delta s}\\ &\times [E_y^n(i, j+1/2, k+1) - E_y^n(i, j+1/2, k) + E_z^n(i, j, k+1/2)\\ &\quad - E_z^n(i, j+1, k+1/2)],\end{aligned} \tag{2.8a}$$

$$
\begin{aligned}
E_x^{n+1}(i+1/2,j,k) &= \exp\left[-\frac{\varepsilon_i(i+1/2,j,k)}{\varepsilon_r(i+1/2,j,k)}\omega\Delta t\right]E_x^n(i+1/2,j,k) \\
&+ \exp\left[-\frac{\varepsilon_i(i+1/2,j,k)}{\varepsilon_r(i+1/2,j,k)}\omega\Delta t/2\right]\frac{\Delta t}{\varepsilon_r(i+1/2,j,k)\Delta s} \\
&\times \Big[H_y^{n+1/2}(i+1/2,j,k-1/2) - H_y^{n+1/2}(i+1/2,j,k+1/2) \\
&+ H_z^{n+1/2}(i+1/2,j+1/2,k) - H_z^{n+1/2}(i+1/2,j-1/2,k)\Big], \qquad (2.8b)
\end{aligned}
$$

where E_x, E_y, E_z and H_x, H_y, H_z denote electric and magnetic components, respectively. To guarantee the numerical stability of the FDTD scheme in the absorbing medium, we use $\Delta t = \Delta s/2c$, where c is the light speed in free space. This is a more strict stability criterion than the Courant–Friedrichs–Levy condition [2]. The coordinates (i, j, k) denote the center positions of the cubic cells in the FDTD grid. The positions of the magnetic and electric field components on a cubic cell are shown in Fig. 3.1.

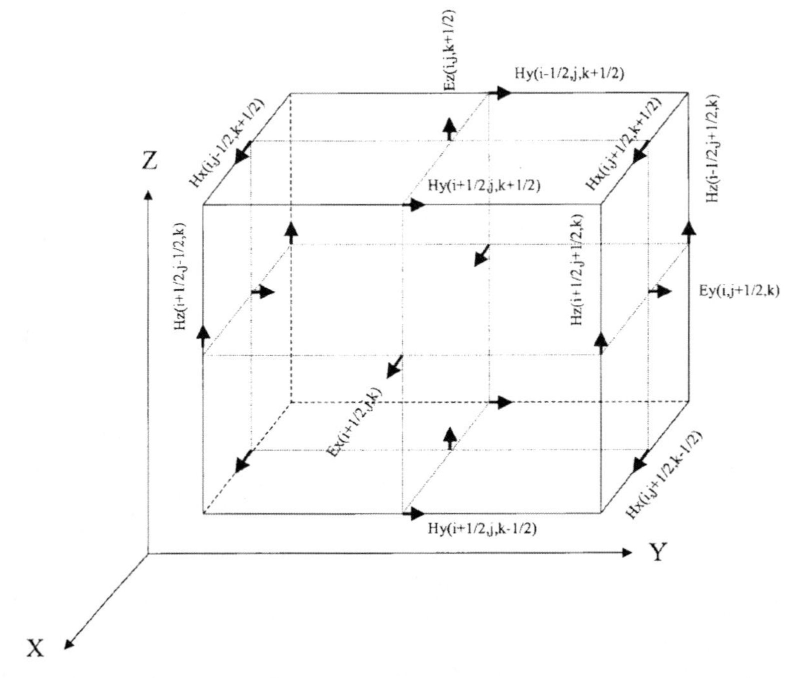

Fig. 3.1. Positions of the electric- and the magnetic-field components in an elementary cubic cell of the FDTD lattice.

Note here that, when the host medium is a free space, the FDTD field updating equations have no difference from those for a non-free host space [5]. The host medium is integrated in the computational domain by assigning the host material properties at the grid points outside the scattering particle.

3.2.2 Incident wave source conditions for open dielectric medium space

On the FDTD grid points, we use the total- and scattered-field formulation [29, 38, 39] to excite the magnetic and electric fields in order to simulate a linearly polarized plane wave propagating in a finite region of a homogeneous dielectric medium. In this formulation, a closed inner surface is assumed in the computational domain. Based on the equivalence theorem [38], the existence of wave-excitation in the spatial domain enclosed by the closed inner surface can be replaced by the equivalent electric and magnetic currents on the inner surface. If there is a scatterer inside the closed surface, the interior fields are the total fields (incident and scattered) and the fields outside are just the scattered fields. In this study, on a rectangular closed interface between the total- and scattered-field zones as shown in Fig. 3.2, the magnetic and electric wave sources are implemented. For example, in a host medium the x components of the magnetic and electric fields on the rectangular faces of the closed interface are as follows [3, 6]:

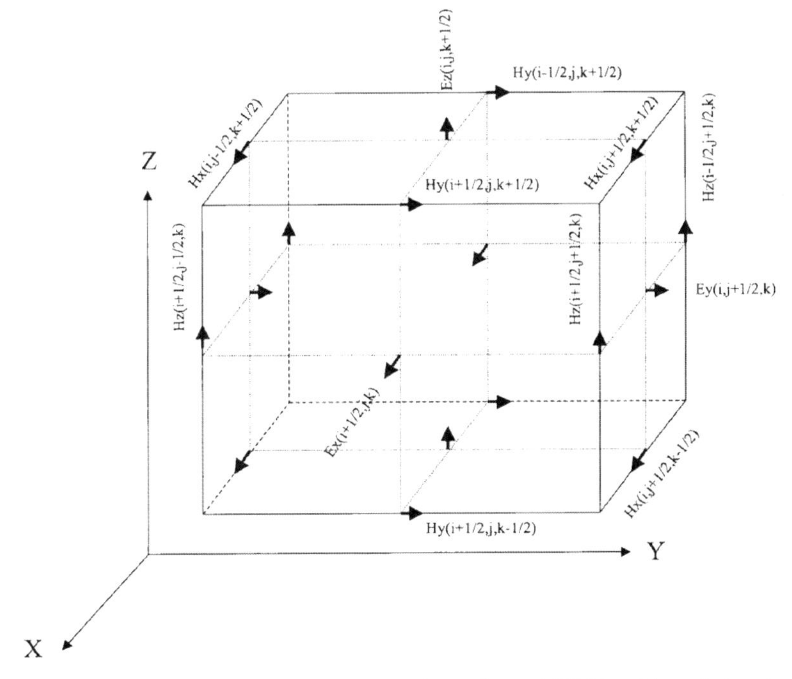

Fig. 3.2. Configuration of the one-dimensional auxiliary FDTD grid and the closed rectangular interface of the total field and scattered field. Here i_a and i_b denote the start and end positions of the total-field space, respectively, in the x direction; j_a and j_b denote the start and end positions of the total-field space, respectively, in the y direction; and k_a and k_b denote the start and end positions of the total-field space, respectively, in the z direction.

At the face $y = j_a - 1/2 (i = i_a, ..., i_b; j = j_a - 1/2; k = k_a + 1/2, ..., k_b - 1/2)$,

$$H_x^{n+1/2}(i, j_a - 1/2, k) = \{H_x^{n+1/2}(i, j_a - 1/2, k)\}_{(2.8a)} + \frac{\Delta t}{\mu(i, j_a - 1/2, k)\Delta s} E_{z,\mathrm{inc}}^{n}(i, j_a, k). \quad (2.9a)$$

At the face $y = j_b + 1/2 (i = i_a, ..., i_b; j = j_b + 1/2; k = k_a + 1/2, ..., k_b - 1/2)$,

$$H_x^{n+1/2}(i, j_b + 1/2, k) = \{H_x^{n+1/2}(i, j_b + 1/2, k)\}_{(2.8a)} - \frac{\Delta t}{\mu(i, j_b + 1/2, k)\Delta s} E_{z,\mathrm{inc}}^{n}(i, j_b, k). \quad (2.9b)$$

At the face $z = k_a - 1/2 (i = i_a, ..., i_b; j = j_a + 1/2, ..., j_b - 1/2; k = k_a - 1/2)$,

$$H_x^{n+1/2}(i, j, k_a - 1/2) = \{H_x^{n+1/2}(i, j, k_a - 1/2)\}_{(2.8a)} - \frac{\Delta t}{\mu(i, j, k_a - 1/2)\Delta s} E_{y,\mathrm{inc}}^{n}(i, j, k_a). \quad (2.9c)$$

At the face $z = k_b + 1/2 (i = i_a, ..., i_b; j = j_a + 1/2, ..., j_b - 1/2; k = k_b + 1/2)$,

$$H_x^{n+1/2}(i, j, k_b + 1/2) = \{H_x^{n+1/2}(i, j, k_b + 1/2)\}_{(2.8a)} + \frac{\Delta t}{\mu(i, j, k_b + 1/2)\Delta s} E_{y,\mathrm{inc}}^{n}(i, j, k_b). \quad (2.9d)$$

At the face $y = j_a (i = i_a + 1/2, ..., i_b - 1/2; j = j_a; k = k_a, ..., k_b)$,

$$E_x^{n+1}(i, j_a, k) = \{E_x^{n+1}(i, j_a, k)\}_{(2.8b)} - \exp\left[-\frac{\varepsilon_i(i, j_a, k)}{\varepsilon_r(i, j_a, k)}\omega\Delta t/2\right]\frac{\Delta t}{\varepsilon_r(i, j_a, k)\Delta s} H_{z,\mathrm{inc}}^{n+1/2}(i, j_a - 1/2, k). \quad (2.10a)$$

At the face $y = j_b (i = i_a + 1/2, ..., i_b - 1/2; j = j_b; k = k_a, ..., k_b)$,

$$E_x^{n+1}(i, j_b, k) = \{E_x^{n+1}(i, j_b, k)\}_{(2.8b)} + \exp\left[-\frac{\varepsilon_i(i, j_b, k)}{\varepsilon_r(i, j_b, k)}\omega\Delta t/2\right]\frac{\Delta t}{\varepsilon_r(i, j_b, k)\Delta s} H_{z,\mathrm{inc}}^{n+1/2}(i, j_b + 1/2, k). \quad (2.10b)$$

At the face $z = k_a (i = i_a + 1/2, ..., i_b - 1/2; j = j_a, ..., j_b; k = k_a)$,

$$E_x^{n+1}(i, j, k_a) = \{E_x^{n+1}(i, j, k_a)\}_{(2.8b)} + \exp\left[-\frac{\varepsilon_i(i, j, k_a)}{\varepsilon_r(i, j, k_a)}\omega\Delta t/2\right]\frac{\Delta t}{\varepsilon_r(i, j, k_a)\Delta s} H_{y,\mathrm{inc}}^{n+1/2}(i, j, k_a - 1/2). \quad (2.10c)$$

At the face $z = k_b (i = i_a + 1/2, ..., i_b - 1/2; j = j_a, ..., j_b; k = k_b)$,

$$E_x^{n+1}(i, j, k_b) = \{E_x^{n+1}(i, j, k_b)\}_{(2.8b)} - \exp\left[-\frac{\varepsilon_i(i, j, k_b)}{\varepsilon_r(i, j, k_b)}\omega\Delta t/2\right]\frac{\Delta t}{\varepsilon_r(i, j, k_b)\Delta s} H_{y,\mathrm{inc}}^{n+1/2}(i, j, k_b + 1/2). \quad (2.10d)$$

Note that in Eqs. (2.9) and (2.10), $\{H_x^{n+1/2}(i,j,k)\}_{(2.8a)}$ and $\{E_x^{n+1}(i,j,k)\}_{(2.8b)}$ denote the magnetic and electric fields directly from Eqs. (2.8a) and (2.8b), respectively. The incident field$E^n_{y,\text{inc}}$,$E^n_{z,\text{inc}}$ and$H^{n+1/2}_{y,\text{inc}}$, $H^{n+1/2}_{z,\text{inc}}$ in Eqs. (2.9) and (2.10) are from the linear interpolation of the fields produced by an auxiliary one-dimensional FDTD scheme. This interpolation treatment numerically can significantly reduce the computational time for obtaining incident fields at the grid points on the total/scattered-field interface. This scheme can simulate the propagation of an incident plane wave on the one-dimensional grid starting at $m = 0$ (origin of the 3D grid) and stretching in the incident direction to a maximum position $m = m_{\max}$, as shown in Fig. 3.1. The incident wave vector $\mathbf{k}_{\text{inc}}$ is oriented with a zenith angle θ and an azimuth angle ϕ. The parameter $m_{\max}$is chosen to be half of the total simulation time steps for the incident wave propagation in a dielectric medium. Because we cannot use a transmitting boundary condition (TBC) in the medium that is absorbing for the truncation of the one-dimensional spatial domain, the selected $m_{\max}$ value is necessary so that no numerical reflection occurs at the forward end of the one-dimensional domain before the 3D FDTD simulation ends. A Gaussian-pulse hard wave source is positioned at the one-dimensional grid point $m = 2$ in the form [3–6]

$$E^n_{\text{inc}}(m=2) = \exp\left[-\left(\frac{t}{30\Delta t} - 5\right)^2\right]. \tag{2.11}$$

Using the hard wave source rather than a soft one [3] at $m = 2$, the field at the grid points $m = 0$ and 1 do not affect the field at the grid points $m \geq 2$, therefore no boundary conditions need to be considered at this grid end. In practice, Eq. (2.11) cannot produce a perfect spatial Gaussian distribution in an absorbing medium. The front part of the pulse becomes steeper, but the rear part becomes flatter in the absorbing medium than in free space. However, the FDTD simulation does not require a perfect spatial shape of the pulse. Any pulse which contains the required frequency should be able to be used.

In a dielectric medium, assuming a plane wave is incident from the coordinate origin to the closed interface of the total and scattered fields as shown in Fig. 3.2, the one-dimensional FDTD equations are

$$\begin{aligned} H^{n+1/2}_{\text{inc}}(m+1/2) &= H^{n-1/2}_{\text{inc}}(m+1/2) \\ &+ \frac{\Delta t}{\mu(m+1/2)\Delta s\left[\dfrac{v_p(\theta=0,\phi=0)}{v_p(\theta,\phi)}\right]} \\ &\times \left[E^n_{\text{inc}}(m) - E^n_{\text{inc}}(m+1)\right], \end{aligned} \tag{2.12a}$$

$$\begin{aligned} E^{n+1}_{\text{inc}}(m) &= \exp\left[-\frac{\varepsilon_i(m)}{\varepsilon_r(m)}\omega\Delta t\right] E^n_{\text{inc}}(m) \\ &+ \exp\left[-\frac{\varepsilon_i(m)}{\varepsilon_r(m)}\omega\Delta t/2\right] \frac{\Delta t}{\varepsilon_r(m)\Delta s\left[\dfrac{v_p(\theta=0,\phi=0)}{v_p(\theta,\phi)}\right]} \\ &\times \left[H^{n+1/2}_{\text{inc}}(m-1/2) - H^{n+1/2}_{\text{inc}}(m+1/2)\right], \end{aligned} \tag{2.12b}$$

where $\varepsilon_i(m)$ and $\varepsilon_r(m)$ denote the imaginary and real permittivity of the host medium at position m, respectively, and $\mu(m+1/2)$ is the permeability of the host medium at position $(m+1/2)$. The equalization factor

$$\left[\frac{v_p(\theta=0,\phi=0)}{v_p(\theta,\phi)}\right] \leq 1$$

is a ratio of numerical phase velocities in the 3D FDTD grid. Using the same spatial and temporal increments in the one-dimensional source and the 3D FDTD simulations, the wave in the source grid would propagate more slowly than the wave in the 3D grid. This is because the wave in the source grid behaves as if it were on axis in the 3D grid. Therefore, if the incident wave is not in the direction of the axis of the 3D grid, it is necessary to use an equalization factor to slightly speed up the wave in the one-dimensional source grid to equalize the numerical phase velocities of the incident wave in the 3D and one-dimensional source grids. The equalization factor

$$\left[\frac{v_p(\theta=0,\phi=0)}{v_p(\theta,\phi)}\right]$$

can be calculated using Newton's method for a solution of the numerical dispersion relation in the 3D FDTD algorithm [3].

3.2.3 Uniaxial perfectly matched layer absorbing boundary condition

When the medium between the target and the absorbing boundary is not free space, the uniaxial PML (UPML) developed by Sacks *et al.* [35] should be used to truncate the computational domain in the FDTD calculation. The UPML is a physical model based on an anisotropic, perfectly matched medium. For a single interface, the anisotropic medium is uniaxial and is composed of both electric permittivity and magnetic permeability tensors. For the scattered/total field formulation of the FDTD method, the conventional UPML formulations can be applied directly; however, for the scattered-field formulation of the FDTD method, the UPML ABC must be modified to account for the incident-wave source terms.

Generally, to match a UPML along a planar boundary to a isotropic half-space characterized by permittivity ε and conductivity σ, the time-harmonic Maxwell's equations can be written in the form [36]:

$$\nabla \times \mathbf{H}(x,y,z) = (i\omega\varepsilon + \sigma)\bar{\bar{s}}\mathbf{E}(x,y,z), \tag{2.13a}$$

$$\nabla \times \mathbf{E}(x,y,z) = -i\omega\mu_0\bar{\bar{s}}\mathbf{H}(x,y,z), \tag{2.13b}$$

where $\bar{\bar{s}}$ is the diagonal tensor defined by

$$\begin{aligned}\bar{\bar{s}} &= \begin{bmatrix} s_x^{-1} & 0 & 0 \\ 0 & s_x & 0 \\ 0 & 0 & s_x \end{bmatrix} \begin{bmatrix} s_y & 0 & 0 \\ 0 & s_y^{-1} & 0 \\ 0 & 0 & s_y \end{bmatrix} \begin{bmatrix} s_z & 0 & 0 \\ 0 & s_z & 0 \\ 0 & 0 & s_z^{-1} \end{bmatrix} \\ &= \begin{bmatrix} s_y s_z s_x^{-1} & 0 & 0 \\ 0 & s_x s_z s_y^{-1} & 0 \\ 0 & 0 & s_x s_y s_z^{-1} \end{bmatrix},\end{aligned} \tag{2.14}$$

where

$$s_x = \kappa_x + \frac{\sigma_x}{i\omega\varepsilon_0}, \quad s_y = \kappa_y + \frac{\sigma_y}{i\omega\varepsilon_0}, \quad \text{and} \quad s_z = \kappa_z + \frac{\sigma_z}{i\omega\varepsilon_0}.$$

Note here that the UPML properties (κ_x, σ_x), (κ_y, σ_y), and (κ_z, σ_z) are independent of the medium permittivity ε and conductivity σ, and are assigned to the FDTD grids in the UPML with the following strategy: (1) at the $x_{\min}$ and $x_{\max}$ boundaries, $\sigma_y = \sigma_z = 0$ and $\kappa_y = \kappa_z = 1$; at the $y_{\min}$ and $y_{\max}$boundaries, $\sigma_x = \sigma_z = 0$ and $\kappa_x = \kappa_z = 1$; at the $z_{\min}$ and $z_{\max}$boundaries, $\sigma_y = \sigma_x = 0$ and $\kappa_y = \kappa_x = 1$; (2) at the $x_{\min}$, $x_{\max}$ and $y_{\min}$, $y_{\max}$ overlapping dihedral corners, $\sigma_z = 0$ and $\kappa_z = 1$; at the $y_{\min}$, $y_{\max}$ and $z_{\min}$, $z_{\max}$ overlapping dihedral corners, $\sigma_x = 0$ and $\kappa_x = 1$; at the $z_{\min}$, $z_{\max}$ and $x_{\min}$, $x_{\max}$ overlapping dihedral corners, $\sigma_y = 0$ and $\kappa_y = 1$; (3) at all overlapping trihedral corners, the complete general tensor in Eq. (2.14) is used. To reduce the numerical reflection from the UPML, several profiles have been suggested for grading (κ_x, σ_x), (κ_y, σ_y) and (κ_z, σ_z) when they do not adopt the aforesaid specific values. In practice, a polynomial grading of the UPML material parameters is used; e.g., we can simply specify (κ_x, σ_x) as [36]

$$\kappa_x(x) = 1 + (x/d)^m(\kappa_{x,\max} - 1), \tag{2.15a}$$

$$\sigma_x(x) = (x/d)^m \sigma_{x,\max}, \tag{2.15b}$$

where x is the depth in the UPML and d is the UPML thickness in this direction. The parameters $\kappa_{x,\max}$ and $\sigma_{x,\max}$ denote the maximum κ_x and σ_x at the outermost layer of the UPML; e.g., considering an x-directed plane wave impinging at angle θ upon a PEC-backed UPML with polynomial grading material properties, the reflection factor can be derived as [33]

$$R(\theta) = \exp\left[-\frac{2\cos\theta}{\varepsilon_0 c}\int_0^d \sigma(x)dx\right] = \exp\left[-\frac{2\sigma_{x,\max} d\cos\theta}{\varepsilon_0 c(m+1)}\right]. \tag{2.16}$$

Therefore, with a reflection factor $R(0)$ for normal incidence, $\sigma_{x,\max}$ can be defined as

$$\sigma_{x,\max} = -\frac{(m+1)\varepsilon_0 c \ln[R(0)]}{2d}. \tag{2.17}$$

As an accurate approach, $R(0)$ can be 10^{-12} to 10^{-5}, and $\kappa_{x,\max}$ can be a real number from 1 to 30.

In this study, we implement the UPML only in the boundary layers to reduce the memory and CPU time requirement. In the non-UPML region, the common FDTD formulations of Eq. (2.8) are used.

To obtain explicitly the updating equations for the magnetic field in the UPML using Eq. (2.13b), the three components of an auxiliary vector field variable B are introduced as [36]

$$B_x(x,y,z) = \mu\left(\frac{s_z}{s_x}\right)H_x(x,y,z), \tag{2.18a}$$

$$B_y(x,y,z) = \mu\left(\frac{s_x}{s_y}\right)H_y(x,y,z), \tag{2.18b}$$

$$B_z(x,y,z) = \mu\left(\frac{s_y}{s_z}\right)H_z(x,y,z). \tag{2.18c}$$

Then Eq. (2.13b) can be expressed as

$$\begin{bmatrix} \frac{\partial E_y(x,y,z)}{\partial z} - \frac{\partial E_z(x,y,z)}{\partial y} \\ \frac{\partial E_z(x,y,z)}{\partial x} - \frac{\partial E_x(x,y,z)}{\partial z} \\ \frac{\partial E_x(x,y,z)}{\partial y} - \frac{\partial E_y(x,y,z)}{\partial x} \end{bmatrix} = i\omega \begin{bmatrix} s_y & 0 & 0 \\ 0 & s_z & 0 \\ 0 & 0 & s_x \end{bmatrix} \begin{bmatrix} B_x(x,y,z) \\ B_y(x,y,z) \\ B_z(x,y,z) \end{bmatrix}. \tag{2.19}$$

On the other hand, inserting the definitions of s_x, s_y and s_z into Eq. (2.18) and reformulating yields the following:

$$\left(i\omega\kappa_x + \frac{\sigma_x}{\varepsilon_0}\right) B_x(x,y,z) = \left(i\omega\kappa_z + \frac{\sigma_z}{\varepsilon_0}\right) \mu H_x(x,y,z), \tag{2.20a}$$

$$\left(i\omega\kappa_y + \frac{\sigma_y}{\varepsilon_0}\right) B_y(x,y,z) = \left(i\omega\kappa_x + \frac{\sigma_x}{\varepsilon_0}\right) \mu H_y(x,y,z), \tag{2.20b}$$

$$\left(i\omega\kappa_z + \frac{\sigma_z}{\varepsilon_0}\right) B_z(x,y,z) = \left(i\omega\kappa_y + \frac{\sigma_y}{\varepsilon_0}\right) \mu H_z(x,y,z). \tag{2.20c}$$

Applying the inverse Fourier transform using the identity $i\omega f(\omega) \to \partial f(t)/\partial t$ to Eqs. (2.19) and (2.20) yields the equivalent time-domain differential equations:

$$\begin{bmatrix} \frac{\partial E_y(x,y,z,t)}{\partial z} - \frac{\partial E_z(x,y,z,t)}{\partial y} \\ \frac{\partial E_z(x,y,z,t)}{\partial x} - \frac{\partial E_x(x,y,z,t)}{\partial z} \\ \frac{\partial E_x(x,y,z,t)}{\partial y} - \frac{\partial E_y(x,y,z,t)}{\partial x} \end{bmatrix} = \frac{\partial}{\partial t} \begin{bmatrix} \kappa_y & 0 & 0 \\ 0 & \kappa_z & 0 \\ 0 & 0 & \kappa_x \end{bmatrix} \begin{bmatrix} B_x(x,y,z,t) \\ B_y(x,y,z,t) \\ B_z(x,y,z,t) \end{bmatrix} + \frac{1}{\varepsilon_0} \begin{bmatrix} \sigma_y & 0 & 0 \\ 0 & \sigma_z & 0 \\ 0 & 0 & \sigma_x \end{bmatrix} \begin{bmatrix} B_x(x,y,z,t) \\ B_y(x,y,z,t) \\ B_z(x,y,z,t) \end{bmatrix}, \tag{2.21}$$

$$\kappa_x \frac{\partial B_x(x,y,z,t)}{\partial t} + \frac{\sigma_x}{\varepsilon_0} B_x(x,y,z,t) = \mu\kappa_z \frac{\partial H_x(x,y,z,t)}{\partial t} + \mu\frac{\sigma_z}{\varepsilon_0} H_x(x,y,z,t), \tag{2.22a}$$

$$\kappa_y \frac{\partial B_y(x,y,z,t)}{\partial t} + \frac{\sigma_y}{\varepsilon_0} B_y(x,y,z,t) = \mu\kappa_x \frac{\partial H_y(x,y,z,t)}{\partial t} + \mu\frac{\sigma_x}{\varepsilon_0} H_y(x,y,z,t), \tag{2.22b}$$

$$\kappa_z \frac{\partial B_z(x,y,z,t)}{\partial t} + \frac{\sigma_z}{\varepsilon_0} B_z(x,y,z,t) = \mu\kappa_y \frac{\partial H_z(x,y,z,t)}{\partial t} + \mu\frac{\sigma_y}{\varepsilon_0} H_z(x,y,z,t). \tag{2.22c}$$

After discretizing Eqs. (2.21) and (2.22) on the Yee mesh points [1], we obtain the explicit FDTD formulations for the magnetic field components in the UPML. For example, we have

$$
\begin{aligned}
&B_x^{n+1/2}(i,j+1/2,k+1/2) \\
&\quad = \left(\frac{2\varepsilon_0\kappa_y - \sigma_y \Delta t}{2\varepsilon_0\kappa_y + \sigma_y \Delta t}\right) B_x^{n-1/2}(i,j+1/2,k+1/2) + \left(\frac{2\varepsilon_0 \Delta t/\Delta s}{2\varepsilon_0\kappa_y + \sigma_y \Delta t}\right) \\
&\quad \times [E_y^n(i,j+1/2,k+1) - E_y^n(i,j+1/2,k) + E_z^n(i,j,k+1/2) \\
&\qquad - E_z^n(i,j+1,k+1/2)],
\end{aligned} \tag{2.23}
$$

$$
\begin{aligned}
&H_x^{n+1/2}(i,j+1/2,k+1/2) \\
&\quad = \left(\frac{2\varepsilon_0\kappa_z - \sigma_z \Delta t}{2\varepsilon_0\kappa_z + \sigma_z \Delta t}\right) H_x^{n-1/2}(i,j+1/2,k+1/2) \\
&\quad + \left(\frac{1/\mu}{2\varepsilon_0\kappa_z + \sigma_z \Delta t}\right) \\
&\quad \times [(2\varepsilon_0\kappa_x + \sigma_x \Delta t) B_x^{n+1/2}(i,j+1/2,k+1/2) \\
&\qquad - (2\varepsilon_0\kappa_x - \sigma_x \Delta t) B_x^{n-1/2}(i,j+1/2,k+1/2)].
\end{aligned} \tag{2.24}
$$

Similarly, for the electric field in the UPML, two auxiliary field variables **P** and **Q** are introduced as follows [36]:

$$P_x(x,y,z) = \left(\frac{s_y s_z}{s_x}\right) E_x(x,y,z), \tag{2.25a}$$

$$P_y(x,y,z) = \left(\frac{s_x s_z}{s_y}\right) E_y(x,y,z), \tag{2.25b}$$

$$P_z(x,y,z) = \left(\frac{s_x s_y}{s_z}\right) E_z(x,y,z), \tag{2.25c}$$

$$Q_x(x,y,z) = \left(\frac{1}{s_y}\right) P_x(x,y,z), \tag{2.26a}$$

$$Q_y(x,y,z) = \left(\frac{1}{s_z}\right) P_y(x,y,z), \tag{2.26b}$$

$$Q_z(x,y,z) = \left(\frac{1}{s_x}\right) P_z(x,y,z). \tag{2.26c}$$

Inserting Eq. (2.25) into Eq. (2.13a), and following the steps in deriving Eq. (2.24), we obtain the updating equations for the **P** components, e.g.

$$
\begin{aligned}
&P_x^{n+1}(i+1/2,j,k) = \left(\frac{2\varepsilon - \sigma \Delta t}{2\varepsilon + \sigma \Delta t}\right) P_x^n(i+1/2,j,k) + \left(\frac{2\Delta t/\Delta s}{2\varepsilon + \sigma \Delta t}\right) \\
&\quad \times \left[H_y^{n+1/2}(i+1/2,j,k-1/2) - H_y^{n+1/2}(i+1/2,j,k+1/2) \right. \\
&\quad \left. + H_z^{n+1/2}(i+1/2,j+1/2,k) - H_z^{n+1/2}(i+1/2,j-1/2,k)\right].
\end{aligned} \tag{2.27}
$$

Similarly, inserting Eq. (2.26) into Eq. (2.13b), and following the steps in deriving Eq. (2.24), we obtain the updating equations for the **Q** components, e.g.

$$
\begin{aligned}
&Q_x^{n+1}(i+1/2,j,k) = \left(\frac{2\varepsilon_0\kappa_y - \sigma_y \Delta t}{2\varepsilon_0\kappa_y + \sigma_y \Delta t}\right) Q_x^n(i+1/2,j,k) + \left(\frac{2\varepsilon_0}{2\varepsilon_0\kappa_y + \sigma_y \Delta t}\right) \\
&\quad \times \left[P_x^{n+1}(i+1/2,j,k) - P_x^n(i+1/2,j,k)\right].
\end{aligned} \tag{2.28}
$$

Inserting Eq. (2.25) into Eq. (2.26) and also following the procedure in deriving Eq. (2.24), we can derive the electric field components in the UPML, e.g.

$$E_x^{n+1}(i+1/2,j,k) = \left(\frac{2\varepsilon_0\kappa_z - \sigma_z\Delta t}{2\varepsilon_0\kappa_z + \sigma_z\Delta t}\right) E_x^n(i+1/2,j,k) + \left(\frac{1}{2\varepsilon_0\kappa_z + \sigma_z\Delta t}\right) \times \left[(2\varepsilon_0\kappa_x + \sigma_x\Delta t)Q_x^{n+1}(i+1/2,j,k) - (2\varepsilon_0\kappa_x - \sigma_x\Delta t)Q_x^n(i+1/2,j,k)\right]. \tag{2.29}$$

Note that the absorbing host material properties ε and σ in Eq. (2.27) explicitly show that the host material properties are extended into the UPML. In other words, the UPML matches the electromagnetic material properties of the truncated domain.

Using the UPML for the truncation of the conductive media, 3 auxiliary arrays for the magnetic field and 6 auxiliary arrays for the electric field are needed. By applying the UPML scheme to the truncation of lossless media, we can use a different formulation to calculate the electric field components in the UPML where only 3 auxiliary arrays are needed for the electric field.

3.2.4 Formulation of the single scattering properties

For light scattering by particles in free space, the far-field approximation for the electromagnetic field is usually used to calculate the particle scattering and extinction cross-sections. The far-field approach also has been used by Mundy *et al.* [40], Chylek [41], and Bohren and Gilra [42] to study scattering and absorption by a spherical particle in an absorbing host medium. However, when the host medium is absorptive the scattering and extinction rates based on the far-field approximation depend on the distance from the particle, which do not represent the actual scattering and extinction of the particle. Recently, the single-scattering properties of a sphere in a medium have been derived using the electromagnetic fields on the surface of the particle based on Mie theory [43–46]. In this study we derive the absorption and extinction rates for an arbitrarily shaped particle in an absorbing medium using the electric field inside the particle. This method can generally be applied to a particle in a medium with or without absorption. The absorption and extinction rates calculated in this way depend on the size, shape and optical properties of the particle and the surrounding medium, but they do not depend on the distance from the particle, which is different from the far-field approximation approach.[47, 48]. In non-absorbing host media, the near-field solution and the far-field approximation solution of the scattering and extinction cross-sections have no difference. In a non-absorbing host medium, the scattered waves that flow in non-radial directions in the near field, though they have complicated path, will finally arrive at the far-field zone in radial direction without losing any energy. However, in an absorbing medium the scattered radiation is absorbed in the absorbing medium but it is not necessarily exponentially damped with the radial distance, especially in the region near the particle. Therefore, in an absorbing host medium, the scattering and extinction cross-sections calculated with the far-field approximation have a nonlinear relationship with the near-field ones. Although attention has been paid to this problem [45], further studies are still needed to resolve this issue.

The flow of energy and the direction of the electromagnetic wave propagation are represented by the Poynting vector. In the frequency domain, a complex Poynting vector can be written in the form

$$\mathbf{s} = \mathbf{E} \times \mathbf{H}^*, \tag{2.30}$$

where the asterisk denotes the complex conjugate and $\mathbf{E}$ and $\mathbf{H}$ are the complex electric and magnetic fields, respectively. To derive the absorption and extinction rates of a particle embedded in an absorbing medium, we start from Maxwell's equations in the frequency domain:

$$\nabla \times \mathbf{H} = -i\omega(\varepsilon_r + i\varepsilon_i)\mathbf{E}, \tag{2.31a}$$

$$\nabla \times \mathbf{E} = i\omega\mu\mathbf{H} \tag{2.31b}$$

So we have

$$\begin{aligned}\nabla \cdot s &= \nabla \cdot (\mathbf{E} \times \mathbf{H}^*) = \mathbf{H}^* \cdot (\nabla \times \mathbf{E}) - \mathbf{E} \cdot (\nabla \times \mathbf{H}^*) \\ &= i\omega(\mu\mathbf{H} \cdot \mathbf{H}^* - \varepsilon_r\mathbf{E} \cdot \mathbf{E}^*) - \omega\varepsilon_i\mathbf{E} \cdot \mathbf{E}^*.\end{aligned} \tag{2.32}$$

For the convenience of the following discussion, we define the real and imaginary permittivity for the particle as ε_{tr} and ε_{ti}, and for the host medium as ε_{hr} and ε_{hi}, respectively. The rate of energy absorbed by the particle is [6]

$$\begin{aligned}w_a &= -\frac{1}{2}\mathrm{Re}\left[\oiint_s n \cdot s(\xi)\, d^2\xi\right] \\ &= -\frac{1}{2}\mathrm{Re}\left[\iiint_v \nabla{\cdot}s(\xi)\, d^3\xi\right] \\ &= \frac{\omega}{2}\iiint_v \varepsilon_{ti}(\xi)\mathbf{E}(\xi) \cdot \mathbf{E}^*(\xi)\, d^3\xi,\end{aligned} \tag{2.33}$$

where $\mathbf{n}$ denotes the outward-pointing unit vector normal to the particle surface. The surface and volume integrals are performed over the particle.

The total electric and magnetic field vectors $\mathbf{E}$ and $\mathbf{H}$ are the superposition of the incident and scattered fields. Consequently, the scattered field vectors can be written as

$$\mathbf{E}_s = \mathbf{E} - \mathbf{E}_i, \tag{2.34a}$$

$$\mathbf{H}_s = \mathbf{H} - \mathbf{H}_i, \tag{2.34b}$$

where $\mathbf{E}_i$ and $\mathbf{H}_i$ denote the incident electric and magnetic field vector, respectively. Therefore the rate of energy scattered by the particle can be taken as

$$\begin{aligned}w_s &= \frac{1}{2}\mathrm{Re}\left[\oiint_s n \cdot (\mathbf{E}_s \times \mathbf{H}_s^*)\, d^2\xi\right] \\ &= \frac{1}{2}\mathrm{Re}\left\{\oiint_s n \cdot [(\mathbf{E} - \mathbf{E}_i) \times (\mathbf{H}^* - \mathbf{H}_i^*)]\, d^2\xi\right\}.\end{aligned} \tag{2.35}$$

Because both absorption and scattering remove energy from the incident waves, the rate of the energy extinction can be defined as [6]

$$\begin{aligned} w_e &= w_s + w_a \\ &= \frac{1}{2}\mathrm{Re}\left\{ \oiint_s n \cdot [(\mathbf{E} - \mathbf{E}_i) \times (\mathbf{H}^* - \mathbf{H}_i^*)]\, d^2\xi \right\} - \frac{1}{2}\mathrm{Re}\left[\oiint_s n \cdot (\mathbf{E} \times \mathbf{H}^*)\, d^2\xi \right] \\ &= \frac{1}{2}\mathrm{Re}\left[\oiint_s n \cdot (\mathbf{E}_i \times \mathbf{H}_i^* - \mathbf{E}_i \times \mathbf{H}^* - \mathbf{E} \times \mathbf{H}_i^*)\, d^2\xi \right] \\ &= \frac{1}{2}\mathrm{Re}\left[\iiint_v \nabla \cdot (\mathbf{E}_i \times \mathbf{H}_i^* - \mathbf{E}_i \times \mathbf{H}^* - \mathbf{E} \times \mathbf{H}_i^*)\, d^3\xi \right]. \end{aligned} \tag{2.36}$$

In a similar derivation to that of Eq. (2.33), we can use Eq. (2.31) and Eq. (2.36) to obtain [6]

$$\begin{aligned} w_e &= w_a + w_s \\ &= \frac{\omega}{2} \iiint_v [\varepsilon_{ti}(\xi) + \varepsilon_{hi}(\xi)]\mathrm{Re}\left[\mathbf{E}_i(\xi) \cdot \mathbf{E}^*(\xi)\right] d^3\xi \\ &- \frac{\omega}{2} \iiint_v [\varepsilon_{tr}(\xi) - \varepsilon_{hr}(\xi)]\mathrm{Im}\left[\mathbf{E}_i(\xi) \cdot \mathbf{E}^*(\xi)\right] d^3\xi \\ &- \frac{\omega}{2} \iiint_v \varepsilon_{hi}(\xi)\left[\mathbf{E}_i(\xi) \cdot \mathbf{E}_i^*(\xi)\right] d^3\xi. \end{aligned} \tag{2.37}$$

Assuming the rate of energy incident on a particle of arbitrary shape is f, then the absorption, scattering and extinction efficiencies are $Q_a = w_a/f$, $Q_s = (w_e - w_a)/f$ and$Q_e = w_e/f$, respectively. The single scattering albedo is consequently $\tilde{\omega} = Q_s/Q_e$.

In an absorbing medium, the rate of energy incident on the particle depends on the position and intensity of the wave source, the optical properties of the host medium, and the particle size parameter and shape. Following Mundy *et al.* [40], if the intensity of the incident light at the center of the computational domain is I_0 when no particle is positioned in the absorbing medium, the rate of energy incident on a spherical scatterer centered within the 3D computational domain is

$$f = \frac{2\pi a^2}{\eta^2} I_0\left[1 + (\eta - 1)e^{\eta}\right], \tag{2.38}$$

where a is the radius of the spherical particle, $\eta = 4\pi a m_{hi}/\lambda_0$ and $I_0 = \frac{1}{2}(m_{hr}/c\mu)|\mathbf{E}_0|^2$, λ_0 is the incident wavelength in free space, m_{hr} and m_{hi} are the real and imaginary refractive index of the host medium, respectively, and $|E_0|$ denotes the amplitude of the incident electric field at the center of the 3D computational domain. For non-spherical particles, the rate of energy incident on the particle may only be calculated numerically.

The scattering phase function represents the angular distribution of scattered energy at a large distance from the scatterer, which can be derived using far-field approximation. The scattering phase function and asymmetry factor are calculated by using the 4 elements of the amplitude scattering matrix [49].

The amplitude scattering matrix for a particle immersed in an absorbing dielectric medium is derived as follows. For electromagnetic waves with time dependence $\exp(-i\omega t)$ propagating in a charge-free dielectric medium, Maxwell's equations in the frequency domain are

$$\nabla \cdot \mathbf{D} = 0, \tag{2.39a}$$
$$\nabla \cdot \mathbf{H} = 0, \tag{2.39b}$$
$$\nabla \times \mathbf{E} = i\omega\mu\mathbf{H}, \tag{2.39c}$$
$$\nabla \times \mathbf{H} = -i\omega\mathbf{D}, \tag{2.39d}$$

In this study, we define the material properties of the host medium as the background permittivity and permeability. So the electric displacement vector is defined as

$$\mathbf{D} = \varepsilon_h\mathbf{E} + \boldsymbol{P} = \varepsilon\mathbf{E}, \tag{2.40}$$

where $\mathbf{P}$ is the polarization vector. Therefore, Eqs. (2.39a) and (2.39d) can be rewritten as

$$\nabla \cdot \mathbf{E} = -\frac{1}{\varepsilon_h}\nabla \cdot \boldsymbol{P}, \tag{2.41a}$$
$$\nabla \times \mathbf{H} = -i\omega(\varepsilon_h\mathbf{E} + \boldsymbol{P}). \tag{2.41b}$$

Using Eqs. (2.39c), (2.41a) and (2.41b) in evaluating $\nabla \times (\nabla \times \mathbf{E})$ yields a source-dependent form of the electromagnetic wave equation

$$(\nabla^2 + k_h^2)\mathbf{E} = -\frac{1}{\varepsilon_h}\left[k_h^2\mathbf{P} + \nabla(\nabla \cdot \mathbf{P})\right], \tag{2.42}$$

where $k_h = \omega\sqrt{\mu\varepsilon_h}$ is the complex wave number in the host medium. Using the unit dyad $\mathbf{II} = \mathbf{xx} + \mathbf{yy} + \mathbf{zz}$ (where $\mathbf{x}, \mathbf{y}$ and $\mathbf{z}$ are unit vectors in x, y and z direction, respectively), we can rewrite the polarization vector $\mathbf{P}$ as $\mathbf{II}\cdot\mathbf{P}$, therefore Eq. (2.42) can be rewritten as

$$(\nabla^2 + k_h^2)\mathbf{E} = -\frac{1}{\varepsilon_h}(k_h^2\mathbf{II} + \nabla\nabla) \cdot \mathbf{P}. \tag{2.43}$$

Note that based on the definition in Eq. (2.40), we have $\mathbf{P} = (\varepsilon - \varepsilon_h)\mathbf{E}$, which means $\mathbf{P}$ is nonzero only within the particle. The general solution for Eq. (2.43) is given by a volume integral equation in the form[50]

$$\mathbf{E}(\mathbf{R}) = \mathbf{E}_0(\mathbf{R}) + \iiint_v G(\mathbf{R}, \xi)(k_h^2\mathbf{II} + \nabla_\xi\nabla_\xi) \cdot (\mathbf{P}/\varepsilon_h)\, d^3\xi, \tag{2.44}$$

where $\mathbf{E}_0(\mathbf{R})$ mathematically can be any solution of $(\nabla^2 + k_h^2)\mathbf{E} = 0$, while physically the only nontrivial solution is the incident field in the host medium. The domain of the integration v is the region inside the particle, and $G(\mathbf{R}, \xi)$ is the 3D Green function in the host medium

$$G(\mathbf{R}, \xi) = \frac{\exp(ik_h|\mathbf{R} - \xi|)}{4\pi|\mathbf{R} - \xi|}. \tag{2.45}$$

The scattered field in the far-field region can be derived from Eq. (2.44), i.e.

$$\mathbf{E}_s(\vec{R})|_{k_h\mathbf{R}\to\infty} = \frac{k_h^2\exp(ik_h\mathbf{R})}{4\pi\mathbf{R}}\iiint_v\left[\frac{\varepsilon(\xi)}{\varepsilon_h} - 1\right]\{\mathbf{E}(\xi) - \mathbf{r}[\mathbf{r}\cdot\mathbf{E}(\xi)]\}\exp(-ik_h\mathbf{r}\cdot\xi)\, d^3\xi. \tag{2.46}$$

To calculate the amplitude scattering matrix elements, the scattered field is decomposed into the components parallel and perpendicular to the scattering plane as shown in Fig. 3.3 [4]. So we have

$$\mathbf{E}_s(\mathbf{R}) = \alpha \mathbf{E}_{s,\alpha}(\mathbf{R}) + \beta \mathbf{E}_{s,\beta}(\mathbf{R}). \tag{2.47}$$

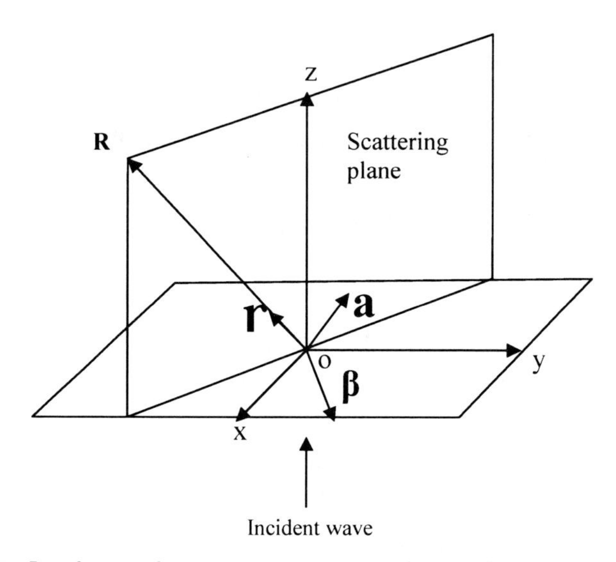

Fig. 3.3. Incident and scattering geometries for the far-field formulation.

Based on the definition of the amplitude scattering matrix, we can write the transformation equation for light scattering by particles as

$$\begin{pmatrix} E_{s,\alpha}(\mathbf{R}) \\ E_{s,\beta}(\mathbf{R}) \end{pmatrix} = \frac{\exp(ik_h R)}{-ik_h R} \begin{bmatrix} s_2 & s_3 \\ s_4 & s_1 \end{bmatrix} \begin{pmatrix} E_{0,\alpha} \\ E_{0,\beta} \end{pmatrix}, \tag{2.48}$$

where $E_{0,\alpha}$ and $E_{0,\beta}$ are the parallel and perpendicular incident field components with respect to the scattering plane. $E_{0,\alpha}$ and $E_{0,\beta}$ are related to the x-polarized and y-polarized incident fields used in the FDTD simulation with

$$\begin{pmatrix} E_{0,\alpha} \\ E_{0,\beta} \end{pmatrix} = \begin{bmatrix} \boldsymbol{\beta}\cdot\mathbf{x} & -\boldsymbol{\beta}\cdot\mathbf{y} \\ \boldsymbol{\beta}\cdot\mathbf{y} & \boldsymbol{\beta}\cdot\mathbf{x} \end{bmatrix} \begin{pmatrix} E_{0,y} \\ E_{0,x} \end{pmatrix}. \tag{2.49}$$

Rewriting Eq. (2.46) in a matrix form yields

$$\begin{pmatrix} E_{s,\alpha}(\mathbf{R}) \\ E_{s,\beta}(\mathbf{R}) \end{pmatrix} = \frac{k_h^2 \exp(ik_h R)}{4\pi R} \iiint_v \left[\frac{\varepsilon(\xi)}{\varepsilon_h} - 1\right] \begin{pmatrix} \boldsymbol{\alpha}\cdot\mathbf{E}(\xi) \\ \boldsymbol{\beta}\cdot\mathbf{E}(\xi) \end{pmatrix} \exp(-ik_h r\cdot\xi)\, d^3\xi. \tag{2.50}$$

From Eqs. (2.48)–(2.50), we can derive the amplitude scattering matrix as [3]

$$\begin{bmatrix} s_2 & s_3 \\ s_4 & s_1 \end{bmatrix} = \begin{bmatrix} F_{\alpha,y} & F_{\alpha,x} \\ F_{\beta,y} & F_{\beta,x} \end{bmatrix} \begin{bmatrix} \boldsymbol{\beta}\cdot\mathbf{x} & \boldsymbol{\beta}\cdot\mathbf{y} \\ -\boldsymbol{\beta}\cdot\mathbf{y} & \boldsymbol{\beta}\cdot\mathbf{x} \end{bmatrix}, \tag{2.51}$$

where $\boldsymbol{\beta}$ denotes the unit vector perpendicular to the scattering plane, and $\mathbf{x}$ and $\mathbf{y}$ are the unit vector in the x and y directions, respectively, as shown in Fig. 3.2. The quantities $F_{\alpha,x}$, $F_{\beta,x}$ and $F_{\alpha,y}$, $F_{\beta,y}$ are calculated for x-polarized incident light and y-polarized incident light, respectively, in the forms [50]:

(1) for x-polarized incident light,

$$\begin{pmatrix} F_{\alpha,x} \\ F_{\beta,x} \end{pmatrix} = \frac{ik_h^3}{4\pi} \iiint_v \left[1 - \frac{\varepsilon(\xi)}{\varepsilon_h}\right] \begin{pmatrix} \boldsymbol{\alpha} \cdot \mathbf{E}(\xi) \\ \boldsymbol{\beta} \cdot \mathbf{E}(\xi) \end{pmatrix} \exp(-ik_h r \cdot \xi)\, d^3\xi; \tag{2.52a}$$

(2) for y-polarized incident light,

$$\begin{pmatrix} F_{\alpha,y} \\ F_{\beta,y} \end{pmatrix} = \frac{ik_h^3}{4\pi} \iiint_v \left[1 - \frac{\varepsilon(\xi)}{\varepsilon_h}\right] \begin{pmatrix} \boldsymbol{\alpha} \cdot \mathbf{E}(\xi) \\ \boldsymbol{\beta} \cdot \mathbf{E}(\xi) \end{pmatrix} \exp(-ik_h r \cdot \xi)\, d^3\xi. \tag{2.52b}$$

In Eq. (2.52), $k_h = \omega\sqrt{\mu\varepsilon_h}$ and ε_h is the complex permittivity of the host medium. $\mathbf{r}$ is the unit vector in the scattering direction, α is the unit vector parallel to the scattering plane, and $\beta \times \alpha = r$. When $\varepsilon_h = \varepsilon_0$ and $\mu = \mu_0$, Eq. (2.52) will degenerate to a formulation for light scattering by particles in free space.

3.2.5 Numerical results

In principle the FDTD method can be accurately applied to particles of arbitrary shapes. However, there are numerical errors involved in the FDTD technique. These errors can be attributed to the numerical dispersion of the finite-difference analog, the approximation of a specific particle shape by a pseudostructure constructed by cubic grid cells, the representation of the near field by the discretized data that do not account for the field variation within each cell, and the reflection from the truncation boundaries. Conventionally, the accuracy of the FDTD method is evaluated by comparing the FDTD results with exact solutions for spherical particles†[5]. This has been standard practice for numerical techniques like the FDTD and DDA algorithms because an exact solution exists for these particles by which a comparison can be made. Since these particles have morphology that both excites strong resonances and is not easily replicated by a Cartesian grid, the errors in such comparisons are significantly higher than are achieved for non-spherical particles [51]. Figure 3.4 shows the phase-matrix elements for a spherical particle with a size parameter of $2\pi a/\lambda_0 = 56.55$, where a is the radius of the spherical particle and λ_0 is the incident wavelength in free space. The black curves are computed with Mie theory and the red curves are the FDTD results. The host medium is free space and the refractive index of the particle is $1.0893 + 0.18216i$. In the FDTD calculations, a FDTD cell size of $\Delta s = \lambda_0/30$ is used. The extinction efficiency Q_e, scattering efficiency Q_s, absorption efficiency Q_a, asymmetry factor g and the errors of the FDTD results are listed in Fig. 3.4. We can see that the FDTD results are very close to the exact Mie solutions.

To demonstrate the applicability of the UPML FDTD scheme for light scattering by a non-spherical particle embedded in an absorbing medium. We assume a phytoplankton cell with a shape of a spheroid which can be described by a surface function

$$\frac{x^2}{a^2} + \frac{y^2}{b^2} + \frac{z^2}{c^2} = 1,$$

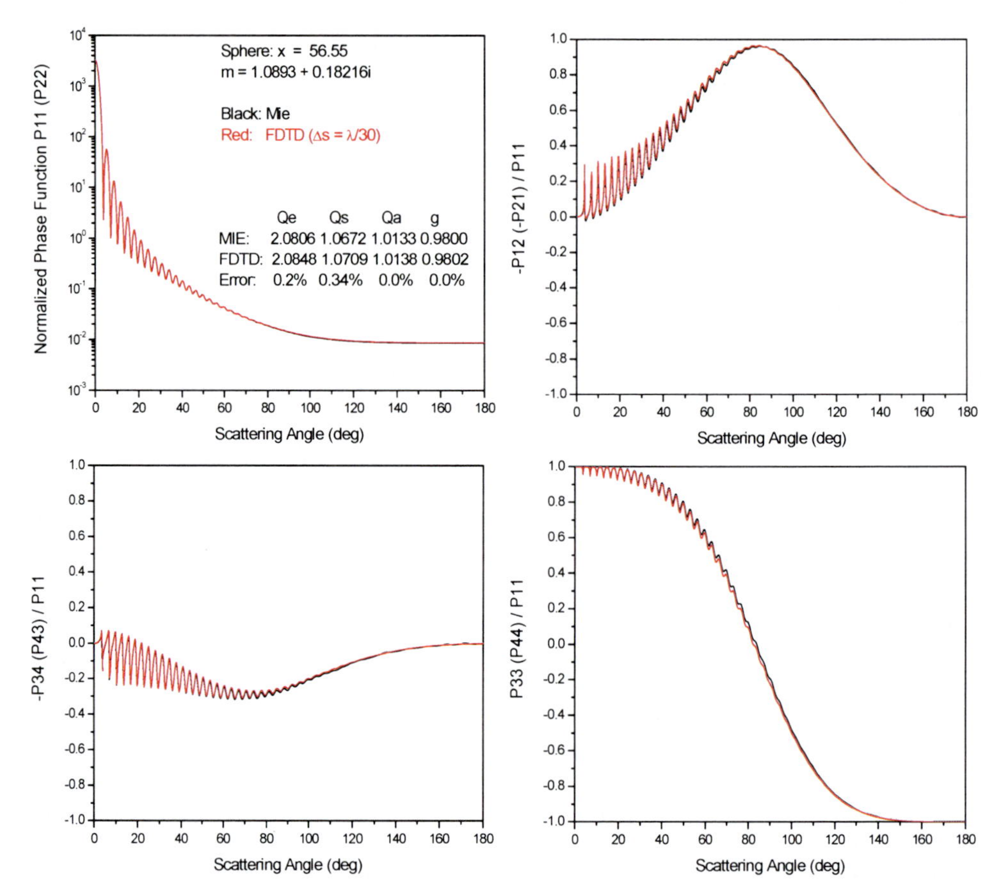

Fig. 3.4. The nonzero phase matrix elements for a spherical particle with a size parameter of $2\pi a/\lambda_0 = 56.55$, where a is the radius of the spherical particle and λ_0 is the incident wavelength in free space. The host medium is free space and the refractive index of the particle is $1.0893 + 0.18216\mathrm{i}$.

where a, b, and c are the half axes of the spheroid with $a = 3b = 6c$. As shown in Fig. 3.5, the light is incident on the spheroid in a direction with a zenith angle of $45°$ and an azimuth angle of $45°$. The size parameter of the spheroid is defined by its half axis a, and $2\pi a/\lambda_0 = 30$. The host medium is assumed to have a refractive index of 1.34 and $1.34 + 0.05\mathrm{i}$. The refractive index of the phytoplankton cell is assumed to be 1.407. Figure 3.6 shows some of the azimuthally averaged elements of the scattering phase matrix for the spheroid and incidence geometries illustrated in Fig. 2.5. A FDTD cell size of $\Delta s = \lambda_0/30$ is used in this simulation. We can see that the absorption by the host medium affects the phase function of the non-spherical particle significantly, similar to what was found in Fu and Sun [45] for spherical particles.

Numerical techniques like the FDTD are ideally suited for solving the light scattered by irregularly shaped particles. Figure 3.7 shows sample Gaussian particle shapes [52] with a correlation angle of $\Gamma = 30°$ and relative standard deviation σ

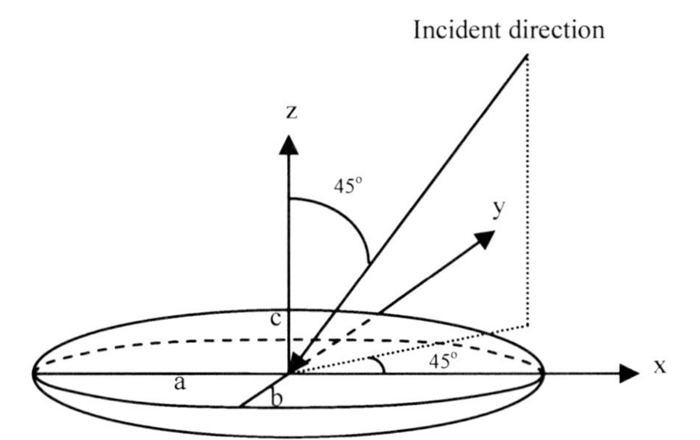

Fig. 3.5. Configuration of a spheroid with half axes a, b, and c. The light is incident on the spheroid in a direction with a zenith angle of 45° and an azimuth angle of 45°.

as: (a) 0.0, (b) 0.05, (c) 0.1, (d) 0.2, (e) 0.4, and (f) 0.8. The nonzero elements of the scattering phase matrix as functions of scattering angle calculated with the FDTD technique for these single, randomly oriented Gaussian ice particles are given in Fig. 3.8 [52]. The incident wavelength $\lambda = 0.55\,\mu$m and the refractive index of ice at this wavelength is 1.311. The particle size parameter $x = 2\pi a/\lambda = \pi$, where a denotes the mean half-size of the particle. In the FDTD calculation, a spatial cell size of $\lambda/20$ is used. We can see that with the increase of the surface deformation, the strong oscillations in the conventional phase function (P11) gradually disappear. This is a very important finding that can greatly simplify the modeling of the radiative transfer in ice clouds. The degree of linear polarization P12/P11 also shows a smooth trend with increasing surface deformation, which suggests the linear polarization properties of irregular ice crystals are significantly different from those of spherical cloud particles. The differences in linear polarization properties between spherical and non-spherical particles may help in the development of methods to detect cloud phase using polarized sensors. The ratios of other elements to P11 also demonstrate a reduced oscillatory pattern when surface deformation increases, except in element P22/P11. Most significant is the shape of the normalized polarization matrix elements (P12/P11, P33/P11, P43/P11, and P44/P11) when the particle shape becomes extremely irregular ($\sigma = 0.8$). These elements in this case bear a very strong resemblance to those of a Rayleigh sphere.

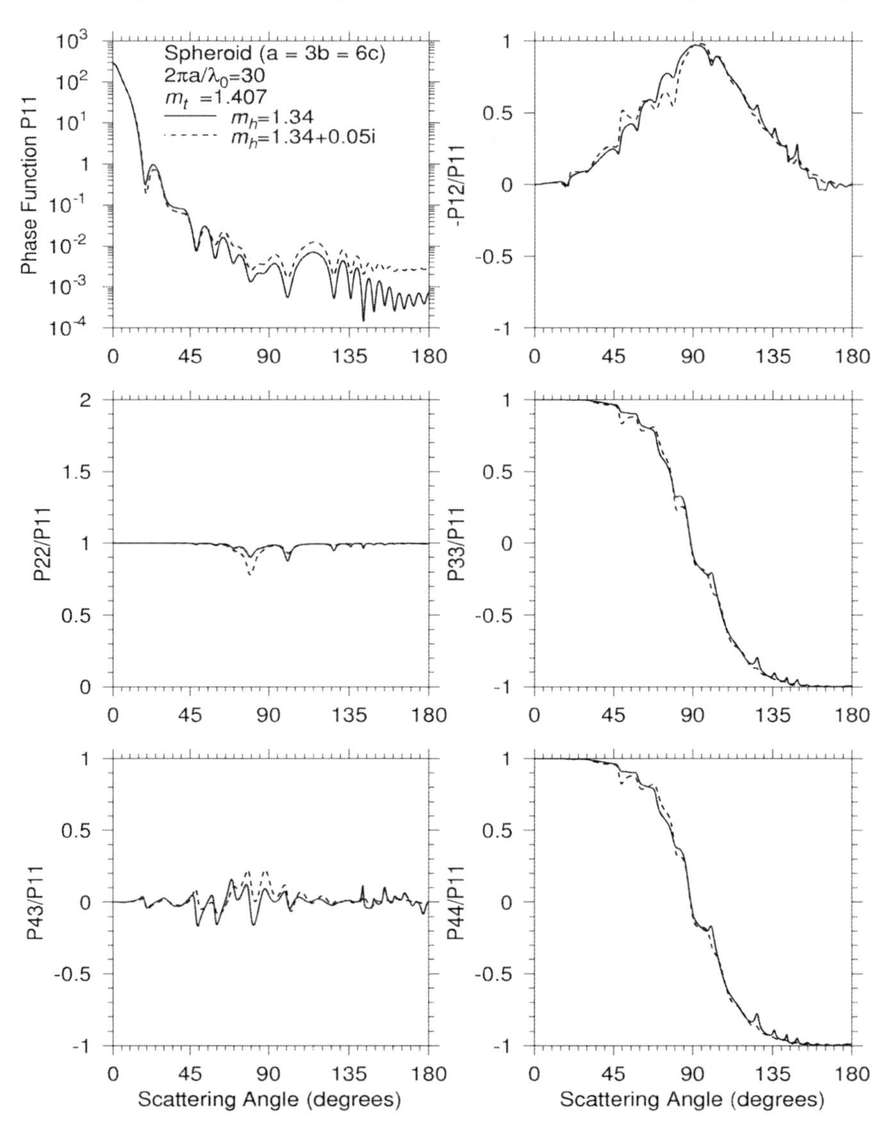

Fig. 3.6. The azimuthally (around the incident direction) averaged elements of the scattering phase matrix for the spheroid and incidence geometries illustrated in Fig. 3.5.

3.3 FDTD method for electromagnetic beam interaction with surfaces

The interaction of electromagnetic radiation with arbitrary dielectric surfaces is one of the major issues for terrestrial or extraterrestrial remote-sensing applications [53]. It is of significant importance in various other fields of applied physics and detection engineering, such as measurement of surface defects or contamination [54], characterization of particles over a rough surface background [55], and non-invasive detection of concealed targets such as improvised explosive devices (IEDs)

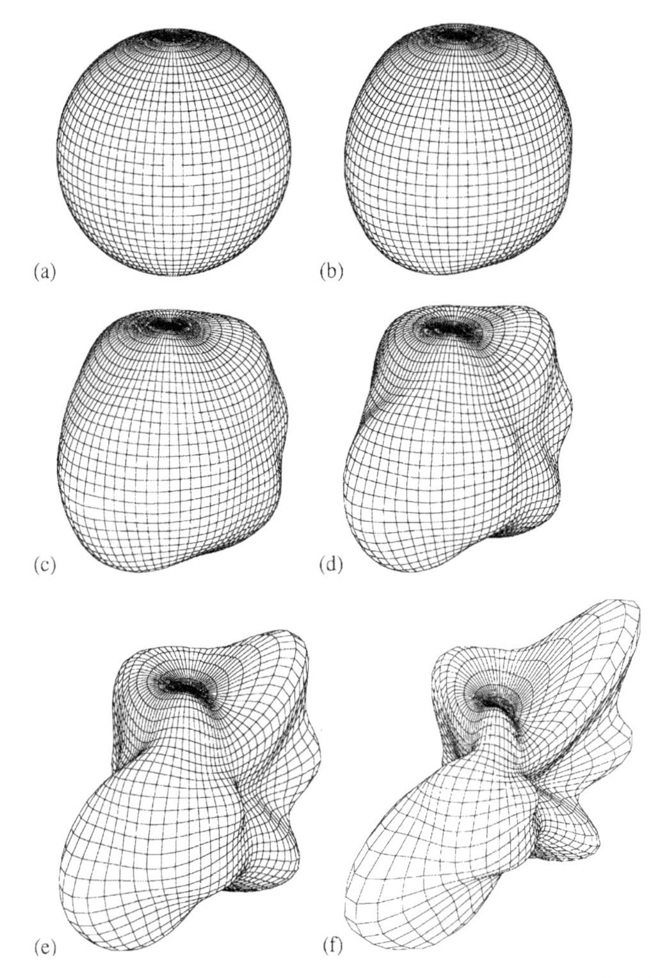

Fig. 3.7. The sample Gaussian particle shapes. The correlation angle $\Gamma = 30^\circ$ and the relative standard deviation σ is chosen as: (a) 0.0, (b) 0.05, (c) 0.1, (d) 0.2, (e) 0.4, and (f) 0.8.

[56]. Some modern detection techniques of biological particles, based on surface enhanced Raman scattering (SERS) obtained as they pass near nanostructures on surfaces, require an accurate analysis of the light scattering from the particle near the nanostructure for signal optimization. Light illuminating these particles emits SERS at chemically specific wavelengths, allowing accurate identification. The signal intensity is dependent on the aerosol particle and surface composition and morphology. Because SERS signals are relatively weak, precise measurement or modeling of the elastic scattering fields are extremely important for obtaining accurate Raman scattering signals.

One striking aspect of SERS-related research is that the great bulk of it is performed experimentally. The reason for this is that no accurate method exists to calculate the interaction of electromagnetic radiation with an arbitrary surface. The frequently used algorithms on wave scattering from rough surfaces are pri-

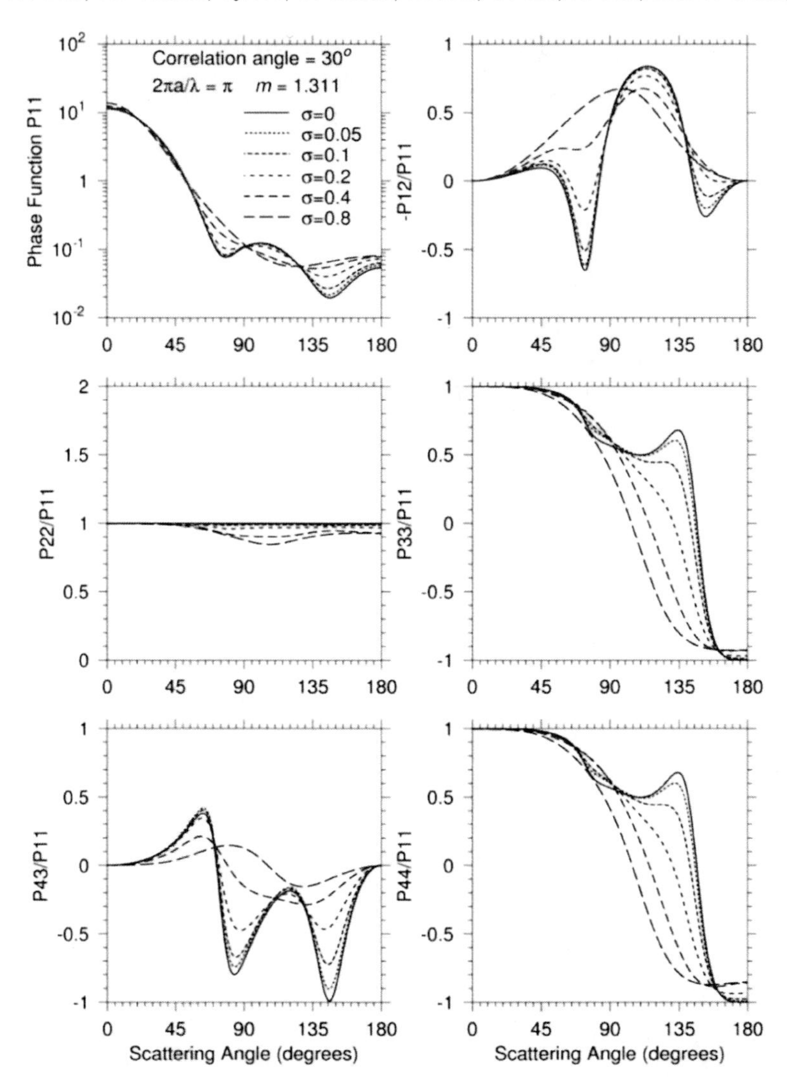

Fig. 3.8. The nonzero elements of the scattering phase matrix as functions of scattering angle calculated with the FDTD technique for randomly oriented Gaussian ice particles.

marily analytical approximate approaches, such as the small perturbation model (SPM) [57–60], the Kirchhoff approximation (KA) [60–62], the integral equation method (IEM) [64–67], and other methods [68]. Such approximate approaches for light scattering by rough surfaces have limited applicability when a precise description of the near field is required. For example, the SPM is valid only for surfaces with roughness scale much smaller than the incident wavelength; whereas, the KA is only valid when the surface roughness level is much greater than the wavelength. When the surface roughness scale is similar to the wavelength, neither the SPM nor the KA are applicable. Although the IEM model was developed to bridge the gap between the SPM and the KA models, as an approximate method, its accuracy is still not sufficient to calculate the near fields accurately [68–72]. With the advancement of computational capabilities, numerical light-scattering solutions such as the finite element method (FEM) [73,74], the DDA [26,75,76], and the FDTD technique [1–6,77,78] play important roles in the accurate calculation of electromagnetic wave scattering. While some analytical methods based on the T-matrix algorithm are extremely efficient at calculating light scattering from specific, relatively regular morphologies such as spheres [79–84], cylinders [85–87], and other particles near surfaces [88–92], these generally have limited applicability to real-world, non-ideal systems. The DDA has been developed to calculate the light scattered from arbitrary surface structures. However, DDA has two shortcomings [93, 94]: First, the Green functions used in this algorithm are for dipoles placed above a planar substrate. As such, the system is less general as it cannot be used to simulate substrates that may have heterogeneities or grooves carved into them. And second, there can be difficulties with convergence when the particles have very high complex refractive indices [95]. Since SERS applications generally address metallic structures on metallic or semi-conducting substrates, this can be a problem. The discrete sources method (DSM) also has been used successfully on surface irregularities, but it also is ideally suited for surface structures on a planar substrate [96]. The FDTD algorithm has been developed to study the interaction of electromagnetic waves with arbitrarily shaped particles on surfaces [76]. While this algorithm has been optimized to handle large complex refractive indices, like those used to generate SERS, it is accurate only when the surface is illuminated normally; i.e., the direction of the incident electromagnetic field is perpendicular to that of the substrate [97]. When this condition is broken, substantial errors result. These errors are especially large in the near-field region, the region of most relevance to SERS studies.

In previous work [7], a novel absorbing boundary condition was developed to truncate the computational domain of the scattered-field FDTD method [98–100], which also can be applied accurately to any FDTD algorithms having source terms in the field-updating equations. For light scattering by particles in free space, the source terms exist only inside the particles, so a regular source-free boundary condition is good enough. However, if the host medium is not free space, the source terms in the scattered-field FDTD updating equations exist everywhere in the computational domain, including the absorbing boundary. To maintain the continuity of the electromagnetic fields crossing the interface between the regular computational domain and the absorbing boundary, the field updating equations inside the absorbing boundary must also have source terms. With use of this novel boundary condition, a scattered-field FDTD algorithm is developed for surface scattering studies. Because the scattered-field FDTD equations include incident-wave source terms, this FDTD algorithm allows the inclusion of an arbitrary incident source beam. Therefore, this FDTD algorithm removes the normally incident plane wave condition that restricts previously developed FDTD methods in surface studies. The incident field can be a plane wave beam incident at non-normal angles with respect to the surface or it can be a Gaussian beam like those emitted by lasers commonly used in surface science, etc. This development provides a robust numerical algorithm for arbitrary surface studies including the calculation of light scattering from particles near nanostructures used for SERS detection.

3.3.1 Scattered-field finite-difference time-domain method

The primary objective of this section is to review the FDTD computer algorithm that can be used to calculate accurately the electromagnetic field in and around a rough surface that may include heterogeneities. This FDTD algorithm includes source terms, so that the incident electromagnetic field may be completely arbitrary; e.g., it will provide accurate scattered near fields from an arbitrary rough surface illuminated by a non-perpendicularly incident, arbitrarily shaped beam. To allow more flexibility in the specification of the form of the incident fields, such as creating an incident beam to have arbitrary amplitude and phase at a given grid point and propagating that beam to the material surface at any preferred angle, we use the pure scattered field formulation of the FDTD equations rather than the total field/scattered field formulation.

The total electromagnetic fields ($\mathbf{E}$ and $\mathbf{H}$) can be decomposed as the sum of incident fields ($\mathbf{E}^i$ and $\mathbf{H}^i$) and scattered fields ($\mathbf{E}^s$ and $\mathbf{H}^s$) as

$$\mathbf{H} = \mathbf{H}^s + \mathbf{H}^i, \tag{3.1a}$$

$$\mathbf{E} = \mathbf{E}^s + \mathbf{E}^i. \tag{3.1b}$$

In a medium with a permittivity ε, conductivity σ, and permeability μ_0, Maxwell's equations for the total electromagnetic fields ($\mathbf{E}$ and $\mathbf{H}$) and incident fields ($\mathbf{E}^i$ and $\mathbf{H}^i$) can be expressed as

$$\mu_0 \frac{\partial \mathbf{H}}{\partial t} = -\nabla \times \mathbf{E}, \tag{3.2a}$$

$$\varepsilon \frac{\partial \mathbf{E}}{\partial t} + \sigma \mathbf{E} = \nabla \times \mathbf{H}; \tag{3.2b}$$

and

$$\mu_0 \frac{\partial \mathbf{H}^i}{\partial t} = -\nabla \times \mathbf{E}^i, \tag{3.2c}$$

$$\varepsilon_0 \frac{\partial \mathbf{E}^i}{\partial t} = \nabla \times \mathbf{H}^i. \tag{3.2d}$$

From Eqs. (3.1) and (3.2) we can derive the scattered-field equations as follows:

$$\mu_0 \frac{\partial \mathbf{H}^s}{\partial t} = -\nabla \times \mathbf{E}^s, \tag{3.3a}$$

$$\varepsilon \frac{\partial \mathbf{E}^s}{\partial t} + \sigma \mathbf{E}^s = \nabla \times \mathbf{H}^s - \sigma \mathbf{E}^i - (\varepsilon - \varepsilon_0) \frac{\partial \mathbf{E}^i}{\partial t}. \tag{3.3b}$$

The pure scattered field FDTD method is simply a discretization of Eqs. (3.3a) and (3.3b). The continuous space and time (x, y, z, t) is replaced by discrete spatial and temporal points, and the field components are calculated only at these discrete points. In a Cartesian grid system the x components of the magnetic and electric fields, e.g., are in the forms [100]

$$H_x^{s,n+1/2}(i,j+1/2,k+1/2) = H_x^{s,n-1/2}(i,j+1/2,k+1/2) + \frac{\Delta t}{\mu_0 \Delta s} \times [E_y^{s,n}(i,j+1/2,k+1) - E_y^{s,n}(i,j+1/2,k) + E_z^{s,n}(i,j,k+1/2) - E_z^{s,n}(i,j+1,k+1/2)], \tag{3.4a}$$

$$\begin{aligned} E_x^{s,n+1}(i+1/2,j,k) &= \left(\frac{2\varepsilon - \sigma\Delta t}{2\varepsilon + \sigma\Delta t}\right) E_x^{s,n}(i+1/2,j,k) \\ &+ \left(\frac{2\Delta t/\Delta s}{2\varepsilon + \sigma\Delta t}\right) [H_y^{s,n+1/2}(i+1/2,j,k-1/2) - H_y^{s,n+1/2}(i+1/2,j,k+1/2) \\ &+ H_z^{s,n+1/2}(i+1/2,j+1/2,k) - H_z^{s,n+1/2}(i+1/2,j-1/2,k)] \\ &- \left(\frac{\sigma\Delta t}{2\varepsilon + \sigma\Delta t}\right) [E_x^{i,n+1}(i+1/2,j,k) + E_x^{i,n}(i+1/2,j,k)] \\ &- \left(\frac{2\varepsilon - 2\varepsilon_0}{2\varepsilon + \sigma\Delta t}\right) [E_x^{i,n+1}(i+1/2,j,k) - E_x^{i,n}(i+1/2,j,k)], \end{aligned} \tag{3.4b}$$

where Δs and Δt denote the cubic cell size and time increment, respectively. To guarantee the numerical stability of the FDTD scheme in the medium, we use $\Delta t = \Delta s/(2c)$, where c is the light speed in free space. The indices (i, j, k) denote the central positions of the cubic cells in the FDTD grid. The time step is denoted by integer n. Note that the electric and magnetic fields are on different spatial positions and are evaluated at alternating half time steps. The positions of the magnetic and electric field components on a cubic cell are identical to those illustrated in Sun *et al* [5].

In the calculation of the scattering patterns in Sun *et al.* [5,6,37,52,77,101,102], the complex form of the uniform plane wave incident field is $\mathbf{A}_0 \exp(-i\mathbf{k}\cdot\mathbf{r})$, where $\mathbf{A}_0$ is a constant vector, $\mathbf{k}$ is the wavenumber vector, and $\mathbf{r}$ denotes the spatial position vector. In this study, we define an arbitrary incident beam as

$$\mathbf{E}^i = \mathbf{A}(\mathbf{r}) \exp(-i\mathbf{k}\cdot\mathbf{r}), \tag{3.5}$$

where $\mathbf{A}(\mathbf{r})$ is an arbitrary complex vector with functional dependence on spatial position $\mathbf{r}$. $\mathbf{A}(\mathbf{r})$ can be defined as any beam shape modulated in both amplitude and phase as functions of location in the computational domain, such as a Gaussian beam [103], Bessel beam [104], etc., depending on the application. In this study, for simplicity and calculation efficiency, we define an artificial beam with

$$\mathbf{A}(\mathbf{r}) = \mathbf{A}_0 \exp\left[-\left(\frac{\rho}{\delta}\right)\right], \tag{3.6}$$

where ρ is the distance from a spatial point perpendicular to the central axis of the incident beam, and δ is the half-width of the beam. The FDTD program requires the incident field to be specified in the time domain, which is accomplished using

$$\mathbf{E}^i(\mathbf{r},t) = \mathrm{Re}\,[\mathbf{A}(\mathbf{r}) \exp(i\omega t - i\mathbf{k}\cdot\mathbf{r})], \tag{3.7}$$

where Re denotes the real value operation. The total field is the superposition of the incident and scattered fields: $\mathbf{E} = \mathbf{E}^s + \mathbf{E}^i$. The simulation is run either until steady state if a continuous incident field is applied or until the pulsed response fades within the computational domain if an incident pulse is used. The frequency-domain values of the total fields and scattered fields are obtained from the discrete Fourier transform (DFT) of the times series of the fields.

3.3.2 Scattered-field uniaxial perfectly matched layer absorbing boundary condition

For applications in the scattered-field FDTD algorithm, Eqs. (2.6a) and (2.6b) also must be modified into their scattered-field forms. Within the absorbing boundary layers, Eqs. (2.6a) and (2.6b) yield the following:

$$\nabla \times [\mathbf{H}^s(x,y,z) + \mathbf{H}^i(x,y,z)] = (i\omega\varepsilon + \sigma)\bar{\bar{s}}[\mathbf{E}^s(x,y,z) + \mathbf{E}^i(x,y,z)], \tag{3.8a}$$

$$\nabla \times [\mathbf{E}^s(x,y,z) + \mathbf{E}^i(x,y,z)] = -i\omega\mu_0\bar{\bar{s}}[\mathbf{H}^s(x,y,z) + \mathbf{H}^i(x,y,z)]. \tag{3.8b}$$

For the incident fields only,

$$\nabla \times \mathbf{H}^i(x,y,z) = i\omega\varepsilon_0\bar{\bar{s}}\mathbf{E}^i(x,y,z), \tag{3.8c}$$

$$\nabla \times \mathbf{E}^i(x,y,z) = -i\omega\mu_0\bar{\bar{s}}\mathbf{H}^i(x,y,z); \tag{3.8d}$$

thus,

$$\nabla \times \mathbf{H}^s(x,y,z) = (i\omega\varepsilon + \sigma)\bar{\bar{s}}\mathbf{E}^s(x,y,z) + [i\omega(\varepsilon - \varepsilon_0) + \sigma]\bar{\bar{s}}\mathbf{E}^i(x,y,z), \tag{3.9a}$$

$$\nabla \times \mathbf{E}^s(x,y,z) = -i\omega\mu_0\bar{\bar{s}}\mathbf{H}^s(x,y,z). \tag{3.9b}$$

To obtain the updating equations for the scattered electric field in the UPML using Eq. (3.9a), we introduce three components of an auxiliary vector field variable P^s, three components of an auxiliary vector field variable Q^s, three components of an auxiliary vector field variables P^i, and three components of an auxiliary vector field variable $\mathbf{Q}^i$, respectively, in the form [6, 7, 36]

$$P_x^s(x,y,z) = \left(\frac{s_y s_z}{s_x}\right) E_x^s(x,y,z), \tag{3.10a}$$

$$P_y^s(x,y,z) = \left(\frac{s_x s_z}{s_y}\right) E_y^s(x,y,z), \tag{3.10b}$$

$$P_z^s(x,y,z) = \left(\frac{s_x s_y}{s_z}\right) E_z^s(x,y,z), \tag{3.10c}$$

$$Q_x^s(x,y,z) = \left(\frac{1}{s_y}\right) P_x^s(x,y,z), \tag{3.11a}$$

$$Q_y^s(x,y,z) = \left(\frac{1}{s_z}\right) P_y^s(x,y,z), \tag{3.11b}$$

$$Q_z^s(x,y,z) = \left(\frac{1}{s_x}\right) P_z^s(x,y,z). \tag{3.11c}$$

$$P_x^i(x,y,z) = \left(\frac{s_y s_z}{s_x}\right) E_x^i(x,y,z), \tag{3.12a}$$

$$P_y^i(x,y,z) = \left(\frac{s_x s_z}{s_y}\right) E_y^i(x,y,z), \tag{3.12b}$$

$$P_z^i(x,y,z) = \left(\frac{s_x s_y}{s_z}\right) E_z^i(x,y,z), \tag{3.12c}$$

$$Q_x^i(x,y,z) = \left(\frac{1}{s_y}\right) P_x^i(x,y,z), \tag{3.13a}$$

$$Q_y^i(x,y,z) = \left(\frac{1}{s_z}\right) P_y^i(x,y,z), \tag{3.13b}$$

$$Q_z^i(x,y,z) = \left(\frac{1}{s_x}\right) P_z^i(x,y,z). \tag{3.13c}$$

Then Eq. (3.9a) can be expressed as

$$\begin{bmatrix} \frac{\partial H_y^s(x,y,z)}{\partial z} - \frac{\partial H_z^s(x,y,z)}{\partial y} \\ \frac{\partial H_z^s(x,y,z)}{\partial x} - \frac{\partial H_x^s(x,y,z)}{\partial z} \\ \frac{\partial H_x^s(x,y,z)}{\partial y} - \frac{\partial H_y^s(x,y,z)}{\partial x} \end{bmatrix} = (i\omega\varepsilon + \sigma) \begin{bmatrix} P_x^s(x,y,z) \\ P_y^s(x,y,z) \\ P_z^s(x,y,z) \end{bmatrix} + [i\omega(\varepsilon - \varepsilon_0) + \sigma] \begin{bmatrix} P_x^i(x,y,z) \\ P_y^i(x,y,z) \\ P_z^i(x,y,z) \end{bmatrix}. \tag{3.14}$$

Applying the inverse Fourier transform to Eq. (3.14) and using the identity $i\omega f(\omega) \to \partial f(t)/\partial t$, we find its equivalent time-domain differential equation as

$$\begin{bmatrix} \frac{\partial H_y^s(x,y,z)}{\partial z} - \frac{\partial H_z^s(x,y,z)}{\partial y} \\ \frac{\partial H_z^s(x,y,z)}{\partial x} - \frac{\partial H_x^s(x,y,z)}{\partial z} \\ \frac{\partial H_x^s(x,y,z)}{\partial y} - \frac{\partial H_y^s(x,y,z)}{\partial x} \end{bmatrix} = \varepsilon \begin{bmatrix} \frac{\partial P_x^s(x,y,z)}{\partial t} \\ \frac{\partial P_y^s(x,y,z)}{\partial t} \\ \frac{\partial P_z^s(x,y,z)}{\partial t} \end{bmatrix} + \sigma \begin{bmatrix} P_x^s(x,y,z) \\ P_y^s(x,y,z) \\ P_z^s(x,y,z) \end{bmatrix} + (\varepsilon - \varepsilon_0) \begin{bmatrix} \frac{\partial P_x^i(x,y,z)}{\partial t} \\ \frac{\partial P_y^i(x,y,z)}{\partial t} \\ \frac{\partial P_z^i(x,y,z)}{\partial t} \end{bmatrix} + \sigma \begin{bmatrix} P_x^i(x,y,z) \\ P_y^i(x,y,z) \\ P_z^i(x,y,z) \end{bmatrix}. \tag{3.15}$$

The discretization of Eq. (3.15) yields the updating equations of $\mathbf{P}^s$. As an example, the x component of $\mathbf{P}^s$ can be expressed as

$$\begin{aligned} P_x^{s,n+1}(i+1/2,j,k) &= \left(\frac{2\varepsilon - \sigma\Delta t}{2\varepsilon + \sigma\Delta t}\right) P_x^{s,n}(i+1/2,j,k) + \left(\frac{2\Delta t/\Delta s}{2\varepsilon + \sigma\Delta t}\right) \\ &\times \left[H_y^{s,n+1/2}(i+1/2,j,k-1/2) - H_y^{s,n+1/2}(i+1/2,j,k+1/2)\right. \\ &+ \left. H_z^{s,n+1/2}(i+1/2,j+1/2,k) - H_z^{s,n+1/2}(i+1/2,j-1/2,k)\right] \\ &- \left(\frac{\sigma\Delta t}{2\varepsilon + \sigma\Delta t}\right)[P_x^{i,n+1}(i+1/2,j,k) + P_x^{i,n}(i+1/2,j,k)] \\ &- \left(\frac{2\varepsilon - 2\varepsilon_0}{2\varepsilon + \sigma\Delta t}\right)[P_x^{i,n+1}(i+1/2,j,k) - P_x^{i,n}(i+1/2,j,k)]. \end{aligned} \tag{3.16}$$

Similarly, from Eqs. (3.11a), (3.11b), and (3.11c), we can derive the updating equations for $\mathbf{Q}^s$, e.g.,

$$\begin{aligned} Q_x^{s,n+1}(i+1/2,j,k) &= \left(\frac{2\varepsilon_0\kappa_y - \sigma_y\Delta t}{2\varepsilon_0\kappa_y + \sigma_y\Delta t}\right) Q_x^{s,n}(i+1/2,j,k) + \left(\frac{2\varepsilon_0}{2\varepsilon_0\kappa_y + \sigma_y\Delta t}\right) \\ &\times \left[P_x^{s,n+1}(i+1/2,j,k) - P_x^{s,n}(i+1/2,j,k)\right]. \end{aligned} \tag{3.17}$$

Inserting Eqs. (3.10a), (3.10b), and (3.10c) into Eqs. (3.11a), (3.11b), and (3.11c), respectively, we can derive the electric field components in the UPML, e.g.,

$$\begin{aligned} E_x^{s,n+1}(i+1/2,j,k) &= \left(\frac{2\varepsilon_0\kappa_z - \sigma_z\Delta t}{2\varepsilon_0\kappa_z + \sigma_z\Delta t}\right) E_x^{s,n}(i+1/2,j,k) + \left(\frac{1}{2\varepsilon_0\kappa_z + \sigma_z\Delta t}\right) \\ &\times \left[(2\varepsilon_0\kappa_x + \sigma_x\Delta t)Q_x^{s,n+1}(i+1/2,j,k) - (2\varepsilon_0\kappa_x - \sigma_x\Delta t)Q_x^{s,n}(i+1/2,j,k)\right]. \end{aligned} \tag{3.18}$$

In a manner similar to the derivations of Eqs. (3.16), (3.17), and (3.18), we can derive the updating equations for $\mathbf{P}^i$ used in Eq. (3.16) from the known incident field $\mathbf{E}^i$ through the ancillary vector $\mathbf{Q}^i$. For example, from Eqs. (3.12a) and (3.12b), the x component of $\mathbf{Q}^i$ can be expressed as

$$Q_x^i(x,y,z) = \left(\frac{1}{s_y}\right) P_x^i(x,y,z) = \frac{s_z}{s_x} E_x^i(x,y,z), \tag{3.19}$$

which yields

$$\left(\kappa_x + \frac{\sigma_x}{i\omega\varepsilon_0}\right) Q_x^i(x,y,z) = \left(\kappa_z + \frac{\sigma_z}{i\omega\varepsilon_0}\right) E_x^i(x,y,z). \tag{3.20}$$

Eq. (3.20) can be expressed as

$$(i\omega\varepsilon_0\kappa_x + \sigma_x) Q_x^i(x,y,z) = (i\omega\varepsilon_0\kappa_z + \sigma_z) E_x^i(x,y,z). \tag{3.21}$$

Applying the inverse Fourier transform to Eq. (3.21) and using the identity $i\omega f(\omega) \to \partial f(t)/\partial t$ yield

$$\varepsilon_0\kappa_x \frac{\partial Q_x^i}{\partial t} + \sigma_x Q_x^i = \varepsilon_0\kappa_z \frac{\partial E_x^i}{\partial t} + \sigma_z E_x^i. \tag{3.22}$$

Eq. (3.22) can be discretized and reformulated as

$$Q_x^{i,n+1}(i+1/2,j,k) = \left(\frac{2\varepsilon_0\kappa_x - \sigma_x\Delta t}{2\varepsilon_0\kappa_x + \sigma_x\Delta t}\right) Q_x^{i,n}(i+1/2,j,k)$$
$$+ \left(\frac{\sigma_z\Delta t}{2\varepsilon_0\kappa_x + \sigma_x\Delta t}\right)\left[E_x^{i,n+1}(i+1/2,j,k) + E_x^{i,n}(i+1/2,j,k)\right]$$
$$+ \left(\frac{2\kappa_z\varepsilon_0}{2\varepsilon_0\kappa_x + \sigma_x\Delta t}\right)\left[E_x^{i,n+1}(i+1/2,j,k) - E_x^{i,n}(i+1/2,j,k)\right]. \tag{3.23}$$

On the other hand, Eq. (3.13a) can be rewritten as

$$P_x^i(x,y,z) = s_y Q_x^i(x,y,z) = \left(\kappa_y + \frac{\sigma_y}{i\omega\varepsilon_0}\right) Q_x^i(x,y,z), \tag{3.24}$$

from which we can derive

$$\varepsilon_0 \frac{\partial P_x^i}{\partial t} = \varepsilon_0\kappa_y \frac{\partial Q_x^i}{\partial t} + \sigma_y Q_x^i. \tag{3.25}$$

Discretizing Eq. (3.25), we obtain

$$P_x^{i,n+1}(i+1/2,j,k) = P_x^{i,n}(i+1/2,j,k)$$
$$+ \left(\frac{\sigma_y\Delta t}{2\varepsilon_0}\right)\left[Q_x^{i,n+1}(i+1/2,j,k) + Q_x^{i,n}(i+1/2,j,k)\right]$$
$$+ \kappa_y\left[Q_x^{i,n+1}(i+1/2,j,k) - Q_x^{i,n}(i+1/2,j,k)\right]. \tag{3.26}$$

In summary, once the incident field is specified, we can calculate Q_x^i using Eq. (3.23) and then calculate P_x^i using Eq. (3.26). After P_x^i is obtained, P_x^s can be found using Eq. (3.26). Using Eqs. (3.16) and (3.17),E_x^scan be calculated up to the absorbing boundary.

Using Eq. (3.9b), the updating equations of the scattered magnetic field in the absorbing boundary can be derived in the same way as for Eqs. (2.16) and (2.17). For example,

$$B_x^{s,n+1/2}(i,j+1/2,k+1/2) = \left(\frac{2\varepsilon_0\kappa_y - \sigma_y\Delta t}{2\varepsilon_0\kappa_y + \sigma_y\Delta t}\right) B_x^{s,n-1/2}(i,j+1/2,k+1/2)$$
$$+ \left(\frac{2\varepsilon_0\Delta t/\Delta s}{2\varepsilon_0\kappa_y + \sigma_y\Delta t}\right)$$
$$\times [E_y^{s,n}(i,j+1/2,k+1) - E_y^{s,n}(i,j+1/2,k) + E_z^{s,n}(i,j,k+1/2)$$
$$- E_z^{s,n}(i,j+1,k+1/2)], \tag{3.27}$$

$$H_x^{s,n+1/2}(i,j+1/2,k+1/2) = \left(\frac{2\varepsilon_0\kappa_z - \sigma_z\Delta t}{2\varepsilon_0\kappa_z + \sigma_z\Delta t}\right) H_x^{s,n-1/2}(i,j+1/2,k+1/2)$$
$$+ \left(\frac{1/\mu_0}{2\varepsilon_0\kappa_z + \sigma_z\Delta t}\right)$$
$$\times [(2\varepsilon_0\kappa_x + \sigma_x\Delta t) B_x^{s,n+1/2}(i,j+1/2,k+1/2)$$
$$- (2\varepsilon_0\kappa_x - \sigma_x\Delta t) B_x^{s,n-1/2}(i,j+1/2,k+1/2)]. \tag{3.28}$$

Using the equations reported in this section in the time-domain, a scattered-field formulation of the FDTD that includes the source term can be implemented inside the UPML. The scattered-field UPML (SF-UPML) provides the first accurate absorbing boundary condition for a source-dependent FDTD algorithm, which has great flexibility in simulations of electromagnetic wave scattering and transmission by arbitrary dielectric surface under an arbitrarily incident beam.

3.3.3 Numerical results

We examine the accuracy of the algorithm by comparing the resulting scattered field intensities with those calculated using the Fresnel equations for a smooth planar surface illuminated by a plane wave at normal incidence. The observation point is located on the central axis of the incident beam, one Δs from the material surface in free space. The beam half-width δ is set to be $30\Delta s$, where $\Delta s = \lambda/30$. We choose an observation point close to the material surface and use a relatively wide beam to reduce the effect of using a non-planar wave in the numerical simulations. Table 3.1 shows the comparison of the reflectivity from the two methods for different refractive indices of the material space. The errors of the FDTD results are within approximately 1% of that obtained using analytical Fresnel solution. To demonstrate the capability of the algorithm in calculating the near-field of the scattered wave from a surface, we design a simple system as illustrated in Fig. 3.9, which shows the FDTD computational domain enclosed by the SF-UPML with a beam incident on a dielectric material half-space with a refractive index of $n_2 = 3$. The computational domain size parameter is

$$\frac{\pi d_x}{\lambda} \times \frac{\pi d_y}{\lambda} \times \frac{\pi d_z}{\lambda} = 15 \times 15 \times 5,$$

where d_x, d_y, and d_z are the domain sizes in the $\mathbf{x}$, $\mathbf{y}$, and $\mathbf{z}$ directions of a right-hand Cartesian coordinate system. We use a FDTD spatial cell size $\Delta s = \lambda/30$ in the simulations; thus, the computational domain is a three-dimensional grid space with $143 \times 143 \times 47$ cubic cells bounded by the 6-Δs-thick SF-UPML. The interface between the material and free space (an x–z plane) passes through the center of the computational domain. The central axis of the incident beam passes through the center of the computational domain within the x–y plane. The polarization

Table 3.1. Comparison of the reflectivity from a planar substrate calculated using the FDTD method and the Fresnel equation.

Material space refractive index (n_r)	Reflectivity	
	FDTD	Fresnel $\left[\left(\frac{n_r - 1}{n_r + 1}\right)^2\right]$
1.1	0.0223922	0.0226757
1.2	0.0082158	0.0082644
1.3	0.0170212	0.0170132
1.4	0.0279624	0.0277777
1.5	0.0405093	0.0400000

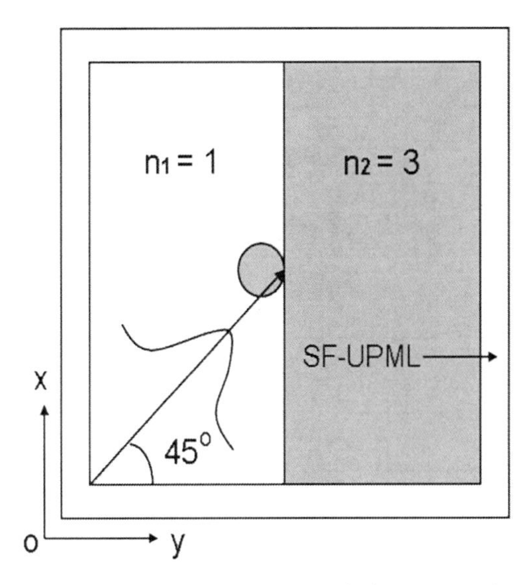

Fig. 3.9. Illustration of the FDTD computational domain enclosed by the SF-UPML and the beam incident on a dielectric surface. The interface between the material and free space passes through the center of the computational domain. The central axis of the incident beam passes through the center of the computational domain within the x–y plane.

direction of the incident electric field is normal to the central axis of the beam and is located within the x–y plane. The incident beam is a continuous wave described by Eqs. (3.5) and (3.6) with a half-width δ of $12\Delta s$.

Figure 3.10 shows the total electric field intensity ($|\mathbf{E}|^2$) for an electromagnetic beam incident on the smooth surface at 45° when the particle refractive index is set to be $n_{\text{particle}} = 1.0$. The fields shown here are those on the x–y plane passing through the center of the computational domain. The upper panel and lower panel are from the 300th and 900th FDTD time step, respectively. The incident electric field intensity at the central axis of the beam is normalized to 1. From this figure, we can see clearly the refracted waves and reflected waves passing smoothly into the SF-UPML absorbing boundary (i.e. at the edges of these figures) without distorting their shapes. This means the artificial reflection from the truncation of the computational domain is very small. In principle, the novel SF-UPML for the scattered-field FDTD algorithm should have a numerical accuracy similar to that of the conventional UPML for source-free FDTD equations. When there is a particle located near the surface, a very different scattering pattern emerges.

Figure 3.11 is the same as Fig. 3.10, but there is a spherical particle with a size parameter of $2\pi a/\lambda = \pi$, where a is the sphere radius, and a refractive index of $n_{\text{particle}} = 3$ on the smooth surface. As in Fig. 3.10, the upper panel and lower panel are from the 300th and 900th time step, respectively. The presence of the sphere changes the near fields significantly. We can see clearly that the presence of the sphere scatters the fields in all directions. We present these results to illustrate the capabilities of the FDTD. More importantly, the algorithm developed is completely general and can accommodate any surface irregularities or medium heterogeneities

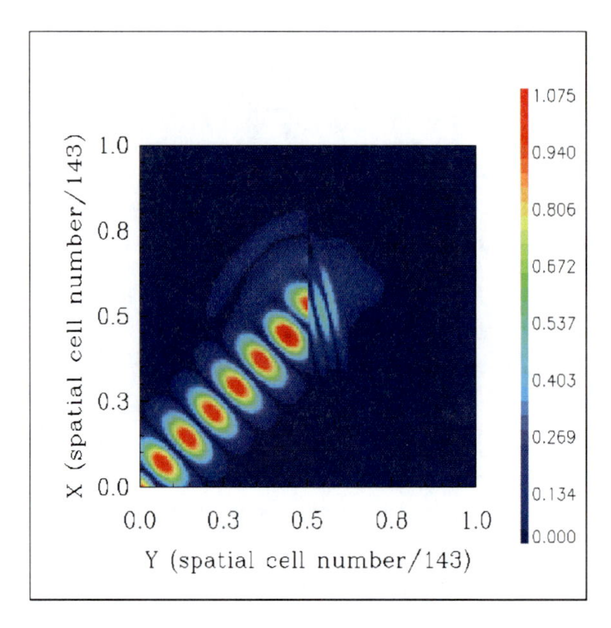

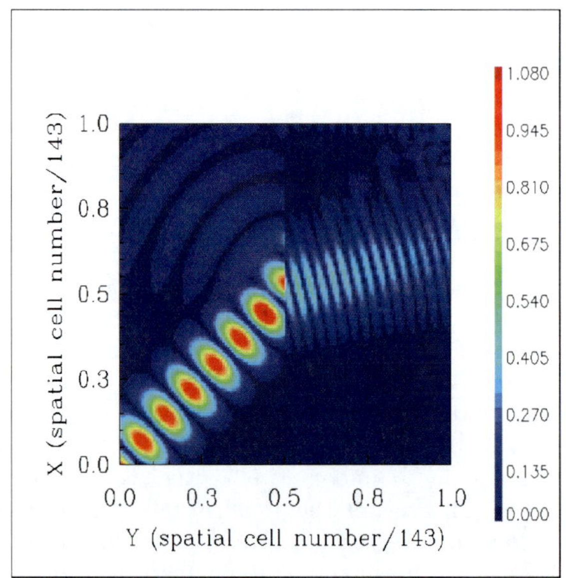

Fig. 3.10. Total electric field intensity ($|\mathbf{E}|^2$) for an electromagnetic beam incident on a smooth surface at 45°. The refractive index of the material space is $n_2 = 3$. The incident beam is a continuous wave described by Eqs. (3.5) and (3.6) with a half-width δ of $12\Delta s$, where $\Delta s = \lambda/30$. The fields shown here are those on the x–y plane passing through the center of the computational domain. The central axis of the incident beam and the polarization direction of the incident electric field are also in this x–y plane. The upper panel and lower panel are from the 300th and 900th FDTD time step, respectively. The incident electric field intensity at the central axis of the beam is normalized to 1.

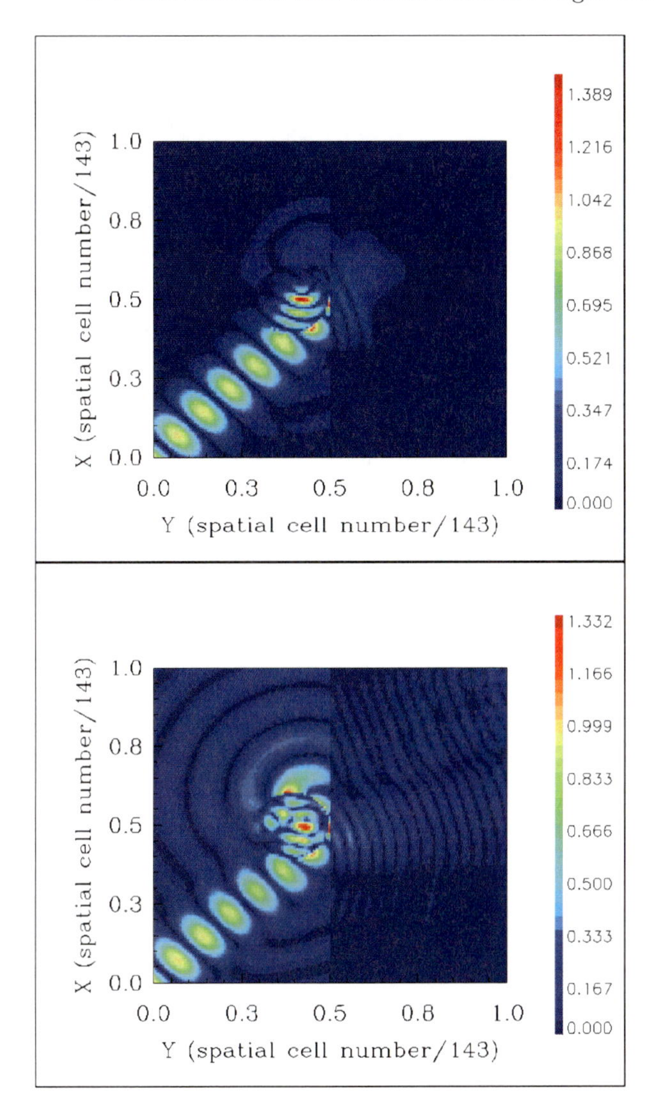

Fig. 3.11. Same as in Fig. 3.10, but with a spherical particle with a size parameter of $2\pi a/\lambda = \pi$ and a refractive index of $n_{\text{particle}} = 3$ resting on the smooth surface.

that are described by their refractive indices. This could have potential applications in many research fields such as surface remote sensing, nanotechnology and IED detection. At present the algorithm is not optimized for computational efficiency and additional work is required to implement practical incident fields, like Gaussian and Bessel beams.

3.4 Summary

The 3D FDTD techniques with uniaxial perfectly matched layer (UPML) absorbing boundary condition (ABC) to simulate light scattering and absorption by arbitrarily shaped particles embedded in a dielectric medium with or without absorption have been reviewed. We have formulated a numerical scheme to simulate the propagation of a plane wave in a finite region of a homogeneous dielectric medium. We also have derived scattering phase functions, extinction and absorption rates by using a volume integration of the electric field inside the particle for particles embedded in a dielectric medium. It is found that the errors in the extinction and absorption efficiencies from the UPML FDTD are smaller than $\sim$2%. The errors in the scattering phase functions are typically smaller than $\sim$5%. The errors in the asymmetry factors are smaller than $\sim$0.1%. For light scattering by particles in free space, the UPML FDTD scheme has a similar accuracy as the PML FDTD model [5]. The UPML FDTD method, which can be applied to both absorbing and free-space hosts, is a more general approach than the PML FDTD. Using the UPML FDTD technique, the problem of light scattering and absorption by non-spherical particles embedded in absorbing media can be accurately solved. Also, using the UPML ABC, the memory requirement for boundary layer is 25% less than using the split-field PML and the simulation can be done on a personal computer for a size parameter as large as 20. This generalized FDTD scheme has applications ranging from remote sensing of aerosols, clouds, ocean water color, modeling target response to ground penetration radar (GPR) wave, and biomedical studies.

In this chapter, we have also reviewed the numerical algorithm to calculate an arbitrary beam's interaction with an arbitrary dielectric surface based on the scattered-field FDTD method. With incident source terms in the FDTD equations, this development enables an arbitrary source to be incident onto an arbitrary dielectric surface or particle. For light scattering by particles in free space, the source terms exist only inside the particles, so a regular source-free boundary condition can work. However, if the host medium is not free space, the source terms in the scattered-field FDTD updating equations exist everywhere in the computational domain, including the absorbing boundary, the field updating equations inside the absorbing boundary must also have source terms. Therefore, a scattered-field uniaxial perfectly matched layer (SF-UPML) absorbing boundary condition (ABC) is developed to truncate the computational domain of the scattered-field FDTD grid. The SF-UPML FDTD algorithm removes the normally incident plane wave condition that restricts previously developed FDTD methods in surface studies. The incident field can be a plane wave beam incident at non-normal angles with respect to the surface or it can be a Gaussian beam like those emitted by lasers commonly used in surface science, etc. The SF-UPML FDTD algorithm is general and can accommodate any type of fabricated system. This development has potential applications in various research fields such as surface and particle characterization, canceled target detection, substrate fabrication, and remote-sensing studies. While the SF-UPML FDTD algorithm provides great flexibility in implementing different wave sources for calculation of light scattering by different objects, because the wave sources are implemented at all grid points where there is not a vacuum, the computational time required by this algorithm is larger than that needed by the total/scattered-field FDTD method.

References

1. K. S. Yee, Numerical solution of initial boundary value problems involving Maxwell's equation in isotropic media, *IEEE Trans. Antennas Propag.* **AP-14**, 302–307 (1966).
2. A. Taflove and M. E. Brodwin, Numerical solution of steady-state electromagnetic scattering problems using the time-dependent Maxwell's equations, *IEEE Trans. Microwave Theory Tech.* **MTT-23**, 623–630 (1975).
3. A. Taflove, *Computational Electrodynamics: The Finite-Difference Time Domain Method* (Artech House, Boston, 1995).
4. P. Yang and K. N. Liou, Finite-difference time domain method for light scattering by small ice crystals in three-dimensional space, *J. Opt. Soc. Am. A* **13**, 2072–2085 (1996).
5. W. Sun, Q. Fu, and Z. Chen, Finite-difference time-domain solution of light scattering by dielectric particles with a perfectly matched layer absorbing boundary condition, *Appl. Opt.* **38**, 3141-3151 (1999).
6. W. Sun, N. G. Loeb, and Q. Fu, Finite-difference time domain solution of light scattering and absorption by particles in an absorbing medium, *Appl. Opt.* **41**, 5728–5743 (2002).
7. W. Sun, H. Pan, and G. Videen, General finite-difference time-domain solution of an arbitrary EM source interaction with an arbitrary dielectric surface, *Appl. Opt.* **48**, 6015–6025 (2009).
8. G. Mie, Beiträge zur Optik trüber Medien, speziell kolloidaler Metallösungen, Leipzig, *Ann. Phys.* **330**, 377–445 (1908).
9. G. Videen, D. Ngo, and M. B. Hart, Light scattering from a pair of conducting, osculating spheres, *Opt. Commun.* **125**, 275–287 (1996).
10. D. Petrov, Y. Shkuratov, G. Videen, Analytic light-scattering solution of two merging spheres using Sh-matrices, **Opt. Comm. 281**, 2411–2423 (2008).
11. S. Asano and G. Yamamoto, Light scattering by a spheroidal particle, *Appl. Opt.* **14**, 29–49 (1975).
12. M. I. Mishchenko, L. D. Travis, and D. W. Mackowski, T-matrix computations of light scattering by non-spherical particles. A review, *J. Quant. Spectrosc. Radiat. Transfer* **55**, 535–575 (1996).
13. Lord Rayleigh, The dispersal of light by a dielectric cylinder, *Philos. Mag.* **36**, 365–376 (1918).
14. J. R. Wait, Scattering of a plane wave from a circular dielectric cylinder at oblique incidence, *Can. J. Phys.* **33**, 189–195 (1955).
15. A. Mugnai and W. J. Wiscombe, Scattering from non-spherical Chebyshev particles, *Appl. Opt.* **25**, 1235–1244 (1986).
16. D. Petrov, Y. Shkuratov, G. Videen, Analytical light-scattering solution for Chebyshev particles, *J. Opt. Soc. Am. A* **24**, 1103–1119 (2007).
17. M. I. Mishchenko, L. D. Travis, and A. Macke, Scattering of light by polydisperse, randomly oriented, finite circular cylinders, *Appl. Opt.* **35**, 4927–4940 (1996).
18. D. Petrov, Y. Shkuratov, G. Videen, The influence of corrugation on light-scattering properties of capsule and finite cylinder particles: *Sh*-matrices analysis, *J. Quant. Spectrosc. Radiat. Transfer* **109**, 650–669 (2008).
19. D. Petrov, Y. Shkuratov, G. Videen, Sh-matrices method as applied to light scattering by circular cylinders, *J. Quant. Spectrosc. Radiat. Transfer* **109**, 1474–1495 (2008).

20. H. Laitinen and K. Lumme, T-matrix method for general star-shaped particles: first results, *J. Quant. Spectrosc. Radiat. Transfer* **60**, 325–334 (1998).
21. D. Petrov, Y. Shkuratov, G. Videen, An analytical solution to the light scattering from cube-like particles using Sh-matrices, *J. Quant. Spectrosc. Radiat. Transfer* **111**, 474–482 (2010).
22. Lord Rayleigh, On the light from the sky, its polarization and colour, *Phil. Mag.* **41**, 107–120, 274–279 (1871).
23. V. G. Farafonov, Light-scattering by spheroidal particles in quasi-static approximation, *Opt. Spektrosk.* **77**, 455–458 (1994).
24. P. Yang and K. N. Liou, Geometric-optics integral-equation method for light scattering by non-spherical ice crystals, *Appl. Opt.* **35**, 6568–6584 (1996).
25. P. Barber and C. Yeh, Scattering of electromagnetic waves by arbitrarily shaped dielectric bodies, *Appl. Opt.* **14**, 2864–2872 (1975).
26. B. T. Draine, The discrete-dipole approximation and its application to interstellar graphite grains, *Astrophys. J.* **333**, 848–872 (1988).
27. B. Engquist and A. Majda, Absorbing boundary conditions for the numerical simulation of waves, *Math. Comp.* **31**, 629–651 (1971).
28. A. Bayliss and E. Turkel, Radiation boundary conditions for wave-like equations, *Commun. Pure Appl. Math.* **33**, 707–725 (1980).
29. G. Mur, Absorbing boundary condition for the finite-difference approximation of the time-domain electromagnetic-field equations, *IEEE Trans. Electromagn. Compat.* **EMC-23**, 377–382 (1981).
30. Z. Liao, H. L. Wong, B. Yang, and Y. Yuan, A transmitting boundary for transient wave analyses, *Sci. Sin.* **27**, 1063–1076 (1984).
31. R. L. Higdon, Absorbing boundary conditions for difference approximations to the multi-dimensional wave equation, *Math. Comp.* **47**, 437–459 (1986).
32. C. E. Reuter, R. M. Joseph, E. T. Thiele, D. S. Katz, and T. Taflove, Ultrawideband absorbing boundary condition for termination of wave guide structures in FD-TD simulations, *IEEE Microwave and Guided Wave Lett.* **4**, 344–346 (1994).
33. J. P. Berenger, A perfectly matched layer for the absorption of electromagnetic waves, *J. Comp. Phys.* **114**, 185–200 (1994).
34. D. S. Katz, E. T. Thiele, and A. Taflove, Validation and extension to three dimensions of the Berenger PML absorbing boundary condition for FD-TD meshes, *IEEE Microwave and Guided Wave Lett.* **4**, 268–270 (1994).
35. Z. S. Sacks, D. M. Kingsland, R. Lee, and J. F. Lee, A perfectly matched anisotropic absorber for use as an absorbing boundary condition, *IEEE Trans. Antennas Propaga.* **43**, 1460–1463 (1995).
36. S. D. Gedney, An anisotropic perfectly matched layer absorbing media for the truncation of FDTD lattices, *IEEE Trans. Antennas Propag.* **44**, 1630–1639 (1996).
37. W. Sun and Q. Fu, Finite-difference time-domain solution of light scattering by dielectric particles with large complex refractive indices, *Appl. Opt.* **39**, 5569–5578 (2000).
38. D. E. Merewether, R. Fisher, and F.W. Smith, On implementing a numeric Huygen's source in a finite difference program to illustrate scattering bodies, *IEEE Trans. Nucl. Sci.* **NS-27**, 1829–1833 (1980).
39. K. Umashanker and A. Taflove, A novel method to analyze electromagnetic scattering of complex objects, *IEEE Trans. Electromagn. Compat.* **EMC-24**, 397–405 (1982).
40. W. C. Mundy, J. A. Roux, and A. M. Smith, Mie scattering by spheres in an absorbing medium, *J. Opt. Soc. Am.* **64**, 1593–1597 (1974).
41. P. Chylek, Light scattering by small particles in an absorbing medium, *J. Opt. Soc. Am.* **67**, 561–563 (1977).

42. C. F. Bohren and D. P. Gilra, Extinction by a spherical particle in an absorbing medium, *J. Colloid Interface Sci.* **72**, 215–221 (1979).
43. M. Quinten and J. Rostalski, Lorenz-Mie theory for spheres immersed in an absorbing host medium, *Part. Part. Syst. Charact.* **13**, 89–96 (1996).
44. A. N. Lebedev, M. Gartz, U. Kreibig, and O. Stenzel, Optical extinction by spherical particles in an absorbing medium: application to composite absorbing films, *Eur. Phys. J. D* **6**, 365–373 (1999).
45. Q. Fu and W. Sun, Mie theory for light scattering by a spherical particle in an absorbing medium, *Appl. Opt.* **40**, 1354–1361 (2001).
46. I. W. Sudiarta and P. Chylek, Mie-scattering formalism for spherical particles embedded in an absorbing medium, *J. Opt. Soc. Am. A* **18**, 1275–1278 (2001).
47. Q. Fu and W. Sun, Apparent optical properties of spherical particles in absorbing medium, *J. Quan. Spectro. Rad. Transfer* **100**, 137–142 (2006).
48. M. I. Mishchenko, Electromagnetic scattering by a fixed finite object embedded in an absorbing medium, *Opt. Express* **15**, 13188–13202 (2007).
49. C. F. Bohren and D. R. Huffman, *Absorption and Scattering of Light by Small Particles* (John Wiley, New York, 1983).
50. G. H. Goedecke and S. G. O'Brien, Scattering by irregular inhomogeneous particles via the digitized Green's function algorithm, *Appl. Opt.* **27**, 2431–2438 (1988).
51. E. Zubko, D. Petrov, Y. Grynko, Y. Shkuratov, H. Okamoto, K. Muinonen, T. Nousiainen, H. Kimura, T. Yamamoto, and G. Videen, Validity criteria of the discrete dipole approximation, *Appl. Opt.* **49**, 1267–1279 (2010).
52. W. Sun, T. Nousiainen, K. Muinonen, Q. Fu, N. G. Loeb, and G. Videen, "Light scattering by Gaussian particles: A solution with finite-difference time domain technique", *J. Quant. Spectrosc. Radiat. Transfer*, **79–80**, 1083–1090 (2003).
53. D. L. Schuler, J.-S. Lee, D. Kasilingam, and G. Nesti, Surface roughness and slope measurements using polarimetric SAR data, *IEEE Transactions on Geoscience and Remote Sensing*, **40**, 687–698 (2002).
54. S. Gomez, K. Hale, J. Burrows, and B. Griffiths, Measurements of surface defects on optical components, *Meas. Sci. Technol.* **9**, 607–616 (1998).
55. H. Lin and J. Zhu, Characterization of nanocrystalline silicon films, *Proc. SPIE*, **4700**, 354–356 (2002).
56. A. Angell and C. Rappaport, Computational modeling analysis of radar scattering by clothing covered arrays of metallic body-worn explosive devices, *Progress In Electromagnetics Research* **PIER 76**, 285–298 (2007).
57. Lord Rayleigh, *The Theory of Sound* (MacMillan, London, 1896).
58. U. Fano, The theory of anomalous diffraction gratings and of quasi-stationary waves on metallic surfaces (Sommerfeld's waves), *J. Opt. Soc. Am.* **31**, 213–222 (1941).
59. S. O. Rice, Reflection of electromagnetic waves from slightly rough surfaces, *Commun. Pure Appl. Math.* **4**, 351–378 (1951).
60. S. O. Rice, *Reflection of EM from Slightly Rough Surfaces* (Interscience, New-York, 1963).
61. C. Eckart, The scattering of sound from the sea surface, *J. Acoust. Soc. Am.* **25**, 66–570 (1953).
62. H. Davies, The reflection of electromagnetical waves from rough surfaces, *Proc. IEE (London)* **101**, 209–214 (1954).
63. P. Beckmann and A. Spizzichino, *The Scattering of Electromagnetic Waves from Rough Surfaces* (Pergamon Press, Oxford, England, 1963).
64. A. K. Fung and G. W. Pan, An integral equation method for rough surface scattering, in *Proceedings of the International Symposium on multiple scattering of waves in random media and random surfaces*, 701–714 (1986).

65. A. K. Fung, Z. Li, and K. S. Chen, Backscattering from a randomly rough dielectric surface, *IEEE Trans. Geosci. and Remote Sens.* **30**, 356–369 (1992).
66. A. K. Fung, *Microwave Scattering and Emission Models and their Applications* (Artech House, Norwood, MA, 1994).
67. L. Tsang, J. A. Kong, K. H. Ding, and C. O. Ao, *Scattering of Electromagnetic Waves: Numerical Simulations* (John Wiley, New York, 2001).
68. M. Saillard and A. Sentenac, Rigorous solutions for electromagnetic scattering from rough surfaces, *Waves in Random Media* **11**, 103–137 (2001).
69. C. Y. Hsieh, A. K. Fung, G. Nesti, A. J. Siber, and P. Coppo, A further study of the IEM surface scattering model, *IEEE Trans. Geosci. and Remote Sens.* **35**, 901–909 (1997).
70. A. K. Fung, Z. Li, and K. S. Chen, An improved IEM model for bistatic scattering from rough surfaces, *J. Electromagn. Waves and Appl.* **16**, 689–702 (2002).
71. K. S. Chen, T. D. Wu, and A. K. Fung, A study of backscattering from multi-scale rough surface, *J. Electromagn. Waves and Appl.* **12**, 961–979 (1998).
72. F. Mattia, Backscattering properties of multi-scale rough surfaces, *J. Electromagn. Waves and Appl.* **13**, 493–527 (1999).
73. P. P. Silvester and R. L. Ferrari, *Finite Elements for Electrical Engineers* (Cambridge Univ. Press, Cambridge, U.K., 1990).
74. J. M. Jin, *The Finite Element Method in Electromagnetics* (John Wiley, New York, 1993).
75. E. M. Purcell and C. R. Pennypacker, Scattering and absorption of light by non-spherical dielectric grains, *Astrophys. J.* **186**, 705–714 (1973).
76. S. B. Singham and G. C. Salzman, Evaluation of the scattering matrix of an arbitrary particle using the coupled dipole approximation, *J. Chem. Phys.* **84**, 2658–2667 (1986).
77. W. Sun, G. Videen, B. Lin, and Y. Hu, Modeling light scattered from and transmitted through dielectric periodic structures on a substrate, *Appl. Opt.* **46**, 1150–1156 (2007).
78. D. Wu and Y. Zhou, Forward scattering light of droplets containing different size inclusions, *Appl. Opt.* **48**, 2957–2965 (2009).
79. G. Videen, Light scattering from a sphere on or near a surface, *J. Opt. Soc. Am. A* **8**, 483–489 (1991); *Errata. J. Opt. Soc. Am. A* **9**, 844–845 (1992).
80. E. Fucile, P. Denti, F. Borghese, R. Saija, and O. I. Sindoni, Optical properties of a sphere in the vicinity of a plane surface, *J. Opt. Soc. Am. A* **14**, 1505–1514 (1997).
81. B. R. Johnson, Calculation of light scattering from a spherical particle on a surface by the multipole expansion method, *J. Opt. Soc. Am. A* **13**, 326–337 (1996).
82. G. Videen, Light scattering from a sphere behind a surface, *J. Opt. Soc. Am. A* **10**, 110–117 (1993).
83. G. Videen, M. G. Turner, V. J. Iafelice, W. S. Bickel, and W. L. Wolfe, Scattering from a small sphere near a surface, *J. Opt. Soc. Am. A* **10**, 118–126 (1993).
84. G. Videen, M. M. Aslan, and M. P. Mengüç, Characterization of metallic nano-particles via surface wave scattering: A. Theoretical framework and formulation, *J. Quant. Spectrosc. Radiative Transfer* **93**, 195–206 (2005).
85. R. Borghi, F. Gori, M. Santarsiero, F. Frezza, and G. Schettini, Plane-wave scattering by a perfectly conducting circular cylinder near a plane surface: Cylindrical-wave approach, *J. Opt. Soc. Am. A* **13**, 483–493 (1996).
86. G. Videen and D. Ngo, Light scattering from a cylinder near a plane interface: Theory and comparison with experimental data, *J. Opt. Soc. Am. A* **14**, 70–78 (1997).
87. G. Videen, Light scattering from a particle on or near a perfectly conducting surface, *Opt. Commun.* **115**, 1–7 (1995).

88. G. Videen, Light scattering from an irregular particle behind a plane interface, *Opt. Commun.* **128**, 81–90 (1996).
89. P. G. Venkata, M. M. Aslan, M. P. Menguc, and G. Videen, Surface plasmon scattering by gold nanoparticles and two-dimensional agglomerates, *J. Heat Transfer–Trans. ASME* **129**, 60–70 (2007).
90. T. Wriedt and A. Doicu, Light scattering from a particle on or near a surface, *Opt Commun.* **152**, 376-384 (1998).
91. P. Denti, F. Borghese, R. Saija, E. Fucile, and O. I. Sindoni, Optical properties of aggregated spheres in the vicinity of a plane surface, *Appl. Opt.* **16**, 167–175 (1999).
92. D. W. Mackowski, Exact solution for the scattering and absorption properties of sphere clusters on a plane surface, *J. Quant. Spectrosc. Radiative Transfer* **109**, 770–788 (2008).
93. R. Schmehl, B. M. Nebeker, and E. D. Hirleman, Discrete-dipole approximation for scattering by features on surfaces by means of a two-dimensional fast Fourier transform technique, *J. Opt. Soc. Am. A* **14**, 3026–3036 (1997).
94. P. Albella, F. Moreno, J. M. Saiz, and F. González, Surface inspection by monitoring spectral shifts of localized Plasmon resonances, *Opt. Exp.* **16**, 12,872–12,879 (2008).
95. M. A. Yurkin, A. G. Hoekstra, R. S. Brock, and J. Q. Lu, Systematic comparison of the discrete dipole approximation and the finite difference time domain method for large dielectric scatterers, *Opt. Express* **15**, 17902–17911 (2007).
96. A. Doicu, Y. Eremin, and T. Wriedt, *Acoustic and Electromagnetic Scattering Analysis Using Discrete Sources* (Academic Press, San Diego, 2000).
97. N. J. Cassidy, A review of practical numerical modeling methods for the advanced interpretation of ground-penetrating radar in near-surface environments, *Near Surface Geophysics* **5**, 5–21 (2007).
98. R. Holland, Threde: A free-field EMP coupling and scattering code, *IEEE Trans. Nuclear Sci.* **24**, 2416–2421 (1977).
99. R. Holland, R. L. Simpson, and K. S. Kunz, Finite-difference analysis of EMP coupling to lossy dielectric structures, *IEEE Trans. Electromagn. Compat.* **22**, 203–209 (1980).
100. A. Taflove and S. C. Hagness, *Computational Electrodynamics: The Finite-Difference Time Domain Method* (Artech House, Boston, 2005).
101. W. Sun, N. G. Loeb, G. Videen, and Q. Fu, Examination of surface roughness on light scattering by long ice columns by use of a two-dimensional finite-difference time-domain algorithm, *Appl. Opt.* **43**, 1957–1964 (2004).
102. W. Sun, B. Lin, Y. Hu, Z. Wang, Y. Fu, Q. Feng, and P. Yang, Side-face effect of a dielectric strip on its optical properties, *IEEE Transac. Geosci. Remote Sens.* **46**, doi: 10.1109/TGRS.2008.916984 (2008).
103. B. Saleh and M. Teich, *Fundamentals of Photonics* (John Wiley, New York, 1991).
104. J. Durnin, Exact solutions for nondiffracting beams. I. The scalar theory, *J. Opt. Soc. Am. A* **4**, 651–654 (1987).

4 Advances in finite-difference time-domain calculation methods

James B. Cole, Naoki Okada, and Saswatee Banerjee

Introduction

Although the finite-difference time-domain (FDTD) method was developed in the 1960s, beginning with Yee's famous algorithm [1], and many advances have been made since then, FDTD is still an active field of research.

One reason that FDTD is so appealing is that it derives directly from the original Maxwell's equations without using any 'black box' mathematical artifices. The algorithm is simple and easy to program. Nevertheless when one tries to actually use FDTD to solve practical problems, one soon finds many 'devils in the details.' In this chapter we introduce some new advances in the FDTD methodology, and directly address some of the devilish details.

In section 4.1 (mainly written by J. B. Cole), we introduce our recent work in nonstandard finite difference models, which have led to great improvements in the accuracy of the basic FDTD algorithm.

In section 4.2 (mainly written by N. Okada), we have applied our methods to compute whispering gallery modes with very high accuracy, and derive a high-accuracy, low-cost NS-FDTD algorithm for the coupled wave equation.

In section 4.3 (mainly written by S. Banerjee), an improved version of FDTD recursion convolution (RC) algorithm for computing propagation in dispersive materials is introduced. Numerical instability has long been a major drawback of the RC method. Here rigorous stability criteria are given. Also extensions of RC-FDTD to the wave equation are given.

The overall goal is to lay out our methods clearly and in detail with a minimum of obfuscating formalism.

4.1 Advances in nonstandard finite-difference time-domain theory and its implementation

4.1.1 Standard versus nonstandard FDTD

The conventional finite-difference time-domain (FDTD) algorithms to solve the wave equation and Maxwell's equations [2] are simple and easy to program, but accuracy is low because second-order finite difference (FD) approximations are used.

Although accuracy can be improved by using higher-order FD approximations, this approach not only complicates the algorithm, but it is often numerically unstable. Mickens [3] has shown that when the order of the FD approximation exceeds the order of the original differential equation, the difference equation has unstable spurious solutions that have nothing to do with those of the original differential equation. Using what is called a nonstandard (NS) finite-difference model [3] of the wave equation and Maxwell's equations it is possible construct high-accuracy FDTD algorithms without using higher-order FD approximations [4]–[6].

NS-FD model of the derivative

The conventional, or standard (S), second-order central FD approximation to the first derivative is,

$$\frac{d}{dt}\psi(t) \cong \frac{d_t\psi(t)}{\Delta t}, \tag{1.1}$$

where the difference operator, d_t, is defined by

$$d_t\psi(t) = \psi(t + \Delta t/2) - \psi(t - \Delta t/2). \tag{1.2}$$

The most general NS-FD model of the first derivative at $t = \tau$ $(t \leq \tau \leq t + \Delta t)$, is

$$\left.\frac{d}{dt}\psi(t)\right|_{t=\tau} \cong \frac{\psi(t + \Delta t) - \sigma(\Delta t)\psi(t)}{s(\Delta t)}, \tag{1.3}$$

where σ and s are chosen to minimize the error with respect to a certain class of functions. For our purposes it is sufficient to consider a NS-FD model of the form

$$\frac{d}{dt}\psi(t) \cong \frac{d_t\psi(t)}{s(\Delta t)}. \tag{1.4}$$

It might seem that choosing

$$s(\Delta t) = \frac{d_t\psi(t)}{\psi'(t)} \tag{1.5}$$

would yield an exact FD expression for the derivative, but (1.5) is not always valid. To guarantee that

$$\lim_{\Delta t \to 0} \frac{d_t\psi(t)}{s(\Delta t)} = \frac{d}{dt}\psi(t) \tag{1.6}$$

s must satisfy the constraints,

$$\lim_{\Delta t \to 0} s(\Delta t) = 0, \tag{1.7}$$

$$s'(0) = 1. \tag{1.8}$$

In addition, s must not depend on t; if it did s would not be useful in practical algorithms.

For example, let $\psi(t) = e^{\pm i\omega t}$ (ω complex). We find that s as given by (1.5) satisfies (1.7) and (1.8). Defining

$$s(\omega, \Delta t) = \frac{2}{\omega}\sin(\omega\Delta t/2), \tag{1.9}$$

and inserting the choice $s(\Delta t) = s(\Delta t, \omega)$ into (1.4) yields an exact NS-FD expression for $\psi'(t)$.

NS-FD model of the one-dimensional wave equation

Let us now seek an exact NS-FD model for the one-dimensional wave equation,

$$\left(\partial_t^2 - v^2\partial_x^2\right)\psi(x,t) = 0, \tag{1.10}$$

where v is the phase velocity. The S-FD approximation to the second derivative is

$$\frac{d^2}{dt^2}\psi(t) \cong \frac{d_t^2\psi(t)}{\Delta t^2}, \tag{1.11}$$

where $d_t^2 = d_t d_t$. It is easy to show that

$$d_t^2\psi(t) = \psi(t+\Delta t) + \psi(t-\Delta t) - 2\psi(t). \tag{1.12}$$

Replacing the derivatives in (1.10) with S-FD expressions yields the S-FD model

$$\left(d_t^2 - \frac{v^2\Delta t^2}{\Delta x^2}d_x^2\right)\psi(x,t) = 0. \tag{1.13}$$

Using (1.12) to expand $d_t^2\psi(x,t)$ and solving for $\psi(x, t+\Delta t)$ gives the conventional or standard (S)-FDTD algorithm,

$$\psi(x, t+\Delta t) = -\psi(x, t-\Delta t) + \left[2 + \frac{v^2\Delta t^2}{\Delta x^2}d_x^2\right]\psi(x,t). \tag{1.14}$$

Solutions to the free space one-dimensional wave equation are $\varphi(x,t) = e^{i(kx \mp \omega t)}$, where $v = \omega/k$. Inserting φ, into the S-FD model (1.13) we find

$$\left(d_t^2 - \frac{v^2\Delta t^2}{\Delta x^2}d_x^2\right)\varphi(x,t) = \varepsilon\varphi(x,t), \tag{1.15}$$

where

$$\varepsilon = 4\left[-\sin^2(\omega\Delta t/2) + \frac{v^2\Delta t^2}{\Delta x^2}\sin^2(k\Delta x/2)\right]. \tag{1.16}$$

The fact that $\varepsilon \neq 0$ (except for the special case $v^2\Delta t^2/\Delta x^2 = 1$) means that φ is not generally a solution of (1.13), so the wave equation and its FD model have different solutions.

Let us now regard $v\Delta t/\Delta x$ as a free parameter – call it u. It is easy to show that $\varepsilon = 0$ for the choice

$$u = \frac{\sin(\omega\Delta t/2)}{\sin(k\Delta x/2)}. \tag{1.17}$$

Thus the difference equation

$$\left(d_t^2 - u^2 d_x^2\right)\psi(x,t) = 0 \tag{1.18}$$

has the same solutions as the wave equation (1.10). This is an example of a NS-FD [3] model.

We have extended NS-FD models to the wave equation and Maxwell's equations in two and three dimensions [4]–[6].

NS-FD model of the two- and three-dimensional wave equation

In two and three dimensions the homogeneous wave equation is given by

$$\left(\partial_t^2 - \frac{v_0^2}{n^2}\nabla^2\right)\psi(\mathbf{x},t) = 0, \tag{1.19}$$

where v_0 is the vacuum phase velocity, and $n = n(\mathbf{x})$ is the refractive index at $\mathbf{x} = (x, y, z)$. Taking $\Delta x = \Delta y = \Delta z = h$, the S-FD approximations for $\nabla\psi(\mathbf{x})$ and $\nabla^2\psi(\mathbf{x})$ are

$$\nabla\psi(\mathbf{x}) \cong \frac{\mathbf{d}\psi(\mathbf{x})}{h}, \tag{1.20}$$

$$\nabla^2\psi(\mathbf{x}) \cong \frac{\mathbf{d}^2\psi(\mathbf{x})}{h^2}, \tag{1.21}$$

where $\mathbf{d} = (d_x, d_y, d_z)$ is a vector difference operator, and $\mathbf{d}^2 = \mathbf{d}\cdot\mathbf{d} = d_x^2 + d_y^2 + d_z^2$. The S-FD model of (1.19) becomes

$$\left(d_t^2 - \frac{v_0^2\Delta t^2}{n^2h^2}\mathbf{d}^2\right)\psi(\mathbf{x},t) = 0. \tag{1.22}$$

In each region of constant n a general monochromatic solution of the wave equation is a superposition of plane wave waves of the form,

$$\psi(\mathbf{x},t) = e^{-i\omega_0 t}\sum_{\hat{\mathbf{k}}} c_{\hat{\mathbf{k}}}\varphi_{\hat{\mathbf{k}}}(\mathbf{x}), \tag{1.23}$$

where $\varphi_{\hat{\mathbf{k}}}(\mathbf{x}) = e^{ink_0\hat{\mathbf{k}}\cdot\mathbf{x}}$, k_0 is the vacuum wavenumber, $\omega_0 = k_0 v_0$, and the $c_{\hat{\mathbf{k}}}$ are constants. The summation is over all propagation directions ($\hat{\mathbf{k}}$ is a unit vector). There is no NS-FD expression for $\nabla^2\varphi_{\hat{k}}$ that is exact for all $\hat{\mathbf{k}}$ directions, but a nearly exact one can be formed with the replacements $\mathbf{d}^2 \to \mathbf{d}_0^2$, $h \to s(k,h)$ in (1.21), where

$$s(k,h) = (2/k)\sin(kh/2), \tag{1.24}$$

$k = nk_0$, and $\mathbf{d}_0^2$ is a new operator defined by

$$\mathbf{d}_0^2 = \mathbf{d}^2 + \gamma_1 d_x^2 d_y^2, \tag{1.25}$$
$$\mathbf{d}_0^2 = \mathbf{d}^2 + \gamma_1\left(d_x^2 d_y^2 + d_x^2 d_z^2 + d_y^2 d_z^2\right) + \gamma_2 d_x^2 d_y^2 d_z^2, \tag{1.26}$$

in two dimensions, and three dimensions, respectively. The parameters γ_1 and γ_2 are defined in section 4.1.7, and the details of the derivation are given in [4] and [6]. Assuming solutions in the form of (1.23), we substitute the exact NS-FD expression,

$$\partial_t^2 \psi = \frac{d_t^2 \psi}{s(\omega_0, \Delta t)^2}, \tag{1.27}$$

and the nearly exact one,

$$\nabla^2 \psi \cong \frac{\mathbf{d}_0^2(n)\psi}{s(nk_0, h)^2}, \tag{1.28}$$

into (1.19) and obtain NS-FD model,

$$\left[d_t^2 - \frac{u_0^2}{\tilde{n}^2}\mathbf{d}_0^2(n)\right]\psi(\mathbf{x},t) = 0, \tag{1.29}$$

where $u_0 = \sin(\omega_0 \Delta t/2)/\sin(k_0 h/2)$, and

$$\tilde{n} = \frac{\sin(nk_0 h/2)}{\sin(k_0 h/2)}. \tag{1.30}$$

The quantity $\tilde{n}$ can be regarded as the effective local refractive index, and $u = u_0/\tilde{n}$ as the effective phase velocity of waves on the grid. Since γ_1 and γ_2 are functions of n, so is $\mathbf{d}_0^2$. Solving (1.29) for $\psi(\mathbf{x}, t+\Delta t)$ gives the NS-FDTD algorithm,

$$\psi(\mathbf{x}, t+\Delta t) = -\psi(\mathbf{x}, t-\Delta t) + \left[2 + \frac{u_0^2}{\tilde{n}^2}\mathbf{d}_0^2(n)\right]\psi(\mathbf{x},t). \tag{1.31}$$

Notice that that (1.28) is exact with respect to $\varphi_{\hat{\mathbf{k}}}$ for $\hat{\mathbf{k}} = \hat{\mathbf{x}}$, $\hat{\mathbf{y}}$, or $\hat{\mathbf{z}}$.

4.1.2 Wave equation for the scattered field and its NS-FD model

Scattered-field wave equation

Let a scatterer of refractive index n be illuminated by an incident field, ψ_0. The total field, ψ, is the sum of the incident (ψ_0) and the scattered field (ψ_{s}), thus

$$\psi_{\mathrm{s}} = \psi - \psi_0. \tag{1.32}$$

Whereas the propagation of ψ is governed by (1.19), by definition ψ_0 propagates according to

$$\left(\partial_t^2 - v_0^2\nabla^2\right)\psi_0(\mathbf{x},t) = 0. \tag{1.33}$$

In other words, the incident field propagates as if there were no scatterer. Subtracting (1.33) from (1.19) we obtain

$$\left[\partial_t^2 - \frac{v_0^2}{n^2}\nabla^2\right]\psi_s(\mathbf{x},t) = -\left(\frac{n^2-1}{n^2}\right)\partial_t^2\psi_0(\mathbf{x},t). \tag{1.34}$$

The term on the right of (1.34) is an effective source term which produces the scattered field, and which vanishes outside the scatterer.

Standard finite-difference model of the scattered field

The S-FD model for the scattered field is obtained by replacing the derivatives on the left of (1.34) with SFD approximations,

$$\left[d_t^2 - \frac{v_0^2\Delta t^2}{n^2h^2}\mathbf{d}^2\right]\psi_s(\mathbf{x},t) = -\Delta t^2\left(\frac{n^2-1}{n^2}\right)\partial_t^2\psi_0(\mathbf{x},t). \tag{1.35}$$

Up until now we constructed the NS-FD model with the replacements $\Delta t \to s(\omega,\Delta t)$, $h \to s(nk,h)$, and $\mathbf{d}^2 \to \mathbf{d}_0^2(n)$. The NS-FDTD algorithm derived from this model is much more accurate than the S-FDTD algorithm, but it is not quite correct because of the way that the source term in (1.34) is modeled. We now proceed to derive a better NS-FD model of the scattered field wave equation.

Nonstandard finite-difference model of the scattered field

The NS-FD model for the propagation of ψ_0 is given by (1.29) with $n = 1 \Rightarrow \tilde{n} = 1$. Let the incident field be

$$\psi_0 = e^{i\left(k_0\hat{\mathbf{k}}\cdot\mathbf{x} - \omega_0 t\right)}. \tag{1.36}$$

Subtracting

$$\left[d_t^2 - u_0^2\mathbf{d}_0^2(1)\right]\psi_0 = 0 \tag{1.37}$$

from (1.29), and using $d_t^2\psi_0 = -4\sin^2\left(\omega_0\Delta t/2\right)\psi_0$, we obtain

$$\left[d_t^2 - \frac{u_0^2}{\tilde{n}^2}\mathbf{d}_0^2(n)\right]\psi_s(\mathbf{x},t) = S_{\mathrm{NS}}(\mathbf{x},t), \tag{1.38}$$

where the NS source term is

$$S_{\mathrm{NS}}(\mathbf{x},t) = \frac{1}{\tilde{n}^2}\left[4\left(\tilde{n}^2-1\right)\sin^2\left(\omega_0\Delta t/2\right) + u_0^2\,\Delta\mathbf{d}_0^2(n)\right]\psi_0(\mathbf{x},t), \tag{1.39}$$

and $\Delta\mathbf{d}_0^2(n) = \mathbf{d}_0^2(n) - \mathbf{d}_0^2(1)$. The NS-FDTD algorithm for the scattered field is thus,

$$\begin{aligned}\psi_{\mathrm{s}}(\mathbf{x},t+\Delta t) &= -\psi_{\mathrm{s}}(\mathbf{x},t-\Delta t) \\ &+ \left[2+\frac{u_0^2}{\tilde{n}^2}\mathbf{d}_0^2(n)\right]\psi_{\mathrm{s}}(\mathbf{x},t)+S_{\mathrm{NS}}(\mathbf{x},t)\end{aligned} \quad (1.40)$$

Since the incident field is known, $\Delta\mathbf{d}_0^2(n)\,\psi_0$ can be analytically evaluated. In two dimensions

$$\Delta\mathbf{d}_0^2(n)=\left[\gamma_1(nk_0)-\gamma_1(k_0)\right]d_x^2 d_y^2. \quad (1.41)$$

From equation (1.106) we find $\gamma_1(nk_0)-\gamma_1(k_0)=(n^2-1)k_0^2h^2/180+\cdots$. Obviously, the larger the refractive index contrast between the scatterer and the environment in which it is immersed the larger $\Delta\mathbf{d}_0^2(n)$. If the incident field direction is $\hat{\mathbf{k}}=\hat{\mathbf{x}}$, $\hat{\mathbf{y}}$, or $\hat{\mathbf{z}}$ it is easy to show that $\Delta\mathbf{d}_0^2(n)\,\psi_0=0$ in both two and three dimensions; and, furthermore (1.37) is exact in this case because (1.28) is an exact expression for $\psi=\psi_0$.

Difference from previously published NS-FD algorithm

Finally, let us note that in [4]–[6] we implicitly assumed that

$$S_{\mathrm{NS}}(\mathbf{x},t)=\left[4\left(\frac{n^2-1}{n^2}\right)\sin^2\left(\omega_0\Delta t/2\right)\right]\psi_0(\mathbf{x},t). \quad (1.42)$$

In the new formulation, (1.39), n is replaced by $\tilde{n}$ and the $\Delta\mathbf{d}_0^2(n)$-term is taken into account.

4.1.3 Extension to the absorbing wave equation

The above methodology is easily extended to the absorbing wave equation,

$$\left(\partial_t^2-\frac{v_0^2}{n^2}\nabla^2+2\alpha\partial_t\right)\psi(\mathbf{x},t)=0, \quad (1.43)$$

where $\alpha \geq 0$ is the absorption (factor of 2 is customary).

S-FD model

Replacing the derivatives of (1.43) with S-FD approximations yields the S-FD model,

$$\left(d_t^2-\frac{v_0^2}{n^2}\mathbf{d}^2+\alpha\Delta t\,d_t'\right)\psi(\mathbf{x},t)=0. \quad (1.44)$$

Because ψ is computed only at $t=0,\Delta t,2\Delta t,\cdots$, we use $\partial_t\psi\cong d_t'\psi/2\Delta t$, where $d_t'\psi(\mathbf{x},t)=\psi(\mathbf{x},t+\Delta t)-\psi(\mathbf{x},t-\Delta t)$. Solving for $\psi(\mathbf{x},t+\Delta t)$ yields the S-FDTD algorithm

$$\psi(\mathbf{x},t+\Delta t)=-\frac{A_-^{(\mathrm{S})}}{A_+^{(\mathrm{S})}}\psi(\mathbf{x},t-\Delta t)+\frac{1}{A_+^{(\mathrm{S})}}\left[2+\frac{v_0^2\Delta t^2}{n^2h^2}\mathbf{d}^2\right]\psi(\mathbf{x},t), \quad (1.45)$$

where $A_\pm^{(\mathrm{S})}=1\pm\alpha\Delta t$.

In [4], we introduced a high-accuracy NS-FDTD algorithm. In each region of constant n and α, (1.43) has solutions of the form

$$\psi(\mathbf{x},t) = e^{-i\omega' t}\, e^{-\alpha t} \sum_{\hat{\mathbf{k}}} c_{\hat{\mathbf{k}}} \varphi_{\hat{\mathbf{k}}}(\mathbf{x}), \tag{1.46}$$

where $\omega'^2 = \omega_0^2 - \alpha^2$. The NS-model of the absorbing wave equation is

$$\left(d_t^2 - \frac{u_0^2}{\tilde{n}^2} \mathbf{d}_0^2(n) + a d_t' \right) \psi(\mathbf{x},t) = 0, \tag{1.47}$$

where

$$a = \tanh(\alpha \Delta t), \tag{1.48}$$

$$u^2 = \frac{\sin^2(\omega' \Delta t/2) + \sinh^2(\alpha \Delta t/2)}{\sin^2(n k_0 h/2) \cosh(\alpha \Delta t)}, \tag{1.49}$$

$$\tilde{n} = \frac{u}{u_0}. \tag{1.50}$$

The effective index of refraction, $\tilde{n}$ given by (1.50), reduces to (1.30) for $\alpha = 0$. The NS-FDTD algorithm is

$$\psi(\mathbf{x},t+\Delta t) = -\frac{A_-^{(\mathrm{NS})}}{A_+^{(\mathrm{NS})}} \psi(\mathbf{x},t-\Delta t) + \frac{1}{A_+^{(\mathrm{NS})}} \left[2 + \frac{u_0^2}{\tilde{n}^2} \mathbf{d}_0^2 \right] \psi(\mathbf{x},t), \tag{1.51}$$

where $A_\pm^{(\mathrm{NS})} = 1 \pm a$.

Scattered-field wave equation with absorption

Decomposing ψ into scattered and incident fields, the incident field propagates according to (1.33). Subtracting (1.33) from (1.43) yields

$$\left(\partial_t^2 - \frac{v_0^2}{n^2} \nabla^2 + 2\alpha \partial_t \right) \psi_{\mathrm{s}}(\mathbf{x},t) = -\left[2\alpha \partial_t + \left(\frac{n^2-1}{n^2} \right) \partial_t^2 \right] \psi_0(\mathbf{x},t). \tag{1.52}$$

S-FD model

Replacing the derivatives in (1.52) with S-FD approximations gives the standard model for the scattered field

$$\left(d_t^2 - \frac{v_0^2 \Delta t^2}{h^2 n^2} \mathbf{d}^2 + \alpha \Delta t d_t' \right) \psi(\mathbf{x},t) = -\Delta t^2 \left[2\alpha \partial_t + \left(\frac{n^2-1}{n^2} \right) \partial_t^2 \right] \psi_0(\mathbf{x},t). \tag{1.53}$$

NS-FD model

We derive the NS model by subtracting the NS model for the incident field (1.37) from that for the total field (1.47). Taking the incident field to be (1.36) we find,

$$\left(d_t^2 - \frac{u_0^2}{\tilde{n}^2}\mathbf{d}_0^2(n) + ad_t' \right) \psi_s(\mathbf{x},t) = S_{\mathrm{NS}}(\mathbf{x},t), \tag{1.54}$$

where the source term which gives rise to the scattered field is

$$S_{\mathrm{NS}}(\mathbf{x},t) = \left[4\left(\frac{\tilde{n}^2-1}{\tilde{n}^2}\right) \sin^2(\omega_0 \Delta t/2) + 2ia\sin(\omega_0\Delta t) + \frac{u_0^2}{\tilde{n}^2}\Delta\mathbf{d}_0^2(n) \right] \psi_0(\mathbf{x},t). \tag{1.55}$$

When $\alpha = 0$, (1.55) reduces to (1.39), and $\Delta\mathbf{d}_0^2(n)\,\psi_0 = 0$ if the incident field direction is along one of the coordinate axes. The NS-FDTD algorithm for the scattered field becomes

$$\begin{aligned} \psi(\mathbf{x},t+\Delta t) &= -\frac{A_-^{(\mathrm{NS})}}{A_+^{(\mathrm{NS})}}\psi(\mathbf{x},t-\Delta t) \\ &+ \frac{1}{A_+^{(\mathrm{NS})}}\left[2 + \frac{u_0^2}{\tilde{n}^2}\mathbf{d}_0^2\right]\psi(\mathbf{x},t) + S_{\mathrm{NS}}(\mathbf{x},t). \end{aligned} \tag{1.56}$$

4.1.4 Maxwell's equations for the scattered-field and improved NS–Yee algorithm

In a linear non-dispersive non-conducting medium Maxwell's equations are,

$$\mu\partial_t\mathbf{H}(\mathbf{x},t) = -\nabla\times\mathbf{E}(\mathbf{x},t), \tag{1.57}$$

$$\varepsilon\partial_t\mathbf{E}(\mathbf{x},t) = \nabla\times\mathbf{H}(\mathbf{x},t), \tag{1.58}$$

where μ is the magnetic permeability, and ε the relative electric permittivity.

Standard Yee algorithm

Replacing the derivatives with the S-FD approximations we obtain the S-FD model,

$$d_t\mathbf{H}(\mathbf{x},t) = -\frac{1}{\mu}\frac{\Delta t}{h}\mathbf{d}\times\mathbf{E}(\mathbf{x},t), \tag{1.59}$$

$$d_t\mathbf{E}(\mathbf{x},t+\Delta t/2) = \frac{1}{\varepsilon}\frac{\Delta t}{h}\mathbf{d}\times\mathbf{H}(\mathbf{x},t+\Delta t/2), \tag{1.60}$$

where $h = \Delta x = \Delta y = \Delta z$. In order to use central finite difference approximations, each electromagnetic field component is evaluated at a different position on the numerical grid, so there is one ε array for each $\mathbf{E}$ component and one μ array for each $\mathbf{H}$ component. For the sake of notational simplicity, this complication is suppressed in the notation. The standard (S) Yee algorithm, derives from (1.59) and (1.60). In many materials μ is almost equal to its free-space value, μ_0. To simplify the following developments, we henceforth assume $\mu = \mu_0$ everywhere, while ε may vary with position. Furthermore, let ε denote relative electric permittivity, so the vacuum wave velocity is $v_0^2 = 1/\mu_0$. To develop the new NS-FD model let us first construct an alternative S-FD model. Defining $\mathbf{H}' = \mu h \mathbf{H}/\Delta t$, we obtain

$$d_t \mathbf{H}'(\mathbf{x}, t) = -\mathbf{d} \times \mathbf{E}(\mathbf{x}, t), \tag{1.61}$$

$$d_t \mathbf{E}(\mathbf{x}, t + \Delta t/2) = \frac{v_0^2}{\varepsilon} \frac{\Delta t^2}{h^2} \mathbf{d} \times \mathbf{H}'(\mathbf{x}, t + \Delta t/2). \tag{1.62}$$

Improved NS-Yee algorithm

We previously introduced [4] (also see section 4.1.7) the vector difference operator $\mathbf{d}_0$ with the property that

$$\mathbf{d} \cdot \mathbf{d}_0 = \mathbf{d}_0 \cdot \mathbf{d} = \mathbf{d}_0^2. \tag{1.63}$$

Note that while $\mathbf{d} \cdot \mathbf{d} = \mathbf{d}^2$, $\mathbf{d}_0 \cdot \mathbf{d}_0 \neq \mathbf{d}_0^2$; thus $\mathbf{d}_0^2$ is merely a symbol. In two dimensions $\mathbf{d}_0 = \left(d_x^{(0)}, d_y^{(0)} \right)$, where

$$d_x^{(0)} = d_x \left[1 + \frac{\gamma_1}{2} d_y^2 \right], \tag{1.64}$$

$$d_y^{(0)} = d_y \left[1 + \frac{\gamma_1}{2} d_x^2 \right]. \tag{1.65}$$

In three dimensions $\mathbf{d}_0 = \left(d_x^{(0)}, d_y^{(0)}, d_z^{(0)} \right)$, where

$$d_x^{(0)} = d_x \left[1 + \frac{\gamma_1}{2} \left(d_y^2 + d_z^2 \right) + \frac{\gamma_2}{3} d_y^2 d_z^2 \right], \tag{1.66}$$

$$d_y^{(0)} = d_y \left[1 + \frac{\gamma_1}{2} \left(d_x^2 + d_z^2 \right) + \frac{\gamma_2}{3} d_x^2 d_z^2 \right], \tag{1.67}$$

$$d_z^{(0)} = d_z \left[1 + \frac{\gamma_1}{2} \left(d_y^2 + d_z^2 \right) + \frac{\gamma_2}{3} d_y^2 d_z^2 \right], \tag{1.68}$$

$\mathbf{d}_0 = \mathbf{d}_0(\sqrt{\varepsilon})$, and $\varepsilon = n^2$ ($\mu =$ constant).

Here we diverge from our previous papers [4]–[6] to introduce an improved version of the NS–Yee algorithm for $\mu = \mu_0$. Making the substitutions $\mathbf{d} \to \mathbf{d}_0$ and $\Delta t^2 v_0^2 / \left(h^2 \varepsilon \right) \to u_0^2 / \tilde{\varepsilon}$ in (1.62) we obtain a new NS-FD model of Maxwell's equations. The $\mathbf{H}'$ field is given by (1.61), and $\mathbf{E}$ by

$$d_t \mathbf{E}(\mathbf{x}, t + \Delta t/2) = \frac{u_0^2}{\tilde{\varepsilon}} \mathbf{d}_0(\sqrt{\varepsilon}) \times \mathbf{H}'(\mathbf{x}, t + \Delta t/2). \tag{1.69}$$

The effective value of ε on the numerical grid is

$$\tilde{\varepsilon} = \frac{\sin^2\left(\sqrt{\varepsilon} k_0 h/2\right)}{\sin^2\left(k_0 h/2\right)}. \tag{1.70}$$

Solving (1.61) and (1.69) for $\mathbf{H}'(\mathbf{x}, t + \Delta t/2)$ and $\mathbf{E}(\mathbf{x}, t + \Delta t)$, yields the new NS–Yee algorithm

$$\mathbf{H}'(\mathbf{x}, t + \Delta t/2) = \mathbf{H}'(\mathbf{x}, t - \Delta t/2) - \mathbf{d} \times \mathbf{E}(\mathbf{x}, t), \tag{1.71}$$

$$\mathbf{E}(\mathbf{x}, t + \Delta t) = \mathbf{E}(\mathbf{x}, t) + \frac{u_0^2}{\tilde{\varepsilon}} \mathbf{d}_0(\sqrt{\varepsilon}) \times \mathbf{H}'(\mathbf{x}, t + \Delta t/2). \tag{1.72}$$

This algorithm is different from the one introduced in [4]. We find that it gives better results when μ is constant, as discussed in section 4.1.7.

Maxwell's equations for the scattered field

Assuming uniform μ, let a scatterer of relative permittivity ε be illuminated by an incident electromagnetic field $(\mathbf{H}_0,\mathbf{E}_0)$. By definition the incident fields obey

$$\mu \partial_t \mathbf{H}_0 = -\nabla \times \mathbf{E}_0, \tag{1.73}$$

$$\partial_t \mathbf{E}_0 = \nabla \times \mathbf{H}_0. \tag{1.74}$$

The scattered fields are $\mathbf{H}_\mathrm{s} = \mathbf{H} - \mathbf{H}_0$ and $\mathbf{E}_\mathrm{s} = \mathbf{E} - \mathbf{E}_0$. Subtracting (1.73) from (1.57), and (1.74) from (1.58) yields Maxwell's equations for the scattered fields,

$$\mu \partial_t \mathbf{H}_\mathrm{s} = -\nabla \times \mathbf{E}_\mathrm{s}, \tag{1.75}$$

$$\varepsilon \partial_t \mathbf{E}_\mathrm{s} = \nabla \times \mathbf{H}_\mathrm{s} - (\varepsilon - 1)\, \partial_t \mathbf{E}_0. \tag{1.76}$$

The last term on the right of (1.76) is an effective source current which gives rise to the scattered electromagnetic field.

NS-FD model of the scattered field

The NS-FD model for the scattered field is constructed by subtracting the NS model for the incident field,

$$d_t \mathbf{H}_0'(\mathbf{x}, t) = -\mathbf{d} \times \mathbf{E}_0(\mathbf{x}, t), \tag{1.77}$$

$$d_t \mathbf{E}_0(\mathbf{x}, t + \Delta t/2) = u_0^2 \mathbf{d}_0(1) \times \mathbf{H}_0'(\mathbf{x}, t + \Delta t/2), \tag{1.78}$$

from the NS model for the total field, (1.71) and (1.72). Let the incident electric field to be an infinite plane wave of unit amplitude polarized in the $\hat{\mathbf{e}}_0$-direction and propagating in the $\hat{\mathbf{k}}_0$- direction,

$$\mathbf{E}_0 = \hat{\mathbf{e}}_0\, e^{i\left(k_0 \hat{\mathbf{k}}_0 \cdot \mathbf{x} - \omega_0 t\right)}, \tag{1.79}$$

where $\hat{\mathbf{k}}_0 \cdot \hat{\mathbf{e}}_0 = 0$. From (1.57) we have $\mu_0 \omega_0 \mathbf{H}_0 = k_0 \hat{\mathbf{k}}_0 \times \hat{\mathbf{e}}_0$, hence

$$\mathbf{H}_0' = \frac{h}{v_0} \hat{\mathbf{h}}_0 \psi_0, \tag{1.80}$$

where $\hat{\mathbf{h}}_0 = \hat{\mathbf{k}}_0 \times \hat{\mathbf{e}}_0$ and ψ_0 is given by (1.36). The NS-FD model for the scattered field now becomes

$$d_t \mathbf{H}'_s(\mathbf{x}, t) = -\mathbf{d} \times \mathbf{E}_s(\mathbf{x}, t), \tag{1.81}$$

$$\begin{aligned} d_t \mathbf{E}_s(\mathbf{x}, t + \Delta t/2) &= \frac{u_0^2}{\tilde{\varepsilon}} \mathbf{d}_0(\sqrt{\varepsilon}) \times \mathbf{H}'_s(\mathbf{x}, t + \Delta t/2) \\ &- \mathbf{J}_{\mathrm{NS}}(\mathbf{x}, t + \Delta t/2), \end{aligned} \tag{1.82}$$

where

$$\mathbf{J}_{\mathrm{NS}} = \left(\frac{\tilde{\varepsilon} - 1}{\tilde{\varepsilon}} \right) d_t \mathbf{E}_0 - \frac{u_0^2}{\tilde{\varepsilon}} \left[\mathbf{d}_0(\sqrt{\varepsilon}) - \mathbf{d}_0(1) \right] \times \mathbf{H}'_0 \tag{1.83}$$

is the current that gives rise to the scattered field. Since $\mathbf{J}_{\mathrm{NS}}$ vanishes outside the scatterer, the scatterer is the effective source of the scattered field.

It is also possible to derive another expression for the source current of the scattered field in the form,

$$\mathbf{J}_{\mathrm{NS}} = u_0^2 \left[\frac{\mathbf{d}_0(\sqrt{\varepsilon})}{\tilde{\varepsilon}} - \mathbf{d}_0(1) \right] \times \mathbf{H}'_0. \tag{1.84}$$

While (1.84) is not exactly equivalent to (1.83) the two expressions converge in the limit $h \to 0$. Numerical experiments show that (1.83) is slightly more accurate than (1.84) on a coarse grid.

The NS–Yee algorithm for the scattered field now becomes

$$\mathbf{H}'_s(\mathbf{x}, t + \Delta t/2) = \mathbf{H}'_s(\mathbf{x}, t - \Delta t/2) - \mathbf{d} \times \mathbf{E}_s(\mathbf{x}, t), \tag{1.85}$$

$$\begin{aligned} \mathbf{E}_s(\mathbf{x}, t + \Delta t) &= \mathbf{E}_s(\mathbf{x}, t) + \frac{u_0^2}{\tilde{\varepsilon}} \mathbf{d}_0(\sqrt{\varepsilon}) \times \mathbf{H}'_s(\mathbf{x}, t + \Delta t/2) \\ &- \mathbf{J}_{\mathrm{NS}}(\mathbf{x}, t + \Delta t/2). \end{aligned} \tag{1.86}$$

Both to clarify our condensed notation in the previous developments and to provide a useful example, we now give the NS–Yee algorithm for the TE mode in two-dimensions where $\mathbf{E} = (E_x, E_y, 0)$ and $\mathbf{H} = (0, 0, H_z)$. Taking the incident fields to be $\mathbf{E}^0 = (-\sin\theta_0, \cos\theta_0, 0)\,\psi_0$ and $\mathbf{H}^0 = (0, 0, \psi_0/v_0)$. We find

$$H_z'^{\mathrm{s}}(t + \Delta t/2) = H_z'^{\mathrm{s}}(t - \Delta t/2) - d_y E_x^{\mathrm{s}}(t) - d_x E_y^{\mathrm{s}}(t), \tag{1.87}$$

$$\begin{aligned} E_x^{\mathrm{s}}(t + \Delta t) &= E_x^{\mathrm{s}}(t) + \frac{u_0^2}{\tilde{\varepsilon}_x} d_y^{(0)}(\sqrt{\varepsilon_x}) H_z'^{\mathrm{s}}(t + \Delta t/2) \\ &- J_x^{\mathrm{NS}}(t + \Delta t/2), \end{aligned} \tag{1.88}$$

$$\begin{aligned} E_y^{\mathrm{s}}(t + \Delta t) &= E_y^{\mathrm{s}}(t) - \frac{u_0^2}{\tilde{\varepsilon}_y} d_x^{(0)}(\sqrt{\varepsilon_y}) H_z'^{\mathrm{s}}(t + \Delta t/2) \\ &- J_y^{\mathrm{NS}}(t + \Delta t/2), \end{aligned} \tag{1.89}$$

where the electromagnetic fields and source currents are all evaluated at position $\mathbf{x}$, and the 'S' superscript denotes the scattered field.

Evaluating $\mathbf{J}_{\mathrm{NS}}$ for $\hat{\mathbf{k}}_0 = \left(k_x^{(0)}, k_y^{(0)}, 0\right) = (\cos\theta_0, \sin\theta_0, 0)$ we find

$$\begin{aligned} J_x^{\mathrm{NS}} &= 2i\Big[\sin\left(\omega_0 \Delta t/2\right)\sin\theta_0\ \Delta\tilde{\varepsilon}_x \\ &+ 2\frac{u_0^2}{v_0}\sin\left(k_y^{(0)}h/2\right)\sin^2\left(k_x^{(0)}h/2\right)\Delta\gamma_{1x}\Big]\psi_0 \end{aligned} \tag{1.90}$$

$$\begin{aligned} J_y^{\mathrm{NS}} &= -2i\Big[\sin\left(\omega_0 \Delta t/2\right)\cos\theta_0\ \Delta\tilde{\varepsilon}_y \\ &+ 2\frac{u_0^2}{v_0}\sin\left(k_x^{(0)}h/2\right)\sin^2\left(k_y^{(0)}h/2\right)\Delta\gamma_{1y}\Big]\psi_0, \end{aligned} \tag{1.91}$$

where $\Delta\tilde{\varepsilon}_x = (\tilde{\varepsilon}_x - 1)/\tilde{\varepsilon}_x$, $\Delta\gamma_{1x}(\sqrt{\varepsilon_x}) = \left[\gamma_1(\sqrt{\varepsilon_x}) - \gamma_1(1)\right]/\tilde{\varepsilon}_x$, and similarly for $\Delta\tilde{\varepsilon}_y$ and $\Delta\gamma_{1y}(\sqrt{\varepsilon_y})$.

4.1.5 Extension to the linearly conducting Maxwell's equations

In a linear medium of conductivity σ Maxwell's equations are given by (1.57) and

$$\varepsilon\partial_t\mathbf{E}(\mathbf{x},t) = \nabla\times\mathbf{H}(\mathbf{x},t) - \sigma\mathbf{E}(\mathbf{x},t). \tag{1.92}$$

Alternative SFD model

Following section 4.4, we derive the alternative SFD model analogous to (1.61) and (1.62) for the conducting case. The $\mathbf{H}'$ field is given by (1.61). Since $\mathbf{E}$ is computed only at $t = 0, \Delta t, 2\Delta t, \cdots$, we use the approximation $\mathbf{E}(\mathbf{x}, t+\Delta t/2) \cong [\mathbf{E}(\mathbf{x},t) + \mathbf{E}(\mathbf{x},t+\Delta t)]/2$ and find that $\mathbf{E}$ is given by

$$\begin{aligned} d_t\mathbf{E}(\mathbf{x},t+\Delta t/2) &= \frac{1}{\varepsilon\mu}\frac{\Delta t^2}{h^2}\ \mathbf{d}\times\mathbf{H}'(\mathbf{x},t+\Delta t/2) \\ &- \frac{\sigma}{2\varepsilon}\left[\mathbf{E}(\mathbf{x},t) + \mathbf{E}(\mathbf{x},t+\Delta t)\right]. \end{aligned} \tag{1.93}$$

In [5] we derived a NS–Yee algorithm for the conducting Maxwell's equations. Following section 4.4 we introduce an improved version for uniform μ.

Improved NS-FD model for the linearly conducting Maxwell equations

The equation for $\mathbf{H}'$ is given by (1.61), while for $\mathbf{E}$ we have

$$\begin{aligned} d_t\mathbf{E}(\mathbf{x},t+\Delta t/2) &= \frac{u_0^2}{\tilde{\varepsilon}}\mathbf{d}_0(\sqrt{\varepsilon})\times\mathbf{H}'(\mathbf{x},t+\Delta t/2) \\ &- a\left[\mathbf{E}(\mathbf{x},t) + \mathbf{E}(\mathbf{x},t+\Delta t)\right], \end{aligned} \tag{1.94}$$

where $\tilde{\varepsilon} = \tilde{n}^2$ ($\tilde{n}$ is given by (1.50) with $\alpha\to\sigma/2\varepsilon$), and $a = \tanh(\sigma\Delta t/2\varepsilon)$. Here equations (5.6a,b) of ref. [5] have been replaced with (1.61) and (1.94), respectively.

The update for $\mathbf{H}'$ is given by (1.71). Solving (1.94) for $\mathbf{E}(\mathbf{x}, t + \Delta t)$, the new NS–Yee algorithm becomes

$$\mathbf{E}(\mathbf{x}, t + \Delta t) = \frac{A_-^{(\mathrm{NS})}}{A_+^{(\mathrm{NS})}} \mathbf{E}(\mathbf{x}, t) + \frac{1}{A_+^{(\mathrm{NS})}} \frac{u_0^2}{\tilde{\varepsilon}} \mathbf{d}_0(\sqrt{\varepsilon}) \times \mathbf{H}'(\mathbf{x}, t + \Delta t/2). \quad (1.95)$$

For uniform μ, this version of the NS-Yee algorithm for the conducting Maxwell's equations is simpler, easier to implement, and more accurate than the one given in [5].

Scattered field for linearly conducting Maxwell's equations

Next let us derive Maxwell's equations for the scattered fields. The propagation of $\mathbf{H}_\mathrm{s}$ is governed by (1.75), while the propagation of $\mathbf{E}_\mathrm{s}$ is found by subtracting (1.74) from (1.92) to obtain

$$\varepsilon \partial_t \mathbf{E}_\mathrm{s} = \nabla \times \mathbf{H}_\mathrm{s} - [(\varepsilon - 1)\, \partial_t + \sigma]\, \partial_t \mathbf{E}_0. \quad (1.96)$$

New NS-FD Model for the scattered field

We can now derive the new NS-FD model for the scattered fields. The $\mathbf{H}_\mathrm{s}'$ field is given by (1.81), while $\mathbf{E}_\mathrm{s}$ is found by subtracting (1.78) from (1.94). We obtain

$$\begin{aligned} d_t \mathbf{E}_\mathrm{s}(\mathbf{x}, t + \Delta t/2) &= \frac{u_0^2}{\tilde{\varepsilon}} \mathbf{d}_0(\sqrt{\varepsilon}) \times \mathbf{H}_\mathrm{s}'(\mathbf{x}, t + \Delta t/2) \\ &- a\,[\mathbf{E}_\mathrm{s}(\mathbf{x}, t) + \mathbf{E}_\mathrm{s}(\mathbf{x}, t + \Delta t)] - \mathbf{J}_{\mathrm{NS}}(\mathbf{x}, t + \Delta t/2), \end{aligned} \quad (1.97)$$

where

$$\begin{aligned} \mathbf{J}_{\mathrm{NS}} = \left(\frac{\tilde{\varepsilon} - 1}{\tilde{\varepsilon}}\right) d_t \mathbf{E}_0 &- \frac{u_0^2}{\tilde{\varepsilon}} \left[\mathbf{d}_0(\sqrt{\varepsilon}) - \mathbf{d}_0(1)\right] \times \mathbf{H}_0' \\ &+ 2a \cos(\omega_0 \Delta t/2)\, \mathbf{E}_0 \end{aligned} \quad (1.98)$$

is the effective source current. Since the time dependence of the incident field is harmonic, we have used $\mathbf{E}_0(\mathbf{x}, t) + \mathbf{E}_0(\mathbf{x}, t + \Delta t) = 2\cos(\omega_0 \Delta t/2)\, \mathbf{E}_0(\mathbf{x}, t + \Delta t/2)$ in (1.98). When $\alpha = 0$, (1.98) reduces to (1.83). Solving (1.97) for $\mathbf{E}_\mathrm{s}(\mathbf{x}, t + \Delta t)$ we obtain the NS–Yee algorithm for the scattered fields. The $\mathbf{H}_\mathrm{s}'$ update is given by (1.85), while the $\mathbf{E}_\mathrm{s}$ update is

$$\begin{aligned} \mathbf{E}_\mathrm{s}(\mathbf{x}, t + \Delta t) &= \frac{A_-^{(\mathrm{NS})}}{A_+^{(\mathrm{NS})}} \mathbf{E}_\mathrm{s}(\mathbf{x}, t) + \frac{u_0^2}{\tilde{\varepsilon}} \mathbf{d}_0(\sqrt{\varepsilon}) \times \mathbf{H}_\mathrm{s}'(\mathbf{x}, t + \Delta t/2) \\ &- \frac{1}{A_+^{(\mathrm{NS})}} J_{\mathrm{NS}}(\mathbf{x}, t + \Delta t/2). \end{aligned} \quad (1.99)$$

4.1.6 Verifications and practical tests

In Figs. 4.1–4.3, we compare the S-FDTD and NS-FDTD calculations with Mie theory [7]. An infinite plane wave (vacuum wavelength = λ_0) propagates in the $+x$ direction and scatters from an infinite dielectric cylinder (radius = $0.65\lambda_0$, refractive index $n = 1.8$). The cylinder is parallel to the z-axis, and the incident electric field is parallel to $\hat{z}$ (TM polarization). In the TM mode it can be shown [8] that Maxwell's equations for **E** (but not **H**) reduce to the wave equation when μ is constant.

In Fig. 4.1 the scattered electric field intensity, $|\mathbf{E}_s|^2$, is visualized in shades of red (black = 0) for the analytic solution, the NS-FDTD, and the S-FDTD calculations. For both FDTD calculations the spatial discretization was $\lambda_0/h = 8$ outside the cylinder, and $\lambda_0/nh \cong 4.4$ inside.

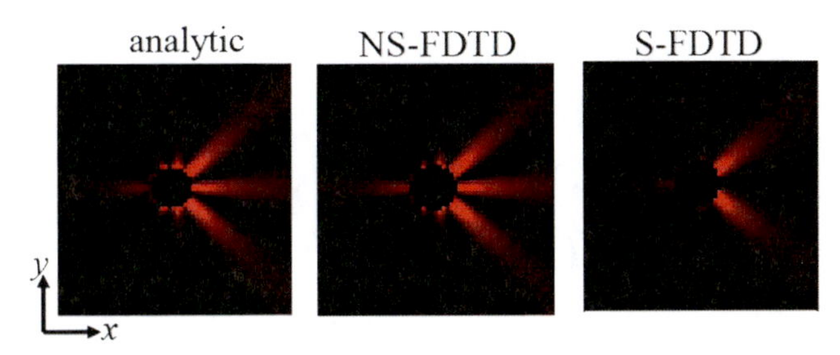

Fig. 4.1. Mie scattering from an infinite dielectric cylinder. Scattered electric field intensity visualized in shades of red, incident electric field parallel to the cylinder axis; cylinder radius = $0.65\lambda_0$, λ_0= vacuum wavelength, refractive index = 1.8, $\lambda_0/h = 8$.

In Fig. 4.2(a) $|\mathbf{E}_s|^2$ (vertical axis) is plotted as a function of scattering angle (horizontal axis) from the $+x$-axis on a circular contour, C, of radius λ_0 centered on the cylinder axis. For $\lambda_0/h = 8$ the root mean square errors for the NS-FDTD and S-FDTD calculations (NSFD-8 and SFD-8, respectively) are $\varepsilon_{NS-8} = 0.04$ and $\varepsilon_{S-8} = 0.20$, respectively. In Fig. 4.2(b) the S-FDTD calculation using $\lambda_0/h = 24$ (SFD-24) is compared with the analytic solution. The root mean square error is $\varepsilon_{S-24} = 0.04$ the same as the NS-FDTD algorithm at $\lambda_0/h = 8$. Thus for $\lambda_0/h = 8$ the NS-FDTD algorithm delivers the same accuracy as the S-FDTD one does at $\lambda_0/h = 24$, but the computational cost of NSFD-8 is only 1/27th that SFD-24.

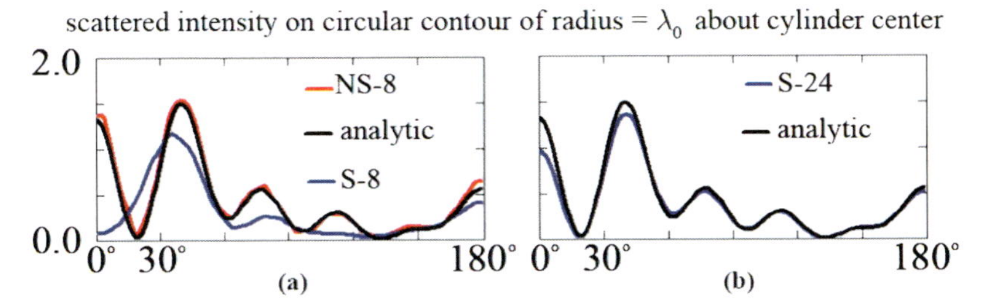

Fig. 4.2. Angular distribution of scattered electric field intensity (data of Fig. 4.1) (a) Analytic solution (black) compared with NSFD-8 (red) and SFD-8 (blue). FDTD calculations use $\lambda_0/h = 8$. (b) Analytic solution (black) compared with SFD-24 (blue, $\lambda_0/h = 24$).

In Fig. 4.3, using $\lambda_0/h = 8$, we compare the new NS-FDTD algorithm with earlier version described in [4]–[6]. In the old version S_{NS} is given by (1.42), while in the present version it is given by (1.39). The greater the refractive index contrast between the scatterer and the surrounding medium, the greater the advantage of using (1.39). Taking $\lambda_0/h = 8$, for S_{NS} given by (1.39) the root mean square deviation of the NS-FDTD calculations from the analytic solution along C is $\varepsilon_{\text{NS}-8} = 0.04$, while for S_{NS} given by (1.42) $\varepsilon_{\text{ONS}-8} = 0.13$. Thus the incorrect model of S_{NS} causes considerable error, since $\varepsilon_{\text{S}-8} = 0.20$.

We carried out similar comparisons for the TE mode [8].

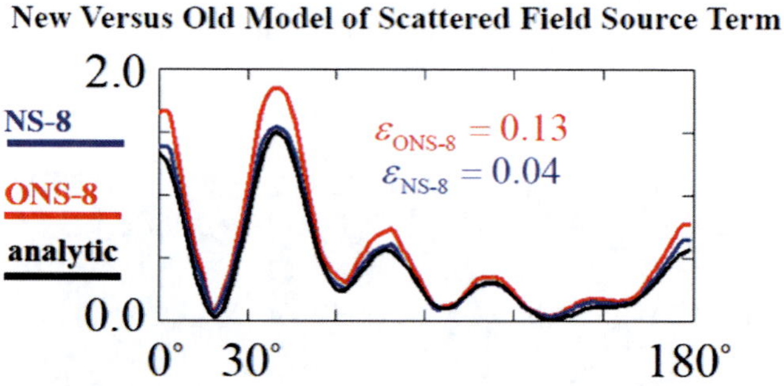

Fig. 4.3. Comparison of old scattered field source model, eqn. (1.42), with the new one, eqn. (1.39). Angular distribution of scattered intensity electric field about a circular contour of radius λ_0, centered on cylinder axis, cylinder radius $= 0.65\lambda_0$, $\lambda_0 =$ vacuum wavelength, refractive index $= 1.8$, $\lambda_0/h = 8$.

4.1.7 Supplementary derivations

NS difference operators

In previous papers we have defined the nonstandard finite difference operators. Here we give more compact expressions that both yield more insight into their mathematical meaning and are easier to implement.

- *Laplacian difference operators*

Defining d_x^2 and d_y^2 analogously to (1.12), the S-FD approximation for $\nabla^2\psi$ in two dimensions is given by (1.21). There is a second FD approximation to $\nabla^2\psi$, $\nabla^2\psi \cong \mathbf{d}_2^2\psi/h^2$, where $\mathbf{d}_2^2$ is defined by

$$\begin{aligned} 2\mathbf{d}_2^2\psi(x,y) &= \psi(x+h,y+h) + \psi(x-h,y+h) \\ &+ \psi(x+h,y-h) + \psi(x-h,y-h) - 4\psi(x,y). \end{aligned} \tag{1.100}$$

Using the fact that $\psi(x+h) + \psi(x-h) = d_x^2\psi(x) + 2\psi(x)$ it is easily shown that

$$\mathbf{d}_2^2 = \mathbf{d}^2 + \frac{1}{2}d_x^2 d_y^2. \tag{1.101}$$

Let us now construct a weighted superposition of the two Laplacian difference operators, $\mathbf{d}_0^2 = \gamma \mathbf{d}^2 + (1-\gamma)\mathbf{d}_2^2$. Let $\varphi_{\mathbf{k}}(\mathbf{x}) = e^{i\mathbf{k}\cdot\mathbf{x}}$, where $\mathbf{x} = (x, y)$ and $\mathbf{k} = k(\cos\theta, \sin\theta)$. We choose γ to minimize the approximation error

$$\epsilon_{\mathrm{NS}} = \frac{1}{\varphi_{\mathbf{k}}} \left(\frac{\mathbf{d}_0^2}{s(k,h)^2} - \nabla^2 \right) \varphi_{\mathbf{k}}. \tag{1.102}$$

It can be shown [4], [6] that the optimal value of γ is

$$\gamma = \frac{2}{3} - \frac{1}{90}(kh)^2 - \frac{1}{15120}(kh)^4 \left(11 - 5\sqrt{2}\right) - \cdots, \tag{1.103}$$

and for this value of γ the Taylor expansion of ϵ_{NS} is

$$\epsilon_{\mathrm{NS}} = \frac{1}{24192}(kh)^6 \left[\left(\sqrt{2} - 1\right) \sin^2(2\theta) - \frac{1}{2}\sin^4(2\theta) \right] + \cdots. \tag{1.104}$$

The factor of 1/24192 corrects an error in [4]–[6]. On the other hand the error of the S-FD approximation (1.2) is

$$\begin{aligned} \epsilon_{\mathrm{S}} &= \frac{1}{\varphi_{\mathbf{k}}} \left(\frac{\mathbf{d}^2}{h^2} - \nabla^2 \right) \varphi_{\mathbf{k}}, \\ &= -\frac{1}{12}k^2h^2 + \frac{1}{24}k^2h^2 \sin^2(2\theta) + \cdots. \end{aligned} \tag{1.105}$$

Using (1.101) is easily shown that γ_1 in (1.25) is given by

$$\begin{aligned} \gamma_1 &= \frac{1-\gamma}{2} \\ &\cong \frac{1}{6} + \frac{1}{180}(kh)^2 + \frac{1}{7698}(kh)^4 + \cdots. \end{aligned} \tag{1.106}$$

In three dimensions, besides $\mathbf{d}^2$, there are two additional FD operators for $\nabla^2\psi$, $\mathbf{d}_2^2$ and $\mathbf{d}_3^2$, given by

$$\begin{aligned} 4\mathbf{d}_2^2\psi(x,y,z) &= \psi(x+h, y+h, z+h) + \psi(x+h, y+h, z-h) \\ &+ \psi(x+h, y-h, z+h) + \psi(x+h, y-h, z-h) \\ &+ \psi(x-h, y+h, z+h) + \psi(x-h, y+h, z-h) \\ &+ \psi(x-h, y-h, z+h) + \psi(x-h, y-h, z-h) - 8\psi(x,y,z). \end{aligned} \tag{1.107}$$

$$\begin{aligned} 4\mathbf{d}_3^2\psi(x,y,z) &= \psi(x, y+h, z+h) + \psi(x, y-h, z+h) \\ &+ \psi(x, y+h, z-h) + \psi(x, y-h, z-h) \\ &+ \psi(x+h, y, z+h) + \psi(x-h, y, z+h) \\ &+ \psi(x+h, y, z-h) + \psi(x-h, y, z-h) \\ &+ \psi(x+h, y+h, z) + \psi(x-h, y+h, z) \\ &+ \psi(x+h, y-h, z) + \psi(x-h, y-h, z) - 12\psi(x,y,z). \end{aligned} \tag{1.108}$$

These operators can be expressed in the form,

$$\begin{aligned}\mathbf{d}_2^2 &= \mathbf{d}^2 + \frac{1}{2}\left(d_x^2 d_y^2 + d_x^2 d_z^2 + d_y^2 d_z^2\right) + \frac{1}{4} d_x^2 d_y^2 d_z^2, \quad (1.109)\\ \mathbf{d}_3^2 &= \mathbf{d}^2 + \frac{1}{4}\left(d_x^2 d_y^2 + d_x^2 d_z^2 + d_y^2 d_z^2\right). \quad (1.110)\end{aligned}$$

Thus a superposition of all three Laplacian difference operators can be expressed in the general form

$$\mathbf{d}_0^2 = \mathbf{d}^2 + \gamma_1'\left(d_x^2 d_y^2 + d_x^2 d_z^2 + d_y^2 d_z^2\right) + \gamma_2 d_x^2 d_y^2 d_z^2. \quad (1.111)$$

When $\partial_z \varphi_{\hat{\mathbf{k}}} = 0$, the three-dimensional version of $\mathbf{d}_0^2$ must reduce to the two-dimensional form (1.25), thus $\gamma_1' = \gamma_1$. We now determine γ_2 such that $\varepsilon_{\mathrm{NS}}$ (1.102) is minimized in three dimensions. We find that

$$\begin{aligned}\gamma_2 &= \frac{\gamma(1-\eta)}{12}\\ &\cong \frac{1}{30} - \frac{1}{905}(kh)^2 - \frac{1}{7698}(kh)^4 + \cdots, \quad (1.112)\end{aligned}$$

where

$$\begin{aligned}\eta &= \frac{2}{5} + \left(\frac{1913}{50400} - \frac{5}{252}\sqrt{2}\right)(kh)^2 \quad (1.113)\\ &+ \left(\frac{1457}{151200} - \frac{17}{30240}\sqrt{2}\right)(kh)^4 + \cdots.\end{aligned}$$

This rather complicated definition of γ_2 is contrived to maintain consistency with previous publications. It can be shown that

$$\epsilon_{\mathrm{NS}} \cong -(kh)^6\left[\frac{1}{16277}\sin^2\theta + \frac{1}{3714}\sin^4\theta\right] + \cdots, \quad (1.114)$$

whereas

$$\begin{aligned}\epsilon_{\mathrm{S}} &= -\frac{(kh)^2}{12} - \frac{(kh)^2}{3}\left(\sin^2\theta - \frac{\sin^2(2\phi)}{12}\right)\\ &+ \frac{(kh)^2}{24}\sin^2(2\theta)\left(1 - \frac{\sin^2(2\phi)}{4}\right) + \cdots. \quad (1.115)\end{aligned}$$

- *Partial difference operators*

In two dimensions, besides $\partial_x \psi \cong d_x \psi / h$, there is a second FD approximation given by $\partial_x \psi \cong d_x^{(2)} \psi / h$, where

$$\begin{aligned}2d_x^{(2)}\psi(x,y) &= \psi(x+h/2, y+h) + \psi(x+h/2, y-h) \quad (1.116)\\ &- \psi(x-h/2, y+h) - \psi(x-h/2, y-h)\end{aligned}$$

It is easy to show that

$$d_x^{(2)} = d_x \left(1 + \frac{1}{2} d_y^2\right), \tag{1.117}$$

and

$$d_y^{(2)} = d_y \left(1 + \frac{1}{2} d_x^2\right). \tag{1.118}$$

Forming the vector difference operator $\mathbf{d}_2 = \left(d_x^{(2)}, d_y^{(2)}\right)$, we define the superposition

$$\mathbf{d}_0 = \alpha \mathbf{d} + (1 - \alpha)\mathbf{d}_2. \tag{1.119}$$

The components of $\mathbf{d}_0$ can be put into the general form

$$d_x^{(0)} = d_x \left[1 + \beta_1 d_y^2\right], \tag{1.120}$$

$$d_y^{(0)} = d_y \left[1 + \beta_1 d_x^2\right]. \tag{1.121}$$

Requiring that $\mathbf{d} \cdot \mathbf{d}_0 = \mathbf{d}_0 \cdot \mathbf{d} = \mathbf{d}_0^2$ we find that

$$\beta_1 = \frac{\gamma_1}{2}. \tag{1.122}$$

In three dimensions, besides d_x, there are two additional FD operators for ∂_x, defined by

$$\begin{aligned} 4d_x^{(2)}\psi(x,y,z) &= \psi(x+h/2, y+h, z+h) + \psi(x+h/2, y+h, z-h) \\ &+ \psi(x+h/2, y-h, z+h) + \psi(x+h/2, y-h, z-h) \\ &- \psi(x-h/2, y+h, z+h) + \psi(x-h/2, y+h, z-h) \\ &- \psi(x-h/2, y-h, z+h) + \psi(x-h/2, y-h, z-h), \end{aligned} \tag{1.123}$$

$$\begin{aligned} 4d_x^{(3)}\psi(x,y,z) &= \psi(x+h/2, y+h, z) && + \psi(x+h/2, y-h, z) \\ &+ \psi(x+h/2, y, z+h) && + \psi(x+h/2, y, z-h) \\ &- \psi(x-h/2, y+h, z) && + \psi(x-h/2, y-h, z) \\ &- \psi(x-h/2, y, z+h) && + \psi(x-h/2, y, z-h). \end{aligned} \tag{1.124}$$

It can now be shown that

$$d_x^{(2)} = d_x \left(1 + \frac{1}{2} d_y^2 + \frac{1}{2} d_z^2 + \frac{1}{4} d_y^2 d_z^2\right), \tag{1.125}$$

$$d_x^{(3)} = d_x \left(1 + \frac{1}{4} d_y^2 + \frac{1}{4} d_z^2\right). \tag{1.126}$$

Similarly for ∂_y and ∂_z, $d_y^{(2)}$ and $d_x^{(2)}$ are given by

$$d_y^{(2)} = d_y \left(1 + \frac{1}{2}d_x^2 + \frac{1}{2}d_z^2 + \frac{1}{4}d_x^2 d_z^2\right), \tag{1.127}$$

$$d_z^{(2)} = d_z \left(1 + \frac{1}{2}d_x^2 + \frac{1}{2}d_y^2 + \frac{1}{4}d_x^2 d_y^2\right), \tag{1.128}$$

while $d_y^{(3)}$ and $d_x^{(3)}$ are given by

$$d_y^{(3)} = d_y \left(1 + \frac{1}{4}d_x^2 + \frac{1}{4}d_z^2\right), \tag{1.129}$$

$$d_z^{(3)} = d_z \left(1 + \frac{1}{4}d_x^2 + \frac{1}{4}d_y^2\right). \tag{1.130}$$

Thus the superpositions $\left(d_x^{(0)}, d_y^{(0)}, d_z^{(0)}\right)$ can be expressed in the general form,

$$\begin{aligned} d_x^{(0)} &= d_x \left[1 + \beta_1' \left(d_y^2 + d_z^2\right) + \beta_2 d_y^2 d_z^2\right], & (1.131)\\ d_y^{(0)} &= d_y \left[1 + \beta_1' \left(d_x^2 + d_z^2\right) + \beta_2 d_x^2 d_z^2\right], & (1.132)\\ d_z^{(0)} &= d_z \left[1 + \beta_1' \left(d_y^2 + d_z^2\right) + \beta_2 d_y^2 d_z^2\right]. & (1.133) \end{aligned}$$

Because the three-dimensional form of $\mathbf{d}_0$ must reduce to the two-dimensional form for $\partial_z \varphi_{\hat{\mathbf{k}}} = 0$, we have $\beta_1' = \gamma_1/2$. Requiring that $\mathbf{d} \cdot \mathbf{d}_0 = \mathbf{d}_0 \cdot \mathbf{d} = \mathbf{d}_0^2$, we find

$$\beta_2 = \frac{\gamma_2}{3}. \tag{1.134}$$

- *Order of the NS Difference Operators*

At first sight it might seem strange to call NS difference operators, with combinations such as $d_x^2 d_y^2$, 'second-order.' With respect to each individual spatial variable, however, they are still second-order. Moreover, in the numerator of (1.28) only $s(nk_0, h)^2 \cong h^2$ appears and, unlike a higher-order FD expression, no higher powers of h appear.

Stability

A detailed analysis of stability was given in [6] which we briefly summarize. Let us write a generalized FD model of the homogeneous wave equation in the form

$$\left(d_t^2 - \bar{u}^2 \bar{\mathbf{d}}^2\right) \psi(\mathbf{x}, t) = 0, \tag{1.135}$$

where $\bar{u}$ stands for either $v\Delta t/h$ (S-FD model) or

$$u = \frac{\sin\left(\omega \Delta t/2\right)}{\sin\left(kh/2\right)}, \tag{1.136}$$

(NS-FD model) and $\bar{\mathbf{d}}^2$ stands for either $\mathbf{d}^2$ (S-FD model) or $\mathbf{d}_0^2$ (NS-FD model). Equation (1.135) is a difference equation; let us postulate solutions of the form $\psi = \varphi(\mathbf{x})\Lambda^t$, where $\varphi = e^{i\mathbf{k}\cdot\mathbf{x}}$, $\mathbf{k} = (k_x, k_y, k_z)$, and Λ is a constant. For mathematical convenience define $\bar{\mathbf{d}}^2\varphi = -2D^2\varphi$, where $D = D(\mathbf{k})$. Inserting this solution into (1.135) and dividing by $\Lambda^{t-\Delta t}\varphi$ we are left with

$$\lambda^2 - 2\lambda\left(1 - \bar{u}^2 D^2\right) + 1 = 0, \tag{1.137}$$

where $\lambda = \Lambda^{\Delta t}$. Solving for λ we find $\lambda = \lambda_\pm$, where

$$\lambda_\pm = 1 - \bar{u}^2 D^2 \pm \sqrt{\left(1 - \bar{u}^2 D^2\right)^2 - 1}. \tag{1.138}$$

We now require that ψ be a stable (non-monotonically increasing) oscillatory solution. For this to be true, Λ^t must be finite for all t, whence the condition $\mid \lambda \mid \leq 1$, and hence the stability condition

$$\mid \lambda_\pm \mid \leq 1. \tag{1.139}$$

It can be shown that (1.139) leads to the condition $\left(1 - \bar{u}^2 D^2\right)^2 \leq 1$, which implies that $\mid \lambda_\pm \mid = 1$. The most conservative stability condition can then be written in the form

$$\bar{u}^2 \leq \frac{2}{\max(D^2)}, \tag{1.140}$$

where $\max(D^2)$ is the maximum value of D^2 over all possible values of $\mathbf{k}$.

- *S-FD stability*

In the S-FD model $\bar{\mathbf{d}}^2 = \mathbf{d}^2$, and we find that $D^2 = D_{\mathrm{S},2}^2$, $D_{\mathrm{S},3}^2$ in two and three dimensions, respectively, where

$$D_{\mathrm{S},2}^2 = 2\left[\sin^2(k_x h/2) + \sin^2(k_y h/2)\right], \tag{1.141}$$

$$D_{\mathrm{S},3}^2 = 2\left[\sin^2(k_x h/2) + \sin^2(k_y h/2) + \sin^2(k_z h/2)\right]. \tag{1.142}$$

Putting $k_x h = k_y h = k_z h = \pi$, we find $\max(D^2) = 2, 4, 6$ in one, two, and three dimensions, respectively. Putting $\bar{u} = v\Delta t/h$ we obtain

$$\frac{v\Delta t}{h} \leq \frac{1}{\sqrt{dim}}, \tag{1.143}$$

where *dim* is the number of spatial dimensions.

• *NS-FD stability*

The NS-FD stability analysis is similar, but evalating $\max(D^2)$ is much more tedious, and putting the result into the form of (1.143) more difficult. We omit the details and simply quote the results:

$$\frac{v\Delta t}{h} \leq c_{\mathrm{NS},dim}, \tag{1.144}$$

where $c_{\mathrm{NS},1} = 1$, $c_{\mathrm{NS},2} = 0.79$, and $c_{\mathrm{NS},3} = 0.74$. These limits are slightly more conservative than those given in [6], and are valid down to the lowest discretization consistent with the Nyquist limit.

Finally it is interesting to note that we can 'design' stability conditions by adjusting the values of γ_1 and γ_2, although they would not be the optimal ones.

Old vs. new NS–Yee algorithm

Taking μ to be spatially constant, the old (O) NS–Yee algorithm (eq. 41 in [4]) for each electromagnetic field component can be derived from the S-FD model of (1.59) and (1.60). In vacuum ($\varepsilon = 1$), $v_0 = 1/\sqrt{\mu}$, thus $1/\mu = (v_0/\sqrt{\varepsilon})\sqrt{\varepsilon/\mu}$ and $1/\varepsilon = (v_0/\sqrt{\varepsilon})\sqrt{\mu/\varepsilon}$. Making the substitution $(v_0\Delta t/h\sqrt{\varepsilon}) \rightarrow u_0\big/\sqrt{\tilde{\varepsilon}}$, yields the ONS–Yee algorithm,

$$d_t H_i = -\frac{u_0}{\sqrt{\tilde{\varepsilon}_?}}\sqrt{\frac{\varepsilon_?}{\mu}}\left[\mathbf{d}_0(?) \times \mathbf{E}\right]_i, \tag{1.145}$$

$$d_t E_i = \frac{u_0}{\sqrt{\tilde{\varepsilon}_i}}\sqrt{\frac{\mu}{\varepsilon_i}}\left[\mathbf{d} \times \mathbf{H}\right]_i, \tag{1.146}$$

where '?' denotes an indeterminate parameter and $i = x, y, z$.

In (1.145) which component of ε should be used to compute $\mathbf{d}_0$, and which one should be used in the denominator? Since $\mathbf{H}$ is given by (1.59), it might seem reasonable to use ε at the grid positions of H_i. This is what we did in the ONS–Yee algorithm. This approach is, however, unsatisfactory because $\mathbf{d}_0$ acts on different components of $\mathbf{E}$, each of which is evaluated at a different grid position and associated with a different value of ε. Moreover $\mathbf{d}_0$ is a complicated computational molecule, so when it acts across boundaries between regions of different ε, $\mathbf{d}_0$ is likely induce more errors than $\mathbf{d}$. By interchanging the positions of $\mathbf{d}_0$ and $\mathbf{d}$ between (1.145) and (1.146) [8] we can circumvent these problems, because $\mathbf{d}$ is a simpler computational molecule and is not a function of ε.

Which value of ε to use in the denominator of (1.145) is still ambiguous. In the ONS–Yee algorithm we used ε at H_i, but this is not necessarily correct. To circumvent this problem we define $\mathbf{H}' = \mu h\mathbf{H}/\Delta t$, so that

$$d_t\mathbf{H}' = -\mathbf{d} \times \mathbf{E} \tag{1.147}$$

in the NS-FD model. This resolves the ambiguity for $\mathbf{H}$. In the alternative S-FD model (1.62) it is clear that

$$d_t E_i = \frac{1}{\varepsilon_i\mu}\frac{\Delta t^2}{h^2}\left[\mathbf{d} \times \mathbf{H}'\right]_i, \tag{1.148}$$

where ε_i is the value of ε at the grid positions of E_i. From this it logically follows that $\Delta t^2/h^2\varepsilon_i\mu \to u_0^2/\tilde{\varepsilon}_i$ in the NS-FD model. Away from boundaries the NS-FD model of Maxwell's equations reduces to the NS-FD model of the wave equation in each $\mathbf{E}$ component,

$$d_t^2 E_i = \frac{u_0^2}{\tilde{\varepsilon}_i}\mathbf{d}_0^2(\sqrt{\varepsilon_i})E_i, \tag{1.149}$$

thus in (1.148) $[\mathbf{d}\times\mathbf{H}']_i \to \left[\mathbf{d}_0(\sqrt{\varepsilon_i})\times\mathbf{H}'\right]_i$. The full form of (1.72) is thus

$$E_i(\mathbf{x}_i, t+\Delta t) = E_i(\mathbf{x}_i, t) + \frac{u_0^2}{\tilde{\varepsilon}_i}\left[\mathbf{d}_0(\sqrt{\varepsilon_i})\times\mathbf{H}'(\mathbf{x}, t+\Delta t/2)\right]_i, \tag{1.150}$$

where $\mathbf{x}_i$ denotes the position of E_i on the grid.

For the scattered fields and the conducting case similar considerations apply. We tested our methodology in numerical experiments by comparing NS-FDTD calculations with analytical solutions in the Mie regime [8].

4.1.8 Summary of nonstandard FDTD methods

Instead of simply substituting finite difference approximations for derivatives in a differential equation (the procedure used to derive the standard FDTD algorithms), in the nonstandard FDTD method we seek a difference equation which has the same (or nearly the same) solutions as the original differential equation. For the wave equation in one dimension, it is sufficient to replace the phase velocity in the the differential equation with something else, but in two and three dimensions and for Maxwell's equations more complicated finite difference operators must be used. Although we have developed the NS-FDTD methodology for monochromatic solutions, it can also be extended to the wide band (but with lower accuracy) see [10].

4.2 High-accuracy simulation of whispering gallery modes

4.2.1 Whispering gallery modes

What is a whispering gallery mode?

Whispering gallery modes (WGMs) are resonances in the interior of highly symmetric structures. The WGMs were first observed as acoustic resonances in the interiors of such structures as cathedral domes (Fig. 4.4), and were analytically described by Lord Rayleigh [11]. Optical WGMs can be excited in dielectric and conducting objects. Since Garrett's experimental work [12], WGMs have been used to measure spherical particle sizes, refractive index, and temperature [13]. In recent years much research effort has focused on micro-resonators, narrow band filters, optical switches, and bio-sensors using the properties of WGMs [14].

Fig. 4.4. Armenian Church in Jerusalem (photographed by Andrew E. Larsen).

Why calculate WGMs?

For simple highly symmetric shapes such as infinite cylinders and spheres, Mie theory provides analytic solutions, including WGMs [7]. But for more complicated shapes, no analytic solutions exist and numerical calculations are necessary. It is important to validate numerical algorithms by comparing with analytic solutions in simple problems.

On the WGM resonance (especially when there are sub-wavelength features), the electromagnetic fields outside the structure are weakly coupled to the inside, and several thousand wave periods are needed to obtain convergence. Error accumulates with each iteration causing accuracy to fall. In addition the WGM calculation is very sensitive to the representation of media interfaces on the numerical grid. Because of this, the computation of WGMs in the Mie regime is a severe test of any numerical algorithm. Although the fact that an algorithm can correctly compute WGMs in the Mie regime for highly symmetric objects is not necessarily a guarantee that it gives accurate results for other problems, it is a good diagnostic.

4.2.2 Infinite cylindrical WGMs

For simplicity, let us consider two-dimensional scattering, in which an infinite plane wave of vacuum wavelength λ impinges upon an infinite dielectric circular cylinder parallel to the z-axis as shown in Fig. 4.5. This problem can be separated into two modes: the transverse magnetic (TM) and transverse electric (TE) modes,

$$\text{TM} \; : \; \mathbf{E} \parallel \text{media interface} \Rightarrow E_x = E_y = H_z = 0, \tag{2.1}$$

$$\text{TE} \; : \; \mathbf{E} \perp \text{media interface} \Rightarrow E_z = H_x = H_y = 0, \tag{2.2}$$

where $\mathbf{E} = (E_x, E_y, E_z)$ is the electric field and $\mathbf{H} = (H_x, H_y, H_z)$ is the magnetic field.

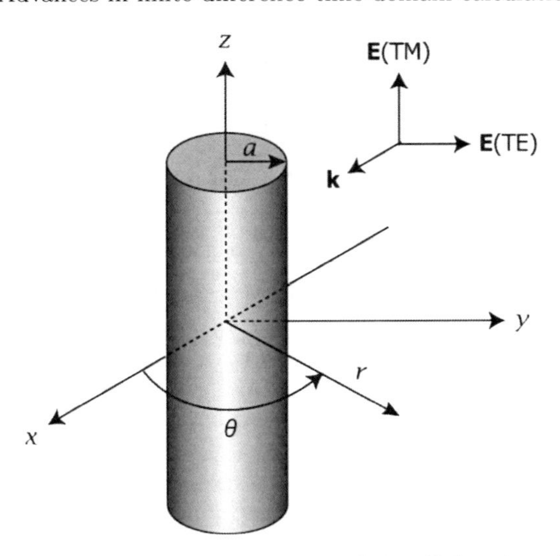

Fig. 4.5. Infinite plane wave impinging on an infinite dielectric cylinder (a = radius, k = wave vector). TM and TE polarizations are shown. Wave propagates along $+x$ axis.

When $\lambda \ll a$ (a = cylinder radius), the resonance wavelength is given by geometrical optics theory,

$$2\pi a = \ell \frac{\lambda}{n}, \tag{2.3}$$

where r is the cylinder radius, n is the refractive index of the cylinder, λ is the wavelength inside the cylinder, and ℓ is the mode number (integer). The light is confined by total internal reflection.

When $\lambda \sim a$, the resonance condition is given by Mie theory. In the TM mode, Maxwell's equations reduce to the Helmholtz equation for E_z,

$$\left(\nabla^2 + n^2 k^2\right) E_z = 0, \tag{2.4}$$

where $k = 2\pi/\lambda$. The fields are independent of z so $E_z = E_z(x, y)$. Outside the cylinder, E_z is the sum of the incident field $E_z^0 = e^{ikx}$, and the outgoing scattered field, E_z^{s}. Taking $(x, y) = r(\cos\theta, \sin\theta)$, E_z^{s} can be expanded in the form,

$$E_z^{\mathrm{s}}(r, \theta) = -\sum_{\ell=-\infty}^{\infty} i^{\ell} b_{\ell} H_{\ell}^{(1)}(kr)\, e^{i\ell\theta}, \tag{2.5}$$

where $H_{\ell}^{(1)}(x)$ is the Hankel function of the first kind and the b_{ℓ} are expansion coefficients to be determined. The electric field inside the cylinder, E_z^{i}, can be expanded in the form,

$$E_z^{\mathrm{i}}(r, \theta) = \sum_{\ell=-\infty}^{\infty} i^{\ell} d_{\ell} J_{\ell}(nkr)\, e^{i\ell\theta}, \tag{2.6}$$

where $J_{\ell}(x)$ is the Bessel function of the first kind, and the d_{ℓ} are expansion coefficients. The expansion coefficients, b_{ℓ} and d_{ℓ}, are determined by the physical conditions that both E_z and its derivative normal to cylinder boundary $\partial_r E_z$ ($\partial_r = \partial/\partial r$), must be continuous on the boundary. Using the fact that

$$E_z^0 = e^{ikx} = \sum_{\ell=-\infty}^{\infty} i^\ell J_\ell(kr)\, e^{i\ell\theta}, \tag{2.7}$$

we obtain

$$E_z^{\rm i}(r,\theta) = E_z^0(a,\theta) + E_z^{\rm s}(a,\theta), \tag{2.8}$$

$$\partial_r E_z^{\rm i}(r,\theta) = \partial_r E_z^0(a,\theta) + \partial_r E_z^{\rm s}(a,\theta). \tag{2.9}$$

Using the identity $Z'_\ell(x) = Z_{\ell-1}(x) - (\ell/x)\, Z_\ell(x)$, where $Z_\ell = J_\ell$ or $H_\ell^{(1)}$, we can determine the expansion coefficients. Abbreviating $x = ka$, we find

$$b_\ell(x,n) = \frac{nJ'_\ell(nx)J_\ell(x) - J_\ell(nx)J'_\ell(x)}{nJ'_\ell(nx)H_\ell^{(1)}(x) - J_\ell(nx){H'}_\ell^{(1)}(x)}, \tag{2.10}$$

$$d_\ell(x,n) = \frac{J_\ell(x) - b_\ell(x)H_\ell^{(1)}(x)}{J_\ell(nx)} = \frac{J_{\ell-1}(x)H_\ell^{(1)}(x) - J_\ell(x)H_{\ell-1}^{(1)}(x)}{nJ'_\ell(nx)H_\ell^{(1)}(x) - J_\ell(nx){H'}_\ell^{(1)}(x)}. \tag{2.11}$$

The WGM resonance occurs when the b_ℓ and d_ℓ become very large. From Eq. (2.10) and Eq. (2.11), we see that the resonance occurs when the b_ℓ denominator, $b_\ell^{\rm d}$, is small. Thus, for $b_\ell^{\rm d} \to 0$, we obtain the resonance condition in the TM mode,

$$\frac{J_{\ell-1}(nka)}{J_\ell(nka)} = \frac{1}{n}\frac{H_{\ell-1}^{(1)}(ka)}{H_\ell^{(1)}(ka)}. \tag{2.12}$$

For example, when $a = \lambda/2$, $b_\ell^{\rm d}$ vanishes at $n \cong 2.745 - i1.506 \times 10^{-3}$, but this is non-physical because it describes a material that is producing energy, not absorbing it. Thus we do not have a perfect resonance. Examples of resonance conditions in the TM mode are given in Table 4.1.

The TE mode can be analyzed in a similar way, and the resonance condition is given by

$$\frac{J_{\ell-2}(nka)}{J_{\ell-1}(nka)} = n\frac{H_{\ell-2}^{(1)}(ka)}{H_{\ell-1}^{(1)}(ka)} - \frac{(\ell-1)(n^2-1)}{nka}. \tag{2.13}$$

Examples of resonance conditions in the TE mode are given in Table 4.1.

In the TM and TE resonance conditions, the radius shift Δa and refractive index shift Δn generate the wavelength shift $\Delta\lambda$. We found that $\Delta\lambda$ is approximated by

$$\Delta\lambda \cong \frac{\lambda_\ell}{a_\ell}\Delta a + \frac{\lambda_{\ell+1}}{n_\ell}\Delta n, \tag{2.14}$$

where λ_ℓ, a_ℓ, n_ℓ are the wavelength, radius, refractive index on the ℓ mode resonance ($a_\ell \to a_\ell + \Delta a$, $n_\ell \to n_\ell + \Delta n \Rightarrow \lambda_\ell \to \lambda_\ell + \Delta\lambda$). The first term in Eq. (2.14) is derived from, $2\pi a_\ell/\lambda_\ell = 2\pi\,(a_\ell + \Delta a)\,/\,(\lambda_\ell + \Delta\lambda)$, where the resonance conditions are maintained. The second term in Eq. (2.14) is more complex, and we numerically approximated it. Eq. (2.14) can be used to estimate mode shift due to manufacturing errors.

Table 4.1. Examples of resonance conditions in TM and TE modes

TM mode		
radius a	refractive index n	mode number ℓ
$0.50\ \lambda$	2.745	6
$0.75\ \lambda$	2.310	8
$1.00\ \lambda$	2.717	10
TE mode		
radius a	refractive index n	mode number ℓ
$0.50\ \lambda$	2.683	6
$0.75\ \lambda$	2.529	9
$1.00\ \lambda$	2.887	11

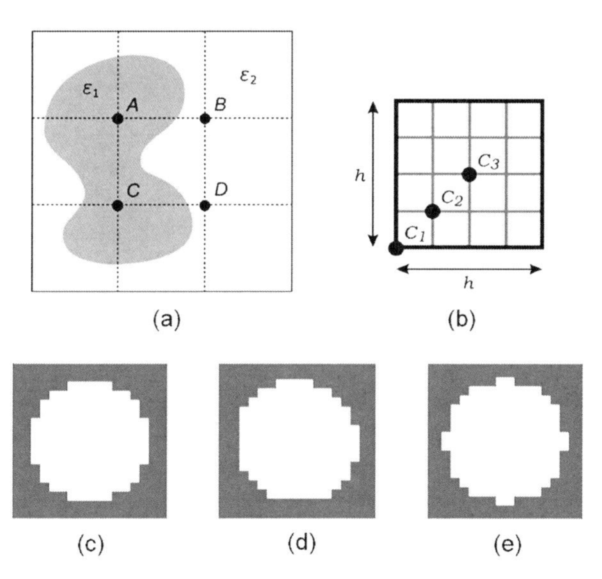

Fig. 4.6. Staircase model. (a) Example grid points. A and C are inside scatterer (gray), B and D are outside (white). (b) Circle center C_1 is on a grid point, C_3 is centered in a cell, C_2 lies between C_1 and C_3. (c), (d), (e) are circle models centered at C_1, C_2, C_3, respectively.

4.2.3 Grid representations

A high-accuracy FDTD algorithm alone does not guarantee a high-accuracy result, because other errors enter into the total calculation. For WGM calculations the largest source of the error is the scatterer representation on the numerical grid.

Staircase model

The most elementary representation is the staircase model. A grid point is either inside or outside the scatterer. In Fig. 4.6(a), a scatterer of permittivity ε_1 (gray

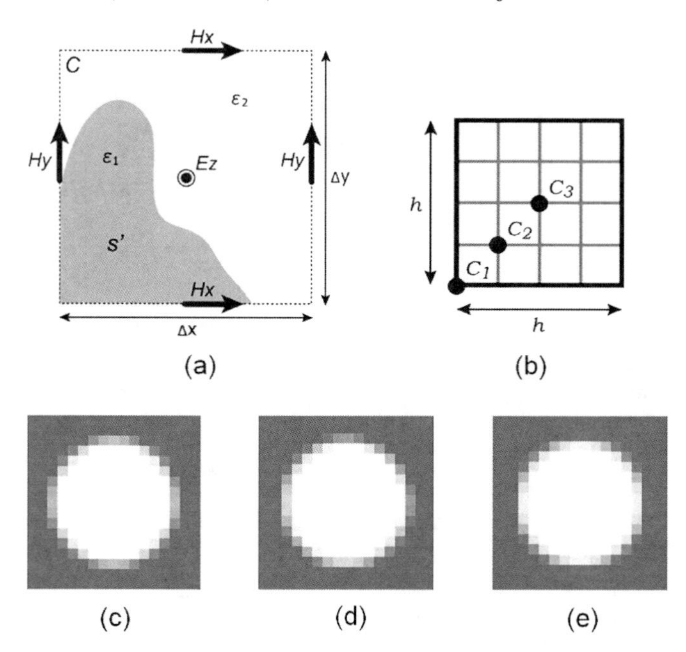

Fig. 4.7. Fuzzy model in the TM mode. (a) Integration of $\mathbf{H}$ on contour about E_z grid point. (b) Circle center C_1 is on a grid point, C_3 is centered in a cell, C_2 lies between C_1 and C_3. (c), (d), (e) are circle models centered at C_1, C_2, C_3, respectively.

region) is immersed in a medium of permittivity ε_2 (white region). In the staircase model $\varepsilon(\mathbf{r}) = \varepsilon_1$ if grid point $\mathbf{r}$ lies within the scatterer (A and C), and $\varepsilon(\mathbf{r}) = \varepsilon_2$ otherwise (B and D).

However, the staircase model obviously cannot include distributions between grid points, and fails to accurately model the shape as shown in Fig. 4.6(c), (d), (e), which gives rise to representation error. For example, consider the circles centered at C_1, C_2, and C_3 on a uniform grid of spacing h (see Fig. 4.6(b)), and their corresponding staircase representation. As we can see the representation various with the position of circle center (Fig. 4.6(c), (d), (e)).

Fuzzy model

The fuzzy model is much better than the staircase model, because the scatterer shape is better represented. The fuzzy model is derived from Ampére's law,

$$\int_S \varepsilon \partial_t \mathbf{E} \cdot d\mathbf{S} = \int_C \mathbf{H} \cdot d\mathbf{s}. \tag{2.15}$$

First let us consider the TM mode. Extracting the E_z component from the left side of Eq. (2.15), we evaluate the integral over the surface shown in Fig. 4.7(a). If Δx and Δy are sufficiently small, we find

$$\int_S \varepsilon \partial_t \mathbf{E} \cdot d\mathbf{S} \cong \partial_t E_z \int \varepsilon \, dx \, dy. \tag{2.16}$$

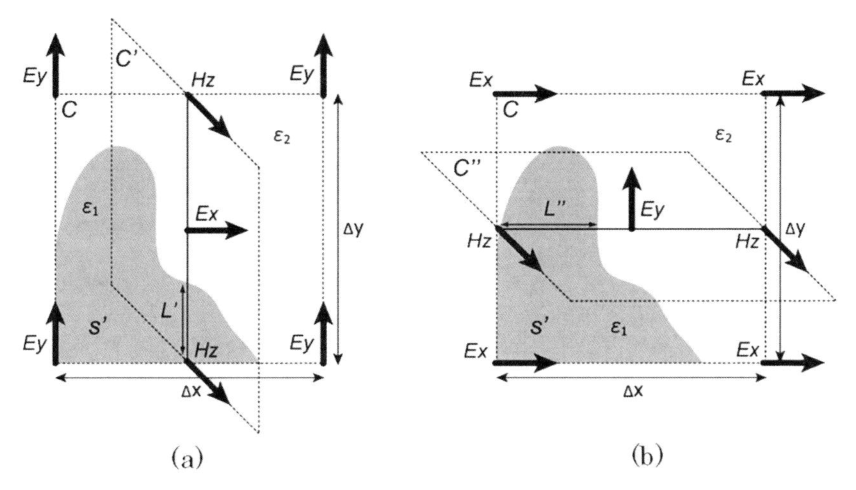

Fig. 4.8. The fuzzy model in the TE mode. (a) Integration of **H** on contour about E_x grid point. (b) Integration of **H** on contour about E_y grid point.

Similarly **H** is essentially constant in Fig. 4.7(a). Using Stoke's theorem, the right side of Eq. (2.15) becomes

$$\int_C \mathbf{H} \cdot ds = \int_S (\nabla \times \mathbf{H}) \cdot d\mathbf{S} \cong \Delta x \Delta y \, (\nabla \times \mathbf{H})_z \,. \tag{2.17}$$

On the other hand, since ε must be a constant at a grid point in FDTD calculations, the differential form of Ampére's law is given by

$$\langle \varepsilon \rangle_{xy} \partial_t E_z = (\nabla \times \mathbf{H})_z \,, \tag{2.18}$$

where $\langle \varepsilon \rangle_{xy}$ means the average of ε on the x-y surface about the grid point. Comparing Eq. (2.18) with Eqs. (2.16) and (2.17), we obtain

$$\begin{aligned} \langle \varepsilon \rangle_{xy} &= \frac{1}{\Delta x \Delta y} \int_S \varepsilon \, dx \, dy \\ &= \varepsilon_1 \left(\frac{S'}{\Delta x \Delta y} \right) + \varepsilon_2 \left(1 - \frac{S'}{\Delta x \Delta y} \right) . \end{aligned} \tag{2.19}$$

Thus in the fuzzy model, in the vicinity of the $\varepsilon_1/\varepsilon_2$ boundary, ε varies continuously between ε_1 and ε_2 depending on its proximity to the interface, whereas in the staircase model ε assumes just one of two possible values. Hence the symmetries are better preserved as shown in Fig. 4.7(c), (d), (e).

In the TE mode, the fuzzy model is derived by integrating E_x on the y-z plane and E_y on the x-z plane. Because the scatterer distributions are constant in the z-direction, ε is replaced by line averages. For example, in Fig. 4.8(a), E_x lies at $\mathbf{r} = (x, y + \Delta y/2)$. Thus $\varepsilon(\mathbf{r})$ on the x-y surface is replaced by the line integral,

$$\langle \varepsilon(\mathbf{r}) \rangle_y = \frac{1}{\Delta y} \int_y^{y+\Delta y} \varepsilon(x, y) \, dy = \varepsilon_1 \left(\frac{L'}{\Delta y} \right) + \varepsilon_2 \left(1 - \frac{L'}{\Delta y} \right) . \tag{2.20}$$

Table 4.2. Example parameters used to simulate the WGM in the TM mode.

wavelength	640 nm
cylinder radius	320 nm
grid spacing	1 0nm
mode	6
refractive index	2.745
computational space	1.5 μm $\times$ 1.5 μm

Similarly, in Fig. 4.8(b), E_y lies at $\mathbf{r} = (x + \Delta x/2, y)$, and $\varepsilon(\mathbf{r})$ is replaced by $\langle\varepsilon(\mathbf{r})\rangle_x = \frac{1}{\Delta x}\int_x^{x+\Delta x} \varepsilon(x, y)\, dx$.

In three dimensions, the fuzzy models at E_x, E_y, E_z are obtained by integration on the y-z, x-z, x-y surfaces, respectively.

Effects of cylinder center

Although the fuzzy model reduces representation error, it does not completely eliminate it. For circular cylinders, numerical experiments show that representation error is minimized when the cylinder center is placed in the middle of a grid cell (position C_3 in Fig. 4.7(b)).

4.2.4 WGM simulation

TM mode

In the TM mode, we simulate a WGM using the standard (S) and nonstandard (NS) FDTD algorithms. Example parameters are listed in Table 4.2, where we used a coarse grid to demonstrate the high accuracy of the NS-FDTD algorithm. The resonance condition can be found by numerically solving Eq. (2.12). We terminate the computational domain with the Berenger's perfectly matched layer (PML) [15]. The scatterer is represented on the numerical grid using the fuzzy model.

We calculated the scattered intensity $|E_z^{\mathrm{s}}|^2$ for 100,000 time steps (about 1000 wave periods) and compared the results with Mie theory in Fig. 4.9. Figs. 4.9(a), 4.9(b), and 4.9(c) visualize $|E_z^{\mathrm{s}}|^2$ distributions of analytic and calculated results at steady state. Fig. 4.9(d) is the angular distribution on a circular contour of radius $1.1a$ around the cylinder center. As shown in Fig. 4.9, the NS-FDTD algorithm is much more accurate than the S-FDTD one.

TE mode

In the TE mode, similarly, we calculated a WGM using the S-FDTD and NS-FDTD algorithms. Example parameters are listed in Table 4.3. The boundary condition and representation model are the same as in the TM mode.

We calculated the scattered intensity $|E_y^{\mathrm{s}}|^2$ at steady state (about 1000 wave periods) and compared the results with Mie theory in Fig. 4.10. Figs. 4.10(a), 4.10(b), and 4.10(c) visualize $|E_z^{\mathrm{s}}|^2$ distributions of analytic and calculated results.

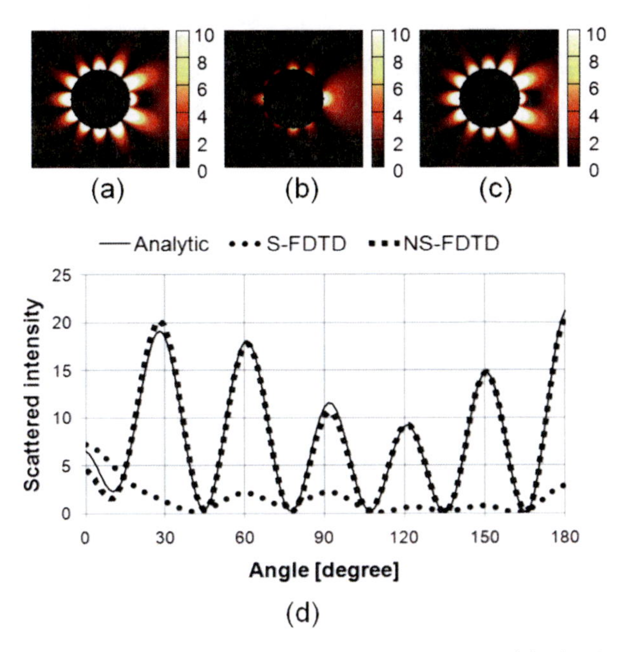

Fig. 4.9. FDTD calculation of a WGM in the TM mode. (a) Analytic solution. (b) S-FDTD result. (c) NS-FDTD result. (d) Angular intensity distributions on a circular contour of radius $1.1a$ (a = cylinder radius).

Table 4.3. Example parameters to simulate the WGM in the TE mode.

wavelength	640 nm
cylinder radius	320 nm
grid spacing	10 nm
mode	6
refractive index	2.683
computational space	1.5 μm $\times$ 1.5 μm

Fig. 4.10(d) is the angular distribution on a circular contour of radius $1.1a$ around the cylinder center. As shown in Fig. 4.10, the accuracy of the NS-FDTD calculation is far superior to the S-FDTD one. But despite the same grid spacing, the accuracy of the NS-FDTD calculation in the TE mode is lower than in the TM mode.

Why does the accuracy fall in the TE mode?

As discussed in section 4.2.3, the TM fuzzy model well represents x-y planar distributions, but the TE fuzzy model is essentially a line average and does not capture the planar structure.

A more fundamental reason for the lower accuracy of the TE calculation can be found by examining Gauss's law. In the TM mode, Gauss's law gives $\nabla \cdot (\varepsilon \mathbf{E}) = \mathbf{E} \cdot \nabla \varepsilon + \varepsilon \nabla \cdot \mathbf{E} = 0$, gives

$$\mathbf{E} \cdot \nabla \varepsilon = \nabla \cdot \mathbf{E} = 0, \tag{2.21}$$

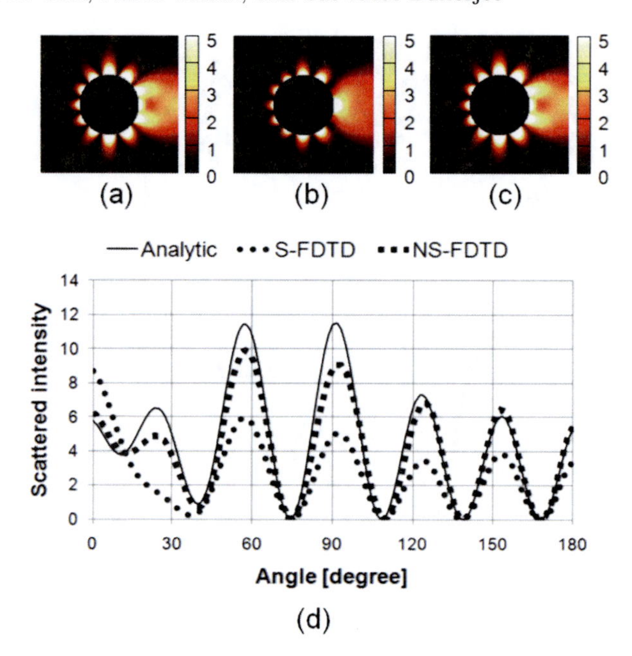

Fig. 4.10. FDTD calculation of a WGM in the TE mode. (a) Analytic solution. (b) S-FDTD result. (c) NS-FDTD result. (d) Angular intensity distributions on a circular contour of radius $1.1a$ (a = cylinder radius).

because $\mathbf{E} \perp \nabla\varepsilon$. In this case Maxwell's equations reduce to the homogeneous wave equation,

$$\left(\partial_t^2 - v^2\nabla^2\right)\mathbf{E} = 0, \tag{2.22}$$

where $v = 1/\sqrt{\varepsilon\mu}$ (ε = permittivity, μ = permeability). In the TE mode, however, because $\mathbf{E} \parallel \nabla\varepsilon \Rightarrow \nabla\cdot\mathbf{E} \neq 0$, and we are left with the coupled wave equation,

$$\partial_t^2\mathbf{E} = v^2\left(\nabla^2\mathbf{E} - \nabla\left(\nabla\cdot\mathbf{E}\right)\right). \tag{2.23}$$

Since the NS-FDTD algorithm is optimized to the homogeneous wave equation, it gives excellent results in the TM mode. But, in the TE mode, errors increase at media interfaces because the boundary information, $\nabla\cdot\mathbf{E} = \mathbf{E}\cdot\nabla\varepsilon/\varepsilon$, is ignored in the optimization.

4.2.5 Coupled wave equation approach

As discussed in section 4.2.4, the NS-FDTD algorithm for Maxwell's equations is optimized to a uniform region and boundaries are ignored. To accurately include the boundary information, we derive a NS-FDTD algorithm for the coupled wave equation (CWE). The CWE approach can also reduce the computing time relative to the Yee algorithm, because $\mathbf{E}$ is only calculated. $\mathbf{H}$ can be easily computed using Maxwell's equations.

Standard FDTD algorithm

The central finite difference (FD) approximation to a derivative is given by

$$\frac{df(x)}{dx} \simeq \frac{d_x f(x)}{\Delta x}, \tag{2.24}$$

where d_x is difference operator defined by $d_x f(x) = f(x + \Delta x/2) - f(x - \Delta x/2)$. For reasons that will soon be obvious we call Eq. (2.24) the standard (S) FD approximation. Analogously we define d_y and d_t. For example, the CWE can be expanded for the E_x component,

$$\partial_t^2 E_x = v^2 \left(\partial_y^2 E_x + \partial_z^2 E_x - \partial_x \partial_y E_y - \partial_x \partial_z E_z\right). \tag{2.25}$$

Replacing the derivatives in Eq. (2.25) with the S-FD approximations, we obtain

$$d_t^2 E_x = \frac{v^2 \Delta t^2}{h^2} \left(d_y^2 E_x + d_z^2 E_x - d_x d_y E_y - d_x d_z E_z\right), \tag{2.26}$$

where $h = \Delta x = \Delta y = \Delta z$. Expanding $d_t^2 E_x$ and solving for $E_x(x, y, t + \Delta t)$ yields the S-FDTD algorithm for the CWE,

$$E_x^{t+\Delta t} = -E_x^{t-\Delta t} + 2E_x^t + \frac{v^2 \Delta t^2}{h^2} \left(d_y^2 E_x^t + d_z^2 E_x^t - d_x d_y E_y^t - d_x d_z E_z^t\right), \tag{2.27}$$

where for simplicity we write $E_x(x, y, t) \to E_x^t$. The E_y, E_z formulations are obtained by exchanging x and y, z in Eq. (2.27). Each electric field component lies at a different position on the numerical grid so that central FD approximations can be used for the spatial derivatives. As shown in Fig. 4.11, our arrangements are

$$E_x^t \to E_x(x, y + h/2, z + h/2, t), \tag{2.28}$$
$$E_y^t \to E_y(x + h/2, y, z + h/2, t), \tag{2.29}$$
$$E_z^t \to E_z(x + h/2, y + h/2, z, t). \tag{2.30}$$

The scattered field formula also can be derived easily. The total electric field $\mathbf{E}$ can be decomposed into sum of the incident field $\mathbf{E}^0$ and the scattered field $\mathbf{E}^\mathrm{s}$. Using $v = v_0$ in the homogeneous wave equation satisfied $\mathbf{E}^0$ and subtracting from the CWE, we obtain the scattered formula of the CWE,

$$\partial_t^2 \mathbf{E}^\mathrm{s} = v^2 \left(\nabla^2 \mathbf{E}^\mathrm{s} - \nabla \left(\nabla \cdot \mathbf{E}^\mathrm{s}\right)\right) + \mathbf{J}, \tag{2.31}$$

where $\mathbf{J}$ is the source term and given by

$$\mathbf{J} = \left(\frac{1}{n^2} - 1\right) \partial_t^2 \mathbf{E}^0. \tag{2.32}$$

The FDTD algorithm for the scattered field also can be derived in a similar way.

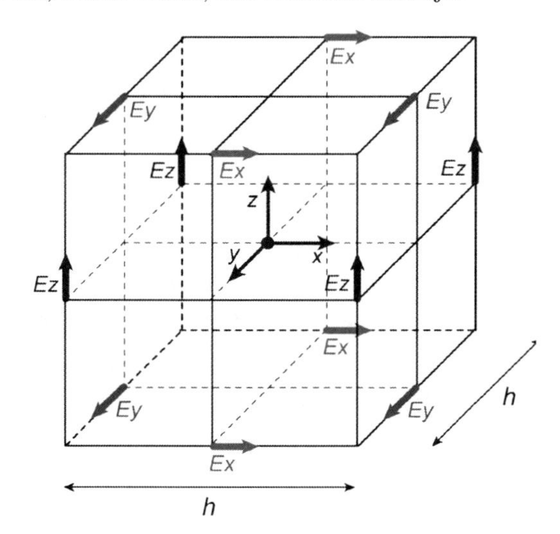

Fig. 4.11. Layout of the electric field on the numerical grid (h = grid spacing).

Nonstandard FDTD algorithm

In one dimension, the nonstandard (NS) FD approximation has the form,

$$\frac{df(x)}{dx} = \frac{d_x f(x)}{s(\Delta x)}, \tag{2.33}$$

where $s(\Delta x) = s(\Delta x) = d_x f(x)/f'(x)$. For monochromatic waves e^{ikx} (k = wave number) an exact FD expression is obtained by putting $s(\Delta x) = s(k, \Delta x)$, where

$$s(k, \Delta x) = \frac{2}{k} \sin\left(\frac{k\Delta x}{2}\right). \tag{2.34}$$

Analogously we find $\partial_t f(t) = d_t f(t)/s(\omega, \Delta t)$ for $f(t) = e^{i\omega t}$ (ω = angular frequency).

In two or three dimensions, however, there is no exact NS-FDTD expressions for the spatial derivatives because $s(\mathbf{k}, h)$ depends on direction of the wave, $\mathbf{k} = (k_x, k_y)$. We found that the spatial error can be greatly reduced by using NS-FD approximations. In the CWE, there are two spatial derivative types, ∂_y^2, $\partial_x \partial_y$ (analogously we can derive NS-FD approximations of ∂_x^2, ∂_z^2, $\partial_y \partial_z$, $\partial_x \partial_z$). First we consider the S-FD approximation of $\partial_y^2 \simeq d_y^2/h^2$. The error is defined by

$$\epsilon_{yy}^{\mathrm{S}} \varphi = \left(h^2 \partial_y^2 - d_y^2\right) \varphi, \tag{2.35}$$

where $\varphi = e^{i(k_x x + k_y y)}$, and $(k_x, k_y) = k\,(\cos\theta, \sin\theta)$. Expanding in a Taylor series, we find

$$\begin{aligned} \epsilon_{yy}^{\mathrm{S}} &= k^4 h^4 \left(-\frac{\sin^2\theta}{12} + \frac{\sin^2(2\theta)}{48}\right) \\ &+ k^6 h^6 \left(\frac{\sin^2\theta}{360} - \frac{\sin^2(2\theta)\left(1+\sin^2\theta\right)}{1440}\right) + O(k^8 h^8). \end{aligned} \tag{2.36}$$

Putting $h \rightarrow s(k,h) = (2/k)\sin(kh/2)$ in Eq. (2.36), the error reduces to

$$\epsilon_{yy}^{\mathrm{NS}} = k^4h^4\frac{\sin^2(2\theta)}{48} - k^6h^6\frac{\sin^2(2\theta)\left(1+\sin^2\theta\right)}{1440} + O(k^8h^8). \tag{2.37}$$

Next we consider the FD approximation of $\partial_x\partial_y \simeq d_x d_y/h^2$. Similarly, the error is defined by

$$\epsilon_{xy}^{\mathrm{S}}\varphi = \left(h^2\partial_x\partial_y - d_x d_y\right)\varphi. \tag{2.38}$$

Expanding in a Taylor series, we find

$$\epsilon_{xy}^{\mathrm{S}} = -k^4h^4\frac{\sin(2\theta)}{48} + k^6h^6\left(\frac{\sin(2\theta)}{3840} + \frac{\sin^3(2\theta)}{11520}\right) + O(k^8h^8). \tag{2.39}$$

Putting $h^2 \rightarrow s'(k,h)^2 = (2h/k)\sin(kh/2)$ in Eq. (2.39), the error can reduce to

$$\epsilon_{xy}^{\mathrm{NS}} = k^6h^6\frac{\sin^3(2\theta)}{11520} + O(k^8h^8). \tag{2.40}$$

Replacing derivatives in the CWE of Eq. (2.26) with the NS-FD approximations, we obtain a nonstandard (NS) finite-difference time-domain (FDTD) algorithm to solve the CWE,

$$\begin{aligned} E_x^{t+\Delta t} &= -E_x^{t-\Delta t} + 2E_x^t \\ &+ u_1\left(d_y^2E_x^t + d_z^2E_x^t\right) - u_2\left(d_xd_yE_y^t + d_xd_zE_z^t\right), \end{aligned} \tag{2.41}$$

$$u_1 = \frac{\sin^2(\omega\Delta t/2)}{\sin^2(nkh/2)}, \tag{2.42}$$

$$u_2 = \frac{2\sin^2(\omega\Delta t/2)}{nkh\sin(nkh/2)}. \tag{2.43}$$

The E_y, E_z formulations are obtained by exchanging x and y, z in Eq. (2.41). Using Eq. (2.32), the NS-FDTD algorithm for the scattered field also can be derived in a similar way.

The stability of the CWE approach in homogeneous medium is the same as Yee algorithm, because $\nabla \cdot \mathbf{E} = 0$ is satisfied. In non-homogeneous medium, the stability and performance are shown in [16], which the FDTD calculation of the CWE approach is accurately included the boundary information, $\nabla \cdot \mathbf{E}$, and twice as fast as the Yee algorithm.

4.3 A quasi-stable FDTD algorithm for dispersive materials to compute optical characteristics of subwavelength metal gratings

4.3.1 Computing light propagation in dispersive materials

In dispersive materials in which the real part of the electric permittivity (ε) is negative, the ordinary FDTD algorithm is numerically unstable. Even for monochromatic light the frequency-dependence of the electrical permittivity of the material must be taken into account. One way to do this is to use a method called recursive convolution (RC) FDTD [2,6].

Two-dimensional structures which are invariant in one dimension (say the z-axis) are of great practical and scientific interest, but many commercially available software packages do not provide FDTD in two dimensions. While useful three-dimensional computations require high-end workstations, meaningful and often more insightful two-dimensional FDTD calculations can be done on desktop and even laptop computers. Furthermore, in two-dimensional calculations, in the TM (transverse magnetic) mode (electric field, **E** is polarized perpendicular to the plane of incidence, x-y plane), Maxwell's equations reduce to the wave equation in E_z (assuming that magnetic permeability is constant, but not the electric permittivity). Thus in the TM mode it is not necessary to compute the magnetic field which greatly reduces the computational cost. Here, we present a new formulation of RC-FDTD for the wave equation that was already introduced in Ref. [18]. In the TE (transverse electric) mode (**E** lies in the x-y plane), the Maxwell curl equations are implemented in the RC-FDTD framework.

RC-FDTD requires an analytical model of dispersion to simulate the material permittivity. Conventionally, RC-FDTD is implemented for pulsed light sources. This requires that the analytical model of permittivity should fit the tabulated permittivities closely over a broadband of frequencies. However, this is not easy to achieve. Various authors have experimented with different permittivity models [19]. It turns out that choice of a particular model depends on the metal being simulated [20]. In this work we choose the first-order Drude model [2]. The broadband Drude parameters are obtained by fitting the analytical model to the tabulated permittivity over a wide range of frequencies. Hence at any particular frequency the Drude permittivity obtained using the fitted parameters might differ substantially from the tabulated values. To be able to use the tabulated permittivity values at all frequencies, we devised a monochromatic version of the RC-FDTD both in TM and TE modes [18]. We computed the Drude parameters at each frequency using the tabulated value of the permittivity and the space-time discretization. In this method, the space-time discretization is chosen to ensure the stability of the algorithm. A stability analysis is presented for our algorithm following Ref. [6], in which the stability constraints are derived as functions of space-time discretization.

One of our objectives is to test the efficacy of the algorithm in solving a real-life problem. To that end we compute the far-field diffraction efficiencies of zero-order subwavelength metal gratings. These gratings are known as wire grid polarizers (WGP) in technical literature. In recent times, the WGPs find application as highly efficient polarizers in liquid crystal based display systems [21]–[27].

In general, a liquid crystal display (LCD) unit consists of a back light unit and a liquid crystal panel unit. The back light unit houses the light source and other optical components to guide light from the source to a flat output surface and to ensure uniformity of brightness over that surface. Both the backlight unit and the liquid crystal panel unit contain multiple polarizing components. Conventionally, dichroic, sheet polarizers are used in the LCD systems [28]. WGPs are fast replacing the conventional sheet polarizers on account of improved performance, such as, higher polarization index, brightness, and light recycling efficiency.

The references reviewed here cover fabrication [21] as well as design and numerical simulation aspects of WGPs [22]–[26]. Three-dimensional FDTD with Lorentz model of dielectric permittivity are used to design finite aperture WGPs [22]. Metallic waveguide theory and finite element method is used to design WGPs meant for liquid crystal on silicon (LCoS) systems [23]. Refs. [24] and [25] use a modal analysis and a differential theory of diffraction gratings respectively to simulate WGPs used in LCoS microdisplays. These studies are significant from the point of view of practical design. However, simulation of two-dimensional WGPs using RC-FDTD and first-order Drude model has not been reported before. Also, how the use of broadband Drude permittivities in FDTD simulations modifies reflection and transmission properties of the WGPs has not been investigated before.

The simulations presented here involve two-dimensional WGPs made of aluminum. The structures are periodic along one direction, finite along a perpendicular direction and invariant along the third orthogonal direction. This fact enables one to simulate these structures in two dimensions reducing the computation time greatly. Also, the chosen grating material being aluminum, the RC-FDTD developed here using first-order Drude model turns out to be a suitable simulation tool.

The reflection, transmission and polarization characteristics of a grating are determined by its spatial period (Λ), groove height (h), fill factor (f) and the grating profile. The grating fill factor is defined as the ratio of the width of one grating line (w) and the grating spatial period. The spatial period of a zero-order WGP is chosen to be smaller than the wavelength of the incident light, λ_p. This particular choice makes the WGPs zero-order i.e., they can only reflect and transmit light in the specular order while blocking the higher diffraction orders.

WGPs when illuminated with unpolarized light transmit the light component polarized in the TE mode and reflects the light polarized in the TM mode. Note that, the light in TE mode is polarized in a plane perpendicular to the grating lines while light in the TM mode is polarized parallel to the grating lines.

In typical LCD units, a WGP can either be used as a stand-alone polarizing component or combined with other components. The second approach facilitates compactness, miniaturization and cost efficiency. In edge-lit display systems appearing frequently in mobile phones or laptop computers, a light guiding plate is employed to guide light emitted from the light sources placed on one side of the plate to the output surface. A few patents including ours have suggested integrating a WGP on the output surface of such a lightguide.

Here we present simulations of two WGPs; the first one for integration on the output surface of the light guide plate and the second one to be used as a stand-alone component in a projection display. The first example deals with the simulation of an ordinary WGP, albeit with a larger fillfactor than usual. The second example deals with the simulation of a double-layered, small fill-factor WGP. The first one is entirely metallic while the second one comprises a metallic part and a weakly conducting dielectric part.

Ordinarily the WGPs suggested for integration on lightguide plates employ low fill factor ($f \leq 0.5$) grating designs that maximize the transmission in the TE mode [24]. Such a WGP mounted on a lightguide would let more light to be transmitted near the light sources and less as the distance from the light sources increase eventually making the brightness distribution to be inhomogeneous on the surface of the lightguide. From simulations we find out that the amount of light that is transmitted through a WGP integrated on the top surface of such a lightguide depends on the grating fillfactor. As the fillfactor increases the TE mode transmission diminishes at any point on the output surface. However, the brightness gradient or the rate at which brightness falls off as the distance of the exit point from the light source increases tend to flatten out as the fillfactor increases. Hence, a large fillfactor WGP integrated on top of the lightguide ensures that polarized light exiting the output surface of the lightguide spreads more homogeneously over the same surface [26].

For some applications, such as, LCD projectors, the reflected light from the WGP can cause ghost images to form. To prevent the ghost images from forming, the reflected light has to be removed from the system. This is why absorbing WGPs are used [27]. Few patents have proposed the use of multilayered WGPs for this purpose. However, none of these works have studied how the absorption in the coating layer affects the reflection and transmission behaviour of an ordinary WGP in quantitative terms. Our simulations indicate that a double-layered WGP consisting of one metal and another weakly conducting dielectric layer is sufficient in many cases to bring down the reflection to an acceptable level. The trick is to ensure that the weakly conducting dielectric layer possess appropriate refractive and absorptive indices. Semiconductors, such as, silicon and various sulphides and phosphides of indium or gallium turn out to be suitable candidates and are easy to coat on the metal surface. Such a double-layered WGP can show low transmission behavior over a narrow band of frequencies and might be suitable for filtering applications. The simulation of weakly conducting dielectric part is carried out using a non-dispersive FDTD suitable for such media [29].

Each of the WGPs is simulated using two sets of permittivity data for aluminum. The first set of permittivities comprises of the tabulated data [30]. The second set of permittivity values is obtained using a set of best-fit Drude parameters. These Drude parameters are determined from a closest fit of the tabulated values of permittivities to the first-order Drude model over the entire visible domain. A comparison between the two sets of results obtained with two sets of permittivities highlights the dependence of the reflection and transmission spectra of the WGPs on the choice of material properties. The results obtained using exact permittivities can be quite different than those obtained using the wide-band, best-fit Drude parameters.

The monochromatic RC-FDTD algorithm is described in section 4.3.2. In this section, besides RC-FDTD, we include the non-dispersive FDTD field update equations. These update mechanisms are required to update fields in non-conducting and weakly conducting dielectrics. The stability analysis is presented in section 4.3.3. The modeling and simulation of gratings is described in section 4.3.4. Section 4.3.5 presents a brief summary of the work presented in this section.

4.3.2 RC-FDTD

In two dimensions the effect of unpolarized light incident on a structure is estimated by the average of optical properties computed in two orthogonal polarization states. In the TE mode, light is polarized parallel to the plane of incidence. In this mode, Maxwell's equations for materials with frequency-dependent permittivities are given by

$$\mu \partial_t \mathbf{H}(\mathbf{r},t) = -\nabla \times \mathbf{E}(\mathbf{r},t), \tag{3.1}$$

$$\partial_t \mathbf{D}(\mathbf{r},t) = \nabla \times \mathbf{H}(\mathbf{r},t). \tag{3.2}$$

In the above μ is the magnetic permeability taken to be constant and equal to the vacuum permeability, μ_0, $\mathbf{H}$ is the magnetic vector, $\nabla \equiv \hat{\mathbf{x}}\partial_x + \hat{\mathbf{y}}\partial_y$ is the vector differential operator, $\times$ stands for vector curl operation. In this notation, ∂_x, ∂_y, ∂_t denote $\partial/\partial x$, $\partial/\partial y$ and $\partial/\partial t$ respectively. $\mathbf{D}$ is the displacement current, given by

$$\mathbf{D}(\mathbf{r},\omega) = \varepsilon(\mathbf{r},\omega)\,\mathbf{E}(\mathbf{r},\omega), \tag{3.3}$$

where $\mathbf{E}$ is the electric vector and ε is the permittivity. The time domain behavior of $\mathbf{D}$ is obtained by taking the inverse Fourier transform on both sides of Eq. (3.3), and using the convolution theorem. We find [2]

$$\mathbf{D}(t) = \int_0^t \varepsilon(\tau)\mathbf{E}(t-\tau)\,d\tau, \tag{3.4}$$

where $\varepsilon(t) = \mathrm{F}^{-1}\{\varepsilon(\omega)\}$, and F^{-1} denotes the inverse Fourier transform operator.

In the first-order Drude model, $\varepsilon(\omega)$ is given by

$$\varepsilon(\omega) = \varepsilon_0\varepsilon_\infty + \frac{\varepsilon_0\omega_{pl}^2}{\omega(i\nu_c - \omega)} = \varepsilon_0\varepsilon_\infty + \varepsilon_0\chi(\omega), \tag{3.5}$$

where ε_0 is the vacuum permittivity, ε_∞ is the infinite frequency dielectric constant, ω_{pl} is the plasma frequency, ν_c is the collision frequency and $\chi(\omega)$ is the susceptibility. ε_∞ is taken to be 1 in all subsequent discussions. Taking the inverse Fourier transform of $\chi(\omega)$ we find [2],

$$\chi(t) = \frac{\omega_{pl}^2}{\nu_c}\left[1 - e^{-\nu_c t}\right]U(t), \tag{3.6}$$

where $U(t)$ is a unit step function defined by

$$U(t) = \begin{cases} 0 & (t \le 0) \\ 1 & (t > 0) \end{cases}$$

In order to evaluate Eq. (3.4), we first replace $\varepsilon(\tau)$ with the Fourier transform of the right-hand side of Eq. (3.5). The resulting form is discretized in keeping with FDTD time-stepping scheme. If Δt is the minimum time step, and if n represents the total number of time steps corresponding to time instant t, then $t = n\Delta t$. Assuming $\mathbf{E}$ to be constant over any single interval $[m\Delta t,\,(m+1)\,\Delta t]$, where $m \in 0, ..., n$, the integral in Eq. (3.4) reduces to a partial summation as follows [2]:

$$\mathbf{D}^n = \varepsilon_0\mathbf{E}^n + \varepsilon_0\sum_{m=0}^{n-1}\mathbf{E}^{n-m}\int_{m\Delta t}^{(m+1)\Delta t}\chi(\tau)\,d\tau, \tag{3.7}$$

Eq. (3.7) can be written in the following compact form

$$\mathbf{D}^n = \varepsilon_0 \mathbf{E}^n + \varepsilon_0 \mathbf{\Psi}^n, \tag{3.8}$$

where,

$$\begin{aligned}\mathbf{\Psi}^n &= \sum_{m=0}^{n-1} \mathbf{E}^{n-m} \int_{m\Delta t}^{(m+1)\Delta t} \chi(\tau)\, d\tau \\ &= \sum_{m=0}^{n-1} \mathbf{E}^{n-m} \frac{\omega_{pl}^2}{\nu_c} \left[\Delta t + \frac{e^{-(m+1)\nu_c \Delta t} - e^{-m\nu_c \Delta t}}{\nu_c} \right].\end{aligned} \tag{3.9}$$

The vector quantity, $\mathbf{\Psi}$, is known as the accumulation field.

TE mode: FDTD for Maxwell's equations

In this mode, we compute the H_z, E_x, E_y, Ψ_x, Ψ_y components. The $\mathbf{H}$ update equation follows from Eq. (3.1) and is given by

$$\mathbf{H}^{n+1/2} = \mathbf{H}^{n-1/2} - \frac{\Delta t}{\mu_0 \Delta s} \mathbf{d} \times \mathbf{E}^n. \tag{3.10}$$

Even in non-dispersive case, the $\mathbf{H}$ update equation retains the same form. We denote the second-order accurate, central finite difference operator by d_x, where $d_x f(x,y) = f(x + \Delta s/2, y) - f(x - \Delta s/2, y)$, thus $\partial_x f \cong d_x f/\Delta s$, and where Δs is the minimum space step. FDTD spatial grid is chosen to be square implying Δs is the same as Δx and Δy, the minimum space steps respectively in x and y directions. The vector difference operator $\mathbf{d}$ is defined by $\mathbf{d} \equiv \hat{\mathbf{x}} d_x + \hat{\mathbf{y}} d_y$, thus $\nabla\times$ is approximated by $(1/\Delta s)\mathbf{d}\times$.

In order to cast Eq. (3.2) into an equivalent standard finite difference form, we approximate $\Delta t \partial_t \mathbf{D}$ by $d_t \mathbf{D}$, where the latter is given by

$$d_t \mathbf{D} = \mathbf{D}^{n+1} - \mathbf{D}^n. \tag{3.11}$$

Now, using Eqs. (3.2), (3.11) and (3.8), the $\mathbf{E}$ update equation becomes

$$\mathbf{E}^{n+1} = \mathbf{E}^n - \left(\mathbf{\Psi}^{n+1} - \mathbf{\Psi}^n \right) + \frac{\Delta t}{\varepsilon_0\, \Delta s} \mathbf{d} \times \mathbf{H}^{n+1/2}, \tag{3.12}$$

where $\mathbf{\Psi}^{n+1} - \mathbf{\Psi}^n$ is given by a recursive relation (see Eq. (3.29)). In Eq. (3.12), $\mathbf{d}$ represents the standard, second-order accurate, central finite difference operator same as in Eq. (3.10).

In the non-dispersive case, let us assume that the material to be simulated has a complex permittivity $\tilde{\varepsilon}$, given by

$$\tilde{\varepsilon} = \varepsilon_0 \varepsilon_r + i \varepsilon_0 \varepsilon_i, \tag{3.13}$$

where ε_r and ε_i are respectively the real and imaginary parts of the relative complex permittivity $(\tilde{\varepsilon}/\varepsilon_0)$. We define weakly conducting dielectrics as materials for which ε_r is positive and $\varepsilon_i \ll \varepsilon_r$. In this case, instead of using Eq. (3.4), we use Eq. (3.13) directly to replace ε in Eq. (3.2). Furthermore, assuming harmonic time dependence for the electric fields we can rearrange Eq. (3.2) as follows:

$$\varepsilon_0 \varepsilon_r \partial_t \mathbf{E}(\mathbf{r},t) = \nabla \times \mathbf{H}(\mathbf{r},t) - \varepsilon_0 \varepsilon_i \omega \mathbf{E}(\mathbf{r},t). \tag{3.14}$$

The update equation for the **E** field corresponding to Eq. (3.14) is given by

$$\mathbf{E}^{n+1} = \left(\frac{1-a}{1+a}\right) \mathbf{E}^n + \frac{\Delta t}{\varepsilon_0 \varepsilon_r \Delta s} \left(\frac{1}{1+a}\right) \mathbf{d} \times \mathbf{H}^{n+1/2}, \tag{3.15}$$

where

$$a = \frac{\omega\, \varepsilon_i}{2\varepsilon_r} \Delta t. \tag{3.16}$$

The **E** update equation given in Eq. (3.15) is stable as long as $\varepsilon_r > 0$. The results are reliable if $\varepsilon_i \ll \varepsilon_r$.

In the case of a non-conducting pure dielectric (relative real permittivity, (ε), putting $a = 0$ in Eq. (3.15) **E** field update equation becomes

$$\mathbf{E}^{n+1} = \mathbf{E}^n + \frac{\Delta t}{\varepsilon_0 \varepsilon \Delta s} \mathbf{d} \times \mathbf{H}^{n+1/2}, \tag{3.17}$$

The nonstandard form of Eqs. (3.15) and Eq. (3.17) are given in section 4.1.

TM mode: FDTD for wave equation

In TM mode, light is polarized normal to the plane of incidence, and using Eqs. (3.1) and (3.2), Maxwell's equations reduce to the wave equation of the following form:

$$\mu_0 \partial_t^2 \mathbf{D} = \nabla^2 \mathbf{E} - \nabla (\nabla \cdot \mathbf{E}). \tag{3.18}$$

Highly conducting metals can be regarded as electrically neutral because any accumulation of net charge is soon canceled, thus $\nabla \cdot \mathbf{D} = 0$ [31]. In a linear medium $\mathbf{D} = \varepsilon \mathbf{E}$, hence $\nabla \cdot \mathbf{D} = 0$ implies that $\varepsilon(\nabla \cdot \mathbf{E}) + \mathbf{E} \cdot (\nabla \varepsilon) = 0$. In a uniform medium where ε is a constant with respect to position, $\nabla \varepsilon = 0$ implying $\nabla \cdot \mathbf{E} = 0$ and (3.18) reduces to

$$\mu_0 \partial_t^2 \mathbf{D} = \nabla^2 \mathbf{E}. \tag{3.19}$$

At the interface between two different media, $\nabla \varepsilon \neq 0$ and hence Maxwell's equations do not reduce to (3.18) in general. In the TM mode, however, $\nabla \varepsilon$ lies in the x-y plane, and **E** is normal to it, thus $\mathbf{E} \cdot (\nabla \varepsilon) = 0 \Rightarrow \nabla \cdot \mathbf{E} = 0$ and (3.18) reduces to (3.19) even when $\nabla \varepsilon \neq 0$.

To construct the finite-difference model in the TM mode, we replace the second derivatives of **E** and **D** in Eq. (3.19) by equivalent finite difference operators in space and time domains respectively. Denoting the equivalent finite difference operator for second time derivative by $d_t^2\mathbf{D}$, we find

$$d_t^2\mathbf{D} = \mathbf{D}^{n+1} + \mathbf{D}^{n-1} - 2\mathbf{D}^n. \tag{3.20}$$

The above form is obtained by evaluating $d_t^2\mathbf{D}$ as $d_t^2\mathbf{D} = d_t(d_t\mathbf{D})$, and by applying Eq. (3.11) twice consecutively.

Finally, the $\mathbf{E}$ update equation is obtained using Eq. (3.8) and is given by

$$\begin{aligned}\mathbf{E}^{n+1} &= 2\mathbf{E}^n - \mathbf{E}^{n-1} \\ &- \left(\boldsymbol{\Psi}^{n+1} + \boldsymbol{\Psi}^{n-1} - 2\boldsymbol{\Psi}^n\right) + \frac{\Delta t^2}{\mu_0\varepsilon_0\Delta s^2}d^2\mathbf{E}^n,\end{aligned} \tag{3.21}$$

where we used the relation $\nabla^2 f \cong d^2 f/\Delta s^2$. The finite-difference operator $d^2 = \mathbf{d}\cdot\mathbf{d} = d_x^2 + d_y^2$, where d_x^2 and d_y^2 obtained using equations similar to Eq. (3.20) in space domain. In the above, $\boldsymbol{\Psi}^{n+1} + \boldsymbol{\Psi}^{n-1} - 2\boldsymbol{\Psi}^n$ is computed from a recursive relation (*see* Eq. (3.30)).

Like the previous section, we use a non-dispersive FDTD to update the $\mathbf{E}$ field in TM mode in a weakly conducting dielectric having a complex permittivity ($\tilde{\varepsilon}$) defined by Eq. (3.13), where ε_r is positive and $\varepsilon_i \ll \varepsilon_r$. In this case, the appropriate form of the wave equation follows from Eqs. (3.1) and (3.14). The corresponding $\mathbf{E}$ update equation is given by

$$\mathbf{E}^{n+1} = \mathbf{E}^n + \left(\frac{1-a}{1+a}\right)\left[\mathbf{E}^n - \mathbf{E}^{n-1}\right] + \frac{\Delta t^2}{\mu_0\varepsilon_0\varepsilon_r\Delta s^2\,(1+a)}\,d^2\mathbf{E}^n, \tag{3.22}$$

a in Eq. (3.22) is the same as that given in Eq. (3.16).

In case of a non-conducting pure dielectric (relative real permittivity, ε), putting $a = 0$ in Eq. (3.22) $\mathbf{E}$ field update equation becomes

$$\mathbf{E}^{n+1} = 2\mathbf{E}^n - \mathbf{E}^{n-1} + \frac{\Delta t^2}{\mu_0\varepsilon_0\varepsilon\Delta s^2}\,d^2\mathbf{E}^n, \tag{3.23}$$

Nonstandard versions of Eqs. (3.22) and Eq. (3.23) are given in section 4.1.

Recursive relation for the accumulation variable

Let us write Eq. (3.9) as

$$\boldsymbol{\Psi}^n = \sum_{m=0}^{n-1}\mathbf{E}^{n-m}\chi^m, \tag{3.24}$$

where

$$\chi^m = \frac{\omega_{pl}^2}{\nu_c^2}\left[\nu_c\Delta t + e^{-m\nu_c\Delta t}\left(e^{-\nu_c\Delta t} - 1\right)\right]. \tag{3.25}$$

From Eq. (3.25), we derive following recursive relation for χ^{m+1},

$$\chi^{m+1} = c_1 + c_2\chi^m, \tag{3.26}$$

where

$$c_1 = \frac{\omega_{pl}^2 \Delta t}{\nu_c}\left(1 - e^{-\nu_c \Delta t}\right), \text{ and } c_2 = e^{-\nu_c \Delta t}, \tag{3.27}$$

$$\mathbf{\Psi}^0 = 0, \mathbf{\Psi}^1 = \mathbf{E}^1 \chi^0, \mathbf{\Psi}^2 = \mathbf{E}^2 \chi^0 + c_1 \mathbf{E}^1 + c_2 \mathbf{\Psi}^1. \tag{3.28}$$

Hence, we find the following recursive relations

$$\begin{aligned} \mathbf{\Psi}^{n+1} - \mathbf{\Psi}^n &= \chi^0 \left[\mathbf{E}^{n+1} - \mathbf{E}^n\right] \\ &+ c_1 \left[\mathbf{E}^n\right] + c_2 \left[\mathbf{\Psi}^n - \mathbf{\Psi}^{n-1}\right] \end{aligned} \tag{3.29}$$

$$\begin{aligned} \mathbf{\Psi}^{n+1} + \mathbf{\Psi}^{n-1} - 2\mathbf{\Psi}^n &= \chi^0 \left[\mathbf{E}^{n+1} + \mathbf{E}^{n-1} - 2\mathbf{E}^n\right] \\ &+ c_1 \left[\mathbf{E}^n - \mathbf{E}^{n-1}\right] + c_2 \left[\mathbf{\Psi}^n + \mathbf{\Psi}^{n-2} - 2\mathbf{\Psi}^{n-1}\right] \end{aligned} \tag{3.30}$$

Now, $\chi^0 = 0$ (from Eq. 3.6), and we set values of $\mathbf{\Psi}$ for all $n \leq 0$ to be 0.

Calculating c_1, c_2 using the complex permittivities of metals

Equation (3.27) shows that c_1 and c_2 are functions of ω_{pl} and ν_c. In simulations presented here we adopted two approaches. In the first approach we computed c_1 and c_2 at each frequency using the tabulated values of ε. In the second approach we determined a set of best-fit Drude parameters by fitting the tabulated permittivities to the first-order Drude model over the entire range of visible frequencies.

Let us assume that $\bar{\bar{\varepsilon}}$ $(= \bar{\varepsilon}_r + i\bar{\varepsilon}_i)$ represents the tabulated value of the permittivity at a given frequency $(\omega = \bar{\omega})$. If $\bar{\omega}_{pl}$ and $\bar{\nu}_c$ represent the values of ω_{pl} and ν_c at $\omega = \bar{\omega}$, then $\bar{\omega}_{pl}$ and $\bar{\nu}_c$ can be expressed in terms of $\bar{\varepsilon}_r$, $\bar{\varepsilon}_i$ using Eq. (3.5). Separating the real and imaginary parts in Eq. (3.5) and rearranging we find following expressions for $\bar{\nu}_c$ and $\bar{\omega}_{pl}$:

$$\bar{\nu}_c = -\frac{\bar{\varepsilon}_i \bar{\omega}}{1 - \bar{\varepsilon}_r} \tag{3.31}$$

$$\bar{\omega}_{pl} = \sqrt{-\frac{\bar{\omega}(\bar{\nu}_c^2 + \bar{\omega}^2)\bar{\varepsilon}_i}{\bar{\nu}_c}} \tag{3.32}$$

Thus at $\omega = \bar{\omega}$, c_1, c_2 are computed using Eq. (3.27) as

$$c_1 = \frac{\bar{\omega}_{pl}^2 \Delta t}{\bar{\nu}_c}\left(1 - e^{-\bar{\nu}_c \Delta t}\right), \text{ and } c_2 = e^{-\bar{\nu}_c \Delta t} \tag{3.33}$$

In the second approach the best-fit Drude parameters ($\hat{\omega}_{pl}$ and $\hat{\nu}_c$) are obtained by minimizing a merit function given by

$$\phi = \frac{1}{N}\sqrt{\sum_{j=1}^{N}\left(\varepsilon_{r,j} - \bar{\varepsilon}_{r,j}\right)^2 + \sum_{j=1}^{N}\left(\varepsilon_{i,j} - \bar{\varepsilon}_{i,j}\right)^2}, \tag{3.34}$$

where $\mathbf{N}$ is the total number of frequency points. $\bar{\varepsilon}_{r,j}$ and $\bar{\varepsilon}_{i,j}$ are respectively the real and imaginary parts of tabulated permittivity value at the jth frequency point.

$\varepsilon_{r,j}$ and $\varepsilon_{i,j}$ are given by

$$\varepsilon_{r,j} = 1 - \frac{\omega_{pl}^2}{\nu_c^2 + \omega_j^2} \tag{3.35}$$

$$\varepsilon_{i,j} = -\frac{\omega_{pl}^2 \nu_c}{(\nu_c^2 + \omega_j^2)\omega_j} \tag{3.36}$$

The minimization is carried out using a genetic algorithm (GA) scheme. GA is a stochastic intelligent search method. In our implementation the algorithm generates a random population of chromosomes and mimics biological evolution process through a given number of generations. The evolution process is facilitated by a combination of a tournament selection scheme and a uniform crossover scheme. The details pertaining to this version of GA can be found in Ref. [32]. The GA parameters chosen are as follows: population size of chromosomes = 500, number of generations = 100, tournament size = 5, tournament probability = 0.8, crossover probability = 0.8, mutation probability = 0.01. With these parameters the minimum value of ϕ comes out to be 0.2958.

c_1, c_2 are computed using Eq. (3.27) as

$$c_1 = \frac{\hat{\omega}_{pl}^2 \Delta t}{\hat{\nu}_c} \left(1 - e^{-\hat{\nu}_c \Delta t}\right), \text{ and } c_2 = e^{-\hat{\nu}_c \Delta t} \tag{3.37}$$

Normalization of FDTD related parameters

The parameters λ, ω, Δs, and Δt used in the FDTD formulations above are normalized in the following manner. Throughout this manuscript we use a subscript p with a symbol to represent the physical value of the parameter associated with the symbol. The same symbol without a subscript represents the normalized value of the same parameter as used in FDTD formulations.

In this normalization scheme $\Delta s = \Delta t = 1$. The physical values of the minimum space and time steps are denoted by symbols Δs_p and Δt_p respectively. The number of grid points needed to represent one wavelength on the grid is obtained by taking the ratio of the physical wavelength (λ_p) to Δs_p, i.e., $\lambda = \lambda_p / \Delta s_p$, λ being the number of grid points needed to represent one wavelength. The speed of the wave v on the FDTD grid is decided by the stability requirement of the algorithm. Hence the normalized frequency ω is given by

$$\omega = \frac{2\pi v}{\lambda}. \tag{3.38}$$

We find that ω is related to ω_p by

$$\omega = \frac{\lambda_p v}{\lambda v_p} \omega_p \tag{3.39}$$

To evaluate Eq. (3.37) we use the physical values of $\hat{\omega}_{pl}$, $\hat{\nu}_c$ and Δt.

4.3.3 Algorithm stability

Following Ref. [6], we show that the algorithm stability critically depends on the choice of c_1 and c_2. The stability analysis shows that choosing c_1 and c_2 using tabulated permittivity values may or may not make the algorithm stable.

TE mode

In TE mode the update equations for $\mathbf{E}$ and $\boldsymbol{\Psi}$ (Eqs. (3.12) and (3.29)) can be rearranged using definitions of finite-difference time operators as follows

$$\mathbf{E}^{n+1} = \mathbf{E}^n - \left(d_t\boldsymbol{\Psi}^{n+1/2}\right) + \frac{\Delta t}{\varepsilon_0 \Delta s}\mathbf{d} \times \mathbf{H}^{n+1/2} \tag{3.40}$$

$$d_t\boldsymbol{\Psi}^{n+3/2} = c_1\left[\mathbf{E}^{n+1}\right] + c_2\left[d_t\boldsymbol{\Psi}^{n+1/2}\right] \tag{3.41}$$

The $\mathbf{H}$ update equation (Eq. (3.10)) together with Eqs. (3.40) and (3.41) give the field updates for the algorithm. We can cast the field update equations into a compact form given by

$$\mathbf{F}^{n+1} = \mathbf{M}\mathbf{F}^n + \mathbf{N}\mathbf{F}^{n+1}, \tag{3.42}$$

where

$$\mathbf{F}^{n+1} = \begin{bmatrix} \mathbf{H}^{n+1/2} \\ \mathbf{E}^{n+1} \\ d_t\boldsymbol{\Psi}^{n+3/2} \end{bmatrix}, \text{ and } \mathbf{F}^n = \begin{bmatrix} \mathbf{H}^{n-1/2} \\ \mathbf{E}^{n} \\ d_t\boldsymbol{\Psi}^{n+1/2} \end{bmatrix}.$$

$\mathbf{M}$ and $\mathbf{N}$ are coefficient matrices. Solving (3.42) for $\mathbf{F}^{n+1}$ we find

$$\mathbf{F}^{n+1} = (\mathbf{I} - \mathbf{N})^{-1}\mathbf{M}\mathbf{F}^n, \tag{3.43}$$

where $\mathbf{I}$ is the identity matrix. The elements of $\mathbf{M}$ and $\mathbf{N}$ are determined below.

For brevity, in Eq. (3.10), we write $\mathbf{p} = \frac{\Delta t}{\Delta s}\frac{1}{\mu_0}(\mathbf{d}\times)$. Here, $\mathbf{p}$ is a vector finite-difference operator. The notation $\mathbf{pE}$ does not indicate multiplication but rather it denotes vector operator $\mathbf{p}$ operating on $\mathbf{E}$.

Similarly, in Eq. (3.40), we write $\mathbf{q} = \frac{1}{\varepsilon_0}\frac{\Delta t}{\Delta s}(\mathbf{d}\times)$. Note that, $\mathbf{q}$ is a vector finite difference operator. As above, the symbol $\mathbf{qH}$ represents the operation of $\mathbf{q}$ on $\mathbf{H}$.

Following Eqs. (3.10), (3.40) and (3.41), we find $\mathbf{M}$ and $\mathbf{N}$ as given below

$$\mathbf{M} = \begin{bmatrix} 1 & -\mathbf{p} & 0 \\ 0 & 1 & -1 \\ 0 & 0 & c_2 \end{bmatrix} \tag{3.44}$$

$$\mathbf{N} = \begin{bmatrix} 0 & 0 & 0 \\ \mathbf{q} & 0 & 0 \\ 0 & c_1 & 0 \end{bmatrix}. \tag{3.45}$$

For brevity, in Eq. (3.43) we write

$$\mathbf{C} = (\mathbf{I} - \mathbf{N})^{-1}\mathbf{M}, \tag{3.46}$$

Using Eq. (3.46), it follows from Eq. (3.43) that

$$\mathbf{F}^n = \mathbf{C}^n \mathbf{F}^0 \tag{3.47}$$

If L_j and l_j are the eigenvectors and eigenvalues of $\mathbf{C}$, where $j = 1, ..., 3$, making use of linear systems theory, we obtain following expressions

$$\mathbf{F}^0 = a_1 L_1 + a_2 L_2 + a_3 L_3 \tag{3.48}$$

$$\mathbf{F}^n = a_1 l_1^n L_1 + a_2 l_2^n L_2 + a_3 l_3^n L_3 \tag{3.49}$$

where a_i are constants. For the stability of the algorithm (3.49) we need to ensure that $|l_j| \leq 1$. Hence, to ensure stability of the algorithm, the values of the constants c_1 and c_2 at each wavelength must be chosen such that the eigenvalues of $\mathbf{C}$, are ≤ 1.

In the above derivation, $\mathbf{F}$ has three components. In general, however, $\mathbf{E}$ and $d_t\mathbf{\Psi}$ both have two components (x and y) each. This is the case when light does not propagate along one of the coordinate axes. In this general case $\mathbf{F}$ has five components and accordingly Eqs. (3.48) and (3.49) have to be modified so as to yield five eigenvalues. All these five eigenvalues must be lesser and equal to unity to ensure the algorithm stability. However, if the light propagates along one of the coordinate axes, for example, in case of normal incidence, transmission and reflection, it is enough to consider only one component each of $\mathbf{E}$ and $d_t\mathbf{\Psi}$. This simplification reduces the number of eigenvalues to just three.

In this mode only the z-components of $\mathbf{H}$ and $(\mathbf{pE})$ are nonzero.
Hence, the z-component of $(\mathbf{pE})$ is given by

$$(\mathbf{pE})_z = \frac{1}{\mu_0} \frac{\Delta t}{\Delta s} \left(d_x E_y - d_y E_x \right). \tag{3.50}$$

In general, $\mathbf{qH}$ has two components in TE mode, which we denote as $(\mathbf{qH})_x$ and $(\mathbf{qH})_y$. $(\mathbf{qH})_x$ and $(\mathbf{qH})_y$ correspond to E_x and E_y components respectively and are given by

$$(\mathbf{qH})_x = \frac{1}{\varepsilon_0} \frac{\Delta t}{\Delta s} \left(\mathbf{d} \times H \right)_x = \frac{1}{\varepsilon_0} \frac{\Delta t}{\Delta s} (d_y H_z), \tag{3.51}$$

$$(\mathbf{qH})_y = \frac{1}{\varepsilon_0} \frac{\Delta t}{\Delta s} \left(\mathbf{d} \times H \right)_y = -\frac{1}{\varepsilon_0} \frac{\Delta t}{\Delta s} (d_x H_z). \tag{3.52}$$

Note that, the quantities $(\mathbf{pE})_z$, $(\mathbf{qH})_x$ and $(\mathbf{qH})_y$ are all scalars.

To compute the eigenvalues (l_j, $j = 1, 2, 3$), we need to evaluate $(\mathbf{pE})_z$, $(\mathbf{qH})_x$ and $(\mathbf{qH})_y$ numerically. For that purpose, we take E_x, E_y and H_z to be infinite plane waves of the form $E_x = E_{0x}\, e^{i(\mathbf{k}\cdot\mathbf{r} - \omega t)}$, $E_y = E_{0y}\, e^{i(\mathbf{k}\cdot\mathbf{r} - \omega t)}$ and $H_z = H_{0z}\, e^{i(\mathbf{k}\cdot\mathbf{r} - \omega t)}$. Evaluating $d_m G$, where $(m = x, y)$ and $G = E_x$, E_y or H_z we find,

$$d_m G = \left(e^{ik_m \Delta s/2} - e^{-ik_m \Delta s/2} \right) G, \tag{3.53}$$

In Eq. (3.53), the x and y components of propagation constant k_x and k_y, are found to be

$$k_x = \frac{2\pi\sqrt{\tilde{\varepsilon}}}{\lambda}\cos(\theta) \tag{3.54}$$

$$k_y = \frac{2\pi\sqrt{\tilde{\varepsilon}}}{\lambda}\sin(\theta), \tag{3.55}$$

where θ is the angle propagation direction makes with x-axis.

TM mode

In TM mode, the update equations for $\mathbf{E}$ and $\boldsymbol{\Psi}$ (Eqs. (3.21) and (3.30)) can be rearranged using the definitions of finite-difference time operators as follows:

$$d_t^2\boldsymbol{\Psi}^n = c_1\left[d_t\mathbf{E}^{n-1/2}\right] + c_2\left[d_t^2\boldsymbol{\Psi}^{n-1}\right] \tag{3.56}$$

$$\mathbf{E}^{n+1} = d_t\mathbf{E}^{n-1/2} - \left(d_t^2\boldsymbol{\Psi}^n\right) + (1+b)\mathbf{E}^n\,. \tag{3.57}$$

For brevity, we write $b = (\Delta t^2/\mu_0\varepsilon_0\Delta s^2)d^2$, in Eq. (3.57). Here, b is a scalar finite-difference operator and the notation bE_z indicates the operation of the operator b on z component of $\mathbf{E}$. Note that Eqs. (3.56) and (3.57) together can not be cast into a convenient form like Eq. (3.42) because of the presence of an unbalanced $\mathbf{E}^n$ on the right-hand side of Eq. (3.57). Hence we add the definition of $d_t\mathbf{E}^{n+1/2}$ with Eqs. (3.56) and (3.57) as follows:

$$d_t\mathbf{E}^{n+1/2} = \mathbf{E}^{n+1} - \mathbf{E}^n \tag{3.58}$$

Now Eqs. (3.56)–(3.58) can be written in a compact form such as Eq. (3.42), where $\mathbf{F}^{n+1}$, $\mathbf{F}^n$ $\mathbf{M}$ and $\mathbf{N}$ are given by

$$\mathbf{F}^{n+1} = \begin{bmatrix} d_t^2\boldsymbol{\Psi}^n \\ \mathbf{E}^{n+1} \\ d_t\mathbf{E}^{n+1/2} \end{bmatrix}, \tag{3.59}$$

$$\mathbf{F}^{n} = \begin{bmatrix} d_t^2\boldsymbol{\Psi}^{n-1} \\ \mathbf{E}^{n} \\ d_t\mathbf{E}^{n-1/2} \end{bmatrix}, \tag{3.60}$$

$$\mathbf{M} = \begin{bmatrix} c_2 & 0 & c_1 \\ 0 & 1+b & 1 \\ 0 & -1 & 0 \end{bmatrix}, \tag{3.61}$$

and

$$\mathbf{N} = \begin{bmatrix} 0 & 0 & 0 \\ -1 & 0 & 0 \\ 0 & 1 & 0 \end{bmatrix}. \tag{3.62}$$

We define a coefficient matrix $\mathbf{C}$ following Eq. (3.46), where $\mathbf{M}$ and $\mathbf{N}$ are given by Eqs. (3.61) and (3.62). To ensure stability of the algorithm, the values of the constants c_1 and c_2 used at each wavelength must be such that the eigenvalues of $\mathbf{C}$, are ≤ 1.

To compute the eigenvalues (l_j, $j = 1, 2, 3$), we need to evaluate bE_z numerically, as in TE mode, we take E_z as an infinite plane wave of the form $E_z = E_{0z}\, e^{i(\mathbf{k}\cdot\mathbf{r}-\omega t)}$. We find

$$d^2E_z = \left[\left(e^{ik_x\Delta s} + e^{-ik_x\Delta s} - 2\right) + \left(e^{ik_y\Delta s} + e^{-ik_y\Delta s} - 2\right)\right] E_z \tag{3.63}$$

Note that, in this mode only the z-component of $\mathbf{E}$ i.e., E_z is nonzero. Hence $\mathbf{E}$ can be replaced by a scalar quantity E_z. Hence, bE_z in Eq. (3.57) can be evaluated as follows:

$$\begin{aligned} bE_z &= \frac{\Delta t^2}{\mu_0\varepsilon_0\Delta s^2} \times \\ &\times \left[\left(e^{ik_x\Delta s} + e^{-ik_x\Delta s} - 2\right) + \left(e^{ik_y\Delta s} + e^{-ik_y\Delta s} - 2\right) \right] E_z\,. \end{aligned} \tag{3.64}$$

In Eq. (3.64), the x and y components of propagation constant k_x and k_y, are computed using Eqs. (3.54) and (3.55).

In the next section, we compute the eigenvalues (l_j, $j = 1, 2, 3$) of $\mathbf{C}$, for various simulation wavelengths in TM and TE modes. These eigenvalues are computed using the c_1 and c_2 values obtained at each wavelength using the corresponding aluminum permittivity. We use Mathcad 11 built-in utilities to compute the eigenvalues.

4.3.4 Simulating subwavelength gratings

In this section, we present the modeling and simulation of subwavelength dielectric–metal or dielectric–metal–semiconductor gratings using a combination of dispersive and non-dispersive FDTD. The semiconductor in the second example is treated as a weakly conducting dielectric. In both cases gratings are realized by mounting a plurality of thin single- or double-layered strips of rectangular cross-sections, parallel to each other on a substrate of transparent material. The material between any two consecutive strips is a medium with permittivity 1. In practice, the substrate layer is made from either glass or transparent polymers like PMMA (polymethylmethacrylate) with permittivity close to 2.25. For simulation purposes, the permittivity of the substrate region is taken to be 2.25. The thickness of the substrate region is very large compared to λ_p. Hence in the simulation we incorporate a substrate with infinite thickness by ending the substrate on the left-hand side with an absorbing boundary condition.

Fig. 4.12 shows a schematic representation of FDTD simulation space employed to simulate the dielectric–metal grating. The figure depicts four periods of a metal grating (an ordinary WGP) on a FDTD spatial grid. Each period contains one rectangular metallic part that represents a cross-section of a metal strip in a plane normal to the long axes of the strips. Each of the metal rectangles is surrounded on three sides by a medium with permittivity 1. On the fourth side the rectangles are in contact with the substrate. The incident beam is generated by a pair of

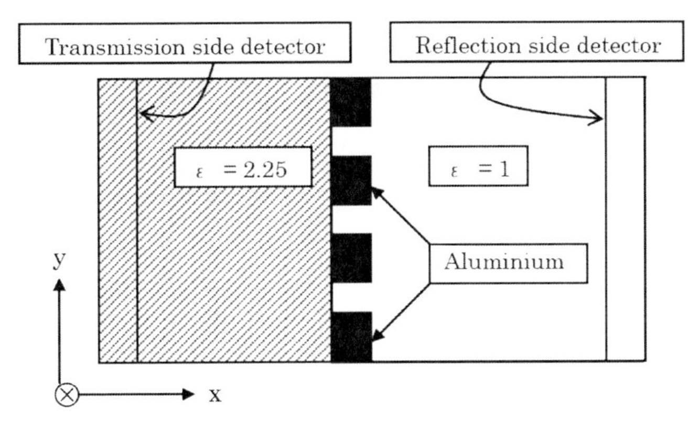

Fig. 4.12. Schematic diagram of the FDTD simulation space for the single-layer (aluminum) WGP.

linear phased current source arrays placed in the medium of permittivity 1. Along the propagation direction ($-x$-axis), the simulation domain is terminated using the nonstandard Mur absorbing boundary condition on both sides [33]. The other two sides of the simulation domain are terminated with a periodic boundary. The gratings and the current sources extend from one end to the other of the simulation domain. Periodic boundary conditions on the top and bottom ends of the source and grating emulate an infinitely periodic grating and infinite plane incident wave for practical purposes. The material of the grating is taken to be aluminum. The tabulated data for complex permittivities of aluminum as a function of frequency are obtained from Ref. [30].

The field values at the grid points on or inside the boundaries of the metal rectangles are updated using RC-FDTD (Eqs. (3.12) and (3.21)). The fields at grid points outside the metal rectangles are updated using the non-dispersive FDTD for non-conducting media (Eqs. (3.17) and (3.23)).

Figure 4.13 shows a schematic diagram of the FDTD simulation space for a double-layered grating (an absorbing WGP). The figure depicts four periods of the grating. Each individual grating line is a composite of two layers; the one in direct contact with the substrate is aluminum while the one on top of the aluminum portion is silicon. The total height of any grating line is h, the thickness of the silicon layer is δ, and the thickness of the aluminum layer is $h - \delta$. The field values at the grid points on or inside the boundary of the silicon layer are updated using non-dispersive FDTD for weakly conducting dielectrics (Eqs. (3.15) and (3.22)). The frequency spectrum of complex permittivities of silicon is obtained from Ref. [34].

The incident light propagates along the $-x$-axis in case of normal incidence. The physical value of the spatial discretization (Δs_p) of the FDTD mesh is 4 nm. The width of the simulation domain along the y-axis is 4 grating periods. Hence the total number of grid points in the y-direction (N_y) is given by $4\Lambda/\Delta s_p$. The length of the simulation domain along the x-direction is fixed so that both the reflected and transmitted fields propagate through a distance of about 7λ–8λ before reaching the detectors where the steady-state field values are recorded and used to compute

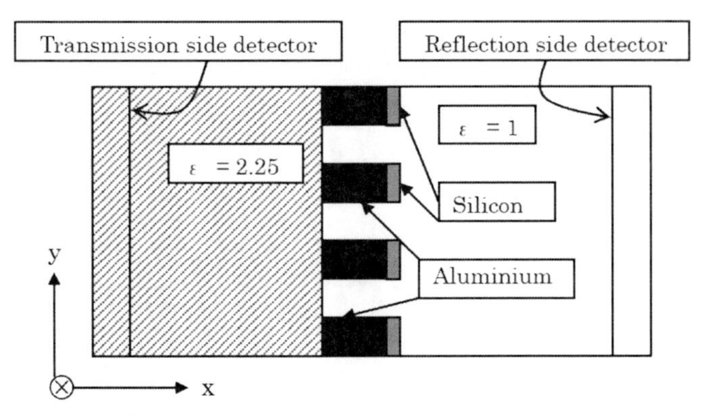

Fig. 4.13. Schematic diagram of the FDTD simulation space for the double-layered (aluminum + silicon) WGP.

the diffraction efficiency. The grid equivalent of the detector to grating surface distance (D) is exactly 7λ when $\lambda = 700$ nm $/\Delta s_p$. D at any other wavelength (λ) is given by

$$D = \frac{7 \times 700}{\lambda \Delta s_p}. \tag{3.65}$$

The choice of D ensures that the near-fields die out completely before the propagating waves reach the detectors.

Both for the single- and double-layered gratings, the reflectivity (R) and transmissivity (T) are computed for 16 wavelengths in the visible range (400 nm–700 nm), sampled at 20 nm interval. In general, the tabulated values of aluminum and silicon permittivities are specified at equal intervals of energy or frequency. Since frequency and wavelength are inversely related, the frequencies at which the permittivities are specified do not correspond to the simulation wavelengths. Hence, we use a cubic spline interpolation algorithm to fit the tabulated data and to determine the permittivity values corresponding to the simulation wavelengths (set 1). In this case the values of c_1 and c_2 at each wavelength are computed using Eq. (3.33) with $\bar{\bar{\varepsilon}}$ representing the permittivity value corresponding to the simulation wavelength. The set 2 permittivity data are obtained using a set of best-fit Drude parameters as described in section 4.3.2. The physical values of the best-fit Drude parameters are $\hat{\omega}_{pl} = 2.2637 \times 10^{16}\ s^{-1}$ and $\hat{\nu}_c = 11.99 \times 10^{13}\ s^{-1}$. c_1 and c_2 are computed using Eq. (3.37). Figure 4.14 shows a comparison between the wavelength spectra of the tabulated aluminum permittivities after interpolation (set 1) and the permittivities obtained using the best-fit Drude parameters (set 2).

The transmissivity (T) and reflectivity (R) computed at any particular wavelength give the total fraction of incident light that is directed in the forward (transmissivity) or backward (reflectivity) directions respectively. T and R are given by

$$T = \frac{I_t}{I_i}, \tag{3.66}$$

$$R = \frac{I_r}{I_i}, \tag{3.67}$$

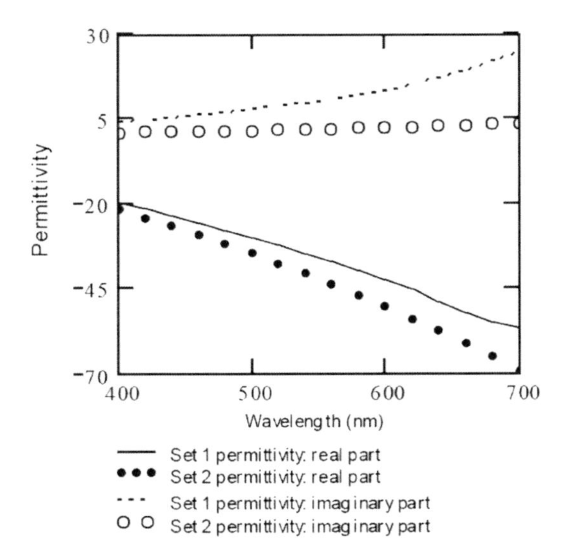

Fig. 4.14. Comparison between the wavelength spectra of two sets of complex permittivities of aluminum; set 1: the tabulated permittivities; set 2: the permittivities obtained using best-fit Drude parameters $\hat{\omega}_{pl} = 2.2637 \times 10^{16} s^{-1}$ and $\hat{\nu}_c = 11.99 \times 10^{13}\ s^{-1}$.

where I_t and I_r are the intensities associated with the zeroth-order transmitted and reflected waves respectively. I_i is the incident intensity. I_t is computed using the total field values recorded at the transmission side detector. However, I_r is obtained from scattered fields computed by subtracting the incident fields from the total field values at each grid point on the reflection side detector.

The incident light intensity (I_i) at each wavelength is computed by carrying out the same simulations without the grating and its substrate with other conditions remaining exactly the same as those done with the gratings present. The complex field data retrieved from the detector with the grating and one without for each wavelength are Fourier transformed separately using a complex fast Fourier transform (CFFT) algorithm (Mathcad 11). The CFFT in both cases show single peaks centred at $\theta = 0$, θ being the diffraction angle. The square modulus of the peak of the Fourier spectrum of the fields obtained with the grating is normalized by the square modulus of the peak of the Fourier spectrum of the fields obtained from simulation without the grating to yield reflectivity and transmissivity.

The real and imaginary parts of the complex field are computed using the nonstandard finite time differences at each grid point. Let us assume that the field $\mathbf{E}$ generated by FDTD has following form: $\mathbf{E} = \mathbf{A}(\cos(\omega t) + i \sin(\omega t)) + \mathbf{B}$. Here $\mathbf{B}$ is the dc part of the signal and $\mathbf{A}$ is the amplitude. To eliminate the dc part, we employ the $\mathbf{E}$ field values generated at three consecutive time steps. In this scheme the real part is given by

$$E_{re} = \frac{\mathbf{E}^{n+1} - \mathbf{E}^{n-1}}{2\,\Delta t_{NS}}, \tag{3.68}$$

where Δt_{NS} is the nonstandard time step and is given by [4]

$$(\Delta t)_{NS} = \frac{\sin(\omega \Delta t/2)}{\omega \Delta t/2}. \tag{3.69}$$

$\mathbf{E}^{n-1}$, $\mathbf{E}^n$ and $\mathbf{E}^{n+1}$ are the electric fields evaluated at $(n-1)$th, nth and $(n+1)$th time steps respectively. Since the imaginary part of the $\mathbf{E}$ field (E_{im}) can be generated by differentiating the real part of the $\mathbf{E}$ field, E_{im} is given by

$$E_{im} = \frac{\mathbf{E}^{n+1} + \mathbf{E}^{n-1} - 2\,\mathbf{E}^n}{[\Delta t_{NS}]^2}. \tag{3.70}$$

The simulations are performed at each wavelength (λ) using following FDTD parameters: $\Delta s_p = 4$ nm, $\lambda = \lambda_p/\Delta s_p$, $v = 0.175$ (Courant limit for standard, non-dispersive FDTD in two dimensions is 0.7), $\varepsilon_0 = 1$, $\mu_0 = 1/v^2$, and $\Delta t_p = 0.00933$ fs. Hence number of grid points varies from 100 at $\lambda_p = 400$nm to 175 at $\lambda_p = 700$ nm. Δs_p is chosen from two considerations. Firstly, Δs_p should be small enough to represent the smallest features of the structure adequately on the grid. Furthermore, since the choice of Δs and v affects the stability of the algorithm, these quantities should be chosen to ensure both stability of the algorithm and the accuracy of the results. As an example, we compute the eigenvalues with $\Delta s_p = 4$ nm, $v = 0.7$ and $v = 0.175$. At each v, eigenvalues are computed for two values of λ_p, namely, 400 nm and 700 nm in both TM and TE polarization modes. Tabulated permittivities (set 1) are employed to compute these eigenvalues. The results of this computation are shown in Tables 4.4 and 4.5. Tables 4.4 and 4.5 show that both in TE and TM modes, $l_1 > 1$ with $v = 0.7$ and l_1 increases with increasing λ_p. The eigenvalues in Tables 4.4 and 4.5 indicate that to make the algorithm perfectly stable, one might need to lower the values of Δs_p and v infinitely. However, choosing Δs_p and v to be sufficiently small we can make the algorithm practically stable, yielding reliable results at the same time.

To find out the total number of time steps needed to reach the steady-state we studied the time variation of the $\mathbf{E}$ field intensity (I) at a point 10 grid units away from the left hand side boundary of the simulation space at $\lambda_p = 540$ nm. The results are plotted in Fig. 4.15. I at the nth point is computed by averaging the $\mathbf{E}$ field intensity over two time periods (τ). I is given by

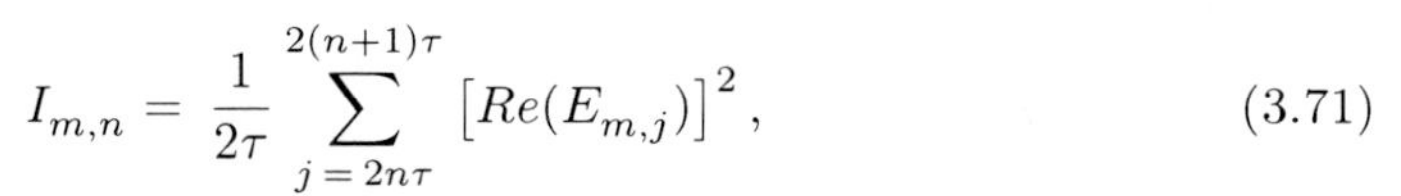

$$I_{m,n} = \frac{1}{2\tau} \sum_{j=2n\tau}^{2(n+1)\tau} \left[Re(E_{m,j})\right]^2, \tag{3.71}$$

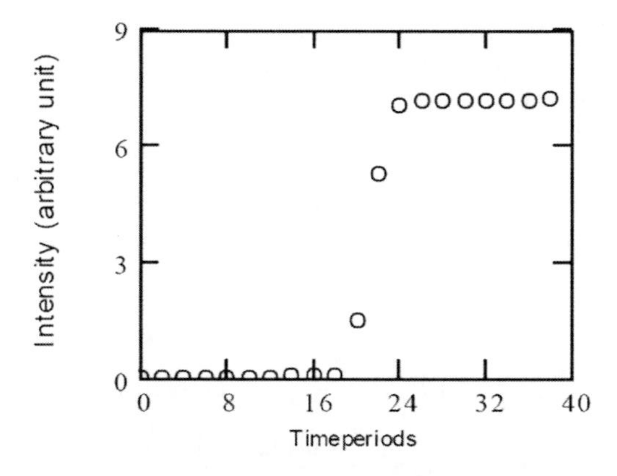

Fig. 4.15. Time variation of intensity over 40 time periods.

Table 4.4. Stability eigenvalues as functions of λ_p and v in TE mode.

v	λ_p (nm)	l_1	l_2	l_3
0.7	400	1.075	0.93	0.999
0.7	700	1.315	0.76	0.999
0.175	400	1.007	1.000	0.992
0.175	700	1.036	1.000	0.965

Table 4.5. Stability eigenvalues as functions of λ_p and v in TM mode.

v	λ_p (nm)	l_1	l_2	l_3
0.7	400	1.026	0.999	0.974
0.7	700	1.128	0.999	0.886
0.175	400	0.997	1.001	1.000
0.175	700	0.988	1.01	1.000

where $n = 0, 1, \ldots, 19$ and $m = y$ in TE mode and $m = z$ in TM mode respectively. τ is given by

$$\tau = \frac{2\pi}{\omega}. \tag{3.72}$$

ω in Eq. (3.72) is defined by Eq. (3.38). The figure indicates that steady state of intensity is reached in 40τ time steps and τ being a normalized quantity, the results hold for all wavelengths. All simulation results presented here are recorded at 40τ time steps.

Figures 4.16–4.19 show the far-field wavelength spectra of reflectivity and transmissivity for the ordinary single-layered aluminum WGP in TE and TM modes

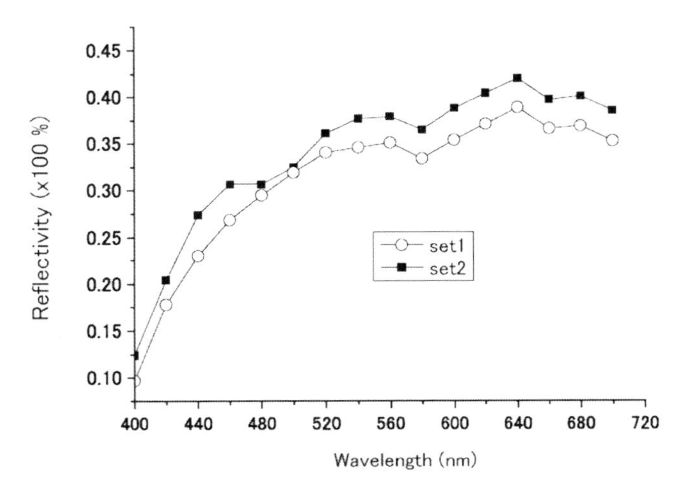

Fig. 4.16. Wavelength spectra of reflectivity for an aluminum WGP (Λ = 170 nm, h = 100 nm, f = 0.6) in TE mode, at normal incidence, simulated using two sets of permittivities for aluminum (set 1 and set 2 in Fig. 4.14).

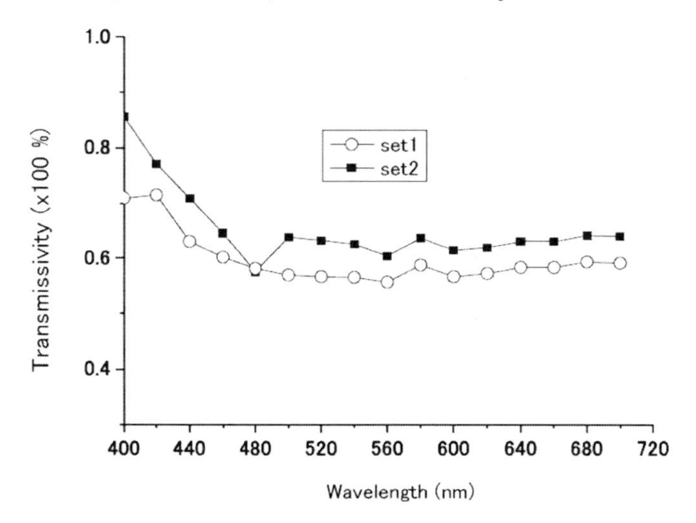

Fig. 4.17. Wavelength spectra of transmissivity for an aluminum WGP (Λ = 170 nm, h = 100 nm, f = 0.6) in TE mode, at normal incidence, simulated using two sets of complex permittivities for aluminum (set 1 and set 2 in Fig. 4.14).

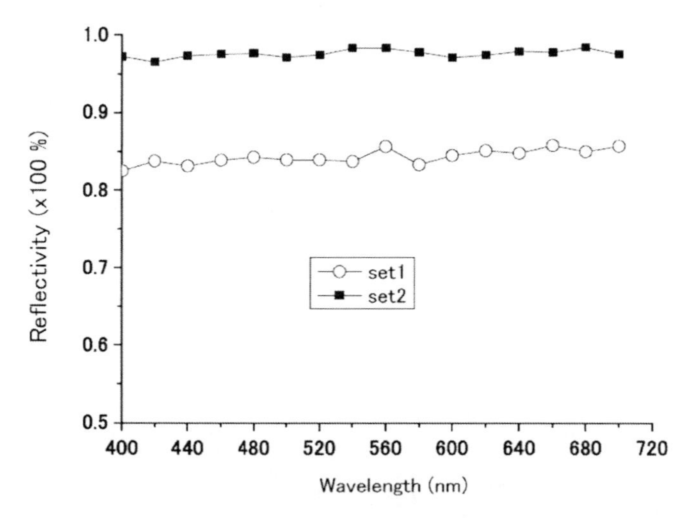

Fig. 4.18. Wavelength spectra of reflectivity for an aluminum WGP (Λ = 170 nm, h = 100 nm, f = 0.6) in TM mode, at normal incidence, simulated using two sets of complex permittivities for aluminum (set 1 and set 2 in Fig. 4.14).

respectively. In all cases rectangular profile gratings are used at normal incidence with following parameters: $\Lambda = 170$ nm, $h = 100$ nm, and $f = 0.6$.

The reflectivity at any wavelength in the TE mode of a large fill-factor WGP (Fig. 4.16) is larger than that of a low fill-factor WGP. The opposite is true for the TE mode transmissivity, that is, TE mode transmissivity of a large fill-factor WGP (Fig. 4.17) is actually lower than that of a low fill-factor WGP. The reflectivity or transmissivity at each wavelength depend on the choice of permittivity, but the overall wavelength dependence is similar for the two chosen sets of permittivities.

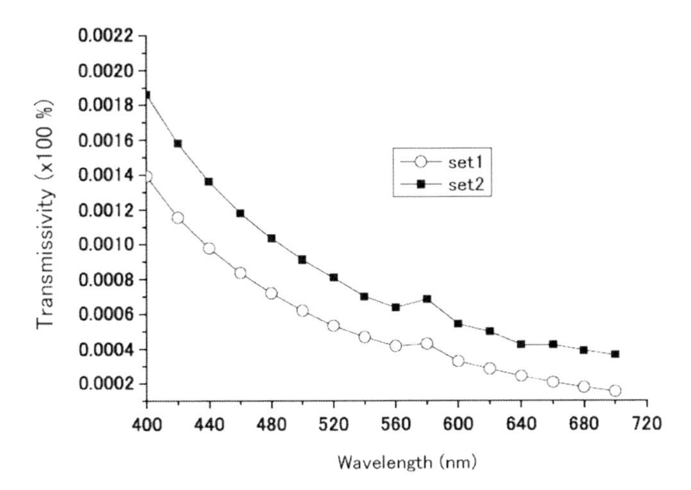

Fig. 4.19. Wavelength spectra of transmissivity for an aluminum WGP ($\Lambda = 170$ nm, $h = 100$ nm, $f = 0.6$) in TM mode, at normal incidence, simulated using two sets of complex permittivities for aluminum (set 1 and set 2 in Fig. 4.14).

The difference between two sets of reflectivity or transmissivity at any particular wavelength is no more than 15%. Similar observations can be made from TM mode reflectivity and transmissivity spectra (Figs. 4.18 and 4.19) except that the difference between the two sets of reflectivity at any wavelength can be as much as 20%.

Figures 4.20–4.23 show the far-field wavelength spectra of reflectivity and transmissivity for the double-layered aluminum–silicon absorbing WGP in TE and TM modes respectively. In all cases rectangular profile gratings are used at normal

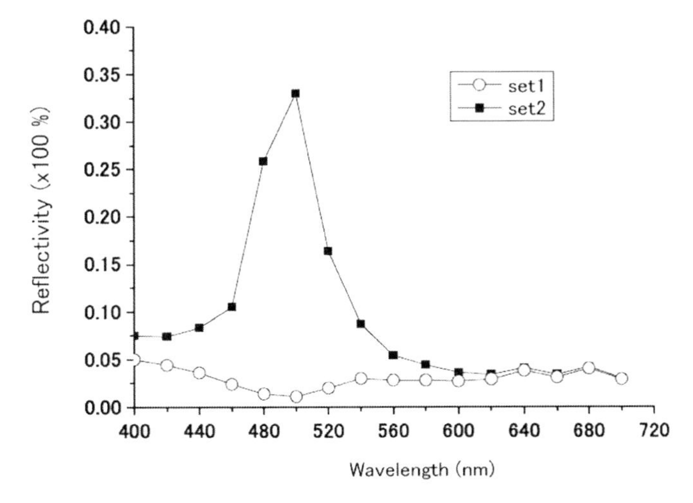

Fig. 4.20. Wavelength spectra of reflectivity for a part aluminum, part silicon WGP ($\Lambda = 155$ nm, $h = 220$ nm, $f = 0.36$, $\delta = 20$ nm) in TE mode, at normal incidence, simulated using two sets of complex permittivities (set 1 and set 2 in Fig. 4.14).

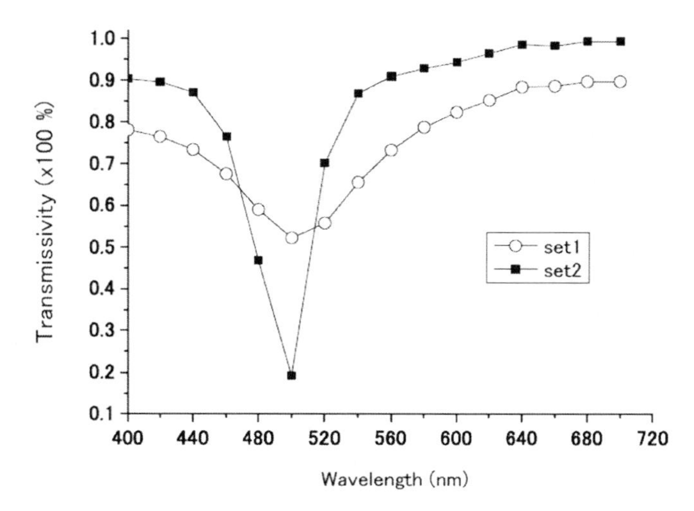

Fig. 4.21. Wavelength spectra of transmissivity for a part aluminum, part silicon WGP (Λ = 155 nm, h = 220 nm, f = 0.36, δ = 20 nm) in TE mode, at normal incidence, simulated using two sets of complex permittivities for aluminum (set 1 and set 2 in Fig. 4.14).

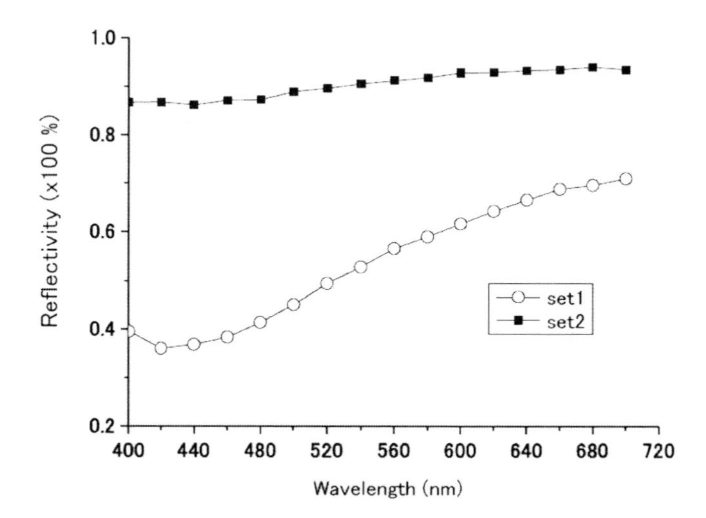

Fig. 4.22. Wavelength spectra of reflectivity for a part aluminum, part silicon WGP (Λ = 155 nm, h = 220 nm, f = 0.36, δ = 20 nm) in TM mode, at normal incidence, simulated using two sets of complex permittivities for aluminum (set 1 and set 2 in Fig. 4.14).

incidence with following parameters: Λ = 155 nm, h = 220 nm, f = 0.36, and δ = 20 nm.

Figures 4.20–4.23 indicate that both in TM and TE modes reflectivity and transmissivity spectra depend strongly on the choice of permittivity. Particularly in TE mode set 2 permittivities derived from the best-fit Drude parameters can show the existence of sharp maximum or minimum, which disappear or become less

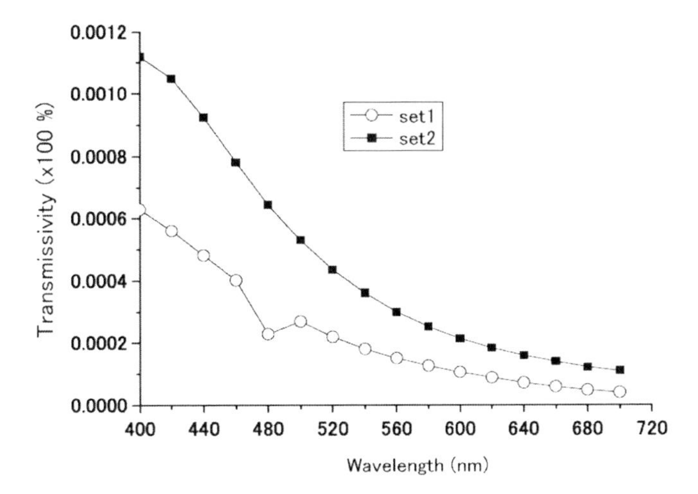

Fig. 4.23. Wavelength spectra of transmissivity for a part aluminum, part silicon WGP (Λ = 155 nm, h = 220 nm, f = 0.36, δ = 20 nm) in TM mode, at normal incidence, simulated using two sets of complex permittivities for aluminum (set 1 and set 2 in Fig. 4.14).

sharp if the exact tabulated values of permittivities are used. However, both TM and TE mode results show that the difference between the two sets of reflectivities (or transmissivities) diminish as the wavelength increases.

4.3.5 Summary

We implemented a recursive convolution FDTD (RC-FDTD) method with first-order Drude model in two dimensions for monochromatic applications. In TM mode we implemented the wave equation using RC-FDTD. In our knowledge, this is the first time that RC-FDTD for the wave equation is implemented. With only one electric field component to compute, the wave equation implementation is inherently faster than implementations that use Maxwell's equations in this mode.

We presented a stability analysis that links RC-FDTD stability to space and time discretization, and the metal permittivity. The stability analysis is valid as long as wave at each point of the simulation space can be approximated by an infinite plane wave. The choice of the minimum spatial grid step and the numerical speed of light on the grid is seen to affect the results profoundly. Choosing the minimum grid step and numerical speed to be sufficiently small it is possible to render the algorithm practically stable and guarantee the reliability of the results.

The RC-FDTD scheme is combined with a complex fast Fourier transform to compute the far-field wavelength spectra of reflectivity and transmissivity for two subwavelength gratings, one, a purely metal grating and another, a double-layered grating formed of one metal layer and one weakly conducting dielectric layer.

The wavelength spectra of the gratings are computed using two sets of complex permittivities: the first set represents the tabulated values of the complex permittivities, while the second set is obtained using a set of best-fit Drude parameters that are determined by fitting the first-order Drude model to the tabulated values of permittivities over the entire visible range of wavelengths. This is done to highlight the effect of choice of permittivities on the optical characteristics of these structures. The results are seen to depend strongly on the choice of permittivity values.

4.4 Summary and Conclusions

In this chapter we have introduced new theoretical and practical developments in the finite-difference time-domain (FDTD) methodology.

In section 4.1, we introduced the nonstandard (NS) FDTD algorithm, and a presented a more useful, more accurate, and more mathematically meaningful description than that presented in previous papers [4]–[6]. We then extended the NS-FDTD methodology to include the effective source terms in scattered field calculations.

The NS-FDTD algorithm derives from nonstandard finite difference models of the wave equation and Maxwell's equations in which model parameters of the difference eqautions are adjusted so that known analytic solutions of the differential equations are also solutions of the corresponding difference equations. Whereas the error of the conventional or standard (S) FDTD algorithm is $\epsilon_{\mathrm{S}} \sim h^2$, where $h = \Delta x = \Delta y = \Delta z$ is the grid spacing, the error of the NS-FDTD algorithm is $\epsilon_{\mathrm{NS}} \sim h^6$. We verified that our methodology is correct by comparing NS-FDTD calculations of scattered field intensity for cylinders and spheres with Mie theory.

The high accuracy of the NS-FDTD algorithm, alone, does not necessarily guarantee high-accuracy results. The accuracy of FDTD calculations depends not only on the FDTD algorithm used, but also on how the computational domain is terminated, and on how objects, such as scatterers, are represented on the numerical grid. Although there is no ideal termination of the computational boundary, there are many good ones, for example [8]. On the other hand, there is no completely satisfactory representation of an object on a coarse grid. In section 4.2.3 we derive the 'fuzzy model.' It gives excellent results in 'most' cases, but to compute the whispering gallery modes discussed in section 4.2 a quite fine grid is required because the object representation gives rise to large errors on resonance. For example, in the TM mode (see Table 4.2) we need a discretization of $\lambda/h = 64$ (λ = vacuum wavelength) in the NS-FDTD calculation to obtain good results, but just off resonance (same cylinder radius, but refractive index = 2.7) the NS-FDTD calculation is in excellent agreement with Mie theory for $\lambda/h = 10$.

While the development of the second-order NS-FDTD algorithms seems nearly complete, further progress is needed in representing objects on a coarse grid.

In section 4.2, we applied the methodology of section 4.1 to compute whispering gallery modes (WGM) in dielectric cylinders in the Mie regime. Using NS-FDTD, we computed the scattered and internal fields due to an infinite plane wave impinging perpendicularly to the cylinder axis. Compared to off-resonance, roughly an order of magnitude more iterations are needed to compute the fields of the WGMs. Besides algorithmic error, such errors as that due to the termination of the computational boundary and the representation of the cylinder on the numerical grid accumulate. Nevertheless we found excellent agreement with Mie theory.

While the fact that NS-FDTD gives the correct result on such a challenging problem, does not necessarily guarantee that it will yield always correct results on all other problems, it is nonetheless the best validation available when there are no analytic solutions. Having validated our algorithms against Mie theory we now plan to investigate resonances in such structures as dielectric cubes, and photonic crystals, where there is no analytic guidance.

Most media are, at least to some extent, dispersive. The FDTD methodology can be extended to include dispersion, using what is called recursive convolution (RC). The computational cost of the RC-FDTD algorithm is not only much higher than that of ordinary FDTD, but also its accuracy is much lower, and it is often numerically unstable. To implement RC-FDTD, an analytic model of the dispersion is needed.

In section 4.3, a monochromatic implementation of RC-FDTD in two dimensions incorporating the Drude model of dispersion is discussed. One advantage of this implementation is that it allows one to use the tabulated values of permittivity at any frequency. A typical broadband implementation does not permit the use of tabulated values of permittivity. Computations presented here indicate that such a broadband implementation might lead to quite different results.

A rigorous numerical stability analysis for the RC-FDTD algorithm, including the Drude parameters is presented.

To obtain reasonable results with RC-FDTD, it is necessary to use a fairly fine grid (typically λ/h is at least 50). It might be possible to increase the accuracy (and hence use a coarser grid) if a suitable NS-FD model could be found. This is a topic for future work.

Acknowledgments

This research was partially supported by Fujikura Corporation (we thank Dr Ogawa) and by the Japanese Ministry of Science and Education. In addition we thank Dr Ogawa, Prof. S. Yamada, and Prof. Y. Katayama for substantive research discussions at our regular meetings. In addition, we thank Sumitomo Chemical Co. Ltd. for ongoing support and Dr Nakatsuka for his advice and encouragement.

References

1. K. S. Yee, Numerical solution of initial boundary value problems involving Maxwell's equations in isotropic media, *IEEE Trans. Antenna Propagation*, **AP-14**, 302–307 (1966).
2. K. S. Kunz and R. J. Luebbers, The Finite Difference Time Domain Method for Electromagnetics, CRC Press, Boca Raton (1993).
3. R. E. Mickens, Nonstandard Finite Difference Models of Differential Equations, World Scientific, Singapore (1994).
4. J. B. Cole, High accuracy Yee algorithm based on nonstandard finite differences: new developments and verification, *IEEE Trans. Antennas and Propagation*, **50**, no. 9, 1185–1191 (2002).
5. J. B. Cole, High accuracy FDTD solution of the absorbing wave equation, and conducting Maxwell's equations based on a nonstandard finite difference model, *IEEE Trans. on Antennas and Propagation*, **53**, no. 2, 725–729 (2004).

6. J. B. Cole, S. Banerjee, M. I. Haftel, High accuracy nonstandard finite-difference time-domain algorithms for computational electromagnetics: applications to optics and photonics, Chapter 4, pp. 89–109 in *Advances in the Applications of Nonstandard Finite Difference Schemes*, R. E. Mickens, ed., Scientific (Singapore, 2005).
7. P. W. Barber, S. C. Hill, *Light Scattering by Particles: Computational Methods*, World Scientific, Singapore (1990).
8. N. Okada, J. B. Cole, Simulation of whispering gallery modes in the Mie regime using the nonstandard finite difference time domain algorithm, *J. Optical Society of America B*, **27**, issue 4, 631–639 (2010).
9. We thank Dr. Till Plewe for this idea; private discussions, unpublished.
10. T. Ohtani, K. Taguchi, T. Kashiwa, T. Kanai, J. B. Cole, Nonstandard FDTD method for wideband analysis, *IEEE Trans. on Antennas and Propagation*, **57**, issue 8, 2386–2396 (2009).
11. L. Rayleigh, The problem of the whispering gallery, *Philos. Mag.*, **20**, 1001–1004 (1910).
12. C. G. B. Garrett, W. Kaiser, W. L. Bond, Stimulated emission into optical whispering gallery modes of spheres, *Phys. Rev.*, **124**, 1807–1809 (1961).
13. P. Chyek, V. Ramaswamy, A. Ashkin, and J. M. Dziedzic, Simultaneous determination of refractive index and size of spherical dielectric particles from light scattering data, *Appl. Opt.*, **22**, 2302–2307 (1983).
14. A. B. Matsko, A. A. Savchenkov, D. Strekalov, V. S. Ilchenko, L. Maleki, Review of applications of whispering gallery mode resonators in photonics and nonlinear optics, IPN Progress Report, pp. 42–162 (2005).
15. J. P. Berenger, A perfectly matched layer for the absorption of electromagnetic waves, *Journal of Computational Physics*, **114**, 185–200 (1994).
16. N. Okada, J. B. Cole, High-accuracy finite-difference time domain algorithm for the coupled wave equation, *J. Opt. Soc. Am. B*, **27**, 7, 1409-1413 (2010).
17. A. Taflove and S. C. Hagness: *Computational Electrodynamics, the Finite Difference Time-Domain Method*, 3rd ed., Chap. 8, p. 329, Chap. 9, p. 355, Artech House, Boston, 2005.
18. S. Banerjee, T. Hoshino and J. B. Cole, Simulation of subwavelength metallic gratings using a new implementation of recursive convolution FDTD, *JOSA A*, **25**, no. 8, 1921 (2008).
19. A. Vial, A. S. Grimault, D. Macias, D. Barchiesi, and M. L. de la Chapelle, Improved analytical fit of gold dispersion: application to the modeling of extinction spectra with a finite-difference time-domain method, *Phys. Rev. B* **71**, 085416-1 (2005).
20. A. Vial, Problems encountered when modeling dispersive materials using the FDTD methodÅ, Proceedings of Workshop, Nano particles, nano structures and near field computation, T. Wriedt, Y. Eremin, W. Hergert, Eds., Bremen, pp. 56, 2010.
21. H. Tamada, T. Doumuki, T. Yamaguchi, and S. Matsumoto, Al wire-grid polarizer using the *s*-polarization resonance effect at the 0.8-mm-wavelength band, *Opt. Lett.*, **22**, no. 6, 419 (1997).
22. M. A. Jensen and G. P. Nordin, Finite-aperture wire grid polarizers, *JOSA A*, **17**, no. 12, 2191 (2000).
23. M. Xu, H. P. Urbach, D. K. G de Boer, and H. J. Cornelissen, Wire-grid diffraction gratings used as polarizing beam splitter for visible light and applied in liquid crystal on silicon, *Opt. Exp.*, **13**, no. 7, 2303 (2005).
24. X. D. Mi, D. Kessler, L. W. Tutt, and L. W. Brophy, Low fill-factor wire grid polarizers for LCD backlighting, *Society for Information Display (SID) Digest 2005*, p. 1004.
25. M. Paukshto, Simulation of sub-100 nm gratings incorporated in LCD stack, *Society for Information Display (SID) Digest 2006*, p. 848.

26. S. Banerjee and K. Nakatsuka, Compact design of light guides using metal grating based polarizing optical controller element, Proc. International Display Workshop 2007, Sapporo, Japan, p. 2087.
27. J. H. Lee, Y. W. Song, J. G. Lee, J. Ha, K. H. Hwang, and D. S. Zang, Optically bifacial thin-film wire-grid polarizers with nano-patterns of a graded metal-dielectric composite layer, *Optics Express*, **16**, no. 21, 16867 (2008).
28. E. H. Land, Some aspects of the development of sheet polarizers, *JOSA*, **41**, no. 12, 957 (1951).
29. S. Banerjee, J. B. Cole and T. Yatagai, Colour characterization of a Morpho butterfly wing-scale using a high accuracy nonstandard finite-difference time-domain method, *Micron*, **38**, 97–103 (2007).
30. C. L. Foiles, Optical properties of pure metals and binary alloys, Chapter 4 of Landolt-Bornstein Numerical Data and Functional Relationships in Science and Technology New Series, Vol. 15, Subvolume b, K.-H. Hellwege and J. L. Olsen, Eds., Springer-Verlag, Berlin, 1985, p. 228.
31. M. Born, and E. Wolf, *Principles of Optics*, 7th (expanded) ed., Chap. XIV. Cambridge University Press, San Francisco, 1999.
32. S. Banerjee and L. N. Hazra, Experiments with a genetic algorithm for structural design of cemented doublets with prespecified aberration targets, *App. Opt.*, **40**, no. 34, 6265 (2001).
33. J. B. Cole and D. Zhu, Improved version of the second-order Mur absorbing boundary condition based on a nonstandard finite difference model, *J. Applied Computational Electromagnetics Society*, **24**, no. 4 (2009).
34. D. E. Aspnes and A. A. Studna, Dielectric functions and optical parameters of Si, Ge, GaP, GaAs, GaSb, InP, InAs, and InSb from 1.5 to 6.0 eV, *Phys. Rev. B*, **27**, 985 (1983).

Part II

Radiative Transfer and Remote Sensing

5 Radiative Transfer in Coupled Systems

Knut Stamnes and Jakob J. Stamnes

5.1 Introduction

In many applications an accurate description is required of light propagation in two adjacent slabs separated by an interface, across which the refractive index changes. Such a two-slab configuration will be referred to as a *coupled* system. Three important examples are atmosphere–water systems [1]–[5], atmosphere–sea ice systems [6, 7], and air–tissue systems [8]. In each of these three examples, the change in the refractive index across the interface between the two slabs must be accounted for in order to model the transport of light throughout a *coupled* system correctly. In the second example, the refractive-index change together with multiple scattering leads to a significant trapping of light inside the strongly scattering, optically thick sea ice medium [6, 7]. For imaging of biological tissues or satellite remote sensing of water bodies an accurate radiative transfer (RT) model for a *coupled* system is an indispensable tool [9]–[12]. In both cases, an accurate RT tool is essential for obtaining satisfactory solutions of retrieval problems through iterative forward/inverse modeling [10]–[18].

In this review, the discussion is limited to applications based on scalar RT models that ignore polarization effects. There are numerous RT models available that include polarization effects (see Zhai et al. [19] and references therein for a list of papers), and the interest in applications relying on polarized radiation is growing.

Section 5.2 provides definitions of inherent optical properties including absorption and scattering coefficients as well as the normalized angular scattering cross-section (scattering phase function). In section 5.3 an overview is given of the scalar RT equation (RTE) applicable to coupled media consisting of two adjacent slabs with different refractive indices. Two different methods of solutions are discussed: the discrete-ordinate method and the Monte Carlo method. The impact of a rough interface between the two adjacent slabs is also discussed. In section 5.4 a few typical applications are discussed including coupled atmosphere–water systems, coupled atmosphere–snow–ice systems, and coupled air–tissue systems. Finally, a summary is provided in section 5.5.

5.2 Inherent optical properties

The optical properties of a medium can be categorized as inherent or apparent. An *inherent optical property* (IOP) depends only on the medium itself, and *not* on the ambient light field within the medium [20]. An *apparent optical property* (AOP) depends also on the illumination, i.e. on light propagating in particular directions inside or outside the medium.

The absorption coefficient α and the scattering coefficient σ are important IOPs, defined as [4]

$$\alpha(s) = \frac{1}{I^i}\left(\frac{dI^\alpha}{ds}\right), \tag{1}$$

$$\sigma(s) = \frac{1}{I^i}\left(\frac{dI^\sigma}{ds}\right). \tag{2}$$

Here I^i is the incident radiance entering a volume element $dV = dA\ ds$ of the medium of cross-sectional area dA and length ds, and $dI^\alpha > 0$ and $dI^\sigma > 0$ are respectively the radiances that are absorbed and scattered in all directions as the light propagates the distance ds, which is the thickness of the volume element dV along the direction of the incident light. If the distance ds is measured in [m], the unit for the absorption or scattering coefficient defined in Eq. (1) or Eq. (2) becomes [m^{-1}].

The angular distribution of the scattered light is given in terms of the *volume scattering function* (VSF), which is defined as

$$\beta(s, \hat{\mathbf{\Omega}}', \hat{\mathbf{\Omega}}) = \frac{1}{I^i}\frac{d^2 I^\sigma}{ds\, d\omega} = \frac{1}{I^i}\frac{d}{ds}\left(\frac{dI^\sigma}{d\omega}\right) \quad [\mathrm{m}^{-1}\mathrm{sr}^{-1}]. \tag{3}$$

Here $d^2 I^\sigma$ is the radiance scattered from an incident direction $\hat{\mathbf{\Omega}}'$ into a cone of solid angle $d\omega$ around the direction $\hat{\mathbf{\Omega}}$ as the light propagates the distance ds along $\hat{\mathbf{\Omega}}'$. The plane spanned by $\hat{\mathbf{\Omega}}'$ and $\hat{\mathbf{\Omega}}$ is called the *scattering plane*, and the *scattering angle* Θ is given by $\cos\Theta = \hat{\mathbf{\Omega}}' \cdot \hat{\mathbf{\Omega}}$. Integration of Eq. (3) over all scattering directions yields

$$\begin{aligned}\sigma(s) &= \frac{1}{I^i}\frac{d}{ds}\int_{4\pi}\left(\frac{dI^\sigma}{d\omega}\right)d\omega = \frac{1}{I^i}\left(\frac{dI^\sigma}{ds}\right)\\ &= \int_{4\pi}\beta(s, \hat{\mathbf{\Omega}}', \hat{\mathbf{\Omega}})d\omega = \int_0^{2\pi}\int_0^{\pi}\beta(s, \cos\Theta, \phi)\sin\Theta\, d\Theta\, d\phi,\end{aligned} \tag{4}$$

where Θ and ϕ are respectively the polar angle and the azimuth angle in a spherical coordinate system in which the polar axis is along $\hat{\mathbf{\Omega}}'$. As indicated in Eq. (4), the VSF [$\beta(s, \cos\Theta, \phi)$] is generally a function of both Θ and ϕ, but for randomly oriented scatterers one may assume that the scattering potential is spherically symmetric implying that there is no azimuthal dependence, so that $\beta = \beta(s, \cos\Theta)$. Then one finds

$$\sigma(s) = 2\pi\int_0^{\pi}\beta(s, \cos\Theta)\sin\Theta d\Theta = 2\pi\int_{-1}^{1}\beta(s, \cos\Theta)d(\cos\Theta). \tag{5}$$

A normalized VSF, denoted by $p(s, \cos\Theta)$ and referred to hereafter as the *scattering phase function*, may be defined as follows

$$p(s, \cos\Theta) = 4\pi \frac{\beta(s, \cos\Theta)}{\int_{4\pi} \beta(s, \cos\Theta)\, d\omega} = \frac{\beta(s, \cos\Theta)}{\frac{1}{2}\int_{-1}^{1} \beta(s, \cos\Theta)\, d\cos\Theta}, \tag{6}$$

so that

$$\frac{1}{4\pi}\int_{4\pi} p(s, \cos\Theta)\, d\omega = 1. \tag{7}$$

The scattering phase function has the following physical interpretation. Given that a scattering event has occurred, $p(s, \cos\Theta)\, d\omega/4\pi$ is the probability that a photon traveling in the direction $\hat{\mathbf{\Omega}}'$ is scattered into a cone of solid angle $d\omega$ around the direction $\hat{\mathbf{\Omega}}$ within the volume element dV with thickness ds along $\hat{\mathbf{\Omega}}'$.

The scattering phase function $[p(s, \cos\Theta)]$ describes the angular distribution of the scattering, while the scattering coefficient $\sigma(s)$ describes its magnitude. A convenient measure of the 'shape' of the scattering phase function is the average over all scattering directions (weighted by $p(s, \cos\Theta)$) of the cosine of the scattering angle Θ, *i.e.*

$$\begin{aligned} g &= \langle \cos\Theta \rangle = \frac{1}{4\pi}\int_{4\pi} p(s, \cos\Theta)\, \cos\Theta\, d\omega \\ &= \frac{1}{2}\int_{0}^{\pi} p(s, \cos\Theta)\, \cos\Theta\, \sin\Theta\, d\Theta = \frac{1}{2}\int_{-1}^{1} p(s, \cos\Theta)\, \cos\Theta\, d(\cos\Theta). \end{aligned} \tag{8}$$

The average cosine g is called the *asymmetry factor* of the scattering phase function. Equation (8) yields complete forward scattering if $g = 1$, complete backward scattering if $g = -1$, and $g = 0$ if $p(s, \cos\Theta)$ is symmetric about $\Theta = 90°$. Thus, isotropic scattering also gives $g = 0$. The scattering phase function $p(s, \cos\Theta)$ depends on the refractive index as well as the size of the scattering particles, and will thus depend on the physical situation and the practical application of interest. Two different scattering phase functions, which are useful in practical applications, are discussed below.

In 1941 Henyey and Greenstein [21] proposed the one-parameter scattering phase function given by [oppressing the dependence on the position s]

$$p(\cos\Theta) = \frac{1 - g^2}{(1 + g^2 - 2g\, \cos\Theta)^{3/2}}, \tag{9}$$

where the parameter g is the asymmetry factor defined in Eq. (8). The Henyey and Greenstein (HG) scattering phase function has no physical basis, but is very useful for describing a highly scattering medium, such as skin tissue or sea ice, for which the actual scattering phase function is unknown. The HG scattering phase function is convenient for Monte Carlo simulations and other numerical calculations because it has an analytical form. In deterministic plane parallel RT models it is also very convenient because the addition theorem of spherical harmonics can be used to expand the scattering phase function in a series of Legendre polynomials [4] as reviewed in the next section. For the HG scattering phase function, the expansion coefficients χ_l in this series (see Eq. (17) below) are simply given by $\chi_l = g^l$. The HG scattering phase function is used for scatterers with sizes comparable to or larger than the wavelength of light.

When the size d of the scatterers is small compared with the wavelength of light ($d < \frac{1}{10}\lambda$), the Rayleigh scattering phase function gives a good description of the angular distribution of the scattered light. The Rayleigh scattering phase function for unpolarized light is given by

$$p(\cos\Theta) = \frac{3}{3+\tilde{f}}(1 + \tilde{f}\cos^2\Theta), \tag{10}$$

where the parameter $\tilde{f}$ is the depolarization factor [22–24]. Originally this scattering phase function was derived for light scattering by an electric dipole [25]. Since the Rayleigh scattering phase function is symmetric about $\Theta = 90°$, the asymmetry factor is $g = 0$.

5.3 Basic Theory

5.3.1 Radiative transfer equation

Consider a *coupled* system consisting of two adjacent slabs separated by a plane, horizontal interface across which the refractive index changes abruptly from a value m_1 in one of the slabs to a value m_2 in the other. If the IOPs in each of the two slabs vary only in the vertical direction denoted by z, where z increases upward, the corresponding vertical optical depth, denoted by $\tau(z)$, is defined by

$$\tau(z) = \int_z^\infty [\alpha(z') + \sigma(z')]\, dz', \tag{11}$$

where the absorption and scattering coefficients α and σ are defined in Eqs. (1) and (2). Note that the vertical optical depth is defined to increase downward from $\tau(z = \infty) = 0$ at the top of the upper slab.

In either of the two slabs, assumed to be in local thermodynamic equilibrium, so that they emit radiation according to the local temperature $T(\tau(z))$, the diffuse radiance distribution $I(\tau, \mu, \phi)$ can be described by the radiative transfer equation (RTE)

$$\begin{aligned} \mu\frac{dI(\tau,\mu,\phi)}{d\tau} &= I(\tau,\mu,\phi) - S^*(\tau,\mu',\phi') - [1 - a(\tau)]B(\tau) \\ &- \frac{a(\tau)}{4\pi}\int_0^{2\pi} d\phi' \int_{-1}^{1} p(\tau,\mu',\phi';\mu,\phi)I(\tau,\mu',\phi')\, d\mu'. \end{aligned} \tag{12}$$

Here μ is the cosine of the polar angle θ, and ϕ is the azimuth angle of the observation direction, $a(\tau) = \sigma(\tau)/[\alpha(\tau) + \sigma(\tau)]$ is the single-scattering albedo, $p(\tau, \mu', \phi'; \mu, \phi)$ is the scattering phase function defined by Eq. (6), where μ' is the cosine of the polar angle θ' and ϕ' is the azimuth angle of the direction of an incident beam that is scattered into the observation direction θ, ϕ, and $B(\tau)$ is the thermal radiation field given by the Planck function. The differential vertical optical depth is [see Eq. (11)]

$$d\tau(z) = -[\alpha(\tau) + \sigma(\tau)]\, dz, \tag{13}$$

where the minus sign indicates that τ increases in the downward direction, whereas z increases in the upward direction as noted above. The scattering angle Θ and the polar and azimuth angles are related by

$$\hat{\boldsymbol{\Omega}}' \cdot \hat{\boldsymbol{\Omega}} = \cos\Theta = \cos\theta\, \cos\theta' + \sin\theta'\, \sin\theta\, \cos(\phi' - \phi).$$

By definition, $\theta = 180°$ is directed toward nadir (straight down) and $\theta = 0°$ toward zenith (straight up). Thus, μ varies in the range $[-1, 1]$ (from nadir to zenith). For cases of oblique illumination of the medium, $\phi = 180°$ is defined to be the azimuth angle of the incident light. The vertical optical depth τ is defined to increase downward with depth from $\tau = 0$ at the top of the upper slab (slab$_1$).

The single-scattering source term $S^*(\tau, \mu', \phi')$ in Eq. (12) in slab$_1$ (with complex refractive index $m_1 = n_1 + in_1'$) is different from that in the lower slab (slab$_2$, with refractive index $m_2 = n_2 + in_2'$). In slab$_1$ it is given by

$$\begin{aligned} S_1^*(\tau, \mu, \phi) &= \frac{a(\tau)F^s}{4\pi} p(\tau, -\mu_0, \phi_0; \mu, \phi)\, e^{-\tau/\mu_0} \\ &+ \frac{a(\tau)F^s}{4\pi} \rho_F(-\mu_0; m_1, m_2) p(\tau, \mu_0, \phi_0; \mu, \phi)\, e^{-(2\tau_1 - \tau)/\mu_0}, \end{aligned} \tag{14}$$

where τ_1 is the vertical optical depth of the upper slab, $\rho_F(-\mu_0; m_1, m_2)$ is the Fresnel reflectance at the slab$_1$–slab$_2$ interface, $\mu_0 = \cos\theta_0$, with θ_0 being the zenith angle of the incident beam of illumination, and where $n_2 > n_1$. Note that the real part of the refractive index of the medium in slab$_1$ has been assumed to be smaller than that of the medium in slab$_2$, as would be the case for air overlying a water body or a skin surface. The first term on the right-hand side of Eq. (14) is due to first-order scattering of the attenuated incident beam of irradiance F^s (normal to the beam) while the second term is due to first-order scattering of the attenuated incident beam that is reflected at the slab$_1$–slab$_2$ interface. In slab$_2$ the single-scattering source term consists of the attenuated incident beam that is refracted through the interface, i.e.

$$\begin{aligned} S_2^*(\tau, \mu, \phi) &= \frac{a(\tau)F^s}{4\pi} \frac{\mu_0}{\mu_{0n}} \mathcal{T}_F(-\mu_0; m_1, m_2) \\ &\times\ p(\tau, -\mu_{0n}, \phi_0; \mu, \phi) e^{-\tau_1/\mu_0} e^{-(\tau - \tau_a)/\mu_{0n}}, \end{aligned} \tag{15}$$

where $\mathcal{T}_F(-\mu_0; m_1, m_2)$ is the Fresnel transmittance through the interface, and μ_{0n} is the cosine of the polar angle θ_{0n} in slab$_2$, which is related to $\theta_0 = \arccos\mu_0$ by Snell's law.

For a two-slab system with source terms as given by Eqs. (14) and (15), a solution based on the discrete-ordinate method [26, 27] of the RTE in Eq. (12) subject to appropriate boundary conditions at the top of slab$_1$, at the bottom of slab$_2$, and at the slab$_1$–slab$_2$ interface, was first developed by Jin and Stamnes [2] (see also Thomas and Stamnes [4]).

Isolation of azimuth dependence

The azimuth dependence in Eq. (12) may be isolated by expanding the scattering phase function in *Legendre polynomials*, $P_l(\cos\Theta)$, and making use of the addition theorem for spherical harmonics [4]

$$p(\cos\Theta) = p(\mu',\phi';\mu,\phi) = \sum_{m=0}^{2N-1}(2-\delta_{0,m})p^m(\mu',\mu)\cos m(\phi'-\phi), \tag{16}$$

where $\delta_{0,m}$ is the Kronecker delta function, *i.e.* $\delta_{0,m}=1$ for $m=0$ and $\delta_{0,m}=0$ for $m\neq 0$, and

$$p^m(\mu',\mu) = \sum_{l=m}^{2N-1}(2l+1)\chi_l\Lambda_l^m(\mu')\Lambda_l^m(\mu). \tag{17}$$

Here $\chi_l = \frac{1}{2}\int_{-1}^{1} d(\cos\Theta)P_l(\cos\Theta)p(\cos\Theta)$ is an expansion coefficient and $\Lambda_l^m(\mu)$ is given by

$$\Lambda_l^m(\mu) \equiv \sqrt{\frac{(l-m)!}{(l+m)!}}P_l^m(\mu), \tag{18}$$

where $P_l^m(\mu)$ is an associated Legendre polynomial of order m. Note that for simplicity we have suppressed the dependence of the phase function on optical depth in Eqs. (16)–(18). Expanding the radiance in a similar way,

$$I(\tau,\mu,\phi) = \sum_{m=0}^{2N-1} I^m(\tau,\mu)\cos m(\phi-\phi_0), \tag{19}$$

where ϕ_0 is the azimuth angle of the direction of the incident light, one finds that each Fourier component satisfies the following RTE (see Thomas and Stamnes [4] for details)

$$\begin{aligned}\frac{dI^m(\tau,\mu)}{d\tau} &= I^m(\tau,\mu) - S^{*m}(\tau,\mu)\\ &\quad -\frac{a(\tau)}{2}\int_{-1}^{1} p^m(\tau,\mu',\mu)\, I^m(\tau,\mu)\, d\mu,\end{aligned} \tag{20}$$

where $m=0,1,2,\ldots,2N-1$ and $p^m(\tau,\mu',\mu)$ is given by Eq. (17).

The interface between the two slabs

When a beam of light is incident upon a plane interface between two slabs of different refractive indices, one fraction of the incident light will be reflected and another fraction will be transmitted or refracted. For unpolarized light incident upon the interface between the two slabs, the Fresnel reflectance ρ_F is given by

$$\rho_F = \frac{1}{2}(R_\perp + R_\parallel), \tag{21}$$

where $R_\perp$ is the reflectance for light polarized with the electric field perpendicular to the plane of incidence, and $R_\parallel$ is the reflectance for light polarized with the electric field parallel to the plane of incidence [4,28]. Thus, one finds

$$\rho_F = \frac{1}{2}\left[\left|\frac{\mu_i - m_r\mu_t}{\mu_i + m_r\mu_t}\right|^2 + \left|\frac{\mu_t - m_r\mu_i}{\mu_t + m_r\mu_i}\right|^2\right], \tag{22}$$

where $\mu_i = \cos\theta_i$, θ_i being the angle of incidence, $\mu_t = \cos\theta_t$, θ_t being the angle of refraction determined by Snell's law ($n_1 \sin\theta_i = n_2 \sin\theta_t$), and $m_r = m_2/m_1$. Similarly, the Fresnel transmittance becomes

$$\mathcal{T}_F = 2m_{rel}\mu_i\mu_t\left[\left|\frac{1}{\mu_i + m_r\mu_t}\right|^2 + \left|\frac{1}{\mu_t + m_r\mu_i}\right|^2\right], \tag{23}$$

where $m_{rel} = n_2/n_1$.

5.3.2 Discrete-ordinate solution of the radiative transfer equation

To solve Eq. (20) for a coupled (two-slab) system one needs to take into account the boundary conditions at the top of slab$_1$ and at the bottom of slab$_2$ as well as the reflection and transmission at the slab$_1$–slab$_2$ interface. In addition, the radiation field must be continuous across interfaces between horizontal layers with different IOPs within each of the two slabs (with constant refractive index). Such horizontal layers are introduced to resolve vertical variations in the IOPs within each slab.

The integro-differential RTE [Eq. (20)] may be transformed into a system of coupled, ordinary differential equations by using the discrete-ordinate approximation to replace the integral in Eq. (20) by a quadrature sum consisting of $2N_1$ terms in slab$_1$ and $2N_2$ terms in slab$_2$, where N_1 terms are used to represent the radiance in the downward hemisphere in slab$_1$ that refracts through the interface into slab$_2$. In slab$_2$, N_2 terms are used to represent the radiance in the downward hemisphere. Note that $N_2 > N_1$ because additional terms are needed in slab$_2$ with real part of the refractive index $n_2 > n_1$ to represent the downward radiance in the region of total internal reflection.

Seeking solutions to the discrete ordinate approximation of Eq. (20), one obtains the Fourier component of the radiance at any vertical position both in slab$_1$ and slab$_2$. The solution for the pth layer of slab$_1$ is given by [4]

$$I_p^m(\tau, \pm\mu_i^u) = \sum_{j=1}^{N_1}\left\{C_{-jp}g_{-jp}^u(\pm\mu_i^u)e^{k_{jp}^u\tau} + C_{+jp}g_{jp}^u(\pm\mu_i^u)e^{-k_{jp}^u\tau} + U_p(\tau, \pm\mu_i^u)\right\}, \tag{24}$$

where $i = 1, \ldots, N_1$ and p is less than or equal to the number of layers in slab$_1$. The solution for the qth layer of slab$_2$ is given by [4]

$$I_q(\tau, \pm\mu_i^\ell) = \sum_{j=1}^{N_2} C_{-jq}g_{-jq}^\ell(\pm\mu_i^\ell)e^{k_{jq}^\ell\tau} + C_{+jq}g_{jq}^\ell(\pm\mu_i^t)e^{-k_{jq}^\ell\tau} + U_q(\tau, \pm\mu_i^\ell), \tag{25}$$

where $i = 1, \ldots, N_2$. The superscripts u and ℓ are used to denote *upper* slab$_1$ and *lower* slab$_2$ parameters, respectively, the plus (minus) sign is used for radiances streaming upward (downward), and $k_{jp}^u, g_{jp}^u, k_{jq}^\ell$, and g_{jq}^ℓ are eigenvalues and eigenvectors determined by the solution of an algebraic eigenvalue problem, which results when one seeks a solution of the homogeneous version of Eq. (20) (with $S^{*m}(\tau, \mu) = 0$) in the discrete-ordinate approximation. The terms $U_p(\pm\mu_i^u)$ and $U_q(\pm\mu_i^\ell)$ are the particular solutions. The coefficients $C_{\pm jp}$ and $C_{\pm jq}$ are determined by boundary conditions at the top of slab$_1$ and at the bottom of slab$_2$, the continuity of the basic radiance (the radiance divided by the square of the real part of the refractive index) at each interface between internal layers in each of the slabs, and Fresnel's equations at the slab$_1$–slab$_2$ interface.

The numerical code C-DISORT [2,3] computes radiances at any optical depth, polar, and azimuth angle by solving the RTE in Eq. (20) for each layer of the two slabs by using the discrete-ordinate method to convert the integro-differential RTE into a system of coupled ordinary differential equations. The C-DISORT method can be summarized as follows:

1. Slab$_1$ and slab$_2$ are separated by a plane interface at which the refractive index changes from $m_1 = n_1 + in_1'$ in slab$_1$ to $m_2 = n_2 + in_2'$ in slab$_2$, where m_2 depends on the wavelength.
2. Each of the two slabs is divided into a sufficiently large number of homogenous horizontal layers to adequately resolve the vertical variation in its IOPs.
3. Fresnel's equations for the reflectance and transmittance are applied at the slab$_1$–slab$_2$ interface, in addition to the law of reflection and Snell's Law to determine the directions of the reflected and refracted beams.
4. Discrete-ordinate solutions to the RTE are computed separately for each layer in the two slabs.
5. Finally, boundary conditions at the top of slab$_1$ and the bottom of slab$_2$ are applied, in addition to continuity conditions at layer interfaces within each of the two slabs.

Fourier components of the radiances at a vertical location given by the pth layer in slab$_1$ or the qth layer in slab$_2$ are computed from Eqs. (24)–(25), and the azimuth-dependent diffuse radiance distribution from Eq. (19). Upward and downward hemispherical irradiances are then calculated by integrating the azimuthally-averaged zeroth-order ($m = 0$) Fourier component $I_p^0(\tau, +\mu_i)$ or $I_p^0(\tau, -\mu_i)$ over polar angles.

The downward irradiance in slab$_1$ consists of a direct component E_{dir}^u, given by

$$E_{dir}^u(\tau) = \mu_0 F^s \, e^{\tau/\mu_0}, \tag{26}$$

and a diffuse component $E_{d,diff}^u$, given by

$$E_{d,diff}^u(\tau) = 2\pi \int_0^1 \langle I_{d,diff}^u(\tau, \mu)\rangle \mu \, d\mu, \tag{27}$$

where $\langle I_{d,diff}^u(\tau, \mu)\rangle$ is the azimuthally-averaged diffuse downward radiance at optical depth $\tau \le \tau_1$ in slab$_1$. Similarly, the upward diffuse irradiance $E_{u,diff}^u$ in slab$_1$ and the average diffuse radiance (mean intensity) are given by

$$E_{u,diff}^u(\tau) = 2\pi \int_0^1 \langle I_{u,diff}^u(\tau, \mu)\rangle \mu \, d\mu \tag{28}$$

$$\bar{I}_{diff}^u(\tau) = \frac{1}{2} \int_0^1 \langle [I_{d,diff}^u(\tau, \mu) + I_{u,diff}^u(\tau, \mu)]\rangle \, d\mu. \tag{29}$$

In slab_2, the downward direct and diffuse irradiances, the upward diffuse irradiance, and the average diffuse radiance become:

$$E_{dir}^{\ell}(\tau) = \mu_0 F^s \, e^{[-\tau_1/\mu_0 - \tau/\mu_{0n}]}, \tag{30}$$

$$E_{d,diff}^{\ell}(\tau) = 2\pi \int_0^1 \langle I_{d,diff}^{\ell}(\tau,\mu)\rangle \mu \, d\mu, \tag{31}$$

$$E_{u,diff}^{\ell}(\tau) = 2\pi \int_0^1 \langle I_{u,diff}^{\ell}(\tau,\mu)\rangle \mu \, d\mu, \tag{32}$$

$$\bar{I}_{diff}^{\ell}(\tau) = \frac{1}{2} \int_0^1 \langle [I_{d,diff}^{\ell}(\tau,\mu) + I_{u,diff}^{\ell}(\tau,\mu)]\rangle \, d\mu, \tag{33}$$

where the downward and upward azimuthally-averaged radiances in slab_2 are given by $\langle I_{d,diff}^{\ell}(\tau,\mu)\rangle$ and $\langle I_{u,diff}^{\ell}(\tau,\mu)\rangle$.

5.3.3 Monte Carlo simulations

The Monte Carlo (MC) method is a numerical approach that is based on the use of random numbers to perform statistical simulations by means of a computer. Computer-generated numbers are not really random, since computers are deterministic. But a random number algorithm, like the one described in [29], has a periodicity of 2.3×10^{18}, and generates numbers that are uncorrelated. Such a random number algorithm is therefore acceptable for practical MC simulations. The concept of the MC method is very simple, and may be used to model light propagation in systems with geometries that are too complicated to be modeled with other numerical methods. The main drawback of the MC method is its heavy demands on computing time. To speed up MC simulations one may use photon packets rather than single photons. Also, there are more advanced methods like the ones described in [30] and [31].

In a Monte Carlo simulation for a coupled two-slab system (C-MC), the solution of the RTE in Eq. (12) with source terms as given in Eqs. (14)–(15) is obtained by following the life histories of a very large number of photons. Each photon enters the coupled system at the top of slab_1, and is followed until it leaves the system either by absorption in either of the slabs (or at the bottom of slab_2) or by backscattering out of slab_1.

From the absorption and scattering coefficients [Eqs. (1) and (2)] two very useful non-dimensional IOPs can be defined. First, the *single-scattering albedo* appearing in the RTE (Eq. (12))

$$a(\tau) \equiv \frac{\sigma(\tau)}{\sigma(\tau) + \alpha(\tau)}, \tag{34}$$

is the probability that a photon will be scattered rather than absorbed in an attenuation event. In a medium where the attenuation of the beam primarily is due to scattering, $a(\tau) \to 1$.

The second non-dimensional IOP, which also appears in the RTE (Eq. (12)), is the *vertical optical depth* τ, which in differential form is given by Eq. (13). Each of the two slabs are divided into a number of of horizontal plane-parallel layers to resolve vertical variations in their IOPs $(\sigma(\tau), \alpha(\tau), p(\tau,\mu',\phi';\mu,\phi))$. Thus, each layer in each of the two slabs has its own individual absorption coefficient α, scattering coefficient σ, asymmetry factor g for the HG scattering phase function (or more generally coefficients χ_ℓ for the Legendre polynomial expansion of the scattering phase function), and depolarization factor $\tilde{f}$ for the Rayleigh scattering phase function.

As an example, if each layer in each of the slabs can be described by three scattering coefficients, denoted by σ_{Ray}, σ_1, and σ_2 for Rayleigh scattering and scattering by non-Rayleigh scatterers of type 1 and 2, respectively, one may define two variables η_1 and η_2, such that $0 \leq \eta_1 \leq \eta_2 \leq 1$, by

$$\eta_1 = \frac{\sigma_1}{\sigma_1 + \sigma_2 + \sigma_{Ray}}, \tag{35}$$

$$\eta_2 = \frac{\sigma_1 + \sigma_2}{\sigma_1 + \sigma_2 + \sigma_{Ray}}. \tag{36}$$

These two variables can be used together with a random number ρ_1 to decide which scattering phase function and scattering coefficient to be used. Let the HG scattering phase function be assumed to be adequate, implying that the asymmetry factors g_1 and g_2 suffice to describe the scattering phase function for scatterers of type 1 and 2, respectively. Then for a coupled system the choice of scattering phase function, scattering coefficient, and asymmetry factor can be made as follows:

- If $\rho_1 \leq \eta_1$, then σ_1 and g_1 are used together with the HG scattering phase function.
- If $\eta_1 < \rho_1 \leq \eta_2$, then σ_2 and g_2 are used together with the HG scattering phase function.
- If $\rho_1 > \eta_2$, then σ_{Ray} and $g_{Ray} = 0$ are used together with the Rayleigh scattering phase function.

In MC simulations, the history of one single photon is tracked from it enters the coupled system at the top of slab$_1$ in a given direction until it reaches the position where it is either absorbed or scattered out of the medium. Then this procedure is repeated for a very large number of photons, each entering the coupled system at the top of slab$_1$ in the same direction as the first one.

The propagation of a photon through the coupled system is tracked one layer at the time. An initial path length s is first calculated, and then the path from the entrance point to the exit point of the first layer is calculated. The initial path length s is determined by noting that light traveling in a medium containing absorbers and scatterers will be attenuated according to the extinction law

$$I = I_0 \, e^{-\tau_s}. \tag{37}$$

Here $\tau_s = cs$ is called the *optical thickness*, and $c = \alpha + \sigma$ is the attenuation coefficient, which is the sum of the absorption coefficient α and the scattering coefficient σ. The optical thickness τ_s is a dimensionless quantity that is related to the *vertical optical depth* τ given previously by $\tau = \tau_s \cos\theta$.

Normalizing the extinction law (so that $\int_0^\infty I \, ds = 1$), one finds that $I_0 = c$ in Eq. (37), so that

$$I_{normalized} = c \, e^{-cs}.$$

The corresponding cumulative function becomes

$$P(s) = \int_0^s c\, e^{-cs'}\, ds' = 1 - e^{-cs},$$

which leads to

$$s = -\frac{\ln(1-\rho_2)}{c}, \tag{38}$$

where $P(s)$ has been replaced by ρ_2. If ρ_2 is a random number uniformly distributed between zero and one, so is $(1-\rho_2)$. Therefore it follows from Eq. (38) that

$$s = -\frac{\ln(\rho_2)}{c} \tag{39}$$

where s is the path length of a photon in a medium with attenuation coefficient c as a function of a random number $\rho_2 \in [0,1]$.

Before the path length in the next layer is calculated, the IOPs are set equal to the IOP values of the new layer, and this process continues until the path length is totally 'consumed', at which time the photon has reached a specific layer. Then the photon interacts with the medium in this layer with single-scattering albedo a, either through absorption or scattering. The single-scattering albedo a gives the probability that a photon is scattered in an attenuation event (Eq. (34)). Thus, to decide if the interaction is absorption or scattering, a random number ρ_3 is drawn. If $\rho_3 > a$, the photon is absorbed and a new photon is allowed to enter the system at the top of the slab. If $\rho_3 \leq a$, the photon is scattered, and the choice of scattering phase function is determined by the size of the random number ρ_1 compared to η_1 and η_2, as described previously. The scattering angle is found from the scattering phase function, and is used to determine the new direction of the photon. This process is then repeated, i.e. one determines the distance to the next interaction, the type of interaction, and if necessary the new direction.

When a photon hits the slab$_1$–slab$_2$ interface at which there is a change in the refractive index, a new random number ρ_4 is drawn to decide if the photon is reflected or refracted. For unpolarized light incident upon the interface between the two slabs, the Fresnel reflectance ρ_F is given by Eq. (22), and the Fresnel transmittance by Eq. (23). The random number ρ_4 is used to decide the fate of the photon. If $\rho_4 \leq \rho_F$, the photon is reflected at the interface, and if $\rho_4 > \rho_F$, it is transmitted into slab$_2$.

5.3.4 Impact of surface roughness on remotely sensed radiances

The bidirectional reflectance distribution function (BRDF) is defined as [4]:

$$\rho(-\hat{\mathbf{\Omega}}', \hat{\mathbf{\Omega}}) = \frac{I_r^+(\hat{\mathbf{\Omega}})}{I^-(\hat{\mathbf{\Omega}}')\cos\theta'\, d\omega'}. \tag{40}$$

Here $I^-(\hat{\mathbf{\Omega}}')\cos\theta'\, d\omega'$ is the radiant energy incident on a flat surface due to an angular beam of radiation with radiance $I^-(\hat{\mathbf{\Omega}}')$ within a cone of solid angle $d\omega'$

around direction $\hat{\boldsymbol{\Omega}}'(\theta', \phi')$, whereas $I_r^+(\hat{\boldsymbol{\Omega}})$ is the radiance of reflected light leaving the surface within a cone of solid and $d\omega$ around the direction $\hat{\boldsymbol{\Omega}}(\theta, \phi)$. θ' is the polar angle between the incident beam direction $\hat{\boldsymbol{\Omega}}'$ and the normal to the surface.

In the preceeding, the interface between the two slabs was assumed to be flat, but natural surfaces are not flat. For example, if the ocean surface were flat, a perfect image of the Sun's disk would be observed in the specular direction. The effect of surface roughness is to spread the specular reflection over a range of angles referred to as the sunglint region in the case of reflections from a wind-roughened surface. If the surface is characterized by a Gaussian random height distribution $z = f(x, y)$ with mean height $\langle z \rangle = \langle f(x, y) \rangle = 0$, and the tangent plane approximation is invoked, according to which the radiation fields at any point on the surface are approximated by those that would be present at the tangent plane at that point [32], the BRDF in Eq. (40) can be expressed as [32–34]

$$\rho_s(\mu', \mu, \Delta\phi) = \frac{\pi P(z_x, z_y) \rho_F(m_1, m_2, \mu', \mu, \Delta\phi)}{4\mu'\mu \cos^4 \beta} \tag{41}$$

where $\Delta\phi = \phi' - \phi$, $\rho_F(m_1, m_2, \mu', \mu, \Delta\phi)$ is the Fresnel reflectance, β is the tilt angle between the vertical and the normal to the tangent plane, and $P(z_x, z_y)$ is the surface slope distribution. For a Gaussian rough surface

$$P(z_x, z_y) = \frac{1}{2\pi\sigma_x\sigma_y} \exp\left[-\frac{z_x^2 + z_y^2}{2\sigma_x\sigma_y}\right] \tag{42}$$

where z_x and z_y are the local slopes in the x and y directions, and σ_x and σ_y are the corresponding mean square surface slopes. Since

$$z_x = \frac{\partial f(x, y)}{\partial x} = \sin\alpha \tan\beta \quad \text{and} \quad z_y = \frac{\partial f(x, y)}{\partial y} = \cos\alpha \tan\beta$$

where α is the azimuth of ascent (clockwise from the Sun), one finds $z_x^2 + z_y^2 = \tan^2\beta$. For an isotropic rough surface, $\sigma_x = \sigma_y$, and $\sigma^2 = \sigma_x^2 + \sigma_y^2 = 2\sigma_x^2 = 2\sigma_x\sigma_y$. Thus, one obtains

$$P(z_x, z_y) = \frac{1}{\pi\sigma^2} \exp\left[-\frac{\tan^2\beta}{\sigma^2}\right]. \tag{43}$$

A corresponding expression for the transmittance $\mathcal{T}$ can be derived as well [32,33].

To analyze remotely sensed radiances obtained by instruments such as the Sea-viewing Wide Field of view Sensor (SeaWiFS, on-board SeaStar), the MODerate-resolution Imaging Spectroradiometer (MODIS, deployed on both the Terra and Aqua spacecraft), and the MEdium Resolution Imaging Spectrometer (MERIS, deployed onboard the European Space Agency (ESA)'s Envisat platform, NASA has developed a comprehensive data analysis software package (SeaWiFS Data Analysis System, SeaDAS), which performs a number of tasks, including cloud screening and calibration, required to convert the raw satellite signals into calibrated top-of-the-atmosphere (TOA) radiances. In addition, the SeaDAS software package has tools for quantifying and removing the atmospheric contribution to the TOA radiance ('atmospheric correction') as well as the contribution due to whitecaps and sunglint due to reflections from the ocean surface [34].

If one ignores effects of shadowing and multiple reflections due to surface facets, the sunglint reflectance can be expressed by Eq. (41) with the distribution $P(z_x, z_y)$ of surface slopes as given in Eq. (43), where $\sigma^2 = 0.003 + 0.00512$ WS, WS being the wind speed in m/s [19, 61, 62].

The directly transmitted radiance (DTR) approach

The sunglint radiance can be expressed as a function of the following variables

$$I_{glint}^{TOA} \equiv I_{glint}^{TOA}(\mu_0, \mu, \Delta\phi, \mathrm{WS}, \mathrm{AM}, \tau_{tot}, \lambda)$$

where the angles μ_0, μ, and $\Delta\phi$ define the sun-satellite geometry, WS is the wind speed, and λ the wavelength. The atmosphere is characterized by its total optical depth τ_{tot}, and the choice of an aerosol model (AM).

In the SeaDAS algorithm a sunglint flag is activated for a given pixel when the reflectance or BRDF, as calculated from Eq. (41) with the slope distribution in Eq. (43), exceeds a certain threshold. If the reflectance for a given pixel is above the threshold, the signal is not processed. If the reflectance is below the threshold, a directly transmitted radiance (DTR) approach is used to calculate the TOA sunglint radiance in the SeaDAS algorithm. Thus, it is computed assuming that the direct beam and its reflected portion only experience exponential attenuation through the atmosphere [35], i.e.

$$I_{glint}^{TOA}(\mu_0, \mu, \Delta\phi) = F^s(\lambda) T_0(\lambda) T(\lambda) I_{GN}, \tag{44}$$

$$T_0(\lambda) T(\lambda) = \exp\left\{ -[\tau_M(\lambda) + \tau_A(\lambda)] \left(\frac{1}{\mu_0} + \frac{1}{\mu} \right) \right\}, \tag{45}$$

where the normalized sunglint radiance I_{GN} is the radiance that would result in the absence of the atmosphere if the incident solar irradiance were $F^s = 1$, and where τ_M and τ_A ($\tau_{tot} = \tau_M + \tau_A$) are the Rayleigh (molecular) and aerosol optical thicknesses. Multiple scattering is ignored in the DTR approach, implying that photons removed from the direct beam path through scattering will not be accounted for at the TOA.

The multiply scattered radiance (MSR) approach

Whereas the DTR approach accounts only for the direct beam (Beam 2 in Fig. 5.1), the multiply scattered radiance (MSR) approach is based on computing the TOA radiance by solving Eq. (12) subject to the boundary condition:

$$\begin{aligned} I(\tau_1, \mu, \phi) &= \frac{\mu F^s}{\pi} e^{-\tau_1/\mu_0} \rho_{glint}(-\mu_0, \phi_0; \mu, \phi) \\ &+ \frac{1}{\pi} \int_0^{2\pi} d\phi' \int_0^1 d\mu' \rho_{glint}(\tau, -\mu', \phi'; \mu, \phi) I(\tau, \mu', \phi') \end{aligned} \tag{46}$$

thereby allowing multiple scattering to be included in the computation. Here τ_1 is the optical thickness of slab$_1$, and $\rho_{glint}(-\mu_0, \phi_0; \mu, \phi) = \rho_s(-\mu_0, \mu, \Delta\phi)$ (see

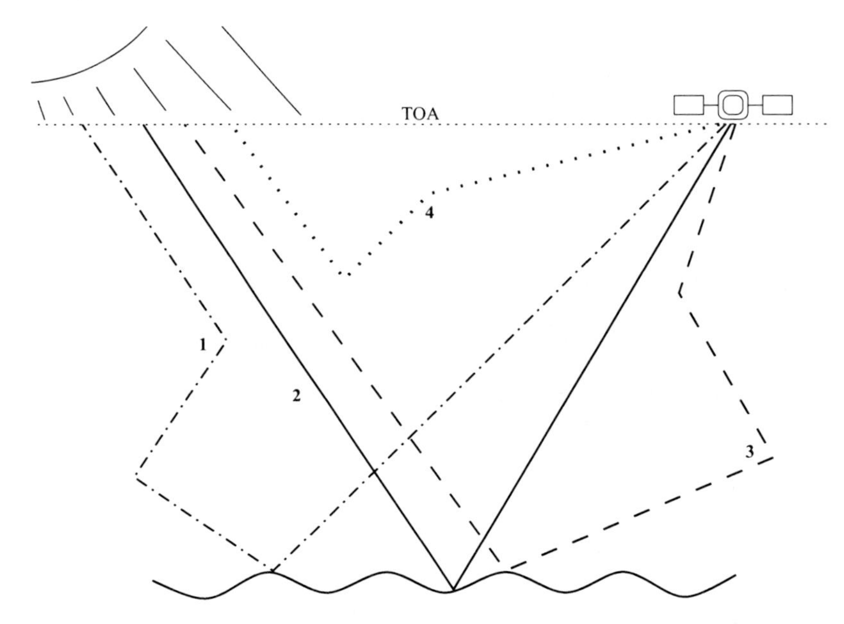

Fig. 5.1. Schematic illustration of various contributions to the TOA radiance in the case of a wind-roughened ocean surface. (1) Diffuse downward component reflected from the ocean surface; (2) direct, ocean-surface reflected beam; (3) beam undergoing multiple scattering after ocean-surface reflection, and (4) (multiply) scattered beam reaching the TOA without hitting the ocean surface (adapted from Ottaviani et al. [36].)

Eq. (41)) is the BRDF of the slab$_1$–slab$_2$ interface. For consistency with the definition of sunglint, radiation reflected from the surface after being scattered on its way down to the ocean surface (Beam 1 in Fig. 5.1) is neglected by not allowing the presence of a downward diffuse term in Eq. (46). The complete solution of Eq. (12) and (46) gives the total TOA radiance $I^{tot}_{TOA}(\mu_0, \mu, \Delta\phi)$, which includes light scattered into the observation direction without being reflected from the ocean surface (Beam 4 in Fig. 5.1). This contribution is denoted by $I^{bs}_{TOA}(\mu_0, \mu, \Delta\phi)$, since it can be computed by considering a black or totally absorbing ocean surface, for which $\rho_{glint} = 0$ in Eq. (46). To isolate the glint contribution one must subtract this 'black-surface' component from the complete radiation field:

$$I^{glint}_{TOA}(\mu_0, \mu, \Delta\phi) = I^{tot}_{TOA}(\mu_0, \mu, \Delta\phi) - I^{bs}_{TOA}(\mu_0, \mu, \Delta\phi). \tag{47}$$

Equation (47) includes multiply scattered reflected radiation, but ignores multiply scattered sky radiation undergoing ocean-surface reflection (Beam 1 in Fig. 5.1). Thus, it guarantees that the difference between the TOA radiances obtained by the DTR and MSR approaches is due solely to that component of the TOA radiance, which is scattered along its path from the ocean surface to the TOA (Beam 3 in Fig. 5.1). In order to quantify the error introduced by the DTR assumption, Ottaviani et al. [36] used a fully coupled atmosphere–ocean discrete ordinate code with a Gaussian (Cox–Munk, Eq. (43)) surface slope distribution [17].

Comparison of DTR and MSR

To correct for the sunglint signal, Wang and Bailey [35] added a procedure to the SeaDAS algorithm based on the DTR assumption, which ignores multiple scattering in the path between the ocean surface and the TOA as well as in the path from the TOA to the ocean surface. To quantify the error introduced by the DTR assumption, Ottaviani et al. [36] neglected the effect of whitecaps as well as the wavelength dependence of the refractive index. Figure 5.2 shows a comparison of DTR and MSR results at 490 nm for several wind speeds and different aerosol types and loads. The incident solar irradiance was set to $F^s = 1$, giving the sun-normalized radiance, and in the computation of the Fresnel reflectance in Eq. (22) the imaginary part of the refractive index n_2' was assumed to be zero. A standard molecular atmospheric model (mid-latitude) with a uniform aerosol distribution below 2 km was used in the computations. Thus, below 2 km the aerosol optical

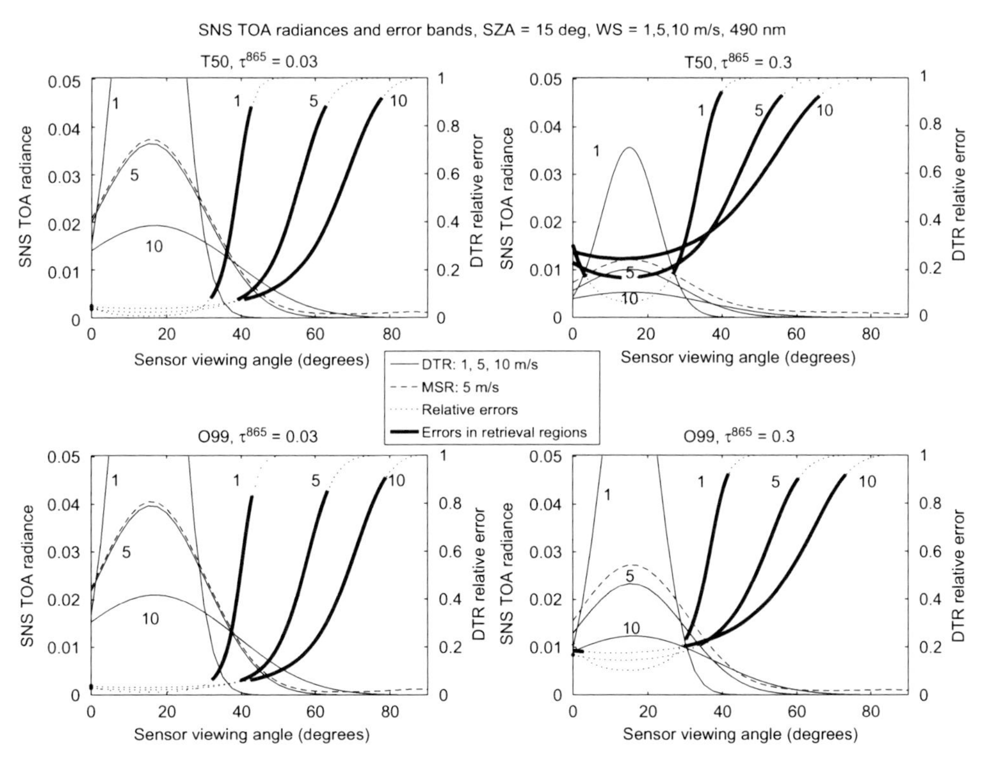

Fig. 5.2. Sun-normalized sunglint TOA radiance (solid and thin curves) at 490 nm for a SZA of 15°, along the principal plane of reflection, and relative error incurred by ignoring multiple scattering along the path from the surface to the TOA (dotted curves). Each plot contains 3 representative wind speeds (1, 5, and 10 m/s). The upper row pertains to small aerosol particles in small amounts ($\tau = 0.03$, left panel) and larger amounts ($\tau = 0.3$, right panel). The bottom row is similar to the top one, but for large aerosol particles. The error curves have been thickened within the angular ranges in which retrievals are attempted (corresponding to $0.0001 \leq I_{TOA}^{tot} \leq 0.001$ in normalized radiance units) (adapted from Ottaviani et al. [36]).

thickness due to scattering (τ_A^s) and absorption (τ_A^a) was added to the molecular optical thickness τ_M:

$$\tau_{tot} = \tau_M + \tau_A = (\tau_M^s + \tau_M^a) + (\tau_A^s + \tau_A^a). \tag{48}$$

The IOPs for aerosols were computed by a Mie code [37], and the IOPs of a multi-component mixture were then obtained as a concentration-weighted average of the IOPs of each aerosol component [38].

The upper panels in Fig. 5.2 pertain to small aerosol particles with optical depths of 0.03 and 0.3, while the lower panels are for large aerosol particles. The DTR curves are shown for wind speeds of 1, 5, and 10 m/s, while only one MSR curve at 5 m/s is shown for clarity. The errors incurred by ignoring multiple scattering in the path from the surface to the TOA are high, typically ranging from 10% to 90% at 490 nm (Fig. 5.2). These error ranges are determined by the radiance threshold values that mark the retrieval region boundaries; the errors are smaller closer to the specular reflection peak (higher threshold). Surface roughness only affects the angular location and extent of the retrieval region where these errors occur. The minimum errors grow significantly in an atmosphere with a heavy aerosol loading, and asymmetries are found close to the horizon, especially in the presence of large (coarse-mode) particles.

Figure 5.2 pertains to the principal plane. Similar computations showed that the errors are azimuth-dependent [36]. Thus, in a typical maritime situation the errors tend to grow as the radiance decreases away from the specular direction, and the high directionality of the radiance peak at low wind speeds causes larger minimum errors away from the principal plane. Correcting for sunglint contamination including multiple scattering effects in future processing of ocean color satellite data is feasible, and would be desirable in view of the magnitude of the errors incurred by the DTR approach.

5.4 Applications

5.4.1 Coupled atmosphere–water systems

Figure 5.3 illustrates the transfer of solar radiation in a coupled atmosphere–ocean system. For this kind of system, Mobley et al. [1] presented a comparison of underwater light fields computed by several different methods including MC methods [39–58], invariant imbedding [59,60], and the discrete-ordinate method [2], demonstrating similar results for a limited set of test cases. However, these comparisons were qualitative rather than quantitative because of the different ways in which the models treated the radiative transfer in the atmosphere, leading to a spread in the downwelling irradiance just above the water surface of 18%, which persisted throughout the underwater column.

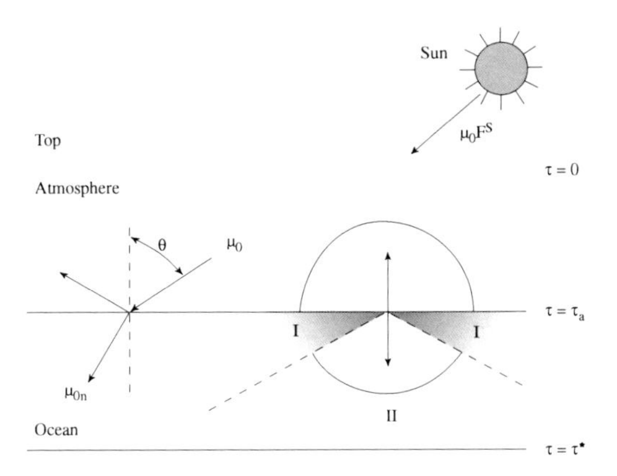

Fig. 5.3. Schematic illustration of the atmosphere and ocean with incident solar irradiance $\mu_0 F^s$ and optical depth, increasing downward from $\tau = 0$ at the top-of-the-atmosphere. The incident polar angle is $\cos^{-1}\mu_0$, which after refraction according to Snell's law changes into the angle $\cos^{-1}\mu_{0n}$. Since the ocean has a larger refractive index than the atmosphere, radiation distributed over 2π sr in the atmosphere will be confined to a cone less than 2π sr in the ocean (region II). Upward radiation in the ocean with directions in region I will undergo total internal reflection at the ocean–air interface (adapted from Thomas and Stamnes [4]).

Comparisons of C-DISORT and C-MC Results

Gjerstad et al. [5] compared irradiances obtained from a MC model for the coupled atmosphere–ocean system (C-MC) with those obtained from a discrete-ordinate method (C-DISORT). By treating the scattering and absorption processes in the two slabs in the same manner in both methods, they were able to provide a more detailed and quantitative comparison than those previously reported [1]. Figure 5.4 shows a comparison of direct and diffuse downward irradiances computed with the C-MC and C-DISORT codes, demonstrating that when precisely the same IOPs are used in the two models, computed irradiances agree to within 1% throughout the coupled atmosphere-ocean system.

One shortcoming of the results discussed above is that the interface between the two media with different refractive indices was taken to be flat. This flat-surface assumption limits the applications of C-DISORT, because a wind-roughened ocean surface is a randomly scattering object. In addition to affecting the backscattered radiation, the surface roughness significantly affects the directional character of the radiation transmitted through the air–water interface. To deal with the surface roughness in the discrete-ordinate method, Gjerstad et al. [5] proposed an ad hoc method of mimicking the irradiances obtained from a C-MC model by adjusting the refractive index in C-DISORT. The limitations incurred by this ad hoc method were removed by Jin et al. [61], who presented a consistent and widely applicable solution of the discrete ordinate RT problem in a coupled atmosphere–ocean system with a rough surface interface having a Gaussian wave slope distribution given by Eq. (43). They concluded that the ocean surface roughness has significant effects on the upward radiation in the atmosphere and the downward radiation in the

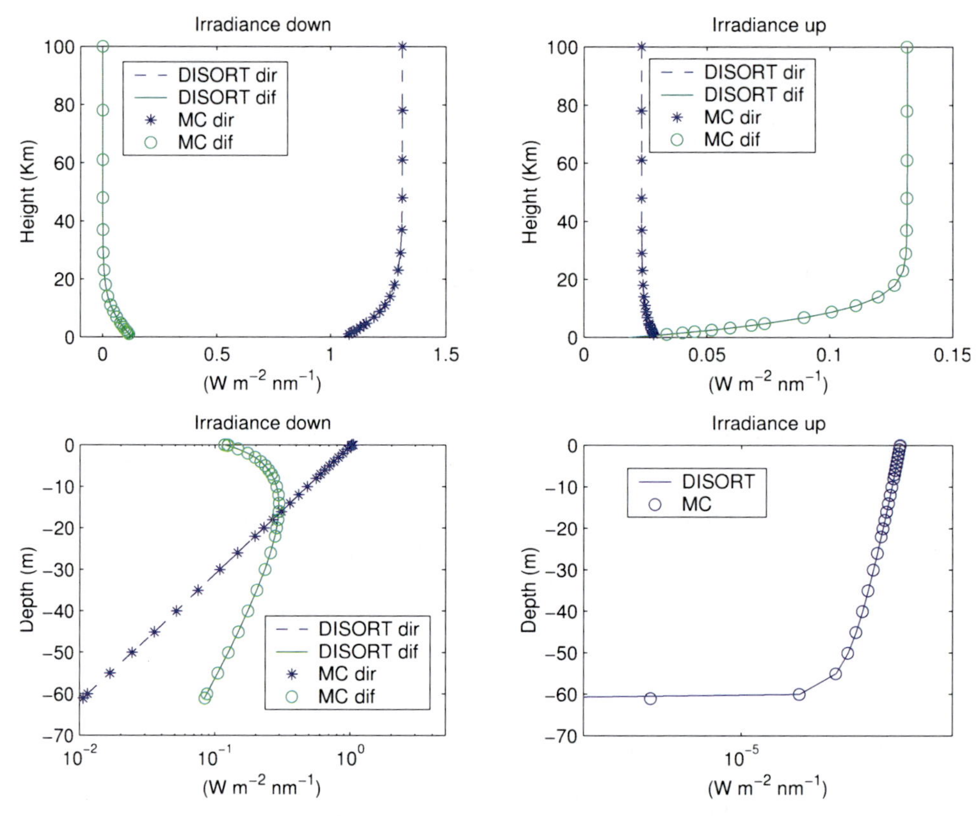

Fig. 5.4. Comparison of irradiance results obtained with C-DISORT and a C-MC code for radiative transfer in a coupled atmosphere–ocean system. The simulations are for an atmosphere containing only molecular absorption and Rayleigh scattering, and for an ocean having no Rayleigh scattering, only absorption and scattering from a chlorophyll concentration of 0.02 mg/m^3, uniformly distributed to a depth of 61 m, below which the albedo is zero (adapted from Gjerstad et al. [5]).

ocean. As the wind speed increases, the angular domain of the sunglint broadens, the surface albedo decreases, and the transmission of radiation through the air–water interface into the ocean increases. The transmitted radiance just below a flat ocean surface is highly anisotropic, but this anisotropy decreases rapidly as the surface wind speed increases. Also, the anisotropy will decrease as the water depth increases because multiple scattering in the ocean interior eventually will make larger contributions to reduce the anisotropy than the surface roughness. The effects of surface roughness on the radiation field depend on both wavelength and angle of incidence (i.e. solar elevation). Although their model predictions of the impact of surface roughness agreed reasonably well with observations, Jin et al. [61] cautioned that the original Cox–Munk surface roughness model [62] adopted in the simulations (Eq. (43)) may be inadequate for high wind speeds.

Simultaneous retrieval of aerosol and marine parameters

Traditional remote sensing ocean color algorithms start by application of an atmospheric correction step to estimate the aerosol optical thickness at a single near infrared channel (865 nm for SeaWiFS), for which the ocean is assumed to be non-scattering (the black-pixel approximation). Based on this atmospheric correction, water-leaving radiances are generated for visible channels [34, 63]. Next, marine constituents are estimated from two or three visible-channel water-leaving radiances, either through regression or by look-up table (LUT) matching based on a suitable bio-optical model. One shortcoming of this approach is that the black-pixel approximation may not be valid [64]. Also, the atmospheric correction step is based on the assumption that the radiation in the atmosphere can be decoupled from that in the ocean, which is potentially a large source of uncertainty, because the oceanic contribution to the total TOA radiance is typically less than 10%. Further, it is difficult if not impossible to quantify systematically error sources in such two-step ad hoc inversion procedures.

To remedy these shortcomings Stamnes et al. [65] devised a one-step iterative inversion scheme, based on simulated radiances stored in LUTs, for simultaneous retrieval of two aerosol parameters and one ocean parameter (chlorophyll concentration). To minimize uncertainties caused by forward model assumptions, an accurate RT model for the coupled atmosphere–ocean system (C-DISORT [2, 3]) was used. Atmospheric correction was not treated separately, since any atmospheric effects were fully integrated in the coupled RT model.

The Stamnes et al. [65] one-step algorithm cannot easily be extended beyond the estimation of three (two aerosol and one ocean) parameters. Therefore, Li et al. [16] developed a new method which employs a linearized version L-CDISORT [18] of the coupled RT code, and simultaneously uses all available visible and near-infrared SeaWiFS measurements (eight channels at 412, 443, 490, 510, 555, 670, 765, and 865 nm). L-CDISORT computes not only radiances, but also Jacobians (radiance partial derivatives) that are required for inversion by standard methods such as the iterative fitting technique based on nonlinear least squares or optimal estimation (OE [66]). According to the new method [16], the retrieval parameters contained in the retrieval state vector includes both boundary-layer aerosol parameters and several marine parameters. At each iteration step, the L-CDISORT forward model is linearized about the current estimate of the retrieval state vector, and used to generate both simulated radiances and Jacobians with respect to the state vector elements and other parameters that are required in the OE fitting.

Atmospheric IOPs

For altitudes up to 2 km, Li et al. [16] used a bimodal aerosol model, in which the IOPs are defined in terms of the optical depth τ_0 at 865 nm and the fractional weighting f between the two aerosol modes:

$$\tau_{aer} \equiv \tau_0 e_{aer} = \tau_0 \left[(1-f)e_1 + f e_2\right], \tag{49}$$

$$\sigma_{aer} \equiv \tau_0[(1-f)\omega_1 e_1 + f\omega_2 e_2]; \quad \omega_{aer} = \frac{\sigma_{aer}}{e_1 + e_2}, \tag{50}$$

$$\chi_{aer,l} = \frac{(1-f)\omega_1 e_1 \beta_{1,l} + f\omega_2 e_2 \beta_{2,l}}{\sigma_{aer}}. \tag{51}$$

Here τ_{aer} is the total extinction aerosol optical depth in the layers containing aerosols, σ_{aer} is the total aerosol scattering optical depth, ω_{aer} the single-scattering albedo for aerosols, and $\chi_{aer,l}$ the total Legendre polynomial expansion coefficients for the bimodal aerosol mixture. The two atmospheric retrieval parameters are τ_0 and f. All other quantities in (49)–(51) are assumed model parameterizations: e_1, ω_1, and $\chi_{1,l}$ are respectively the extinction coefficient normalized to the value at 865 nm, single-scattering albedo, and scattering phase function expansion coefficients for aerosol type 1 ('fine-mode'), and e_2, ω_2, and $\chi_{2,l}$ are the corresponding values for aerosol type 2 ('coarse-mode'). The fine mode (subscript 1) was assumed to be a tropospheric aerosol model with 70% humidity, while the coarse mode (subscript 2) was a coastal aerosol model with 99% humidity. IOPs were calculated for the eight SeaWiFS channels using a Mie program for spherical particles with size and refractive index depending on humidity [37, 67]. A justification for adopting just one large and one small aerosol model (instead of several models of each type) can be found elsewhere [68]. To obtain the total IOPs in the marine boundary layer (MBL) containing aerosols, the Rayleigh scattering coefficient σ_{Ray} and the molecular absorption coefficient α_{gas} are also needed. Rayleigh scattering cross-sections and depolarization ratios were taken from standard sources. Absorption by O_3 (visible channels), O_2 (A-band) and water vapor was included.

Marine IOPs

In the ocean, IOPs can be derived from simple wavelength-dependent parameterizations of the phytoplankton absorption coefficient $a_{ph}(\lambda)$ in terms of the overall chlorophyll concentration CHL in [mg·m^{-3}], and the detrital and colored dissolved material (CDM) absorption coefficient $a_{dg}(\lambda)$ and the constituent backscattering coefficient $b_{bp}(\lambda)$ in terms of their respective values $CDM \equiv a_{dg}(\lambda_0)$ and $BBP \equiv b_{bp}(\lambda_0)$ at some reference wavelength λ_0 [16]:

$$a_{ph}(\lambda) = \alpha_1(\lambda) CHL^{\alpha_2(\lambda)}, \tag{52}$$

$$a_{dg}(\lambda) = CDM \exp^{[-S(\lambda-\lambda_0)]}, \tag{53}$$

$$b_{bp}(\lambda) = BBP \left(\frac{\lambda}{\lambda_0}\right)^{-\eta}. \tag{54}$$

Thus, this bio-optical model is described by the three retrieval elements CHL, CDM, and BBP, and the four model parameters $\alpha_1(\lambda), \alpha_2(\lambda), S$, and η. For α_1 and α_2, wavelength-dependent coefficients are determined by fitting the power-law expression in (52) to field measurements of chlorophyll absorption. From spectral fittings of measurements to the expressions for $a_{dg}(\lambda)$ and $b_{bp}(\lambda)$ in (53) and (54), it was found [16] that $S = 0.012$ and $\eta = 1.0$. All coefficients are in units of [m^{-1}]. Together with the pure water absorption and scattering coefficients $a_w(\lambda)$ and $b_w(\lambda)$ [69, 70] expressed in the same units, the layer total optical depth and total single-scattering albedo, and Legendre polynomial expansion coefficients IOPs for the marine medium become:

$$\tau_{tot} = d\left[a_{ph}(\lambda) + a_{dg}(\lambda) + b_p(\lambda) + a_w(\lambda) + b_w(\lambda)\right], \quad (55)$$

$$\omega = d\frac{b_p(\lambda) + b_w(\lambda)}{\tau_{tot}}, \quad (56)$$

$$\beta_l = \frac{b_p(\lambda)\beta_{l,FF} + b_w(\lambda)\beta_{l,water}}{b_p(\lambda) + b_w(\lambda)}, \quad (57)$$

where d is the layer depth in [m].

The particle size distribution (PSD) function in oceanic water is frequently described by an inverse power law (Junge distribution) $F(r) = \phi/r^{\gamma}$, where F(r) is the number of particles per unit volume per unit bin width, r [μm] is the radius of the assumed spherical particle, ϕ [$\text{cm}^{-3}\cdot\mu\text{m}^{\gamma-1}$] is the Junge coefficient, and γ is the PSD slope, which typically varies between 3.0 and 5.0 [71,72]. Based on this PSD, Forand and Fournier [73] derived an analytic expression for the scattering phase function. Values of $\phi = 1.069$ and $\gamma = 3.38$ were chosen to give a backscattering fraction of 0.005, and are consistent with a certain mixture of living organisms and re-suspended sediments [74]. A moment-fitting code [75] was used to generate Legendre polynomial expansion coefficients for the FF (Forand–Fournier) scattering phase function. Linearized IOPs were obtained by differentiation.

Inverse modeling

In OE the update of retrieval state vector $\mathbf{x}_n$ at iteration step n is given by [66]

$$\mathbf{x}_{n+1} = \mathbf{x}_n + \mathbf{G}_n\{\mathbf{K}_n^T\mathbf{S}_m^{-1}(\mathbf{y}_{\mathbf{meas}} - \mathbf{y}_n) - \mathbf{S}_a^{-1}(\mathbf{x}_n - \mathbf{x}_a)\}, \quad (58)$$

where

$$\mathbf{G}_n = [(1+\gamma_n)\mathbf{S}_a^{-1} + \mathbf{K}_n^T\mathbf{S}_m^{-1}\mathbf{K}_n]^{-1}, \quad (59)$$

and where the superscript ^{T} denotes matrix transpose.

The measurement vector $\mathbf{y}_{meas}$ has covariance error matrix $\mathbf{S}_m$, $\mathbf{y}_n = \mathbf{F}(\mathbf{x}_n)$ are simulated radiances generated by the forward model $\mathbf{F}(\mathbf{x}_n)$, which is a (nonlinear) function of $\mathbf{x}_n$. $\mathbf{K}_n$ is the Jacobian matrix of simulated radiance partial derivatives with respect to $\mathbf{x}_n$. The *a priori* state vector is $\mathbf{x}_a$, with covariance $\mathbf{S}_a$. $\mathbf{G}_n$ is the gain matrix of contribution functions. The Levenberg–Marquardt regularization parameter γ_n is chosen at each step to minimize the cost function. When $\gamma_n \to 0$, this step tends to the Gauss–Newton formula, and when $\gamma_n \to \infty$, it tends to the steepest descent method. One may start with $\gamma_n = 0.01$. The inverse process starts from an initial guess $\mathbf{x}_0$; often set to $\mathbf{x}_a$. Also, one may use the previous pixel's retrieved values as the next pixel's initial values. At each step, a convergence criterion is employed to check progress towards the solution $\mathbf{x}$ that minimizes the cost function. If the error decreases, one updates $\mathbf{x}_n$ and decreases γ_n for the next step. If the error increases, one increases γ_n, keeps $\mathbf{x}_n$ the same, and tries again.

To illustrate the method, Li et al. [16] considered a SeaWiFS image over Santa Barbara Channel obtained on February 28, 2003. The forward and inverse models described above were used for simultaneous retrieval of the 5-element state vector $\{\tau_{865}, f, CHL, CDM, BBP\}$. Figure 5.5 shows retrieved values of the four parameters $\{\tau_{865}, f, CDM, BBP\}$. The most probable value for the aerosol optical depth is about 0.04, with a range between 0.002 and 0.10. The aerosol fraction f ranges

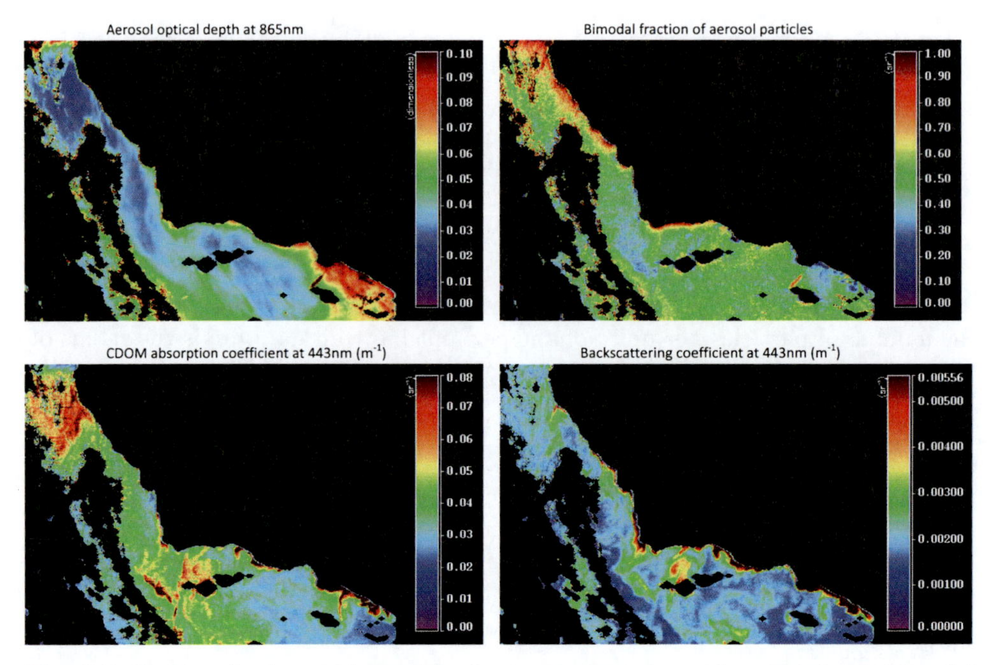

Fig. 5.5. Retrieved values of four of the five parameters: aerosol optical depth, bimodal fraction of aerosols, CDM absorption coefficient at 443 nm, and backscattering coefficient at 443 nm (adapted from Stamnes et al. [76]).

from about 0.2 (predominantly small particles) to 0.9. On average there appears to be equal amounts of small and large particles for this image. The CDM absorption coefficient (Eq. (53)) has a maximum value of around 0.04 m^{-1} and lies between 0.02 and 0.07 m^{-1}. The backscattering coefficient BBP lies between 0.001 and 0.005 m^{-1} with a peak at 0.002 m^{-1}.

The retrieved chlorophyll concentration, shown in Fig. 5.6, ranges from near 0 to about 3.0 mg·m^{-3}. In contrast to the traditional two-step 'atmospheric correction and regression' approach, the simultaneous retrieval described above produces a direct assessment of the error by examining sensor radiance residuals, summarized in the table inserted in Fig. 5.6. For the nearly 35,000 pixels in this SeaWiFS image the residuals are less than 1% for 7 of the 8 SeaWiFS channels, and less than 2% for the remaining 765 nm (O_2 A-band) channel. It may be concluded that this simultaneous forward/inverse retrieval method yielded excellent retrieval capability, and that 8 SeaWiFS channels were sufficient to retrieve 2 atmospheric and 3 marine parameters in coastal waters. In addition to well-calibrated SeaWiFS data the good results are believed to be due to the availability of high-quality field data used to construct a reliable bio-optical model, and an adjustable bimodal fraction of large versus small aerosol particles.

The only drawback with the OE approach, as described above, is the slow speed. The most time-consuming step in the inversion process is the L-CDISORT forward model computations. It is possible, however, to reach operational speed with a fast forward model trained by using neural-network radial-basis functions. It has been

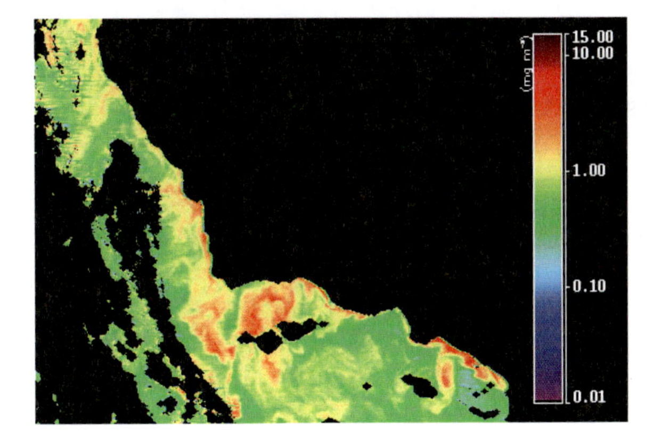

Left panel: Retrieved chlorophyll concentration (mg · m^{-3}) from SeaWiFS image on Feb. 28, 2003 over the Santa Barbara Channel.

Right top panel: The distributions of the other retrieved parameters from the same image. (a) aerosol optical depth; (b) aerosol model fraction; (c) CDOM absorption coefficient at 443nm (m^{-1}); (d) backscattering at 443nm (m^{-1})

Right bottom table: Radiance residuals at all SeaWiFS channels.

Wavelength	Average relative error (%)	Pixels with <2% relative error (%)
412nm	±0.298	99.488
443nm	±0.289	99.625
490nm	±0.555	99.216
510nm	±0.716	98.552
555nm	±0.240	99.168
670nm	±1.050	94.267
765nm	±1.952	64.102
865nm	±0.857	94.745

Fig. 5.6. Retrieved chlorophyll concentration for the same SeaWiFS image as in Fig. 5.5, distribution of the other for parameters, and residuals (adapted from Stamnes et al. [76]).

demonstrated that this approach leads to a performance enhancement of about 1,500 [76].

5.4.2 Coupled atmosphere–snow–ice systems

The C-DISORT coupled RT model can be used to compute the BRDF, defined in Eq. (40), for sea ice as described in [77]. To quantify the BRDF one needs the backscattered radiance distribution as a function of the polar angles θ' and θ of incidence and observation, respectively, as well as the corresponding azimuth-difference angle $\Delta\phi = \phi' - \phi$. According to Eq. (19), the radiance distribution can be expressed as a Fourier cosine series in which the expansion coefficients $I^m(\tau, \mu)$ depend on the polar angles θ' and θ as well as on the sea ice IOPs. Each expansion coefficient satisfies Eq. (20), which is readily solved by C-DISORT to provide $I^m(\tau, \mu)$, and Eq. (19) then yields the complete angular distribution of the radiance as a function of θ', θ, and $\Delta\phi$. With no atmosphere assumed to be present, so that $S_1^*(\tau, \mu, \phi) = 0$ (Eq. (14)), the expansion coefficients $I^m(\tau, \mu)$ for a number of values of θ' (solar zenith angle), θ (observation angle), and sea ice IOPs were computed [77] and stored in a set of LUTs.

To create LUTs for the sea ice BRDF to be used for interpolation, Stamnes et al. [77] assumed the sea ice to float on water with a known albedo A_w, and the IOPs for a slab of sea ice to be characterized in terms of its optical thickness τ, its single-scattering albedo ω, and its asymmetry parameter g. Then C-DISORT was used to tabulate the expansion coefficients $I^m(\tau, \omega, g, \mu_0, \mu, A_w)$, which determine the sea ice BRDF as a function of τ, ω, g, μ_0, μ, and A_w. (Here θ_0 is used instead of θ' and $\mu_0 = \cos\theta_0$.) Stamnes et al. [77] also created a tool [*ISIOP*] for computing sea ice IOPs (τ, ω, and g) for any desired wavelength from sea ice physical parameters: real and imaginary parts of the sea ice refractive index, brine pocket concentration and effective size, air bubble concentration and effective size, volume fraction and absorption coefficient of sea ice impurities, asymmetry parameters for scattering by brine pockets and air bubbles, and sea ice thickness. This approach enabled a reliable computation of the wavelength-dependent BRDF as a function of sea ice IOPs. The BRDF for snow-covered sea ice was readily obtained by including snow as a 'cloud on top of the sea ice'.

A combination of the two different tools developed by Stamnes et al. [77]: (i) *ISIOP* for computing IOPs for ice and snow, as well as (ii) *ISBRDF* for computing the BRDF of sea ice (with or without snow cover), can be used to quantify the BRDF of sea ice in a very efficient manner. An example is shown in Fig. 5.7, where computed albedos of clean snow are compared with laboratory measurements. This figure shows that flux reflectances generated from BRDF values obtained using the *ISIOP* and *ISBRDF* tools agree well with computations done independently [78] as well as with experimental values [79]. In [77] it was shown that sea ice spectral albedo values derived from the ISIOP/ISBRDF tools are consistent with independently computed [80] as well as observed values [81, 82] for a variety of ice types and thicknesses.

In the work reviewed in this paper, it was assumed that snow/ice particles have spherical shapes so that Mie theory could be used to compute their IOPs. However, it should be kept in mind that for BRDF the non-spherical shape of ice crystals may be important [83–85].

Retrieval of snow/ice parameters from satellite data

It is well recognized that snow cover has a strong impact on the surface energy balance in any part of the world. Satellite remote sensing provides a very useful tool for estimating spatial and temporal changes in snow cover, and for retrieving snow optical characteristics. Numerous sensors have been used to retrieve snow optical properties, including Landsat TM (Thematic Mapper) data [86, 87], AVIRIS (the Airborne Visible/Infrared Imaging Spectrometer) data [88–90], and MODIS [91] data. The retrieval of snow grain size and impurity concentration is possible because snow reflectance depends primarily on the impurity concentration (assumed to be soot-contamination) in the visible range, but on snow grain size [92] in the NIR, as shown in Fig. 5.8.

The GLobal Imager (GLI) sensor was launched onboard Japan's Advanced Earth Observing Satellite II (ADEOS-II) on December 14, 2002. GLI was an optical sensor similar to MODIS that observed solar radiation reflected from the Earth's atmosphere and surface including land, oceans, and clouds as well as terrestrial infrared radiation. In addition to atmospheric parameters, GLI like MODIS, was designed to infer information about several other quantities including marine and land parameters such as chlorophyll concentration, dissolved organic matter, surface temperature, vegetation distribution and biomass, distribution of snow and ice, and albedo of snow and ice. The GLI sensor acquired data from April 2 to October 24, 2003. After that date no useful data were retrieved from ADEOS-II due to a power failure. Because the GLI sensor is similar in many respects to the

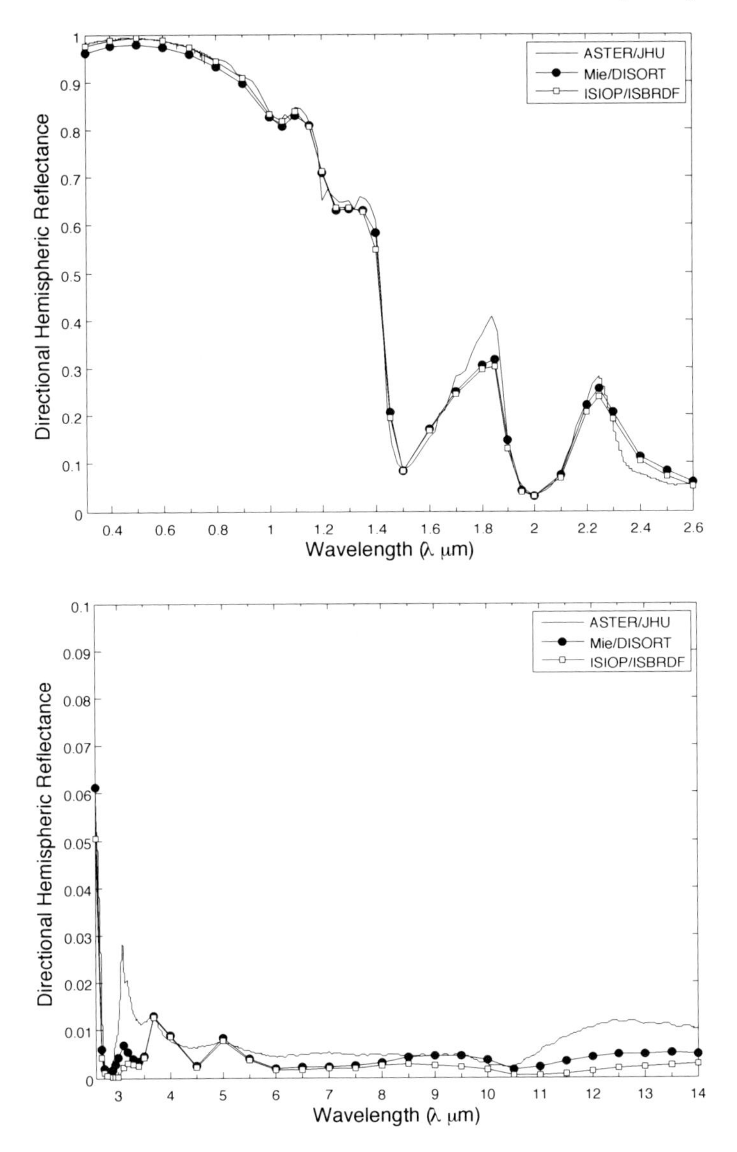

Fig. 5.7. Directional hemispherical reflectance of clean snow for 24 μm snow grain radius from (1) *ISIOP* computed IOPs and *ISBRDF*, (2) Mie computed IOPs and DISCORD [78] and (3) ASTER spectral library observations [79] for 10° solar zenith angle for the visible and near-infrared (upper) and infrared (lower) spectral regions (adapted from Stamnes et al. [77]).

MODIS sensor that was launched prior to ADEOS-II, algorithms developed to retrieve information about the cryosphere from GLI data [94] were tested by the use of MODIS data [95], and are therefore applicable also to data obtained with the MODIS sensor [96].

Snow can be regarded as a mixture of pure ice, air, liquid water, and impurities. Pure ice is highly transparent in the visible, so that an increase in snow grain size

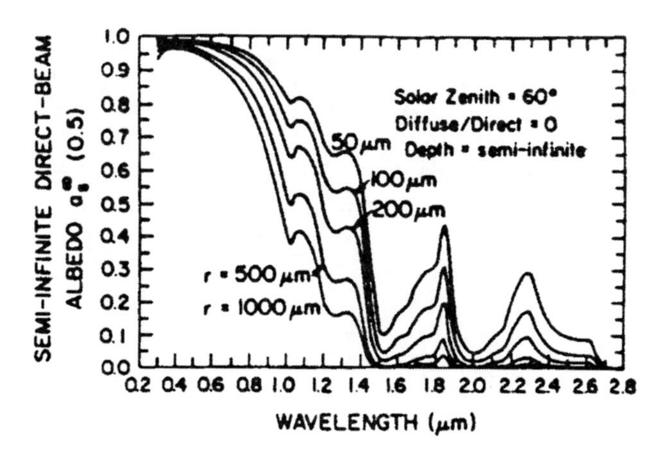

Fig. 5.8. Spectral albedo of snow as a function of wavelength for grain size 50–1000 μm (after Wiscombe and Warren [93]).

has little effect on the reflectance. However, because ice is moderately absorptive in the NIR, reflectance is sensitive to grain size, especially in the wavelength region 0.8–1.3 μm (see Fig. 5.8). For satellite measurements, spectral channels should be selected to lie in wavelength regions where the effect of atmospheric scattering and absorption is small, so that when the radiance values are atmospherically corrected to yield surface reflectance, errors in the characterization of the atmosphere, particularly atmospheric water vapor, are minimized [97–99]. For these reasons, GLI channels 19 (0.86 μm), 24 (1.05 μm), 26 (1.24 μm), and 28 (1.64 μm) would be suitable for retrieval of snow grain size, because the impact of changes in snow grain size is large whereas the effects of the atmosphere and snow impurities are relatively small [94].

Accurate estimates of surface temperature could provide an early signal of climate change, particularly in the Arctic, which is known to be quite sensitive to climate change. The surface temperature in the polar regions controls sea ice growth, snow melt, and surface–atmosphere energy exchange. During the past decade, significant progress was made in estimation of sea-surface temperature [100–102] and snow/ice surface temperature [103–106] from satellite thermal infrared data. Algorithms for surface temperature retrievals in the Arctic based on GLI measurements were developed [94] for retrieval of snow/ice surface temperature (IST), as well as for open-ocean surface temperature (SST). The SST algorithm can be applied to areas consisting of a mixture of snow/ice and melt ponds. GLI channels 35 (10.8 μm) and 36 (12.0 μm) were used in conjunction with RT simulations and a multi-linear regression technique to determine the empirical coefficients in the expression for the surface temperature [94].

Cloud mask and surface classification

In order to infer information about snow and ice properties from visible and IR satellite imagery a cloud mask is required to discriminate between clear and cloudy sky. For scenes that are determined to be cloud-free, the next step is to do a surface classification. Thus, for cloud-free pixels, algorithms are needed to (i) determine whether a given field of view is obstructed by clouds, and (ii) distinguish bare sea ice from snow-covered sea ice. The snow/sea ice discriminator is designed to discriminate bare sea ice from snow-covered sea ice during the bright polar summer. The surface is classified into five possible types: snow, sea ice, cloud shadow, land (tundra), and open ocean. When sea ice is covered by snow (even only a few centimeters), the surface radiative characteristics will be similar to snow [2,6,7]. Thus, sea ice covered by snow will be classified as a snow surface, while only bare sea ice is classified as sea ice.

Snow and sea ice cover and surface temperature

Figure 5.9 shows seasonal variations of the extents of snow and sea ice cover around the northern polar region derived from GLI data. For comparison, seasonal variations of MODIS land snow cover and of the Advanced Microwave Scanning Radiometer (AMSR) sea ice cover are also shown. For the period of April 7–22, 2003,

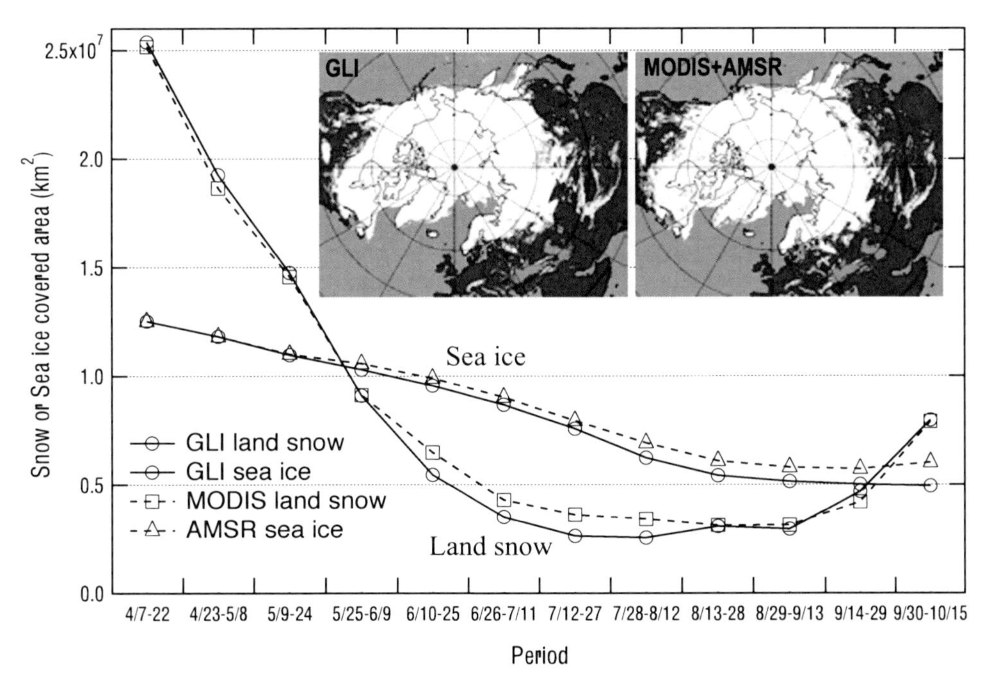

Fig. 5.9. Comparison of the temporal variations (16-day averages from Apr. 7 to Oct. 15) of GLI derived snow-covered land area and sea ice covered area with those derived from MODIS (land snow) and AMSR (sea ice). Images of the extents of snow and sea ice cover for the period of Apr. 7–22 from GLI and MODIS+AMSR are also shown (after Hori et al. [96]).

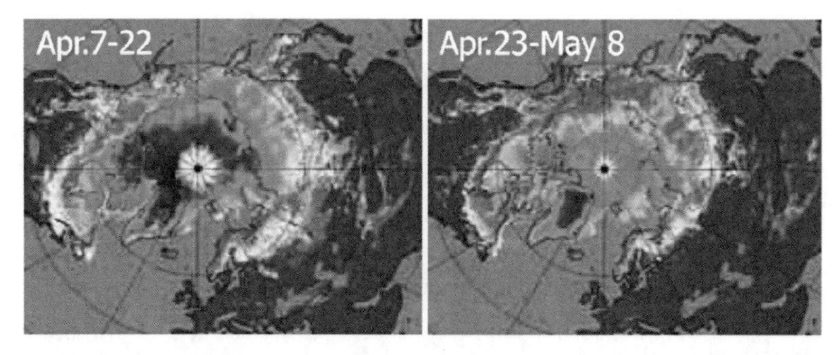

Fig. 5.10. 16-day average GLI snow surface temperature around the northern polar region from April 7 to May 8, 2003 (after Hori et al. [96]).

the snow and sea ice cover maps derived from GLI and MODIS+AMSR data are also shown. The GLI and MODIS snow cover extents are consistent with slight differences in the periods from June 10 to August 12 possibly due to differences in the cloud detection scheme and the ability to detect snow cover. The trend of the GLI sea ice cover also follows closely the variation of the AMSR sea ice cover except for slight negative biases, which become larger in the later 16-day periods. The bias is caused by the loss of valid sea ice pixels in the GLI results partly due to persisting cloudiness over the Arctic Ocean during the 16-day averaging period, particularly for the July to October time frame, and partly due to the drift of the sea ice itself over the averaging period.

Figure 5.10 shows the 16-day average snow surface temperature around the northern polar region from April 7 to May 8, 2003. White areas indicate snow or sea ice covered areas for which no snow physical parameters were determined because at least one of the four snow physical parameters retrieved was beyond the valid range of the analysis. Possible surface types of the white areas can be one of the following: bare ice, spatially inhomogeneous snow (e.g. snow cover contaminated by clouds or vegetation), invalid geometric conditions (e.g. too large solar or sensor zenith angles).

To assess the accuracy of the GLI-derived surface temperature, a comparison between MODIS and GLI snow surface temperatures is shown in Fig. 5.11. The GLI surface temperatures are well correlated with the MODIS temperatures having slight negative biases of about −2.0, −1.0, and −0.5 at 250, 260, and 270 K, respectively. Comparisons between GLI-derived snow grain sizes and surface temperatures also indicate that the GLI derived surface temperatures have about −0.5 K negative bias at around the melting point of ice (273 K), which is estimated from the temperature at which the retrieved snow grain size distribution shifts to a coarser mode due to melting of snow.

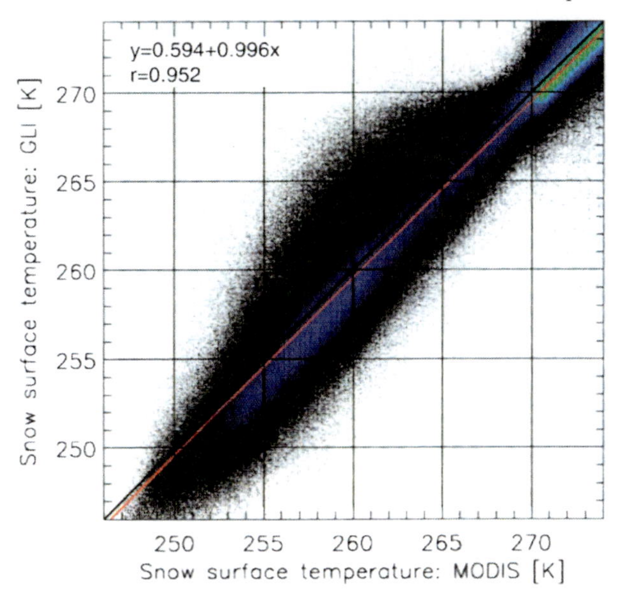

Fig. 5.11. Scatter plot between snow surface temperatures from MODIS and GLI (after Hori et al. [96]).

Snow impurity concentration and grain size

Hori et al. [96] showed that the regional dependence of the retrieved snow impurity is different from that of the snow surface temperature. The snow impurity values are mostly less than 0.3 ppmw over the Arctic sea ice, tundra, polar desert areas, and the Greenland ice sheet. In particular, impurity fractions at the Greenland ice sheet were found to be the lowest (mostly less than 0.05 ppmw) among the snow-covered areas in the Arctic during the 7-month observation period from April 7 to October 15, 2003. Although the retrieved impurity concentrations appear reasonable, their accuracy is uncertain, as discussed in [95].

Figure 5.12 shows the 16-day average spatial distribution of snow grain size of the shallow layer (0–20 cm) retrieved from the NIR channel at 0.875 μm ($R_{s0.9}$). The spatial distributions of the snow grain size exhibit not only a large-scale variation but also several regional patterns. The large-scale variation is the latitudinal dependence similar to that of the snow surface temperature, i.e. the higher the latitude, the smaller the grain size, and vice versa. The regional patterns are related to local weather or the thermal environment (e.g. relatively fine newly fallen snow in the mid-latitude area around the northern prairie in the United States seen in the April 7–22 period and coarse, probably melting snow over sea ice in the Arctic around Baffin Island in April).

The snow grain size of the top surface layer (0–2 cm) can be retrieved from the $\lambda = 1.64$ μm channel ($R_{s1.6}$) [94–96]. When comparing the $R_{s0.9}$ distribution with the $R_{s1.6}$ distribution (not shown), one finds not only that the spatial variability of $R_{s1.6}$ is different from that of $R_{s0.9}$, but also that absolute values of $R_{s1.6}$ are one order of magnitude smaller than those of $R_{s0.9}$. The ratio of $R_{s1.6}$ to $R_{s0.9}$ makes those features more clear as shown for the April 7–22 period in Fig. 5.13. The difference in the spatial distribution between $R_{s0.9}$ and $R_{s1.6}$ may be explained by a possible vertical inhomogeneity of the grain size in the upper several centimeters of the snow cover or by a depth variation in the snow cover taking into account the light penetration depth difference at $\lambda = 0.865$ μm and 1.64 μm.

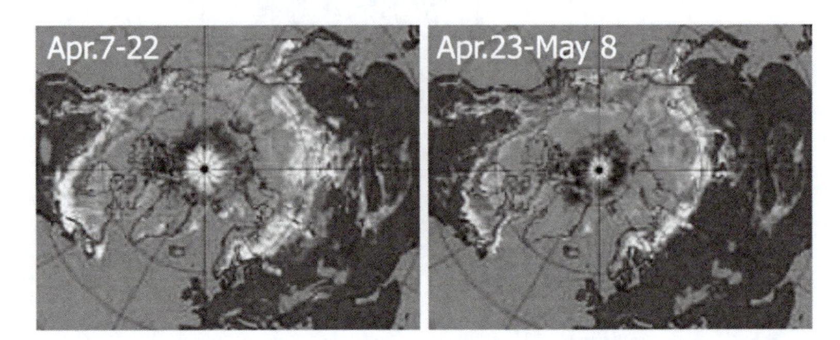

Fig. 5.12. 16-day average of the GLI derived snow grain size of the shallow layer ($R_{s0.9}$) around the northern polar region from April 7 to May 8, 2003 (after Hori et al. [96]).

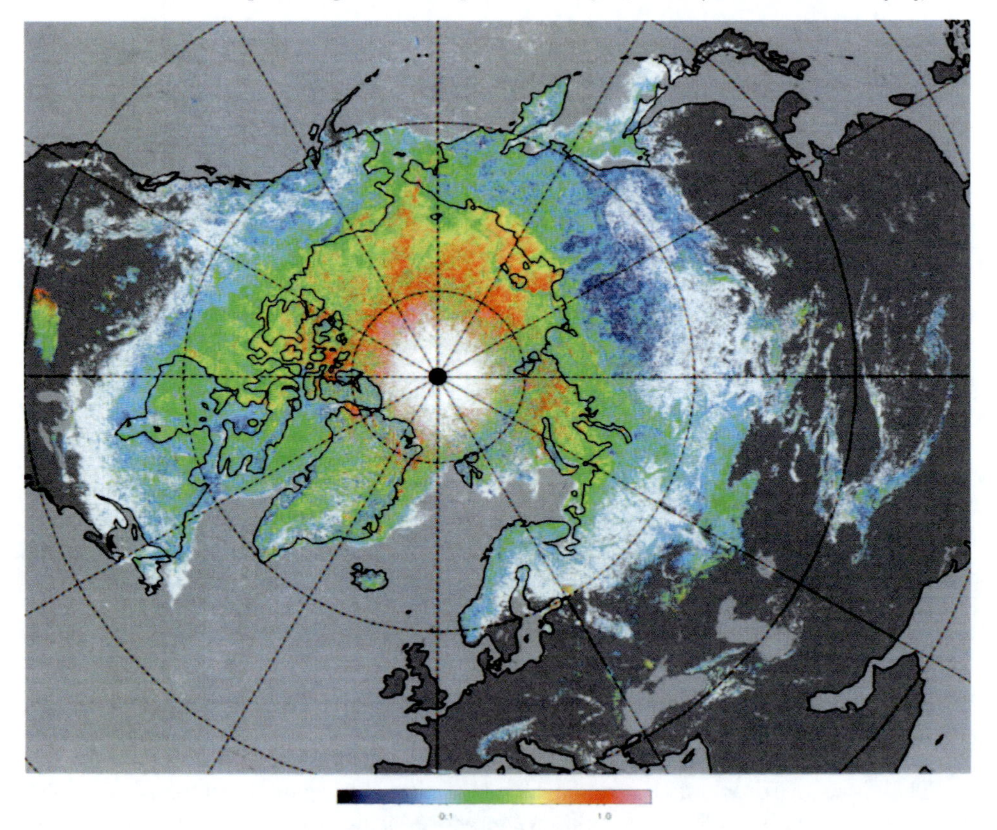

Fig. 5.13. Ratio of the snow grain radius of the top surface ($R_{s1.6}$) to that of the shallow layer ($R_{s0.9}$) for the April 7–22 period. White-colored areas are the same as in Fig. 5.10 (after Hori et al. [96]).

The close relationship between snow grain size ($R_{s0.9}$) and surface temperature (T_s) seen in the melting season is considered as an average feature of the seasonal snow cover on a hemispheric scale. On a local scale, however, the snow cover can shift temporally to different states of the temperature-grain size relationship, e.g. a state with small grain size under warm temperature or coarse size under cold temperature depending on the recent history of the thermal environment to which snow grains were exposed after a snow fall. As an example, Fig. 5.14 shows the spatial distribution of the same snow cover for the period of April 7–22, 2003, but color-coded using the two-dimensional temperature-grain size (T_s–$R_{s0.9}$) relationship. Warm (orange) color denotes small grains under high temperature indicating high potential for metamorphosis into larger grains, whereas cold (blue) color indicates coarse grains under low temperature with sizes that are likely to remain intact for a while. Thus, the map has information about the potential of snow grains to metamorphose in the near future. For example, the snow covers in the two elliptical areas in Fig. 5.14, of which one is shown in orange color, implying high potential for metamorphosis and the other in blue with low potential, have similar grain sizes (around 200 μm) but exist in different temperature regimes (272–273 K in the left orange area and 253 K in the right blue area; see Figs. 5.10 and 5.12). This information will be useful for validation of snow metamorphism models such as CROCUS [107].

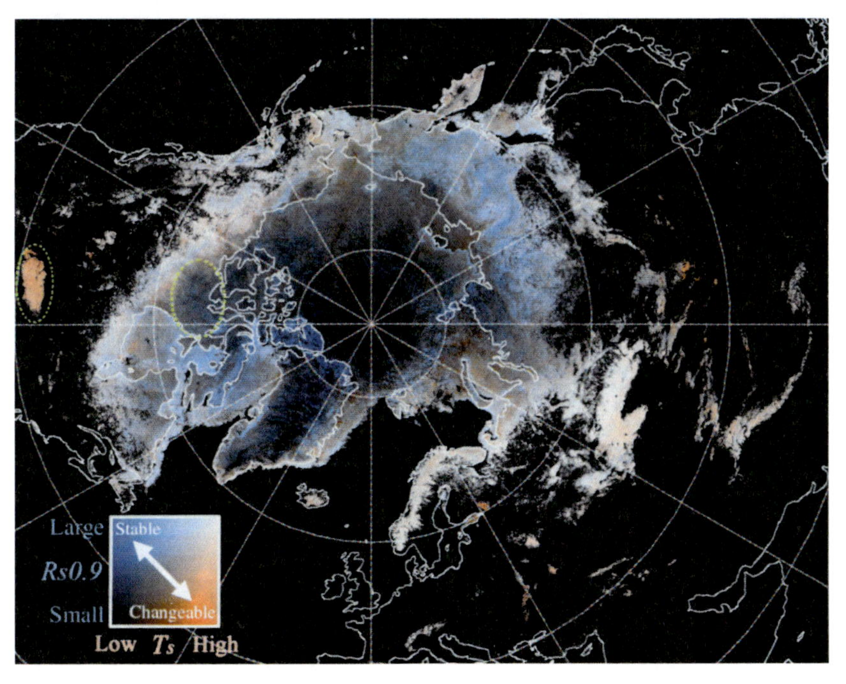

Fig. 5.14. Map of snow metamorphism potential around the northern polar region for the period of April 7–22, 2003 determined from the relation between snow surface temperature (T_s) and snow grain size ($R_{s0.9}$). Warm (orange) color denotes small grains under high temperature whereas cold (blue) color indicates coarse grains under low temperature (after Hori et al. [96]).

Another more practical application of the T_s–$R_{s0.9}$ relationship is for detection of the onset of snow melt at the hemispheric scale. Figure 5.15 shows a map of the date for the onset of snow melt over the Arctic derived from the daily maps of $R_{s0.9}$ and T_s. The melt onset date is defined as the average of the first three days (if determinable) when $R_{s0.9}$ becomes larger than 500 μm in a warm environment, i.e. with T_s higher than 272 K. Black areas in Fig. 5.15 indicate non-melted regions (e.g. the central area of the Greenland ice sheet), or areas where the snow cover evolves from dry to wet under cloudy conditions so that the GLI observation cannot detect the transition in the T_s–$R_{s0.9}$ relationship (e.g. some parts of the Arctic sea ice). The map clearly illustrates the development of the melt zones of snow cover in the northern hemisphere, e.g. Julian Day (JD) 90–150 for snow cover over the continents, JD 110–180 for the sea ice zone of the Arctic Ocean, and JD 150–220 for the marginal Greenland ice sheet. Thus, because of their higher spatial resolution, melt onset maps derived from optical sensor data can be useful in the interpretation of similar maps derived from microwave sensors (SSM/I, AMSR, NSCAT, etc.), which have coarser spatial resolution.

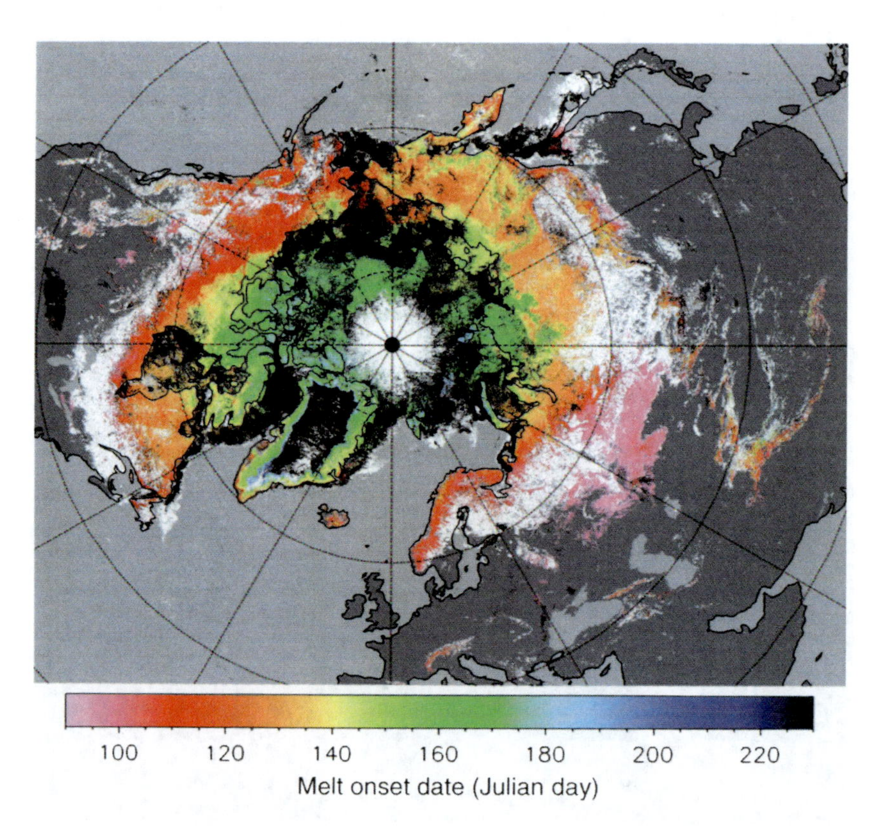

Fig. 5.15. Spatial distribution of melt onset date around the northern polar region in 2003 determined from the relation between snow surface temperature (T_s) and snow grain size ($R_{s0.9}$). Date is indicated by Julian Day (after Hori et al. [96]).

5.4.3 Coupled air–tissue systems

The first models of light propagation in tissue were based on Kubelka–Munk theory [108] and the diffusion approximation to the RTE [109], which give accurate results only if the absorption is negligible ($\sigma \gg \alpha$), or the angular distribution of the scattering is nearly isotropic [9]. For light propagation in skin tissue, the first model that included the air–tissue refractive-index discontinuity, was a coupled Monte Carlo (C-MC) code [110]. C-MC simulations are simple, flexible, and accurate, but very time-consuming. Fortunately, as discussed in section 5.4.1 for a coupled atmosphere-ocean system, the refractive-index discontinuity can be accurately accounted for using a C-DISORT model [2,4,5], which recently was applied for the first time to light propagation in skin tissue [111].

Test cases

Hestenes et al. [8] compared C-MC simulations and C-DISORT computations of radiances in a coupled air–tissue system similar to what was done earlier for irradiances in a coupled atmosphere–ocean system [5]. They considered a bio-optical model of skin having five epidermal layers with a total thickness of 0.05 mm, a 1 mm thick dermal layer, and a 3 mm thick subcutaneous tissue layer [111], and found the C-MC code to be about 1,000 times slower than the C-DISORT code for test cases with incident light at two different angles of incidence (0° and 45°) and at three different wavelengths (280, 540, and 650 nm). These wavelengths were selected because:

- At 280 nm there is strong absorption by proteins in the epidermis, and hence very little backscattering from the skin.
- At 540 nm blood has an absorption band, but the single-scattering albedo of the skin is considerably greater at this wavelength than at 280 nm.
- At 650 nm the absorption coefficients of both blood and melanosome pigments are low, implying that light entering the skin can reach great depths, but most of it is eventually backscattered.

The values used for the optical properties of the skin tissue at these three wavelengths are given in Tables 1, 2, and 3 in [8].

Results

Hestenes et al. [8] compared simulated (C-MC) and computed (C-DISORT) diffuse radiances for the reflected light at polar angles from 0° (straight upward, $\mu = 1$) to 90° (horizontal reflection, $\mu = 0$) and for light at different depths inside the tissue for polar angles from from 0° to 180° (straight downward, $\mu = -1$). Below the skin surface the C-MC curves were found to be noisy at angles close to the horizon because photon irradiances in the C-MC code were weighted by $1/\mu$ to obtain radiances, implying that close to the horizon there was division by a number close to zero.

Hestenes et al. [8] found the percentage error between results from the C-MC and C-DISORT codes to be largest in the direction straight down ($\mu = -1$), at the critical angle, and at $\mu \sim 0$ (i.e. close to the horizon). One reason for these discrepancies was that they did not sample exactly the same μ values in the two codes. Thus, when comparing results from the two codes close to a point with an abrupt change in the slope of the radiance, the relative error might seem larger than it actually was. Also, as noted above, photon irradiances in the C-MC code were weighted by $1/\mu$ to obtain radiances, which may explain why the relative error is larger close to $\mu \sim 0$. For other μ values than those close to the critical angle and close to $\mu \sim 0$ they found the relative error to be within acceptable ranges of 1% and 4% for angles of incidence of $\theta_i = 0°$ and $\theta_i = 45°$, respectively. Also, the relative errors were found to have both positive and negative values, indicating that they were of a stochastic nature, and that the results from the C-MC and C-DISORT codes were essentially the same.

At $\lambda = 280$ nm the radiance values were found to decrease with depth in the skin tissue, due to the large optical depths at $\lambda = 280$ nm (see Table 1 in [8]). At the interface between the epidermis and the dermis (at a depth of 50 μm) the radiance values were found to be very low but to have a maximum in the direction of the refracted light, implying that the radiance distribution was not isotropic. At the layer between the dermis and the subcutaneous tissue (at a depth of 1.05 mm) the radiance values from the C-MC code were found to be zero, and the radiance values from the C-DISORT code to be approximately 10^{-12}, implying that practically no radiation penetrated this deep into the skin tissue.

At $\lambda = 540$ nm and $\lambda = 650$ nm the radiance values were found to first increase with depth in the epidermis, and then to decrease as the epidermal-dermal boundary was approached at a depth of 50 μm. At the interface between the dermis and the subcutaneous tissue at a depth of 1.05 mm the radiance values were found to be significantly smaller than those in the epidermis, and the radiance was found to be almost completely isotropic, particularly for $\lambda = 650$ nm.

The critical angle θ_c at which light propagating upward in the tissue undergoes total internal reflection at the tissue–air (slab_2–slab_1) interface, is given by $\theta_c = \sin^{-1}(n_1/n_2)$, leading to $\theta_c \approx 45°$ for $\lambda = 280$ nm, $\theta_c \approx 47°$ for $\lambda = 540$ nm, and $\theta_c \approx 47°$ for $\lambda = 650$ nm. The corresponding polar angle in the tissue is $\theta_{tc} = \pi - \theta_c$. At $\lambda = 540$ nm and $\lambda = 650$ nm each radiance curve in the epidermis was found to have a sudden discontinuity in its slope at the critical angle for light propagating downward (see Fig. 5.16). As the polar angle increases beyond θ_{tc} ($\mu_{tc} = \cos(\pi - \theta_c) \approx -0.68$ for both $\lambda = 540$ nm and $\lambda = 650$ nm) the radiance is seen to attain a minimum before its value increases again to a maximum. The abrupt change in the radiance at the critical angle illustrates the enhancement effect associated with total internal reflection. For $\mu \in [-0.68, 0]$ light incident upon the tissue–air interface from below is totally reflected and contributes to enhance the radiance at these angles [7].

At $\lambda = 540$ nm the discontinuity in the slope of the radiance at the critical angle was found to disappear at a depth of 50 μm, but still to be present at this depth at $\lambda = 650$ nm, because the scattering optical depth in the epidermal layers was assumed to be greater at $\lambda = 540$ nm than at $\lambda = 650$ nm (see Tables 2 and 3 in [8]).

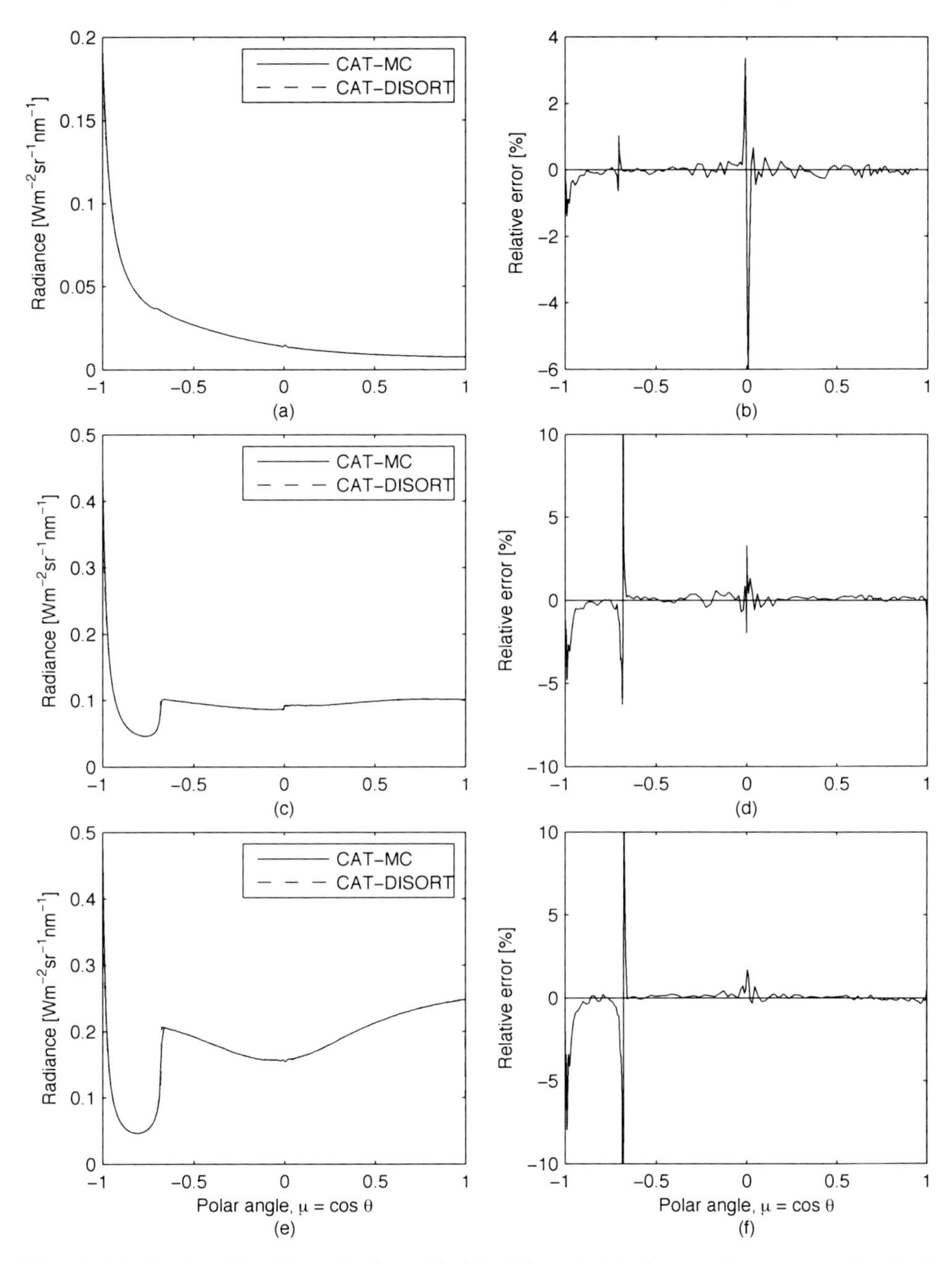

Fig. 5.16. Angle of incidence is $\theta_i = 0°$. (a), (c), and (e) show radiances at a depth of 10 μm for $\lambda = 280$ nm, $\lambda = 540$ nm, and $\lambda = 650$ nm, respectively. (b), (d), and (f) show the relative error between CAT-MC and CAT-DISORT values for (a), (c), and (e), respectively (adapted from Hestenes et al. [8]).

Retrieval of the physiological state of human skin

Nielsen et al. [10] examined the feasibility of using C-DISORT in conjunction with a classic inversion scheme for retrieval of parameters describing the physiological state of human skin. To that end, they analyzed ultraviolet and visible reflectance spectra from human skin measured before, immediately after, and on each day for two weeks after photodynamic treatment with the hexyl ester of ALA and exposure to red light (632 nm), and showed that it is possible to perform a simultaneous retrieval of the melanosome concentration in both the basal and upper layers of the epidermis.

In order to retrieve important physiological properties of skin the following ingredients are needed [10]: (1) A bio-optical model that relates the physiological properties of the skin to its IOPs, i.e. the absorption and scattering coefficients as well as the scattering phase function, each as a function of wavelength and depth in the skin. (2) An accurate RT model for the coupled air–tissue system, such as C-DISORT, which for a given set of IOPs can be used to compute the AOPs, such as the diffuse reflectance spectrum. (3) An iterative inversion scheme that accounts for the nonlinear dependence of the AOPs on the IOPs. Because C-MC simulations, which were considered as the standard for forward RT modeling in the skin at ultraviolet and visible (UV–Vis) wavelengths (e.g. [112–114]), are very time-consuming, Nielsen et al. [10] used the C-DISORT model [2,4,26]. They performed a feasibility study in order to evaluate the potential of employing a bio-optical model [111] together with the C-DISORT forward RT model [8] and inversion based on Bayesian optimal estimation [115] to retrieve important physiological properties of skin. To that end, they analyzed diffuse reflectance spectra measured by Zhao et al. [116]. This feasibility study of retrieving physiological properties of the skin was based on the following premise. Although the bio-optical model contained a larger number of variable parameters than could possibly be retrieved, it would seem reasonable to consider most of them as fixed and only a few of them as free variables. Thus, one might assume the chromophores, such as the epidermal melanosome concentration and the dermal blood concentration, to be free (i.e. retrieval) parameters, since their variability would strongly influence the reflectance spectra of skin. On the other hand, one could fix the parameters describing the optical properties of the cellular matrix into which these chromophores are embedded, since variations in these parameters would have much less impact on the reflectance spectra of skin.

Design of experiments

Reflectance spectra in the wavelength region from 300 nm to 600 nm were measured daily from three test areas and three control areas on a volunteer with skin type III for two weeks during which the skin in the test areas went through erythema and pigmentation. Erythema and pigmentation were induced after 24-hour topical application of a photosensitizer followed by illumination with red light (632 nm) for 2 min on the first day (day 0) of the experiment. The design of the photodynamic experiment, which can be found in Zhao et al. [116], may be summarized as follows. Three test areas (A, B, and C), each 1 cm $\times$ 1 cm with approximately 1.5 cm distance between adjacent areas, were marked on the inner part of the right forearm of the volunteer. Cream was prepared using 10% (w/w) of the hexyl ester of

5-aminolevulinic acid (ALA-Hex) in a standard oil-in-water cream base (Unguentum, Merck, Darmstad, Germany). Freshly prepared cream with approximately (755 ± 10) mg/cm^2 of ALA-Hex was topically applied on each of these three test areas, and subsequently covered with transparent adhesive dressings (OpSite Flexifix, Smith & Nephew Medical Ltd., Hull, UK), in which three openings (1 cm $\times$ 1 cm) had been cut out precisely in the places where the test areas were located. The dressings were intended to prevent the cream from diffusing to adjacent areas. The creams and the dressings were kept for 24 h on the test areas, which were then illuminated with red light (632 nm) for 2 min. Three control areas, which were equal in size to the test areas, were also marked on the volunteer [116]. On the first of these (D) ALA-Hex was applied but it was not illuminated; the second of the control areas (E) was illuminated with red light, but no ALAHex was applied; on the third test site (F) a base-cream without ALA-Hex was applied and it was illuminated with red light. A luminescence spectrometer (Perkin-Elmer LS50B, Norwalk, CT) was employed to record reflectance spectra from each of the test areas and each of the control areas. The spectrometer was equipped with two scanning grating monochromators, one in front of the light source (a pulsed Xenon lamp) and another in front of the detector. Reflectance spectra were measured in synchronous scans in which both gratings were set at the same wavelength and band pass (5 nm) to avoid fluorescence artefacts. The area exposed to the excitation light of the spectrometer was the same as the area from which the reflected light was detected. The geometry of the fiber probe was such that both the directly (Fresnel) reflected and the diffusely reflected irradiances from the skin were collected and recorded. Care was taken not to press the spacer too hard against the skin surface in order to minimize artefacts from pressure-induced reductions in the blood flow.

The coupled air–tissue system can be represented by a turbid, layered medium with fixed IOPs in each layer, so that C-DISORT can be used to compute the diffuse light reflected from it. Each skin layer can be described in terms of its IOPs, which are the absorption coefficient α (mm^{-1}), the scattering coefficient σ (mm^{-1}), the scattering phase function $p(\cos\Theta)$, and its physical thickness Δz (mm). As discussed in section 5.3.1, in terms of α, σ, and Δz, one may define two non-dimensional IOPs given by $\tau = (\alpha + \sigma)\Delta z$ (optical thickness) and $a = \sigma/(\alpha + \sigma)$ (single-scattering albedo), so that the IOPs in each layer of the skin can be adequately described by the two variables τ and a, as well as a third variable g. As discussed previously (see Eq. (8)), the third parameter g is the asymmetry factor of the scattering phase function. Since skin tissue is a complex medium with many different kinds of scattering 'particles' in each layer, the scattering phase function for a particular layer represents a weighted mean of scattering phase functions for several types of particles. Nielsen et al. [10] used the Henyey-Greenstein scattering phase function (Eq. (9)) for 'large particles' with sizes comparable to or larger than the wavelength and the Rayleigh scattering phase function (Eq. (10)) for small particles with diameters $d < \lambda/10$.

To calculate the IOPs for a given set of parameters that describe the physiological state of the skin tissue, Nielsen et al. [10] used a bio-optical model [111]. In order to calculate the AOPs (in this case the diffuse reflectance spectrum of the skin), they employed C-DISORT [2,8,111] to solve the RTE pertaining to a slab of biological tissue stratified into a number of layers, thus properly accounting for the reflection and refraction of the incident radiance at the air–tissue interface (caused by the change in the refractive index), which affect the radiation field significantly [117]. When performing forward and inverse modeling, the bio-optical model and the C-DISORT model were coupled together, implying that the physiological parameters were retrieved directly from the measured AOPs. In order to obtain a unique solution of the inverse modeling problem, some physiological parameters were assumed to be fixed, while the following seven parameters were allowed to vary [10]:

- the dermal blood concentration;
- the relative percentage of oxygenated blood;
- the melanosome concentration in the lower epidermis;
- the thickness of the lower epidermis;
- the melanosome concentration in the upper epidermis;
- the thickness of the upper epidermis;
- the keratin concentration in the upper epidermis.

All other parameters, such as each of the scattering coefficients associated with the non-pigmented constituents of the epidermis and dermis, the optical thickness of the dermis, and the optical properties of the subcutaneous layer, were assumed to be fixed.

Results

Figure 5.17 shows an example of the agreement between measured and simulated reflectances obtained when using the retrieved values for the seven parameters listed above as inputs to the C-DISORT simulations. Figures 5.18–5.21 show the retrieved values of the blood concentration in the dermis (Fig. 5.18), the relative amount of oxygenated blood (Fig. 5.19), the melanosome concentration in the lower layers of the epidermis (Fig. 5.20), and the melanosome concentration in the upper layers of the epidermis (Fig. 5.21), respectively. The three subpanels in the left columns of the figures represent the test areas (A–C), while the three subpanels in the right columns of the figures represent the control areas (D–F).

Discussion

A fundamental problem of biomedical optics is the validation of results obtained in vivo. If the tissue is excised, the physiological properties will change significantly. Therefore, a detailed analysis was carried out of synthetic spectra with realistic noise levels to obtain the the sensitivity and crosstalk for each of the seven parameters that were retrieved [10]. According to Farrell et al. [118], the sensitivity of a parameter in an inversion scheme is the percentage of change in the retrieved parameter for a given change in that same parameter when used as input to compute synthetically generated spectra with realistic noise added, whereas the crosstalk is the retrieved change in a parameter that was not changed in the input to the computations of synthetic spectra. To perform sensitivity and crosstalk calculations, Nielsen et al. [10] first chose an initial model that was representative for the retrieved results: 1% vol. concentration of melanosomes in the upper epidermis,

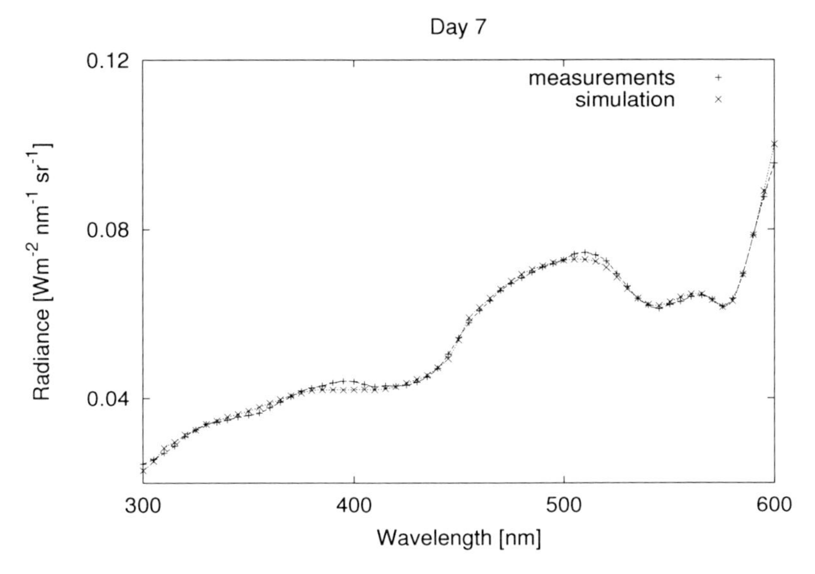

Fig. 5.17. Measured (+) and simulated (×) radiance spectra for test area A on the seventh day after the photodynamic exposure. The simulated spectrum gives the optimal agreement between the measured and simulated spectra obtained when the retrieved physiological parameters for this day were used as inputs to the forward simulations (adapted from Nielsen et al. [10]).

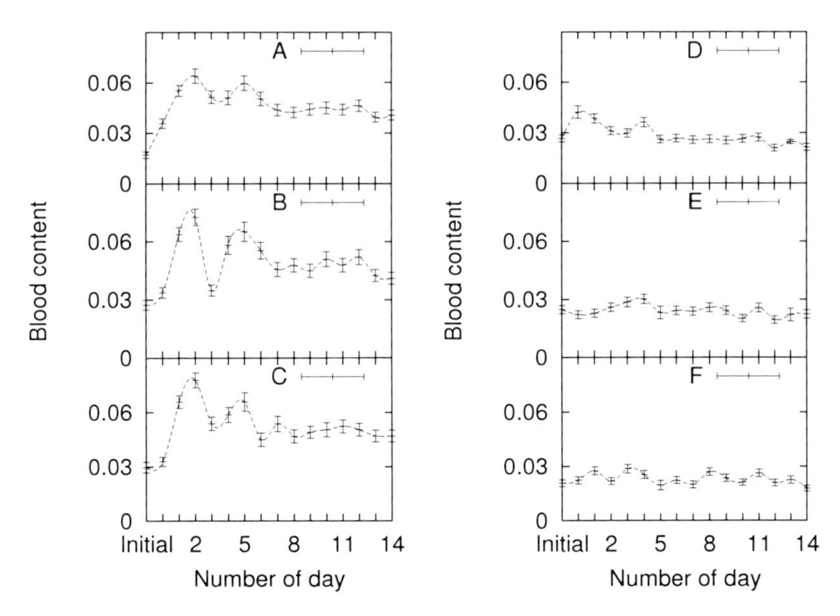

Fig. 5.18. Retrieved dermal blood concentration for each of the measurement areas for the 15 days of measurement. 'Initial' refers to the situation prior to the photodynamic treatment of the skin, while the next tick mark refers to the situation immediately after the photodynamic treatment (adapted from Nielsen et al. [10]).

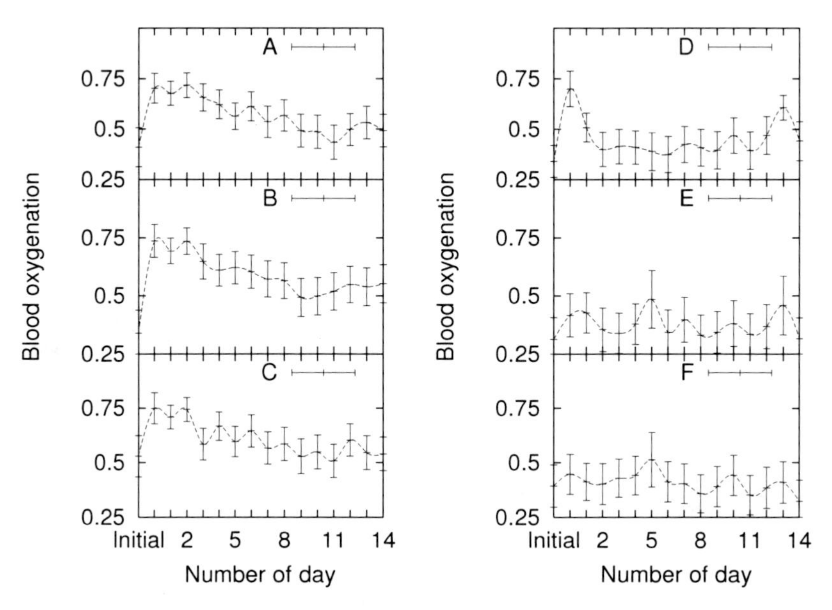

Fig. 5.19. Retrieved percentage of oxygenated blood for each of the measurement areas for the 15 days of measurement. 'Initial' refers to the situation prior to the photodynamic treatment of the skin, while the next tick mark refers to the situation immediately after the photodynamic treatment (adapted from Nielsen et al. [10]).

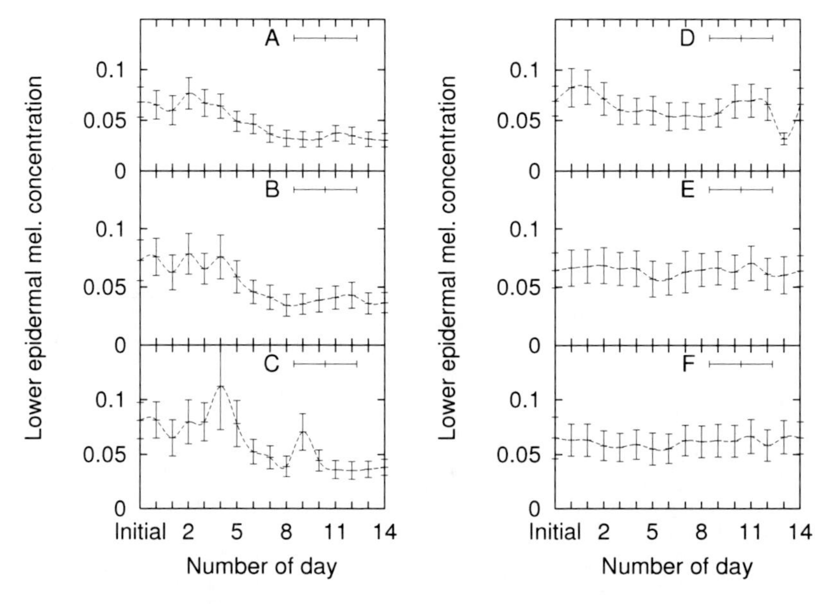

Fig. 5.20. Retrieved melanosome concentration in the lower epidermis for each of the measurement areas for the 15 days of measurement. 'Initial' refers to the situation prior to the photodynamic treatment of the skin, while the next tick mark refers to the situation immediately after the photodynamic treatment (adapted from Nielsen et al. [10]).

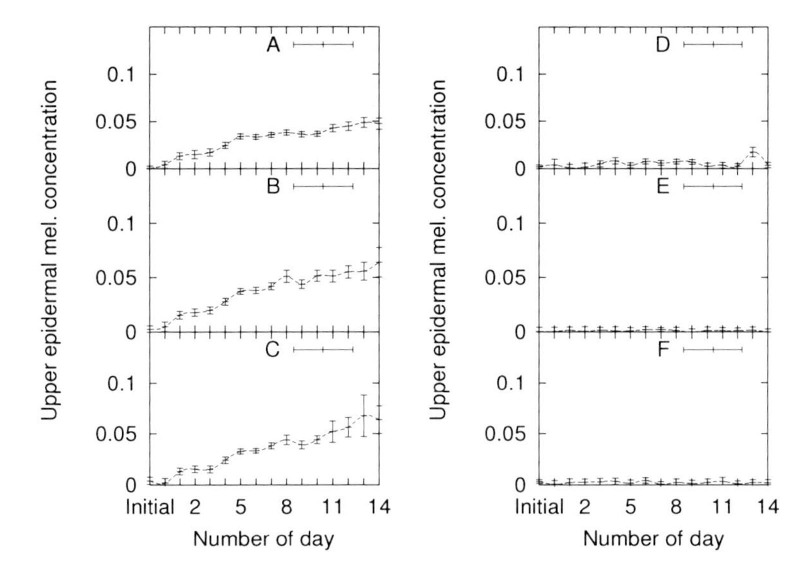

Fig. 5.21. Retrieved melanosome concentration in the upper epidermis for each of the measurement areas for the 15 days of measurement. 'Initial' refers to the situation prior to the photodynamic treatment of the skin, while the next tick mark refers to the situation immediately after the photodynamic treatment (adapted from Nielsen et al. [10]).

5% vol. concentration of melanosomes in the lower epidermis, 5% vol. blood concentration, 75% relative blood oxygenation, 25 μm upper epidermis thickness, and 25 μm lower epidermis thickness. Then they calculated synthetic spectra for which each of the seven parameters were successively perturbed with relative changes of −50%, −40%, ... −30%, −20%, −10%, +10%, +20%, +30%, +40%, and +50%, respectively. Finally, noise (∼N(0, 1%)) was added to each of these 70 synthetic spectra.

From these noisy spectra retrievals were obtained, and sensitivities were calculated as the slope of the retrieved perturbed parameters against the input perturbed parameters. Similarly, crosstalk (which ideally should be zero) was calculated as the slope of each of the the other parameters against each input perturbed parameter. The results [10] showed blood oxygenation to be causing the lowest amount of crosstalk and to have a sensitivity close to 100%. The sensitivities were found to be very high and the crosstalks moderately low also for the upper and lower epidermal melanosome concentrations and the blood concentration. For the upper epidermal keratin concentration the sensitivity was found to be high, but there was considerable crosstalk into the upper epidermal melanosome concentration (43%). For the upper and lower epidermal thicknesses the sensitivities were found to be low, 47% and 64%, respectively, and there were substantial crosstalk into the other epidermal parameters. Thus, it would appear that changes in the thicknesses of the epidermal layers cannot be retrieved accurately from this kind of measurements for this type of skin, but if the thicknesses of the epidermal layers would remain unchanged, reasonable retrievals could be obtained for the other five parameters.

The standard deviations or the error bars in Figs. 5.18–5.21 were calculated during the retrieval procedure from the diagonal elements of the covariance matrix [10].

In general, the temporal variation in the blood concentration of the skin (Fig. 5.18) was found to be in good agreement with expectations. A maximum was reached after 1–2 days, the so-called erythema reaction, typical for a sun-burn, after which the blood concentration slowly decayed. For the three control areas the variations were within the range of the calculated standard deviations. Fig. 5.19 shows the relative amount of oxygenated blood immediately after the photodynamic treatment and during the following two weeks for all six measurement areas. Photodynamic therapy is known to be an oxygen consuming process [119]. Thus, 'immediately after' means several seconds or may be even a minute after the exposure. The marked increase in oxygenation measured at the beginning of the experiment at all three test areas is likely to be a reaction to the photodynamic treatment rather than a direct effect of the treatment itself. The variation in blood oxygenation at the three control areas are of a stochastic nature. The relative change of the percentage of oxygenated blood during the two weeks of measurements seems reasonable. However, the absolute values, which are in the range between 40% and 80%, may be too low (Fig. 5.19). These low oxygenation percentages could be caused by the bandpass of the spectrometer (5 nm) being insufficiently narrow to resolve the spectral fine structure in the 540–580 nm spectral region. The skin reflectance in this spectral region is very sensitive to the percentage of oxygenated blood. Therefore, the percentage of oxygenated blood could be underestimated from these measurements.

The retrieval of the melanosome concentration was less uncertain for the upper epidermis (Fig. 5.21) than for the lower epidermis (Fig. 5.20). Thus, the standard deviations for the lower epidermal melanosome concentration were about twice as large as for the upper epidermal melanosome concentration. For all three test areas, the melanosome concentration in the lower epidermis decreased during the two weeks (Fig. 5.20), while the melanosome concentration in the upper epidermal layers increased, in particular during the first week (Fig. 5.21). This behavior is similar to that caused by pigmentation induced by UVB radiation (with wavelengths shorter than 320 nm). Thus, as a reaction to UVB exposure the melanosome pigment particles tend to be transferred from the melanocytes in the basal layer of the epidermis to the keratinocytes in the upper layers [120]. In the experiment discussed here [116], there was no sign of immediate pigment darkening, a process primarily induced by UVA radiation (with wavelengths longer than 320 nm) [121]. Had immediate pigment darkening occurred, it would have been seen in the measurements taken immediately after the photodynamic treatment. Hence, one may conclude that the photodynamic process induced by topical application of ALA-Hex followed by illumination with red light (632 nm) bears similarities with the photobiological pigmentation process induced by UVB radiation, but that immediate melanin darkening does not take place during photodynamic therapy.

The decrease in the retrieved lower melanosome concentration occurred at the same time as an increase in the retrieved lower epidermal thickness. Thus, the retrieved total lower epidermal melanosome concentration was relatively constant during the two weeks of measurements. At the control areas no significant temporal variations were seen in the retrieved lower or upper epidermal melanosome concentrations (Figs. 5.20 and 5.21), or in the retrieved lower or upper epidermal thicknesses (not shown). The retrieved total thickness of the epidermis was very close to what would be expected. Thus, Sandby-Møller et al. [122] performed measurements on the dorsal forearm of 71 volunteers and found the average epidermal thickness to be 76 μm $\pm$ 15 μm. Nielsen et al. [10] found the retrieved value for the total epidermal thickness to be approximately 70 μm, both for the test sites and the control sites. Given the result from the cross talk analysis discussed above, it was surprising to find that the total epidermal thickness appeared to be well retrieved.

Optical transfer diagnosis (OTD) of pigmented skin lesions

Malignant melanoma is one of the most rapidly increasing cancers in the world; in the United States alone, the estimated incidence for 2008 was 62,480, which would lead to an estimated total of 8,420 deaths [123]. Successful treatment of melanoma depends on early detection by clinicians with subsequent surgical removal of tumors. Visual detection has limitations even when augmented with dermoscopy, especially with less experienced users [124]. Attempts have thus been made to develop automated devices to assist in the screening of pigmented skin lesions for the likelihood of melanoma. Most of these devices have digitalized dermoscopy-related features analyzed by an artificial neural network or support vector machine learning systems [125, 126].

Swanson et al. [11] described a novel melanoma detection system that uses morphologic and physiologic mapping from spectral radiance images of melanocytic lesions. A clinical pilot study carried out to evaluate the novel detection system enrolled 50 patients of the dermatology clinic at Mayo Clinic, Scottsdale, AZ, USA. Referral for optical transfer diagnosis (OTD) typically occurred after examination and evaluation of the pigmented lesion with the naked eye and the dermoscope. Potential participants were identified by the referring dermatologist as having pigmented skin lesions suspected of being melanoma for which skin biopsy was indicated. These included lesions felt to be low risk, likely benign lesions, and high risk, likely melanoma lesions.

The OTD system (Balter Medical AS, Bergen, Norway) used in the study records 30 spectral radiance images (one image set), which constitute one measurement of a lesion. Images were recorded at 10 different wavelengths (365–1,000 nm) from multiple angles of illumination and detection. The OTD device is a spectral radiance meter consisting of a measurement head with 12 fixed diode lamps and 3 IEEE (Institute of Electrical and Electronics Engineers) 1394 FireWire cameras. Each light-emitting diode (LED) is placed at a different angle relative to the skin to enhance the ability to retrieve information on the depth of the lesion. The polar angles of the LEDs vary between 30° and 45°, and the relative azimuth between 34° and 145°. The polar angles for the detectors varied between 0° and 45°, and the relative azimuth between 0° and 180°. An alcohol-based gel interface was used where the measurement head contacted the skin, so that a select area of skin could be illuminated and imaged through a 2.2-cm-diameter circular sapphire plate. Imaging time was approximately 5 s.

On the basis of established absorption and transmission spectra for known skin chromophores and mathematical modeling of skin reflectance [10, 111], the images from each set were used to derive morphologic–physiologic maps of the lesions for the following seven parameters, as described elsewhere [10]: (1) percentage of hemoglobin concentration, (2) percentage of hemoglobin oxygenation, (3) upper epidermal thickness, (4) lower epidermal thickness, (5) percentage of upper melanosome concentration, (6) percentage of lower melanosome concentration, and (7) percentage of keratin concentration. These parameters were assumed to be different for normal and malignant tissue.

From each map, an entropy value was calculated and relative entropy values were calculated among different pairs of 2 maps. The entropy concept used was similar to that used in statistical physics and information theory. For example, from the spatial distribution of the melanosome concentration, the entropy associated with this parameter was computed as the sum of the melanosome concentration multiplied by its logarithm and integrated over the area of the lesion. Simply stated, the entropy provides a measure of the disorder in any one of the maps, and the relative entropy provides a measure of the correlation between two different maps. A positive-definite 7×7 matrix (diagnostic matrix) was constructed and two eigenvectors were selected that corresponded to the 2 largest eigenvalues of the diagnostic matrix. These 2 eigenvectors were used to assign a prediction of malignancy for each lesion that was imaged.

Each suspicious lesion was scanned 3 times with the OTD camera to obtain 3 measurements (i.e. 3 sets of images) per lesion, with the camera repositioned for each image set. The image sets were recorded on a digital video disc and processed independently for the creation of morphologic–physiologic maps of the 7 parameters listed above. Each map set was then used to derive diagnostic matrices as described above.

After each lesion was scanned, the entire lesion was removed *in toto* using a saucerization-type biopsy, and then sent for histopathologic processing and examination. Pathologic specimens were processed with hematoxylin and eosin staining as well as, when indicated, immunohistochemical staining with melan-A. All specimens were reviewed independently by 2 dermatopathologists, who rendered the diagnoses. The histopathologic findings were then compared with the findings from the respective diagnostic matrix calculation. The comparison end points were 'melanoma' and 'nonmelanoma'.

Results

Sixty-three lesions in the 50 patients were imaged with OTD and subsequently removed; all were clinically suspicious for melanoma. Of these 63 lesions, five were melanomas (two *in situ* melanomas, three invasive), 47 were benign melanocytic nevi, two were basal cell cancers, and 11 were other lesions.

The two largest eigenvectors of the diagnostic matrix defined above were used to calculate one index for melanoma (the 'melanoma index') and another index for benign (nonmelanoma) lesions by correlation with the biopsy results. The indices were designed such that the largest value determined the diagnosis. Thus, if the melanoma index was larger than the nonmelanoma index, the lesion would be diagnosed as a melanoma. To avoid false negatives due to inherent uncertainties in the data, the percentage difference between the melanoma and the nonmelanoma indices was required to exceed an empirically defined threshold. The sensitivity of OTD was 100% for identifying melanoma. Depending on the threshold by which the melanoma index was required to exceed the nonmelanoma index, the specificity of OTD for discriminating melanoma from nonmelanoma was 94.8–96.6%.

Dermoscopic images and morphologic–physiologic maps of representative lesions are shown for a typical nonmelanoma lesion (Fig. 5.22) and for a melanoma in situ (Fig. 5.23). The nonmelanoma lesion in Fig. 5.22 was a nevus with mild ar-

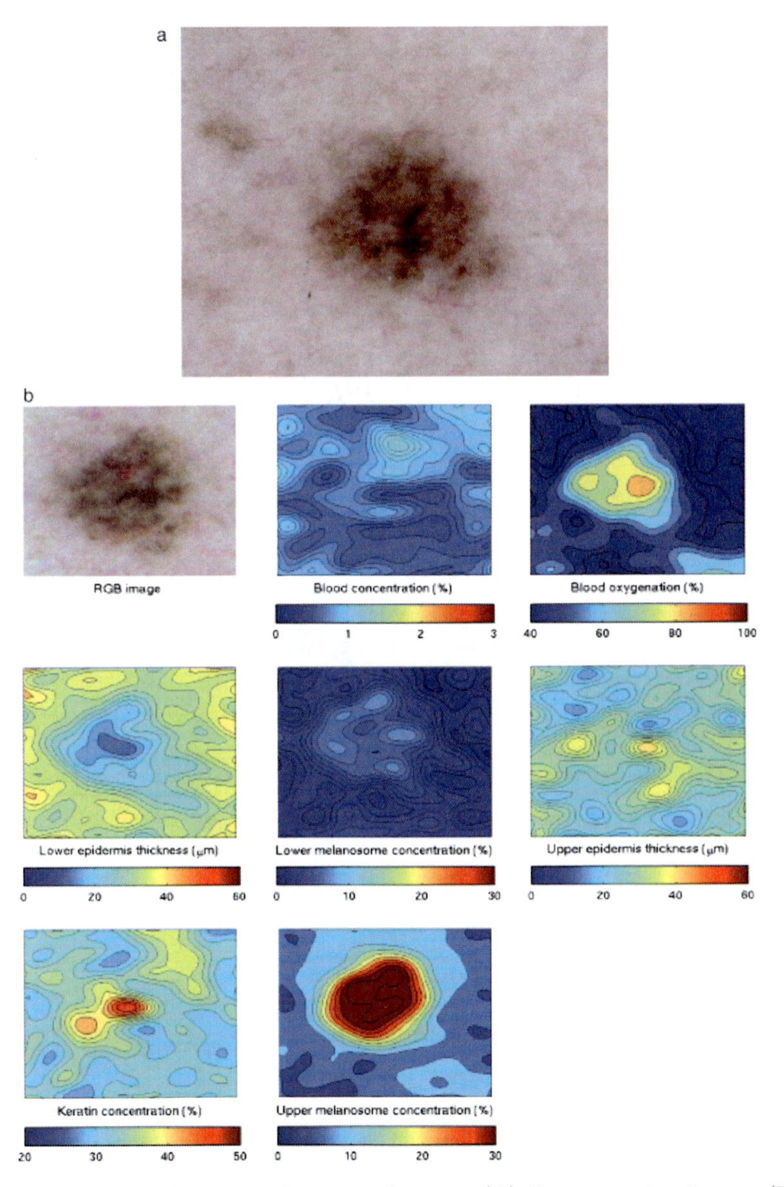

Fig. 5.22. Pathologic diagnosis: Compound nevus. (A) Dermoscopic picture. (B) Red-green-blue (RGB) image (upper left-hand corner) and a set of 7 morphologic–physiologic maps generated by optical transfer diagnosis (adapted from Swanson et al. [11]).

chitectural disorder. In the 7 maps associated with this lesion, there was a relative degree of homogeneity and a lack of entropy (i.e. disorder) compared with those in the equivalent maps in Fig. 5.23, which were for a melanoma *in situ*. These latter maps show a higher level of heterogeneity, especially in the percentage of blood oxygenation and in the percentage of upper epidermal keratin concentration, with an associated higher level of relative entropy between the pairs of maps.

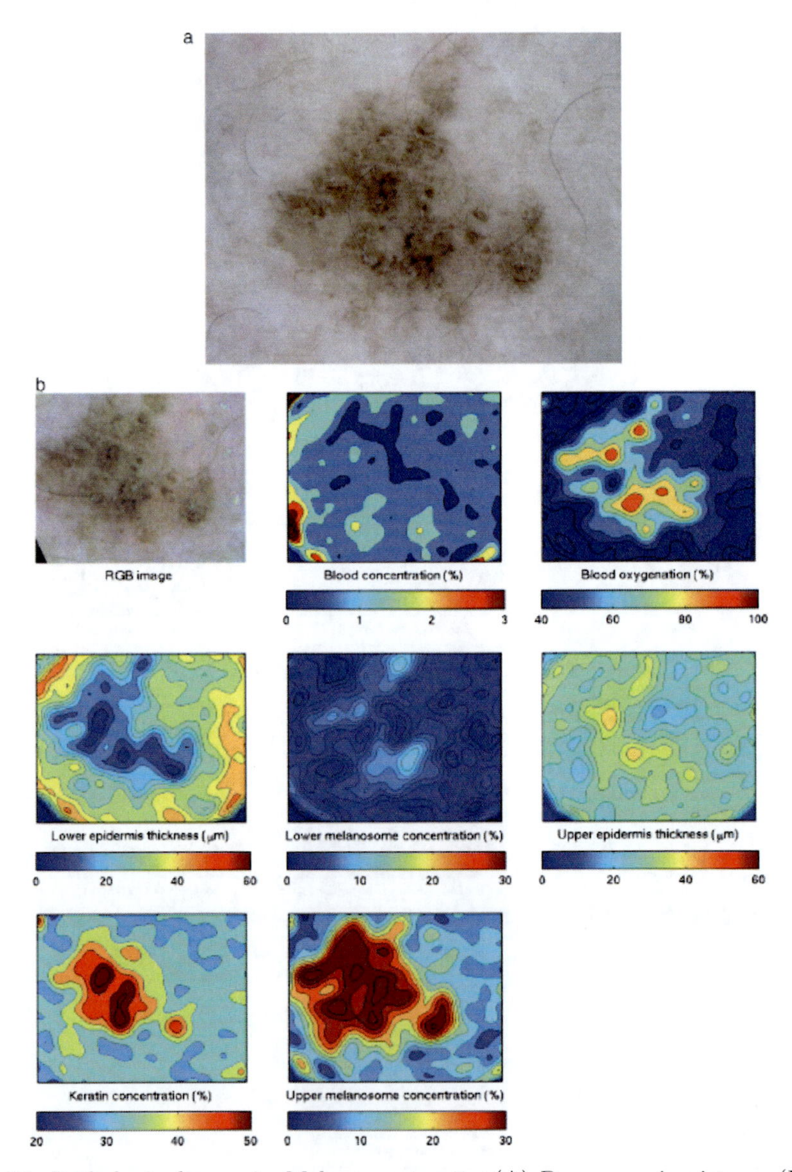

Fig. 5.23. Pathologic diagnosis: Melanoma *in situ*. (A) Dermoscopic picture. (B) Red-green-blue (RGB) image (upper left-hand corner) and a set of 7 morphologic–physiologic maps generated by optical transfer diagnosis (adapted from Swanson et al. [11]).

There were 3 false positives. One was for a seborrheic keratosis that appeared to be a borderline lesion on dermoscopy. OTD imaging of this lesion showed more striking degrees of blood content, keratin concentration, and upper melanin concentration, as well as a high percentage of blood oxygenation. The other 2 lesions were nevi with architectural disorder that showed especially higher levels of percentage of oxygenation heterogeneity. On dermoscopy, both lesions showed atypical vascular features with dot vessels and irregular shades of pink.

Discussion

Automated instruments for the diagnosis of primary melanoma that require no diagnostic input by the operator are currently in various stages of development [127]. The technologies under development typically use data manipulation of digital dermoscopic images by image segmentation for border definition, feature extraction, and subsequent lesion classification by statistical methods [125,126,128]. OTD differs from purely morphometric and color analyses by the additional modeling of physiologic parameters. This modeling is based on known absorption and transmission spectra of melanin, keratin, hemoglobin, and deoxyhemoglobin. In contrast to other technologies, the OTD technology adds an entirely new dimension to *in vivo* analysis by accounting for interactions between the tumor and its environment, such as the relative degree of hemoglobin desaturation in the vicinity of the tumor.

Extended study

In a recent study, Swanson et al. [12] based their findings on a data set that included 47 patients from the pilot study discussed above [11] and 47 patients subsequently studied. The OTD analysis used by Swanson et al. [12] was a revised, more robust version of the analysis used in the previous study [11]. From each physiologic–morphologic map, an entropy value was calculated, and relative entropy values were calculated between different pairs of two maps. Weights were assigned to each entropy and relative entropy value and their respective logarithms and to each pure morphologic parameter and its logarithm. For the seven physiologic–morphologic maps, 28 weights were associated with the entropy and relative entropy values, and 28 weights with their logarithms. Similarly, 10 weights were associated with the 10 pure morphologic parameters and 10 with their logarithms. Thus, in total, 76 weights were assigned.

A proper threshold value for a diagnostic index was developed to differentiate between melanoma and nonmelanoma. This threshold value indicates the desire to discriminate between benign and malignant lesions so that the diagnostic indices for malignant lesions would be well separated from those for benign lesions when inevitable errors in the measurement procedure are properly taken into account. By comparing with the diagnosis of melanoma or nonmelanoma pathology results obtained from clinical data (see below), a cost function was defined consisting of a master term (including pathology), constraints, regularization, and the Occam rule. Constraints were used to take advantage of *a priori* information about the covariance of the measurements, and regularization was used to suppress variations in the measurements, which do not contribute to the correct diagnosis. Some of the physiologic–morphologic maps and the pure morphologic parameters did not contribute substantially to the diagnosis. The Occam rule was designed to exclude inconsequential parameters.

The diagnostic index (D) is defined as $D = \mathbf{w}^T\mathbf{p}$, where the superscript T denotes transpose, $\mathbf{w}$ is a column vector of 76 weights, and $\mathbf{p}$ is a column vector of 76 elements consisting of 28 entropy and relative entropy values with their 28 logarithms and 10 pure morphologic parameters with their 10 logarithms. A value of the diagnostic index was computed for every lesion examined, and the 76 weights were optimized to give the best diagnosis.

To test robustness, the diagnostic procedure was applied to find optimized weights for different subsets of the entire data set. For this purpose, subsets were created by starting with the entire data set of all clinical measurements ($n = 342$) and excluding some of the measurements. One subset was created by dividing the total set into three equal parts, each consisting of every third measurement, and then excluding one-third of the total subset, so that the remaining subset contained two-thirds of the entire data set. The clinical data set had 11 melanomas (see below). Nine different subsets were created by excluding all three measurements that had been performed on one of the 11 lesions – a melanoma according to the pathology report. In the clinical data set described above, several weights were very small and were set equal to 0. Optimization was performed on several subsets, and the criterion for accepting a new set of weights was that it should give a specificity greater than 90% for a sensitivity of 100% for all subsets. In this manner, a new set of weights with fewer nonzero values was tested for acceptance.

Results

In the 94 patients, 118 lesions were imaged with optical transfer diagnosis and subsequently removed; all were clinically suggestive of melanoma; 11 of these were categorized as melanomas (5 *in situ* melanoma, 4 invasive melanoma, and 2 atypical melanocytic hyperplasia consistent with melanoma). Of the 107 benign lesions, 84 were benign melanocytic neoplasms, but 9 of those showed moderate to severe degrees of architectural disorder or cytologic atypia. The remaining lesions were 2 basal cell carcinomas (BCCs) and 21 other lesions. For each lesion, 3 optical transfer diagnosis measurements (each consisting of 30 images) were taken, but some measurements were discarded because of measurement errors, reducing the total number of useful measurements to 342. When the method described above was applied to this clinical data set, the sensitivity was 100% for any specificity less than 91.4%. When the diagnostic procedure was applied to different subsets of the entire data set in the robustness test (see above), after a corrected selection of unimportant weights, the specificity at 100% sensitivity did not change noticeably and was found to be larger than 90% for all subsets.

Dermoscopic images and morphologic–physiologic maps of representative lesions are shown for a typical nonmelanoma lesion (Fig. 5.24) and for a malignant melanoma (Fig. 5.25). The nonmelanoma lesion in Fig. 5.24 was a nevus with mild architectural disorder. In the seven maps associated with this lesion, there was a relative degree of homogeneity and a lack of entropy (disorder) compared with those in the corresponding maps in Fig. 5.25. The maps of the melanoma show a higher

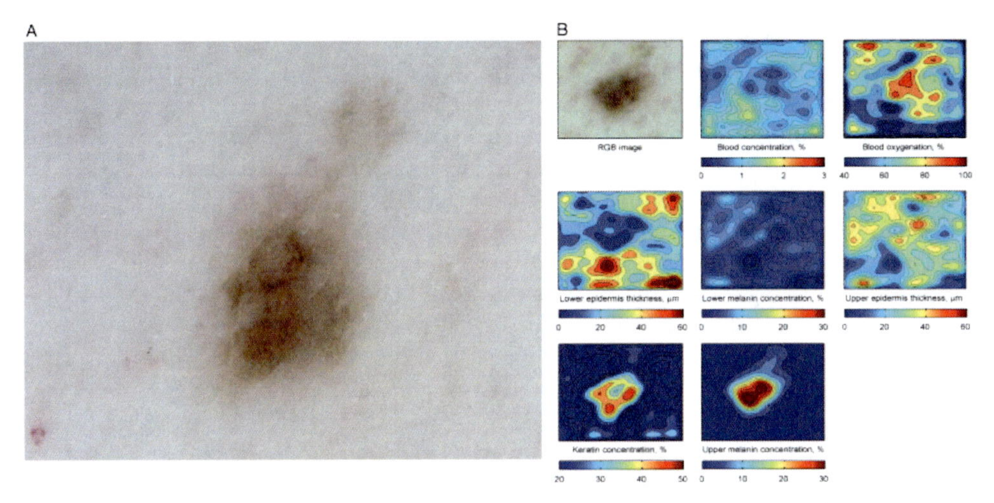

Fig. 5.24. Pathologic diagnosis: nevus with mild architectural disorder. (A) Dermoscopic image. (B) Physiologic maps. The parameters display greater uniformity and higher levels of congruence between maps than the melanoma in Fig. 5.25. RGB, red-green-blue (adapted from Swanson et al. [11]).

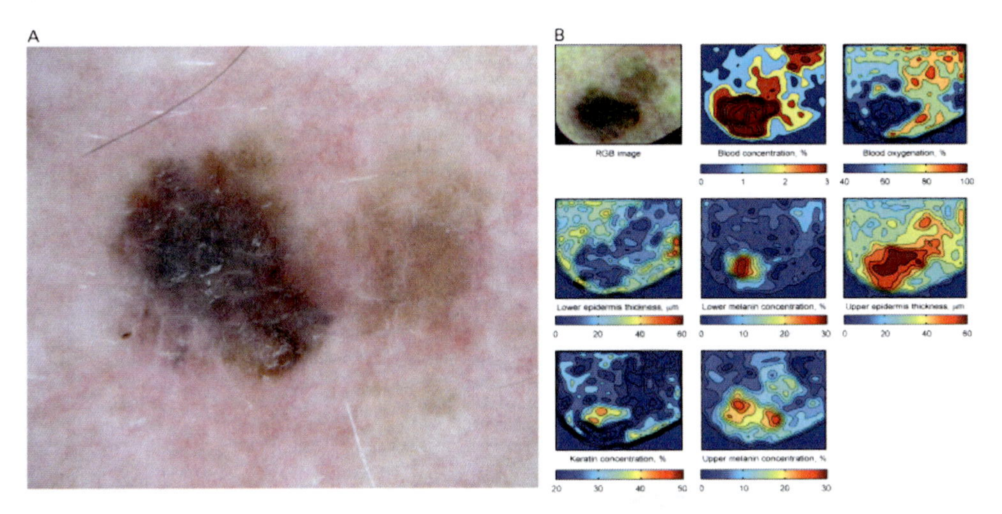

Fig. 5.25. Pathologic diagnosis: malignant melanoma, 1.25 mm thick. (A) Dermoscopic image. (B) Physiologic maps. The physiologic maps display a considerable amount of heterogeneity for the parameters displayed. RGB, red-green-blue (adapted from Swanson et al. [11]).

level of heterogeneity, especially in the percentage of blood oxygenation and in the percentage of upper epidermal keratin concentration, with an associated higher level of relative entropy between each pair of maps. Ten lesions had false-positive results: 4 nevi with mild architectural disorder or mild cytologic atypia (or both), two pigmented actinic keratoses, 1 blue nevus, 1 seborrheic keratosis, 1 dermatofibroma, and 1 large cell acanthoma. Of the lesions identified as not melanoma, there were 2 BCCs and 8 nevi with moderate to severe architectural disorder or cytologic atypia. The BCCs had dermoscopic features of BCC, but were included in the data set because they met the case referral indication.

OTD technology appears to be a promising tool for clinical practice. Additional research needs to be done to validate the ability of OTD to discriminate benign from malignant lesions. Further study is also needed to investigate the ability to discriminate malignant melanoma and bland nevi from benign lesions with moderate to severe degrees of architectural disorder and cytologic atypia. A limitation of optical transfer diagnosis technology is that it did not identify the pigmented BCCs specifically; rather, the algorithm regarded them as nonmelanoma. Further study is needed to determine whether the algorithm can "learn" to discriminate other neoplasms, including BCC. The validity of physiologic mapping as a true reflection of the physiology of human tissue needs to be demonstrated. Physiologic mapping may have applications beyond the screening of melanocytic lesions in inflammatory diseases and other dermatologic disorders.

5.5 Summary

A review was given of scalar radiative transfer theory applicable to a coupled two-slab system consisting of two adjacent slabs with different refractive indices. In section 5.2, a brief description of inherent optical properties (IOPs) was provided, including absorption and scattering coefficients (α and σ) as well as a normalized angular scattering cross-section or scattering phase function. The IOPs depend only on the medium itself, and not on the ambient light field, whereas apparent optical properties (AOPs) depend also on the illumination. The connection between AOPs and IOPs is provided by the RT equation (RTE), described in section 5.3.1 for a coupled two-slab system separated by a plane, horizontal interface across which the refractive index changes abruptly from a value m_1 in one of the slabs to a value m_2 in the other. If one assumes that the IOPs vary only in the vertical direction z, the scalar intensity $I(\tau, \mu, \phi)$ depends on the vertical optical depth $\tau(z) = \int_z^\infty (\alpha(z') + \sigma(z'))\, dz'$ (Eq. (11)), of the medium as well as the polar angle θ ($\mu = \cos\theta$) and the azimuthal angle ϕ. In such a two-slab geometry it is possible to isolate the azimuthal dependence of the intensity. Expansion of the scattering phase function in Legendre polynomials and use of the addition theorem for spherical harmonics led to a Fourier cosine expansion of the scattering phase function. By also expanding the intensity in a Fourier cosine series (Eq. (19)), one finds that each Fourier component $I^m(\tau, \mu)$ satisfies a RTE given by Eq. (20).

Two fundamentally different methods of solution of the RTE for a coupled two-slab system were described. In section 5.3.2 an overview of the discrete-ordinate method was provided. This method consists of approximating the integro-differential RTE (Eq. (20)) by a system of coupled, ordinary differential equations which may be solved by applying methods from linear algebra resulting in a numerical code (C-DISORT) that computes radiances at any optical depth, polar and azimuthal angle. The Monte Carlo (MC) method, described in section 5.3.3, is based on the use of random numbers to perform statistical simulations of the RT process, and may be used to model light propagation in systems with complicated geometries (beyond the coupled two-slab system considered here). In a

Monte Carlo simulation for a coupled two-slab system (C-MC), the solution of the RTE (Eq. (12)) is obtained by following the life histories of a very large number of light beams. Each beam is followed until it leaves the system either by absorption in either of the slabs or by escaping through the bottom of the lower slab or the top of the upper slab.

As discussed in section 5.3.4, the impact of surface roughness on the radiance distribution is to spread the specular reflection over a range of angles, which, in the case of reflections from a wind-roughened water surface, is referred to as the sunglint region. If the surface is characterized by a Gaussian random height distribution, and the tangent plane approximation is invoked, the bidirectional reflection distribution function (BRDF) can be expressed as in Eq. (41), in which the surface slope distribution is given by Eq. (43) for an isotropic, Gaussian rough surface. To correct for sunglint in ocean color remote sensing applications it is customary to use a direct transmittance approach that ignores multiple scattering. As discussed in section 5.3.4, a model for a coupled atmosphere-water system that includes multiple scattering can be used to quantify the error incurred by invoking the direct transmission assumption. These errors typically range from 10% to 90% depending on aerosol loading and sun-satellite geometry.

Sample applications of the theory were discussed in section 5.4, starting with the coupled atmosphere–water system in section 5.4.1. To demonstrate that the two fundamentally different solutions methods give similar results, it was shown (Fig. 5.4) that irradiances computed with the C-MC and C-DISORT codes are in good agreement throughout the coupled atmosphere–ocean system.

An RT code for the coupled atmosphere–water system can be used as a forward model to compute not only radiances, but also Jacobians (radiance partial derivatives with respect to changes in retrieval parameters) that are required for inversion of ocean color data by standard inversion methods such as iterative fitting techniques based on nonlinear least squares or optimal estimation. Such an approach allows for simultaneous retrieval of atmospheric aerosol parameters and marine parameters, and circumvents the 'negative water-leaving radiance problem' encountered in the traditional two-step 'atmospheric correction and regression' approach. At each iteration step, the coupled forward RT model can be used to generate both simulated radiances and Jacobians required in the inversion scheme. It was shown (Figs. 5.5 and 5.6) that this simultaneous forward/inverse retrieval method can provide good retrieval capability, and that 8 SeaWiFS channels were sufficient to retrieve 2 atmospheric and 3 marine parameters in coastal waters.

Applications to coupled atmosphere–snow–ice systems were discussed in section 5.4.2. The C-DISORT coupled RT model can be used to compute the radiation field within sea ice, the BRDF defined in Eq. (40) for bare and snow-covered sea ice, as well as radiances at the top of the atmosphere. Figure 5.7 shows that computed albedos of clean snow agree well with computations done by others as well as with experimental values. Satellite remote sensing can be used to estimate spatial and temporal changes in snow/ice cover, and to retrieve snow/ice optical characteristics and temperatures. To infer information about snow and ice properties from visible and IR satellite imagery a cloud mask is required to discriminate between clear and cloudy sky, and for cloud-free pixels, a snow/sea ice discriminator to distinguish bare sea ice from snow-covered sea ice. Algorithms developed for retrieval of snow ice properties were used to analyze data from the GLI satellite. For example, Fig. 5.11 shows surface temperatures around the northern polar region for April 7 to May 8, 2003, while Fig. 5.11 shows a comparison between temperatures derived from MODIS and GLI data. Figure 5.12 shows the corresponding spatial distribution of snow grain size between 0 and 20 cm retrieved from the NIR channel at 0.875 μm ($R_{s0.9}$). The snow grain size of the top surface layer (0–2 cm) can be retrieved from the $\lambda = 1.64$ μm channel ($R_{s1.6}$). The ratio of $R_{s1.6}$ to $R_{s0.9}$ for the April 7–22 period shown in Fig. 5.13 indicates that the difference in the spatial distribution between $R_{s0.9}$ and $R_{s1.6}$ may be due to vertical inhomogeneity of the grain size in the upper several centimeters of the snow or by different light penetration depths at $\lambda = 0.865$ μm and 1.64 μm. It was also shown that there is a close relationship between the $R_{s0.9}$ snow grain size and surface temperature (T_s), as illustrated in Fig. 5.14, which shows the spatial distribution of the snow cover for the period of April 7–22, 2003, but color-coded using the two-dimensional temperature-grain size ($T_s - R_{s0.9}$) relationship to indicate the potential of snow grains to metamorphose. The snow in the two elliptical areas in Fig. 5.14 has similar grain sizes (around 200 μm) but exists in different temperature regimes (272–273 K in the orange area and 253 K in the blue area). The snow in orange-colored area has a high potential while that in blue-colored area has a low potential of metamorphosis.

In section 5.4.3, C-MC simulations and C-DISORT calculations of spectral radiances in coupled air–tissue systems were compared, and excellent agreement was demonstrated between radiances obtained with the two codes, both above and in the tissue for typical optical properties of skin tissue at the wavelengths 280, 540, and 650 nm. The scattering phase function for internal structures in the skin was represented by the Henyey–Greenstein phase function for large particles and the Rayleigh phase function for small particles. The C-DISORT code was found to be about 1,000 times faster than the C-MC code.

In order to test the feasibility of retrieving important parameters describing the physiologial state of human skin, ultraviolet and visible reflectance spectra from human skin, measured before and on each day for two weeks after photodynamic treatment with the hexyl ester of ALA and exposure to red light (632 nm), were analyzed using C-DISORT for a coupled air–tissue system as a forward model in conjunction with a classic inversion scheme. The results showed that it is possible to perform a simultaneous retrieval of several important physiological parameters, such as the melanin in both the basal and upper layers of the epidermis.

To evaluate the potential of a novel imaging technology, optical transfer diagnosis (OTD), for differentiation of benign from malignant pigmented melanocytic lesions, 50 patients with pigmented lesions suspicious for melanoma were referred for OTD. After OTD scanning, lesions were biopsied for histopathologic examination, each by 2 separate dermatopathologists. Maps of morphological and physiological parameters were derived using forward/inverse modeling by C-DISORT of light absorption and scattering by chromophores such as hemoglobin, keratin, and melanin at different epidermal and dermal depths. Relative entropies were analyzed for output prediction of malignancy vs. nonmalignancy. Sixty-three pigmented suspicious lesions were OTD scanned before being biopsied for histopathologic examination by the 2 dermatopathologists. Of the 63 lesions, 5 were identified as melanoma and

58 were found to be benign. OTD was able to identify the malignant lesions with 100% sensitivity and 94.8–96.6% specificity.

To further evaluate the potential of OTD for distinguishing benign from malignant pigmented melanocytic neoplasms, 49 patients with pigmented lesions suggestive of melanoma were referred for OTD. After lesions were scanned with the OTD camera, they were removed for histopathologic examination by 2 dermatopathologists each. From the recorded images, morphologic–physiologic maps were created, and entropy and relative entropy values derived from the maps and a set of pure morphologic parameters were analyzed for output prediction of melanoma versus nonmelanoma. Of the 118 scanned and biopsied lesions, 11 were identified as melanoma or atypical melanocytic hyperplasia consistent with melanoma. For identification of melanomas, OTD had a sensitivity of 100% and a specificity of 90%.

References

1. C. D. Mobley, B. Gentili, H. R. Gordon, Z. Jin, G. W. Kattawar, A. Morel, P. Reinersman, K. Stamnes, and R. H. Stavn, Comparison of numerical models for computing underwater light fields, *Appl. Opt.* **32**, 7484–7504 (1993).
2. Z. Jin and K. Stamnes, Radiative transfer in nonuniformly refracting layered media: atmosphere–ocean system, *Appl. Opt.* **33**, 431–442 (1994).
3. B. Yan and K. Stamnes, Fast yet accurate computation of the complete radiance distribution in the coupled atmosphere ocean system. *J. Quant. Spectrosc. Radiat. Transfer* **76**, 207–223 (2003).
4. G. E. Thomas and K. Stamnes, *Radiative Transfer in the Atmosphere and Ocean*, Cambridge University Press (1999); second edition (2002).
5. K. I. Gjerstad, J. J. Stamnes, B. Hamre, J. K. Lotsberg, B. Yan, and K. Stamnes, Monte Carlo and discrete-ordinate simulations of irradiances in the coupled atmosphere–ocean system, *Appl. Opt.* **42**, 2609–2622 (2003).
6. B. Hamre, J.-G. Winther, S. Gerland, J. J. Stamnes, and K. Stamnes, Modeled and measured optical transmittance of snow covered first-year sea ice in Kongsfjorden, Svalbard, *J. Geophys. Res.* **109**, d0i:10.1029/2003JC001926 (2004).
7. S. Jiang, K. Stamnes, W. Li, and B. Hamre, Enhanced solar irradiance across the atmosphere–sea ice interface: a quantitative numerical study, *Appl. Opt.* **44**, 2613–2625 (2005).
8. K. Hestenes, K. P. Nielsen, L. Zhao, J. J. Stamnes, and K. Stamnes, Monte Carlo and discrete-ordinate simulations of spectral radiances in the coupled air–tissue system, *Appl. Opt.* **46**, 2333–2350 (2007).
9. B. Chen, K. Stamnes, and J. J. Stamnes, Validity of diffusion approximation in bio-optical imaging, *Appl. Opt.* **40**, 6356–6366 (2001).
10. K. Nielsen, L. Zhao, G. A. Ryzhikov, M. S. Biryulina, E. R. Sommersten, J. J. Stamnes, K. Stamnes, and J. Moan, Retrieval of the physiological state of human skin from UV-VIS reflectance spectra: A feasibility study, *J. Photochem. Photobiol. B* **93**, 23–31 (2008).
11. D. L. Swanson, S. D. Laman, M. Biryulina, K. P. Nielsen, G. Ryzhikov, J. J. Stamnes, B. Hamre, L. Zhao, F. S. Castellana, and K. Stamnes, Optical transfer diagnosis of pigmented lesions: a pilot study, *Skin Res. Technol.* **15**, 330–337 (2009). doi: 10.1111/j.1600-0846.2009.00367.x

12. D. L. Swanson, S. D. Laman, M. Biryulina, K. P. Nielsen, G. Ryzhikov, J. J. Stamnes, B. Hamre, L. Zhao, E. Sommersten, F. S. Castellana, and K. Stamnes, Optical transfer diagnosis of pigmented lesions, *Dermatol. Surg.* **36**, 1–8 (2010). DOI: 10.1111/j.1524-4725.2010.01808.x
13. Ø. Frette, J. J. Stamnes, and K. Stamnes, Optical remote sensing of marine constituents in coastal waters: A feasibility study, *Appl. Opt.* **37**, 8318–8326 (1998).
14. Ø. Frette, S. R. Erga, J. J. Stamnes, and K. Stamnes, Optical remote sensing of waters with vertical structure, *Appl. Opt.* **40**, 1478–1487 (2001).
15. K. Stamnes, W. Li, B. Yan, A. Barnard, W. S. Pegau, and J. J. Stamnes, Accurate and self-consistent ocean color algorithm: simultaneous retrieval of aerosol optical properties and chlorophyll concentrations, *Appl. Opt.* **42**, 939–951 (2003).
16. W. Li, K. Stamnes, R. Spurr, and J. J. Stamnes, Simultaneous retrieval of aerosols and ocean properties: A classic inverse modeling approach. II. SeaWiFS case study for the Santa Barbara channel, *Int. J. Rem. Sens.* **29**, 5689–5698 (2008). DOI: 10.1080/01431160802007632
17. R. J. D. Spurr, LIDORT and VLIDORT: Linearized pseudo-spherical scalar and vector discrete ordinate radiative transfer models for use in remote sensing retrieval algorithms, In A. Kokhanovsky (Ed.), *Light Scattering Reviews*, vol. 3. Springer, Berlin (2008).
18. R. Spurr, K. Stamnes, H. Eide, W. Li, K. Zhang, and J. J. Stamnes, Simultaneous retrieval of aerosol and ocean properties: A classic inverse modeling approach: I. Analytic Jacobians from the linearized CAO-DISORT model, *J. Quant. Spectrosc. Radiat. Transfer* **104**, 428–449 (2007).
19. P.-W. Zhai, Y. Hu, J. Chowdhary, C. R. Trepte, P. L. Lucker, and D. B. Josset, A vector radiative transfer model for coupled atmosphere and ocean systems with a rough interface, *J. Quant. Spectrosc. Radiat. Transfer* **111**, 1025–1040 (2007).
20. C. D. Mobley, *Light and Water*, Cambridge University Press (1994).
21. L. C. Henyey and J. L. Greenstein, Diffuse radiation in the galaxy, *Astrophys. J.* **93**, 70–83 (1941).
22. L. Rayleigh, A re-examination of the light scattered by gases in respect of polarization. I. Experiments on the common gases, *Proc. Roy. Soc.* **97**, 435–450 (1920).
23. L. Rayleigh, A re-examination of the light scattered by gases in respect of polarization. II. Experiments on helium and argon, *Proc. Roy. Soc.* **98**, 57–64 (1920).
24. A. Morel and B. Gentili, Diffuse reflectance of oceanic waters: its dependence on sun angle as influenced by the molecular scattering contribution, *Appl. Opt.* **30**, 4427–4437 (1991).
25. L. Rayleigh, On the light from the sky, its polarization and colour, *Phil. Mag.* **41**, 107–120, 274–279, 447–454 (1871).
26. K. Stamnes, S. C. Tsay, W. J. Wiscombe, and K. Jayaweera, Numerically stable algorithm for discrete-ordinate-method radiative transfer in multiple scattering and emitting layered media, *Appl. Opt.* **27**, 2502–2509 (1988).
27. K. Stamnes, S. C. Tsay, W. J. Wiscombe, and I. Laszlo, DISORT, A General-Purpose Fortran Program for Discrete-Ordinate-Method Radiative Transfer in Scattering and Emitting Layered Media: Documentation of Methodology, ftp://climate.gsfc.nasa.gov/pub/wiscombe/Multiple Scatt/ (2000).
28. M. Born and E. Wolf, *Principles of Optics*, Cambridge University Press (1980)).
29. W. H. Press, S. A. Teukolsky, W. T. Vetterling, and B. P. Flannery. *Numerical Recipes in C*, Cambridge University Press (1992).
30. J. M. Schmitt and K. Ben-Letaief, Efficient Monte Carlo simulation of confocal microscopy in biological tissue, *J. Opt. Soc. Am. A* **13**, 952–961 (1996).

31. A. Bilenca, A. Desjardins, B. Bouma, and G. Tearney, Multicanonical Monte-Carlo simulations of light propagation in biological media, *Opt. Expr.* **13**, 9822–9833 (2005).
32. P. Beckmann and A. Spizzichino, *The Scattering of Electromagnetic Waves from Rough Surfaces*, Macmillan, New York (1963).
33. L. Tsang, J. A. Kong, and R. T. Shin, *Theory of Microwave Remote Sensing*, Wiley, New York (1985).
34. H. R. Gordon, Atmospheric correction of ocean color imagery in the Earth Observing Observation System era, *J. Geophys. Res.* **102**, 17081–17106 (1997).
35. M. Wang and S. Bailey, Correction of sun glint contamination on the SeaWiFS ocean and atmosphere products. *Appl. Opt.* **40**, 4790–4798 (2001)
36. M. Ottaviani, R. Spurr, K. Stamnes, W. Li, W. Su, and W. J. Wiscombe, Improving the description of sunglint for accurate prediction of remotely-sensed radiances, *J. Quant. Spectrosc. Radiat. Transfer*, doi:10.1016/j.jqsrt.2008.05.012, 2008.
37. S.-C. Tsay and G. L. Stephens, A Physical/Optical Model for Atmospheric Aerosols with Application to Visibility Problems, Department of Atmospheric Sciences, Colorado State University, Fort Collins, CO, 1990.
38. B. Yan, K. Stamnes, W. Li, B. Chen, J. J. Stamnes, and S.-C. Tsay, Pitfalls in atmospheric correction of ocean color imagery: How should aerosol optical properties be computed? *Appl. Opt.* **41**, 412–423 (2002).
39. H. Gordon, O. Brown, and M. Jacobs, Computed reletionships between the inherent and apparent optical properties of a flat homogeneous ocean, *Appl. Opt.* **14**, 417–427 (1975).
40. H. Gordon, A bio-optical model describing the distribution of irradiance at the sea surface resulting from a point source embedded in the ocean, *Appl. Opt.* **26**, 4133–4148 (1987).
41. H. Gordon, Can the Lambert–Beer law be applied to the diffuse attenuation coefficient of ocean water?, *Limnol. Oceanogr.* **34**, 1389–1409 (1989).
42. H. Gordon, Dependence of the diffuse reflectance of natural waters on the sun angle, *Limnol. Oceanogr.* **34**, 1484–1489 (1989).
43. H. Gordon, Diffuse reflectance of the ocean: influence of nonuniform phytoplankton pigment profile, *Appl. Opt.* **31**, 2116–2129 (1992).
44. Y. Ge, H. Gordon, and K. Voss, Simulation of inelastic scattering contributions to the irradiance field in the oceanic variation in Fraunhofer line depths, *Appl. Opt.* **32**, 4028–4036 (1993).
45. G. Kattawar and X. Xu, Filling-in of Fraunhofer lines in the ocean by Raman scattering, *Appl. Opt.* **31**, 1055–1065 (1992).
46. G. Kattawar and C. Adams, Stokes vector calculations of the submarine light field in an atmosphere–ocean with scattering according to the Rayleigh phase matrix: Effect of interface refractive index on radiance and polarization, *Limnol. Oceanogr.* **34**, 1453–1472 (1989).
47. G. Kattawar and C. Adams, Errors in radiance calculations induced by using scalar rather than Stokes vector theory in a realistic atmosphere-ocean system, in *Ocean Optics X*, R. W. Spinrad (Ed.), *Proc. Soc. Photo-Opt. Instrum. Eng.* **1302**, 2–12 (1990).
48. C. Adams and G. Kattawar, Effect of volume scattering function on the errors induced when polarization is neglected in radiance calculations in an atmosphere-ocean system, *Appl. Opt.* **32**, 4610–4617 (1993).
49. G. Kattawar and C. Adams, Errors induced when polarization is neglected, in radiance calculations for an atmosphere–ocean system, in *Optics for the Air–Sea Interface*, L. Epstep (Ed.), *Proc. Soc. Photo-Opt. Instrum. Eng.* **1749**, 2–22 (1992).

50. G. Plass and G. Kattawar, Radiative transfer in an atmosphere–ocean system, *Appl. Opt.* **8**, 455–466 (1969).
51. G. Plass and G. Kattawar, Monte–Carlo calculations of radiative transfer in the earth's atmosphere ocean system: I. Flux in the atmosphere and ocean, *Phys. Oceanogr.* **2**, 139–145 (1972).
52. H. Gordon and O. Brown, Irradiance reflectivity of a flat ocean as a function of its optical properties, *Appl. Opt.* **12**, 1549–1551 (1973).
53. G. Plass, G. Kattawar, and J. A. Guinn Jr., Radiative transfer in the earth's atmosphere and ocean: influence of ocean waves, *Appl. Opt.* **14**, 1924–1936 (1975).
54. A. Morel and B. Gentili, Diffuse reflectance of oceanic waters. II. bidirectional aspect, *Appl. Opt.* **32**, 2803–2804 (1993).
55. J. Kirk, Monte Carlo procedure for simulating the penetration of light into natural waters, Techn. Rep., Div. Plant Industry Techn. Paper 36, Commonwealth Scientific and Industrial Research Organization, Canberrea, Australia (1981).
56. W. Blattner, H. Horak, D. Collins, and M. Wells, Monte Carlo studies of the sky radiation at twilight, *Appl. Opt.* **13**, 534 (1974).
57. R. Stavn and A. Weidemann, Optical modeling of clear ocean light fields: Raman scattering effects, *Appl. Opt.* **27**, 4002–4011 (1988).
58. R. Stavn and A. Weidemann, Raman scattering in ocean optics: quantitative assessment of internal radiant emission, *Appl. Opt.* **31**, 1294–1303 (1992).
59. C. Mobley, A numerical model for the computation of radiance distributions in natural waters with wind-roughened surfaces, *Limnol. Oceanogr.* **34**, 1473–1483 (1989).
60. C. Mobley and R. Preisendorfer, A numerical model for the computation of radiance distributions in natural waters with wind-roughened surfaces, NOAA Tech. Meo. ERL PMEL-75 (NTIS PB88-192703), Pacific Marine Environmental Laboratory, Seattle, WA (1988).
61. Z. Jin, T. P. Charlock, K. Rutledge, K. Stamnes, and Y. Wang, An analytical solution of radiative transfer in the coupled atmosphere–ocean system with rough surface, *Appl. Opt.* **45**, 7443–7455 (2006).
62. C. Cox and W. Munk, Measurement of the roughness of the sea surface from photographs of the sun's glitter, *J. Opt. Soc. Am.* **44**, 838–850 (1954).
63. K. Carder, F. Chen, J. Cannizarro, J. Campbell, and B. Mitchell, Performance of MODIS semi analytical ocean colour algorithm for chlorophyll-a, *Adv. Space Res.* **33**, 1152–1159 (2004).
64. D. A. Siegel, M. Wang, S. Maritorena, and W. Robinson, Atmospheric correction of satellite ocean colour imagery: the black pixel assumption. *Appl. Opt.* **39**, 3582–3591 (2000).
65. K. Stamnes, W. Li, B. Yan, H. Eide, A. Barnard, W. S. Pegau, and J. J. Stamnes, Accurate and self-consistent ocean colour algorithm: simultaneous retrieval of aerosol optical properties and chlorophyll concentrations, *Appl. Opt.* **42**, 939–951 (2003).
66. C. Rodgers, *Inverse Methods for Atmospheric Sounding*, World Scientific Press, Singapore (2000).
67. E. P. Shettle and R. W. Fenn, Models for the Aerosols of the Lower Atmosphere and the Effects of Humidity Variations on their Optical Properties, Air Force Geophysics Laboratory, Hanscomb AFB, MA, 1979.
68. K. Zhang, W. Li, K. Stamnes, H. Eide, R. Spurr, and S.-C. Tsay, Assessment of the MODIS algorithm for the retrieval of aerosol parameters over the ocean, *Appl. Opt.* **46**, 1525–1534 (2007).
69. R. M. Pope and E. S. Fry, Absorption spectrum (380–700 nm) of pure water II. Integrating cavity measurements, *Appl. Opt.* **36**, 8710–8723 (1997).

70. R. C. Smith and K. S. Baker, Optical properties of the clearest natural waters, *Appl. Opt.* **20**, 177–184 (1981).
71. P. Diehl and H. Haardt, Measurement of the spectral attenuation to support biological research in a plankton tube experiment, *Oceanol. Acta* **3**, 89–96 (1980).
72. I. N. McCave, Particulate size spectra, behavior, and origin of nephloid layers over the Nova Scotia continental rise, *J. Geophys. Res.* **88**, 7647–7660 (1983).
73. G. R. Fournier and J. L. Forand, Analytic phase function for ocean water, in *Proc. Ocean Optics XII*, SPIE vol. 2558, pp. 194–201 (1994).
74. C. D. Mobley, L. K. Sundman, and E. Boss, Phase function effects on oceanic light fields, *Appl. Opt.* **41**, 1035–1050 (2002).
75. Y. X. Hu, B. Wielicki, B. Lin, G. Gibson, S.-C. Tsay, K. Stamnes, and T. Wong, A fast and accurate treatment of particle scattering phase function with weighted SVD least square fitting, *J. Quant. Spectrosc. Radiat. Transfer* **65**, 681–690 (2000).
76. K. Stamnes, W. Li, Y. Fan, T. Tanikawa, B. Hamre, and J. J. Stamnes, A fast yet accurate algorithm for retrieval of aerosol and marine parameters in coastal waters, Ocean Optics XX, Anchorage, Alaska (2010).
77. K. Stamnes, B. Hamre, J. J. Stamnes, G. Ryzhikov, M. Birylina, R. Mahoney, B. Hauss, and A. Sei, Modeling of radiation transport in coupled atmosphere–snow–ice–ocean systems, *J. Quant. Spectrosc. Radiat. Transfer*, doi:10.1016/j.jqsrt.2010.06.006 (2010).
78. J. Fre and J. Dozier, The image processing workbench - portable software for remote sensing instruction and research, in Proceedings of the 1986 International Geoscience and Remote Sensing Symposium, ESA SP-254, pp. 271–276, European Space Agency, Paris (1986).
79. A. M. Baldridge, S. J. Hook, C. I. Grove, and G. Rivera, The ASTER spectral library version 2.0, *Rem. Sens. Environ.* **113**, 711–715 (2009).
80. T. C. Grenfell, A radiative transfer model for sea ice with vertical structure variations, *J. Geophys. Res.* **96**, 16991–17001 (1991).
81. R. E. Brandt, S. G. Warren, A. P. Worby, and T. C. Grenfell, Surface albedo of the Antarctic sea-ice zone, *J. Climate* **18**, 3606–3622 (2005).
82. B. P. Briegleb and B. Light, A Delta-Eddington multiple scattering parameterization of solar radiation in the sea ice component of the Community Climate System Model, NCAR Technical Note (NCAR/TN-472+STR (2007).
83. Te. Aoki, Ta. Aoki, M. Fukabori, A. Hachikubo, Y. Tachibana, and F. Nishio, Effects of snow physical parameters on spectral albedo and bidirectional reflectance of snow surface, *J. Geophys. Res.* **105**, 10219–10236 (2000).
84. T. Tanikawa, Te. Aoki, M. Hori, A. Hachikubo, and M. Aniya, Snow bidirectional reflectance model using non-spherical snow particles and its validation with field measurements, *EARSel eProceedings*, **5**, 137-145 (2002).
85. A. A. Kohkhanovsy, Te. Aoki, A. Hachikubo, M. Hori, and E. P. Zege, Reflective properties of natural snow: Approximate asymptotic theory versus in situ measurements, IEEE Transactions on Geoscience and Remote Sensing, **43**, 1529-1535, doi:10.1109/TGRS.2005.848414 (2005).
86. M. Fily, B. Bourdelles, J. P. Dedieu, and C. Sergent, Comparison of in situ and Landsat Thematic Mapper derived snow grain characteristics in the Alps, *Rem. Sens. Environ.* **59**, 452–460 (1997).
87. B. Bourdelles and M. Fily, Snow grain-size determination from Landsat imagery over Terre Adelie, *Antarctica, Ann. Glaciol.* **17**, 86–92 (1993).
88. A. W. Nolin and J. Dozier, Estimating snow grain size using AVIRIS data, *Rem. Sens. Environ.* **44**, 231–238 (1993).

89. T. H. Painter, D. A. Roberts, R. O. Green, and J. Dozier, The effect of grain size on spectral mixture analysis of snow-covered area from AVIRIS data. *Rem. Sens. Environ.* **65**, 320–332 (1998).
90. R. O. Green, J. Dozier, D. A. Roberts, and T. H. Painter, Spectral snow reflectance models for grain size and liquid water fraction in melting snow for the solar reflected spectrum, *Ann. Glaciol.* **34**, 71–73 (2002).
91. D. K. Hall, G. A. Riggs, and V. V. Salomonson, Development of methods for mapping global snow cover using Moderate Resolution Imaging Spectroradiometer data, *Rem. Sens. Environ.* **54**, 127–140 (1995).
92. S. G. Warren, Optical properties of snow, *Rev. Geophys. Space Phys.* **20**, 67–89 (1982).
93. W. J. Wiscombe and S. G. Warren, A model for the spectral albedo of snow. I. Pure snow. *J. Atmos. Sci.* **37**, 2712–2733 (1980).
94. K. Stamnes, W. Li, H. Eide, Te. Aoki, M. Hori, and R. Storvold, ADEOS-II/GLI Snow/Ice Products – Part I: Scientific Basis, *Rem. Sens. Environ.* **111**, 258–273 (2007).
95. Te. Aoki, M. Hori, H. Motoyohi, T. Tanikawa, A. Hachikubo, K. Sugiura, T. Yasunari, R. Storvold, H. A. Eide, K. Stamnes, W. Li, J. Nieke, Y. Nakajoma, and F. Takahashi, ADEOS-II/GLI snow/ice products–Part II: Validation Results, *Rem. Sens. Environ.* **111**, 320–336 (2007).
96. M. Hori, Te. Aoki, K. Stamnes, and W. Li, ADEOS-II/GLI snow/ice products – Part III: Retrieved Results, *Rem. Sens. Environ.* **111**, 274–319 (2007).
97. T. Y. Nakajima, T. Nakajima, M. Nakajima, H. Fukushima, M. Kuji, A. Uchiyama, and M. Kishino, Optimization of the advanced earth observing satellite II global imager channels by use of radiative transfer calculations, *Appl. Opt.* **37**, 3149–3163 (1998).
98. J. Nieke, Te. Aoki, T. Tanikawa, H. Motoyoshi, and M. Hori, A satellite cross-calibration experiment, IEEE Geosci. *Rem. Sens. Lett.* **1**, 215–219 (2004).
99. K. Stamnes, W. Li, H. Eide, and J. J. Stamnes, Challenges in atmospheric correction of satellite imagery, *Opt. Eng.* **44**, 041003-1-041003-9 (2005).
100. D. T. Llewellyn-Jones, P. J. Minnett, R. W. Saunders, and A. M. Zavody, Satellite multichannel infrared measurements of sea surface temperature of the northeast Atlantic ocean using AVHRR/2, *Q. J. R. Meteorol. Soc.* **110**, 613–631 (1984).
101. I. J. Barton, Transmission model and ground-truth investigation of satellite-derived sea surface temperatures, *J. Clim. Appl. Meteorol.* **24**, 508–516 (1985).
102. P. J. Minnett, The regional optimization of infrared measurements of sea surface temperature from space, *J. Geophys. Res.* **95**, 13497–13510 (1990).
103. J. Key and M. Haefliger, Arctic ice surface temperature retrieval from AVHRR thermal channels, *J. Geophys. Res.* **97**, 5885–5893 (1992).
104. J. R. Key, J. B. Collins, C. Fowler, and R. S. Stone, High-latitude surface temperature estimates from thermal satellite data, *Rem. Sens. Environ.* **67**, 302–309 (1997).
105. Z. Wan and J. Dozier, A generalized split-window algorithm for retrieving land-surface temperature from space, IEEE Trans. *Geosci. Rem. Sens.* **34**, 892–905 (1996).
106. D. K. Hall, J. R. Key, K. A. Casey, G. A. Riggs, and D. J. Cavalieri, Sea ice surface temperature products from MODIS, *IEEE Trans. Geosci. Rem. Sens.* **42**, 1076–1087 (2004).
107. E. Brun, P. David, M. Sudul, and G. Brugnot, A numerical model to simulate snow-cover stratigraphy for operational avalanche forecasting, *J. Glaciol.* **38**, 13–22 (1992).

108. S. Wan, R. Anderson, and J. A. Parrish, Analytical modeling for the optical properties of the skin with in vitro and in vivo applications, *Photochem. Photobiol.* **34**, 493–499 (1981).
109. L. O. Svaasand, Optical dosimetry for direct and interstitial photoradiation therapy of malignant tumors, in *Porphyrin Localization and Treatment of Tumors*, D. Doiron and C. Gomer (Eds.), John Wiley, New York, pp. 91–114 (1984).
110. S. A. Prahl, M. Keijzer, S. L. Jacques, and A. J. Welch, A Monte Carlo model of light propagation in tissue, in *Proc. Dosim. Laser Rad. Med. Biol.*, G. J. Müller and D. H. Sliney (Eds.), SPIE IS **5**, 102–111 (1989).
111. K. P. Nielsen, L. Zhao, P. Juzenas, K. Stamnes, J. J. Stamnes, and J. Moan, Reflectance spectra of pigmented and non-pigmented skin in the UV spectral region, *Photochem. Photobiol.* **80**, 450–455 (2004).
112. I. V. Meglinski and S. J. Matcher, Computer simulation of the skin reflectance spectra, *Comp. Meth. Program. Biomed.* **70**, 179–186 (2003).
113. T. J. Pfefer, D. Sharma, and L. S. Matchette, Evaluation of a fiberoptic-based system for optical property measurement in highly attenuating turbid media, in: I. Gannot (Ed.), *Optical Fibers and Sensors for Medical Applications V*, SPIE Press, Bellingham, WA, pp. 163–171 (2005).
114. D. Sharma, A. Agrawal, L. S. Matchette, and T. J. Pfefer, Evaluation of a fiberoptic-based system for measurement of optical properties in highly attenuating turbid media, Biomed. Eng. Onl. **5**, http://biomedical-engineeringonline.com/content/5/1/49.
115. A. N. Tikhonov and V. Arsenin, *Solution of Ill-Posed Problems*, Winston, Washington, DC (1977).
116. L. Zhao, K. P. Nielsen, A. Juzeniene, P. Juzenas, V. Iani, L. Ma, K. Stamnes, J. J. Stamnes, and J. Moan, Spectroscopic measurements of photoinduced processes in human skin after topical application of the hexylester of 5-aminolevulinic acid, *J. Environ. Pathol. Toxicol. Oncol.* **25**, 307–320 (2006).
117. A. R. Degheidy, M. S. Abdel Krim, Effects of Fresnel and diffused reflectivities on light transport in a half-space medium, *J. Quant. Spectrosc. Radiat. Transfer* **61**, 751–757 (1999).
118. T. J. Farrell, M. S. Patterson, and M. Essenpreis, Influence of layered tissue architecture on estimates of tissue optical properties obtained from spatially resolved diffuse reflectometry, *Appl. Opt.* **37**, 1958–1972 (1998).
119. B. J. Tromberg, A. Orenstein, S. Kimel, S. J. Barker, J. Hyatt, J. S. Nelson, W. G. Roberts, and M. W. Berns, Tumor oxygen tension during photodynamic therapy, *J. Photochem. Photobiol. B* **5**, 121–126 (1990).
120. R. R. Anderson and J. A. Parrish, Optical properties of human skin, in: J. D. Regan and J. A. Parrish (Eds.), *The Science of Photomedicine*, Plenum Press, New York, NY, pp. 147–194 (1982).
121. H. Hönigsmann, Newer knowledge of immediate pigment darkening (IPD), in: F. Urbach, R. W. Gange (Eds.), *The Biological Effects of UVA Radiation*, Praeger Publishers, CBS Inc., New York, NY, pp. 221–224 (1986).
122. J. Sandby-Møller, T. Poulsen, and H. C. Wulf, Epidermal thickness at different body sites: relation to age, gender, pigmentation, blood content, skin type and smoking habits, *Acta Derm. Venereol.* **83**, 410–413 (2003).
123. A. Jemal, R. Siegel, E. Ward et al., Cancer statistics, CA Cancer J. Clin. **58**, 71–96 (2008). Epub February 20, 2008.
124. M. Binder, M. Schwarz, A. Winkler et al., Epiluminescence microscopy: a useful tool for the diagnosis of pigmented skin lesions for formally trained dermatologists, *Arch. Dermatol.* **131**, 286–291 (1995).

125. M. Carrara, A. Bono, C. Bartoli et al., Multispectral imaging and artificial neural network: mimicking the management decision of the clinician facing pigmented skin lesions, *Phys. Med. Biol.* **52**, 2599–2613 (2007). Epub April 17, 2007.
126. M. E. Celebi, H. A. Kingravi, B. Uddin et al., A methodological approach to the classification of dermoscopy images, *Comput. Med. ImagGraph.* **31**, 362–373 (2007). Epub March 26, 2007.
127. S. W. Menzies, Technologies for the diagnosis of primary melanoma of the skin, *Med. J. Aust.* **185**, 533–534 (2006).
128. M. Elbaum, A. W. Kopf, H. S. Rabinovitz et al., Automatic differentiation of melanoma from melanocytic nevi with multispectral digital dermoscopy: a feasibility study, *J. Am. Acad. Dermatol.* **44**, 207–218 (2001).

6 Airborne measurements of spectral shortwave radiation in cloud and aerosol remote sensing and energy budget studies

Sebastian Schmidt and Peter Pilewskie

6.1 Introduction

Space-borne observations of clouds and aerosols are currently undergoing important developments. A new generation of passive imagers follows in the footsteps of proven instrumentation akin to AVHRR (Advanced Very High Resolution Radiometer: Cracknell, 1997) and MODIS (Moderate Resolution Imaging Spectroradiometer: King et al., 1992). At the same time, novel approaches are diversifying the instrumental infrastructure and thus extending the observable parameter space: radar and lidar explore the vertical distribution of clouds and aerosols; polarimeters help untangle aerosols and clouds, and complement non-polarized imagery for ice and mixed-phase clouds. Curiously, the spectral information in the shortwave (solar) wavelength range has remained largely underutilized for cloud and aerosol remote sensing, whereas the infrared and microwave spectral ranges are extensively used for sounding techniques – particularly for water vapor. Solar spectral imagers such as AVIRIS (Airborne Visible/InfraRed Imaging Spectrometer) are routinely flown in geological surveys, recently in the aftermath of the May 2010 Gulf of Mexico oil spill (Clark et al., 2010). Ecosystem mapping (Pignatti et al., 2009) and ocean color retrievals (Liew and Kwoh, 2003) with Hyperion onboard the NASA satellite EO-1 are examples of space-borne spectral cartography in biology and ocean chemistry. In all of these applications, the atmosphere between the surface and the sensor is a factor that needs to be removed via correction algorithms. The spectral signal from the atmosphere itself is mainly used for fingerprinting trace gases based on differential optical absorption spectroscopy. In addition to gas-phase spectroscopy, the European Space Agency's SCIAMACHY (scanning imaging absorption spectrometer for atmospheric cartography) on ENVISAT and GOME (global ozone monitoring experiment) on ERS-2 provide limited information about aerosols and clouds, which introduce biases in trace gas retrievals due to enhanced scattering and absorption or spatial heterogeneity effects (Wagner et al., 2008). However, derived parameters such as the absorbing and scattering aerosol indices (de Graaf and Stammes, 2005; Penning de Vries et al., 2009) or effective cloud fraction (Grzegorski et al., 2006) remain somewhat quantitative or are limited to certain wavelength bands with strong gas absorption lines (Koelemeijer et al., 2002) or Fraunhofer lines (Ring effect, Joiner and Bhartia, 1995).

Why is cloud and aerosol remote sensing lagging behind exploiting the full information content in solar spectral radiance measurements? One reason is that the large footprint of most space-borne spectral radiometers makes them ill-suited for spatially resolving clouds. Another may be the common belief that spectrally resolved observations add redundant information as far as clouds and aerosols are concerned. Principal component analysis shows that a limited number of independent pieces of information can explain most of the variance in solar spectral radiation measurements (Rabbette and Pilewskie, 2002). However, this information is distributed across the spectrum. While optical thickness and effective radius can be derived from the cloud reflectance at two solar wavelengths (Nakajima and King, 1990), the retrieval accuracy may benefit from the information content in additional wavelengths (Vukicevic et al., 2010).

In fact, the keys to resolving some of the outstanding issues in cloud–aerosol remote sensing and radiative energy budget may lie in the spectral dimension. In this chapter, we will substantiate this claim by giving an overview of recent airborne measurements and modeling of solar spectral radiation that we hope will spur new observational approaches in the future. We start by pointing out the subtle but important differences between multi-channel filter-radiometers and spectrometers, and illustrate instrument setup, calibration, precision and accuracy using as an example the Solar Spectral Flux Radiometer (SSFR: Pilewskie et al., 2003), a spectrometer system that can be used to measure solar spectral radiance or irradiance. In the subsequent sections, we will illustrate how airborne irradiance measurements provide insights into cloud–aerosol remote sensing, where each section brings to bear different aspects of spectrally resolved measurements or model calculations. To date, airborne observations are the only tool for examining spectral phenomena in clouds and aerosols because imagers currently in orbit lack the required spatial resolution or coverage.

One of the distinct advantages of airborne irradiance measurements is that they do not entail an artificial distinction between remote sensing and radiative energy budget, i.e., radiative forcing, albedo, and absorption of cloud or aerosol layers. For satellite observations, cloud or aerosol properties are retrieved from reflected radiances. Irradiance cannot be measured from space and thus needs to be inferred through models. This introduces systematic errors caused by spatial heterogeneity. Upwelling and downwelling irradiance acquired from aircraft observations can be used to retrieve cloud or aerosol properties, while providing layer albedo, forcing, and absorption at the same time. Since an aircraft can fly below layers of interest, the derived bottom-of-layer radiative forcing is more constrained than it would be from a satellite. The layer absorption, impossible to observe from a satellite, can be derived from irradiance measurements without invoking a model.

Spectrally contiguous measurements allow an integration of the spectra to obtain unbiased broadband absorption and radiative forcing, while spectral resolution is invaluable in distinguishing various different contributors to forcing and absorption. Moreover, airborne instrumentation can be regularly calibrated on the ground, and effects of instrument degradation can be tracked. Technology can be outdated at the end of a satellite's lifetime; in an aircraft instrument, it can be kept state-of-the-art. For example, the signal-to-noise ratio of the photodiode arrays used in AVIRIS improved by a factor of ten from 1987 to 1999 (Green and Pavri, 2000).

After introducing instrumentation, relevant radiative quantities and models in section 6.2, we highlight some recent spectral measurements and model results: In section 6.3, we discuss what can be accomplished in remote sensing of homogeneous clouds when transitioning from traditional dual-channel techniques to spectral approaches. Section 6.3.1 introduces three indices based on reflected spectral radiance to retrieve thermodynamic phase and phase ratios in mixed-phase clouds. In section 6.3.2, we capitalize on the intrinsic spectral contrasts of near-infrared liquid water absorption to improve the accuracy of the effective radius retrieval from transmitted radiance, and discuss how cloud retrievals may become less sensitive to absolute radiometric calibration in the future. In section 6.3.3, we present a Bayesian approach for a rigorous error analysis of spectral retrieval techniques that can establish the gain in information content by increasing spectral resolution, sampling and coverage. In section 6.3.4, we show how the accuracy of cloud retrievals from satellite can be assessed with airborne measurements in the presence of aerosol layers overlying homogeneous clouds. In section 6.4, we discuss how to constrain the aerosol direct effect with airborne measurements. We introduce a new approach where aerosol single scattering albedo, asymmetry parameter, and surface albedo are obtained from flight legs above and below an aerosol layer, alongside aerosol top-of-layer and bottom-of-layer forcing and layer absorption.

In Sections 6.5 and 6.6, we address the relationship between cloud heterogeneities and spectral irradiance, forcing, and absorption. We point out that 3D effects alter the spectral signature of heterogeneous clouds (section 6.5) from that of their homogeneous counterparts, and that of cloud–aerosol scenes (section 6.6). In section 6.5, we describe a *spectral consistency* approach that helps identify inconsistencies between cloud properties and spectral cloud albedo, forcing and absorption caused by spatial heterogeneities. We also discuss the spectral shape of measured *apparent absorption*, obtained from the vertical divergence of irradiance in a cloud volume, and we reproduce the spectrum using a 3D radiative transfer model. The radiative forcing of convective boundary layer clouds embedded in an aerosol layer is discussed in section 6.6. We show that the spectral shape of the measured irradiance is reproduced only when the aerosol properties are fully represented in a 3D radiative transfer model.

Finally, we summarize the results of recent measurements and point out the potential of spectrally-resolved measurements for future cloud–aerosol observations.

6.2 Instrumentation, radiative quantities, and models

6.2.1 Spectrometer versus multi-channel filter-radiometer

Radiometers can be categorized by their spatial resolution and viewing geometry, and by their spectral resolution, coverage, and dispersion technique.

Spatial resolution distinguishes imaging from non-imaging instruments. For non-imaging instruments, no spatial information is retained in the measured signal, which originates from either a narrow cone (for example, in sun-photometers), or from any extended directional range, for example, in irradiance measurements, where the signal is integrated over a hemisphere. In imagers, spatial coverage is typically achieved by scanning with a detector array, for example, along-track in *pushbroom* or across-track in *whiskbroom* mode. Most imagers are spatially contiguous because they oversample, and their pixels overlap.

In the spectral domain, a seemingly minor, but important distinction (see section 6.2.2) is the way in which the solar spectrum is dispersed: multi-channel spectroradiometers typically employ filters or dichroic beam-splitters and independent detectors with each filter. They provide measurements in non-contiguous discrete wavelength bands. This is usually referred to as 'multi-spectral'. An example of an imaging spectroradiometer is MODIS (King et al., 1992). The Multi-Filter Rotating Shadowband Radiometer (MFRSR: Harrison et al., 1994) or the sun-photometers used for the Aerosol Robotic Network (AERONET: Holben et al., 1998) are surface-based non-imaging filter-radiometers. Spectrometers, on the other hand, cover parts or the entire solar spectrum continuously. This is sometimes called 'hyperspectral'. In contrast to filter-radiometers, spectrometers use gratings or prisms in conjunction with detector arrays or sequential wavelength scanning devices. The Airborne Visible/InfraRed Imaging Spectrometer (AVIRIS: Green et al., 1998) is composed of four individual spectrometers from the visible to shortwave-infrared range. Each uses a two-dimensional diode array with the spectral dimension along one axis and the spatial dimension along the orthogonal axis. The substrate of the three detector arrays is optimized for maximum response in the respective spectral wavelength range. AVIRIS operates in *whiskbroom* scanning mode; the space-borne Hyperion (two spectrometers), and the Moon-viewing M^3 (Moon Mineralogy Mapper, single spectrometer) both use *pushbroom* scanning. The challenge in building spectral imagers is to optimize the spatial and spectral resolutions and ranges. Increasing signal-to-noise ratios in detector arrays have allowed higher resolution in the spectral and spatial domain while keeping telescopes small enough to be feasible in space.

Multispectral and hyperspectral radiometers exhibit different modes of degradation. Filter-radiometers have multiple optical paths for each band, while spectrometers have a single optical path. In filter-radiometers, degradation of optical components can occur independently for each optical path. In spectrometers, instrument response function variability is *correlated* across the spectral range. Spectrally *uncorrelated* changes can occur only at the detector level. This type of degradation is common to spectrometers and filter-radiometers alike.

Currently, no spectral imager with an adequate combination of resolution and accuracy is in orbit. It was planned for the NASA CLARREO project (Climate Absolute Radiance and Refractivity Observatory), which was recently canceled due to budgetary problems.

6.2.2 Solar spectral flux radiometer

The Solar Spectral Flux Radiometer (Pilewskie et al., 2003) measures upward and downward solar spectral irradiance or radiance. Prior to its first deployment in 1998, airborne irradiance measurements relied mostly on broadband radiometers. The lack of spectral resolution became obvious when Cess et al. (1995) discovered a discrepancy between modeled and measured cloud absorption, thereafter debated in the literature for more than a decade. This discrepancy was one of the motivations for developing SSFR at NASA Ames. A sister instrument, the albedometer by Wendisch et al. (2001), followed soon afterwards. It took until 2007 to achieve better understanding of the so-called *apparent* cloud absorption which is derived from vertical irradiance divergence (section 6.5.2).

SSFR has been deployed on numerous airborne platforms and ground sites (for example at the Southern Great Plains ARM site in Oklahoma, under the name of Shortwave Spectrometer, SWS) and on various research vessels. The albedometer has also been mounted on platforms that were carried by tethered balloons, helicopters and aircraft. SSFR (depicted in Fig. 6.1) covers a wavelength range from 350 nm to 2150 nm, comprising nearly 95% of the total incident solar irradiance. Each SSFR consists of two subsystems, combined in one rack-mounted encasement, and connected by fiber-optic cables to light collectors mounted on top (zenith viewing) and at the bottom (nadir viewing) of the aircraft. The light collectors (on the right in Fig. 6.1) are designed such that they provide a near-ideal cosine weighting of the incoming radiance from the upper and lower hemisphere, defined as irradiance. This is realized by a circular entrance aperture in a miniature integrating sphere, covered by a quartz dome for weather protection. Recent improvements in the cosine-response of the SSFR light collectors are reported in Kindel (2010). For operation on the ship or on the ground, the nadir-viewing hemispheric light collector is replaced with a zenith-viewing narrow-field-of-view collimator. In this setup, SSFR allows simultaneous downward irradiance and radiance measurements. The albedometer (Wendisch et al., 2001) has also been configured to measure upward radiance with a nadir-viewing collimator.

The full spectral range is covered by using pairs of monolithic miniature Zeiss spectrometer modules (MMS-1 for the near-UV, visible, and very-near-infrared from 350 nm to 1000 nm; NIR-PGS 2.2 for the near-infrared from 900 nm to 2200 nm). Partitioning the spectral range was necessary because at the time of its development, no single detector was sensitive enough across the entire solar wavelength range. This design is shared with many spectral imagers. Newer systems (for example, M^3) have employed single focal plane detector arrays. The 350 nm to 1000 nm module is a flat-field grating with a 256-element linear silicon diode array detector. It is temperature-stabilized at 27.0°C ± 0.3°C to minimize dark current fluctuations. Its spectral resolution as indicated by the full-width-half-maximum

Fig. 6.1. Solar Spectral Flux Radiometer (SSFR) with zenith and nadir light collectors.

(FWHM) is 8 nm, with 3 nm sampling. The 900 nm to 2200 nm module utilizes a 256-element InGaAs linear diode array that is thermoelectrically cooled to $-10.0°C \pm 0.1°C$. Recently, the instrument setup was modified to also stabilize the temperature of the electronics (pre-amplifier) for better dark current stability. The FWHM is 12 nm, with 5 nm sampling resolution. In normal operation mode, one spectrum is recorded per second. Different versions of the instrument with higher resolution and different spectral range exist. The instrument is mechanically robust since no moving parts are used, and are therefore optimal for aircraft operations.

Since the SSFR is not only utilized for spectroscopic, but also for radiometric applications, it is essential to ensure high accuracy and stability of its absolute calibration. Therefore, the spectrometers are calibrated before and after each field deployment in the laboratory, using a tungsten-halogen 1000-W lamp that is traceable to a US national laboratory standard. Over the course of each experiment, the stability of the calibration is tracked with a portable field calibration unit. Generally, a calibration stability of 1% to 2% is achieved over the course of a multiple-week field mission. The nominal radiometric accuracy of 3% is dominated by the uncertainty of the calibration light source. The precision of the instrument is 0.1%. This value is driven by dark current fluctuations. Those fluctuations can exceed 0.1%, or even 3% of the signal near the edges of a spectrometer's wavelength range where the signal-to-noise ratio decreases due to decreasing sensitivity of the detector. Figure 6.2 shows a typical SSFR measurement over a vegetated surface in California during a 2010 field experiment, with downwelling irradiance in blue, and upwelling irradiance in red. The wavelength ranges of the InGaAs detector array and the silicon detector array overlap between 900 and 1000 nm and are joined at 940 nm. This wavelength was chosen to maximize the signal-to-noise ratio on either side. The wavelength calibration relies on known emission lines of Hg, Ar, and Xe lamps, and on lasers in the visible and near-infrared wavelength range. It is usually stable to 0.1 nm over several years.

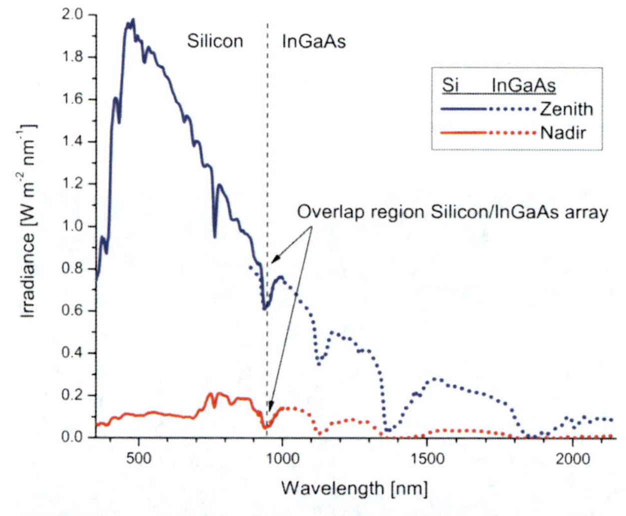

Fig. 6.2. Typical SSFR measurement (blue: downwelling, red: upwelling irradiance; solid lines: silicon spectrometer wavelength range; dotted lines: InGaAs spectrometer wavelength range).

The data are corrected for the angular response of the light collectors which can deviate from an ideal cosine-weighting of the incident radiation (Kindel, 2010). Since irradiance is defined with respect to a horizontal plane, an additional correction is required for aircraft attitude if the light collectors are fix-mounted to the fuselage (Bannehr and Schwiesow, 1993). This can be complicated or impossible if a light collector receives radiation from the opposite hemisphere (see Fig. 6.3(a)). For this reason, Wendisch et al. (2001) introduced a stabilization platform (Fig. 6.3(b)) for the albedometer where attitude changes of the aircraft are counteracted by a two-dimensional tilt stage. In this way, the light collectors are always aligned with the horizon.

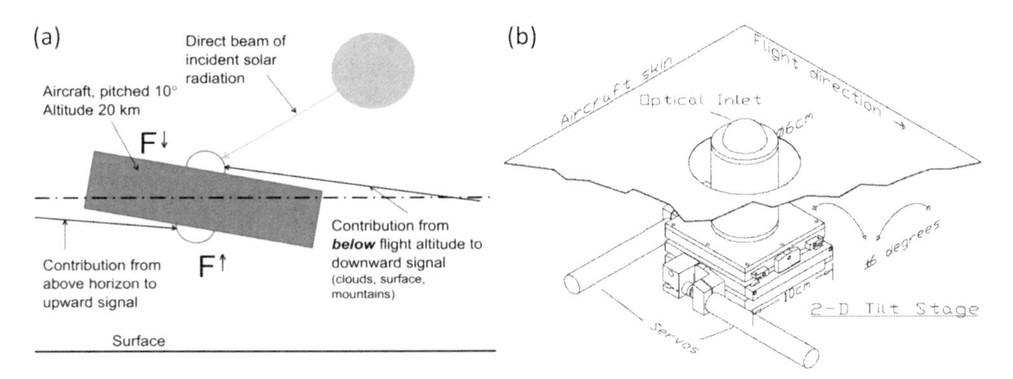

Fig. 6.3. (a) Viewing geometry for irradiance measurements on an airplane. (b) Layout of a tip-tilt stabilizing platform from Wendisch et al. (2001).

6.2.3 Radiative quantities

Optical properties

In the shortwave wavelength range, radiative transfer calculations require the vertical distribution of spectral extinction and absorption coefficients, as well as the scattering phase matrix. Cloud and aerosol layers can be fully described by three parameters:

(1) The optical thickness, τ, is defined as the column-integrated extinction coefficient, β_{ext}. It comprises the scattering and absorption optical thickness, τ_{sca} and τ_{abs}: $\tau_{\mathrm{tot}} = \tau_{\mathrm{sca}} + \tau_{\mathrm{abs}}$. The Ångström parameter describes the wavelength dependence of aerosol optical thickness. Cloud optical thickness has almost no wavelength-dependence.

(2) The single scattering albedo, ϖ, is defined as the ratio between scattering and total optical thickness, $\tau_{\mathrm{sca}}/(\tau_{\mathrm{sca}} + \tau_{\mathrm{abs}})$.

(3) For spherical water droplets, the scattering phase function can be obtained from Mie calculations, whereby drop size distributions are often parameterized by its third moment divided by the second: the effective radius, R_{eff}. Assigning a phase function to ice crystals and aerosol particles is more ambiguous because of their complex shape and internal composition.

Irradiance, radiance, and related properties

Upwelling and downwelling spectral irradiance (or flux density), $F_\lambda^\uparrow$ and $F_\lambda^\downarrow$, are defined as cosine-weighted radiances, I_λ, integrated over the lower and upper hemisphere, respectively. The net irradiance, F_λ, is the difference between downwelling and upwelling irradiance: $F_\lambda = F_\lambda^\downarrow - F_\lambda^\uparrow$. The layer absorption is derived from the difference in net irradiance on top and at the bottom of the layer, ΔF_λ, assuming that the net horizontal divergence vanishes. In real clouds this is rarely achieved except under the most ideal conditions (discussed in section 6.5). The spectral albedo is defined as the ratio between upwelling and downwelling irradiance, $\alpha_\lambda = F_\lambda^\uparrow / F_\lambda^\downarrow$, while layer reflectance pertains to the reflected radiance, $r = \pi I^\uparrow / F_\lambda^\downarrow$, where $F_\lambda^\downarrow$ is the incident irradiance on top of the layer, which depends on the cosine of the solar zenith angle, μ_0. Similarly, transmittance is defined as $t = \pi I^\downarrow / F_\lambda^\downarrow$, and $I^\downarrow$ is the transmitted radiance through the layer. This definition of r and t is slightly different from the satellite bi-directional reflectance function, which involves solar zenith angle and multiple different viewing angles. We will consider nadir-viewing reflectance in section 6.3.1 and zenith-viewing transmittance in section 6.3.2.

6.2.4 Radiative transfer models

For 1D radiative transfer calculations, we used the libRadtran radiative transfer package (http://www.libradtran.org), developed by Mayer and Kylling (2005), and the SSFR-specific radiative transfer code by Coddington et al. (2008), where gas absorption was parameterized with the correlated-k method (Mlawer et al., 1997), based on the HITRAN 2004 data base (Rothman et al., 2005). For the 3D calculations, we used MYSTIC (Monte Carlo code for the physically correct tracing of photons in cloudy atmospheres: Mayer, 2009), which is embedded in libRadtran.

6.3 The value of spectral resolution for cloud retrievals

To date, the most widely used technique for retrieving cloud optical thickness (τ) and effective radius (R_{eff}) from passive imagery relies on only two channels in the solar wavelength range (Arking and Childs, 1985; Twomey and Cocks, 1989; Nakajima and King, 1990; Kokhanovsky et al., 2011). Figure 6.4(a) shows cloud reflectance at a near-infrared channel (r_{nir}) plotted as a function of the reflectance in a visible channel (r_{vis}) for various $\{\tau, R_{\text{eff}}\}$ combinations. With increasing optical thickness, the reflectance in both channels increases monotonically towards its asymptotic limit. In the visible channel, where water does not absorb, the reflectance approaches unity in the limit of large optical thickness. In the near-infrared channel, the maximum reflectance is below 100%, due to liquid water absorption. In the limit of weak absorption and geometric optics, single scattering co-albedo (probability of photon absorption: $1 - \varpi$) depends linearly on drop size, which thus determines the reduction in reflectance in the near-infrared channel. Multiple scattering in optically thick clouds amplifies absorption, which is then proportional to $\sqrt{1 - \varpi}$.

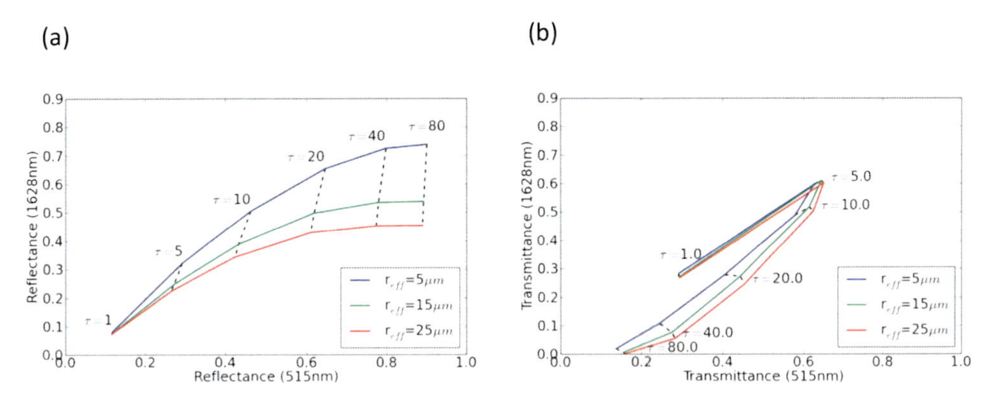

Fig. 6.4. Nakajima–King diagram (lookup table) for (a) reflectance-based and (b) transmittance-based cloud retrievals from McBride et al. (2011).

The retrieval algorithm is based on the simple one-to-one relationship $\{\tau, R_{\text{eff}}\} \leftrightarrow \{r_{\text{vis}}, r_{\text{nir}}\}$. It identifies the closest match between the dual-channel reflectance observations and a pre-calculated reflectance lookup table $\{\tau(r_{\text{vis}}, r_{\text{nir}}), R_{\text{eff}}(r_{\text{vis}}, r_{\text{nir}})\}$. Figure 6.4(b) shows the same lookup table for the transmittance below a cloud. For zero cloud optical thickness molecular scattering is the only contributor to the downward radiance. With an increasing number of cloud droplets, the downward radiance increases up to $\tau \approx 5$, beyond which the signal decreases toward the limit of zero for infinite thickness. The poor separation of individual R_{eff} lines in the transmittance lookup table translates into large uncertainties in the retrieved $\{\tau, R_{\text{eff}}\}$ pairs, especially when $\tau > 40$ where the dual-channel approach is not feasible (section 6.3.2). Further problems arise in the presence of ice or mixed-phase clouds (section 6.3.1), absorbing aerosols above clouds (section 6.3.4), and cloud heterogeneity (section 6.5).

6.3.1 Reflectance from mixed-phase clouds

For water clouds, Mie calculations provide the single scattering properties for forward radiative transfer calculations. For the more complicated shapes encountered in ice clouds, ray-tracing codes provide distinct spectra of single scattering properties for different crystal habits or habit mixtures (Baum et al., 2011). Therefore, the retrieved values for τ and R_{eff} depend on the assumed habit mix and can vary by as much as 50% (Kalesse et al., 2011), introducing uncertainties in the derived absorption and radiative forcing of these clouds.

Mixed-phase clouds are composed of multiple layers with varying drop and crystal sizes. Treating these clouds as liquid in satellite retrievals can result in an underestimation of their absorption in the near-infrared wavelength range. Likewise, the vertical distribution of ice throughout the cloud can alter its reflectance and absorptance considerably. Yoshida and Asano (2005) find that even for fixed liquid water path (LWP) and ice water path (IWP), a varying vertical distribution of liquid/ice microphysics can result in up to 10% changes in 700–2500 nm absorption.

Ehrlich et al. (2008, 2009) used albedometer measurements of spectral reflectance and albedo of Arctic mixed-phase clouds during ASTAR (Arctic Study of Tropospheric Aerosol, Clouds and Radiation, 2007) to explore the impact of vertical structure on near-infrared ice indices. As shown by Pilewskie and Twomey (1987), differences in the bulk absorption between ice and liquid water around 1600 nm and 2000 nm can be exploited for cloud phase discrimination. For example, ice absorption peaks near 1500 nm, while liquid water absorption reaches a maximum near 1440 nm (Fig. 6.5(b)). Due to a water vapor absorption band that extends to 1550 nm, it is impractical to use these maxima for phase discrimination. Instead, reflectance contrasts at longer wavelengths have been used (Knap et al., 2002; Kokhanovsky et al., 2006). Comparing near-infrared and thermal infrared cloud phase detection algorithms suggests that discrepancies between the two methods depend on the choice of near-infrared wavelengths and cloud microphysics, as well as the underlying surface (Chylek et al., 2006). Since the results of the near-infrared retrieval depend on the choice of wavelengths, the need arose to examine ice and water absorption spectrally. Ehrlich et al. (2008) introduced three ice indices; two for the near-infrared and one for visible wavelengths. The first one, I_S, is based on the spectral slope of cloud reflectance between 1550 nm and 1700 nm. Within this wavelength range, the reflectance of pure ice or liquid clouds is fairly linear (Fig. 6.5(a)), and the slope can be assigned by fitting a straight line to the measured reflectance. The second index, I_P, is based on identifying the principal components associated with cloud absorption between 1500 nm and 1800 nm. With I_P, the pre-calculated spectral shape of ice and water cloud reflectance (using 50 combinations of optical thickness and crystal/drop effective radii) is explicitly taken into account, rather than assuming linearity. A third index, I_A, is a monochromatic indicator of cloud phase, based on the differing anisotropy of ice and water cloud

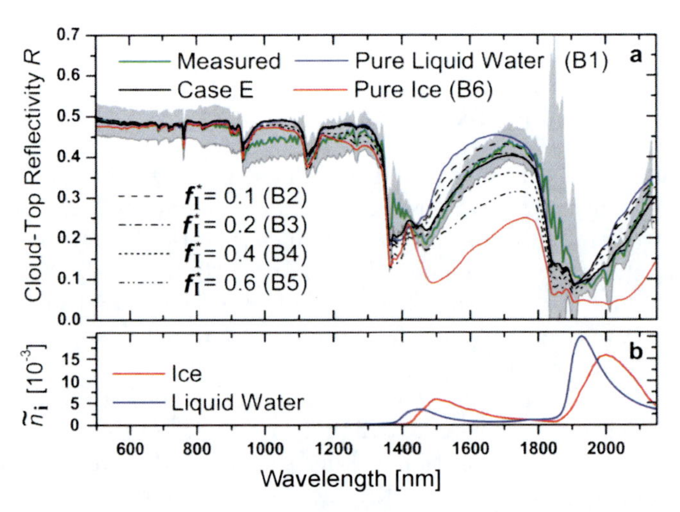

Fig. 6.5. Adapted from Ehrlich et al. (2009). (a) Measured reflectance spectrum above mixed-phase clouds (green), with measurement uncertainty in gray. The lines labeled B1–B6 show model results with different ice optical ratios from pure water to pure ice. The line labeled Case E shows model results for a mixed-phase cloud, capped by an ice cloud. (b) Spectrum of the imaginary part of the refractive index of ice and water.

phase functions. It is defined as ratio of the reflectance to albedo at 645 nm. While the reflectance is sensitive to the scattering phase function at a certain angle if single scattering dominates, the albedo incorporates scattering contributions over the entire hemisphere. The distinct angular patterns in ice crystal phase functions were previously exploited by McFarlane et al. (2005) to retrieve crystal habit from MISR (Multiangle Imaging SpectroRadiometer). Since photons from deeper cloud layers are more likely to have experienced a higher order of scattering, the sensitivity of the angular distribution of reflectance to crystal habit, and the sensitivity of I_{A} to phase, are limited to conditions near cloud top.

For a specific case, Ehrlich et al. (2008) compared I_{S}, I_{P}, and I_{A} with *in situ* phase measurements of a polar nephelometer (Gayet et al., 2007). The anisotropy index, I_{A}, classified mixed-phase clouds as liquid when they were capped by a liquid cloud layer. The other indices, I_{S} and I_{P}, indicated higher ice concentrations than suggested by the *in situ* measurements. To examine the effect of coexisting crystals and water drops, Ehrlich et al. (2009) introduced the ice optical fraction, f_{I}^{*}, the ratio of ice to total optical thickness, and compared modeled spectra with measurements (Fig. 6.5(a)). With this approach, however, no single ice optical fraction could reproduce the measurements consistently across the wavelength range from 1550 nm to 1700 nm. Applying the spectral vertical weighting functions by Platnick (2000) for this specific case showed that the photon penetration depth corresponding to 50% of the measured reflectance gradually increased from 160 m at 1500 nm to 195 m at 1700 nm. Since the measured signal resembled the predicted reflectance for a liquid water cloud at 1700 nm, but showed more similarity with an ice cloud spectrum at shorter wavelengths, this suggested that adding an ice layer within the cloud top layer could reconcile the model with the measurement. Indeed, only the multi-layer cloud structure (solid black line in Fig. 6.5(a), labeled 'Case E') reproduced the observations (green line) across the wavelength range from 1550 nm to 1700 nm.

It should be noted that while the slope index I_{S} could be replaced by the reflectance contrast between two separate channels, the principal component index, I_{P}, can only be derived from spectrally resolved observations. Since principal component analysis bears similarity to spectral matching in differential optical absorption spectroscopy (Platt and Stutz, 2008), I_{P} could be regarded as one of the first results of a newly emerging 'cloud spectroscopy', with the understanding that by comparison with gas absorption lines, spectral changes of cloud parameters occur more steadily since condensed-phase absorption spectra are continua. Also, due to the spectral dependence of photon penetration depth, an extended spectral range provides some insight into the vertical cloud structure. Therefore, profiling of the first few hundred meters below cloud top could be regarded as a further application of spectrally resolved observations in the future.

6.3.2 Cloud spectroscopy with transmitted radiance

As shown in Fig. 6.4 and discussed in Marshak et al. (2004), it is not straightforward to apply a reflectance-based cloud retrieval to transmittance observations. For a given transmittance pair at a visible and a near-infrared wavelength, any effective radii in Fig. 6.4(b) could be retrieved if the radiometric uncertainty exceeds a certain threshold, with the exception of the optical thickness range from 15 to 40 where the transmittance values for different effective radii are fairly well separated. The fundamental difference between cloud reflectance and transmittance originates from the fact that for large optical thickness values, the transmittance is completely attenuated while the reflectance approaches an asymptotic limit determined by single scattering albedo or, equivalently, effective radius. Also, an increasing effective radius usually leads to more pronounced forward scattering and smaller single scattering albedo, both of which lower cloud reflectance, but have opposing effects in transmitted radiance. Therefore, the size-related effects partially cancel each other out in transmittance.

The problem is that nearly indistinguishable effective radii lines in Fig. 6.4(b) translate into large uncertainties, too large to provide statistically significant cloud retrievals from the ground. The solution comes from 'cloud spectroscopy' where slowly changing cloud optical properties across a limited wavelength range are exploited.

McBride et al. (2011) examined the slowly changing asymmetry parameter (Fig. 6.6(a)) and single scattering albedo in water clouds around 1600 nm (Fig. 6.6(b)). Plotted are the spectral shapes of asymmetry parameter and co-albedo for 5 and 20 μm droplets, and for 20 μm ice crystals for comparison. The latter are taken from Baum et al. (2011); smooth and severely roughened droxtals are shown. As mentioned above, the water vapor band around 1400 nm reduces the utility of this wavelength range for cloud retrievals. McBride et al. (2011) therefore used the wavelength range from 1565 nm to 1634 nm to fit a line to the observed transmitted radiance, normalized by its value at 1565 nm. In this wavelength range, the co-albedo has a distinct spectral slope, which results in a slope in cloud transmittance. Figure 6.7 shows modeled cloud transmittance for a set of optical thickness and effective radii, along with a typical boundary layer cloud spectrum, measured by SWS at one of the ground sites of the Department of Energy's Atmospheric Radiation Measurement (ARM) program. With the exception of $\tau = 75$, the effective radii from 5 (solid line) to 25 μm (dashed line) are fairly

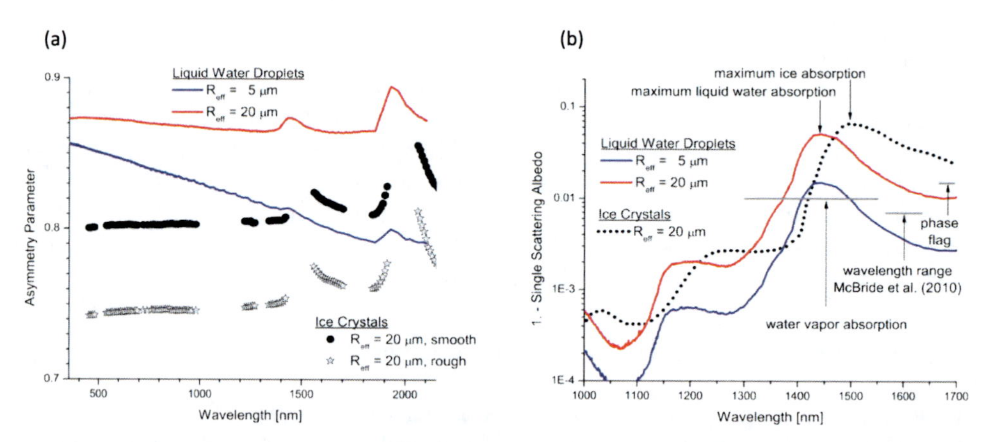

Fig. 6.6. Spectral shape of (a) asymmetry parameter and (b) single scattering co-albedo (1 – single scattering albedo) for water droplets and ice crystals (from Baum et al., 2011).

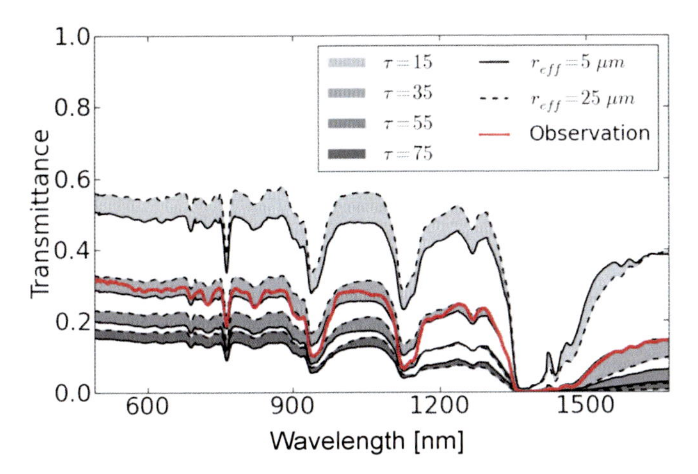

Fig. 6.7. Modeled cloud transmittance spectra, for different optical depths (increasing shades of gray), for effective radii ranging from 5 (solid line) to 25 μm (dashed line). The red spectrum shows a typical measurement.

well separated by their transmittance. Note that in the visible wavelength range, larger transmittance values are associated with larger effective radii due to more pronounced forward scattering and negligible absorption at these wavelengths. At near-infrared wavelengths, this behavior reverses, and larger effective radii are associated with lower transmittance values, because absorption outweighs forward scattering. The crossover wavelength occurs between 1100 nm and 1400 nm.

In the new retrieval, the near-infrared transmittance value in the lookup table is replaced with a spectral shape parameterization, the slope of transmittance in a spectral sub-range. Figure 6.8(a) shows normalized transmittance spectra for two optical thickness values. The error bars within the wavelength range from 1565 nm to 1634 nm reflect 3% radiometric uncertainty. When fitting a line to the spectra, the radiometric error propagates into the uncertainty of the slope. If the correlation between the errors of neighboring wavelengths is large as in spectrometers, the slope

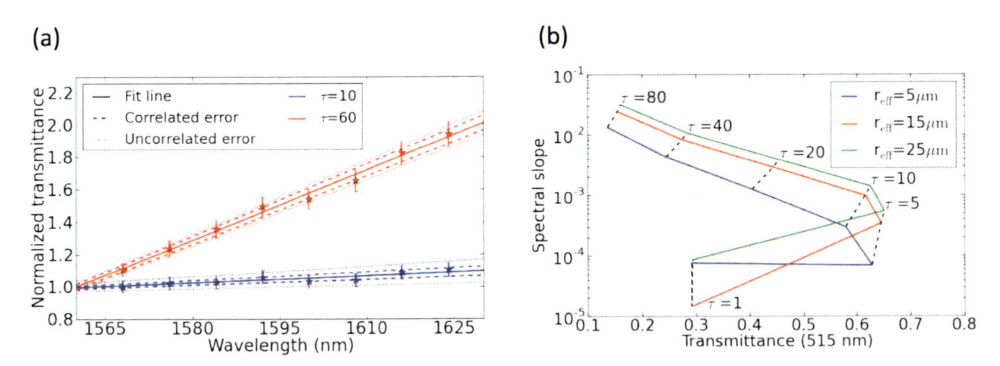

Fig. 6.8. (a) Spectral cloud transmittance, normalized to its value at 1565 nm, for two different effective radii and fixed optical depth. (b) Slope–Transmittance lookup table.

error is smaller than it would be for uncorrelated errors as in filter-radiometers (section 6.1). This is indicated by two sets of lines around the fitted line. The inner (dashed lines) and outer envelopes (dotted lines) reflect the ranges of uncertainty that result from correlated and uncorrelated errors, respectively. Nonlinearities of the normalized radiance increase the error of the slope; the inner envelope would be much closer to the fitted line if the radiance were a more linear function of wavelength. It is important to note that the uncertainty of the slope also depends on the spectral resolution of the spectrometer; an increase from 8 nm to 1 nm spectral sampling decreases the slope error from 60% to 25% (uncorrelated errors) and from 25% to 10% (correlated errors). These errors were obtained for the blue spectrum in Fig. 6.8(a); larger slopes (red spectrum) result in smaller relative errors. In the future, a principal component analysis index as discussed by Ehrlich et al. (2009) could replace the slope and decrease the uncertainty by taking into account the nonlinear spectral shape in the analyzed wavelength range.

The slope around 1565 nm, S_{1565}, replaces the transmittance at 1628 nm, T_{1628}, in Fig. 6.4(b), and a slope-transmittance grid is created with $\{S_{1565}, T_{515}\}$ (Fig. 6.8(b)), where the effective radii are better separated than in the transmittance–transmittance grid. Since the slopes are derived from *normalized* transmittance, they are not affected by the radiometric calibration if instrument degradation changes the radiometric response of all wavelengths from 1565 nm to 1634 nm equally. The algorithm has been applied to liquid-water clouds only. Ice clouds were filtered out using the slope around 1680 nm which is close to zero for ice clouds and negative for water clouds (see Fig. 6.6(b)).

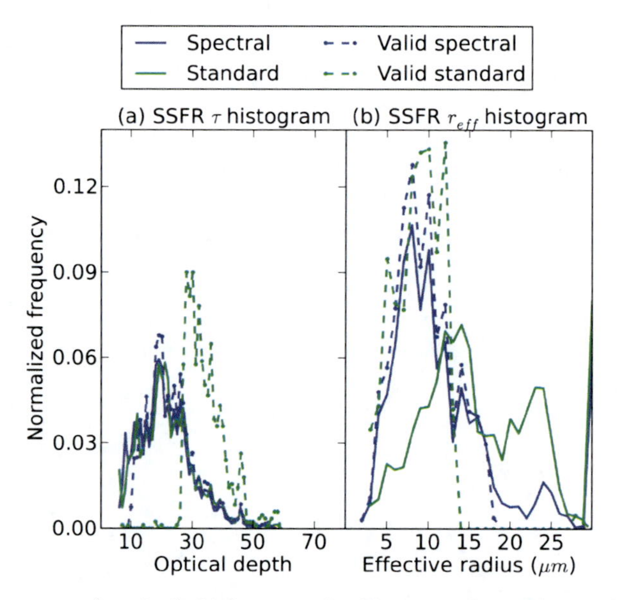

Fig. 6.9. Histograms of optical thickness and effective radius, obtained from the dual-channel (standard) technique (shown in green) and from the new spectral technique (shown in blue). The dashed histograms show the quality-filtered retrievals where only data with an effective radius uncertainty below 2 μm is accepted as valid (dashed lines).

The potential of this technique and new applications are under investigation. For example, the sensitivity to cloud vertical structure and horizontal heterogeneities has not been explored. One promising application is the implementation of additional spectral sub-ranges, for example, using $\{S_{1200}, S_{1565}, S_{2100}\}$.

The retrieval was tested with data from two field experiments: for ICEALOT 2008 (International Chemistry Experiment in the Arctic Lower Troposphere) and CalNex 2010 (Research at the Nexus of Air Quality and Climate Change, California), SSFR was mounted onboard research vessels during cruises across the northern Atlantic and along the coast of California. In addition, data was collected by SWS at the ARM ground site in Oklahoma. Figure 6.9 shows histograms of optical thickness and effective radius from March 20, 2008. The effective radius uncertainty of the dual-channel (standard) retrieval is substantially larger than for the new spectral technique. The dashed histograms only include retrievals with an effective radius absolute uncertainty below 2 μm. Retrievals with higher uncertainty are flagged as invalid. Using this quality filter, the standard retrieval does not provide valid retrievals when the optical thickness decreases below 25, whereas the new technique is sensitive to effective radius even for thin clouds. Figure 6.10

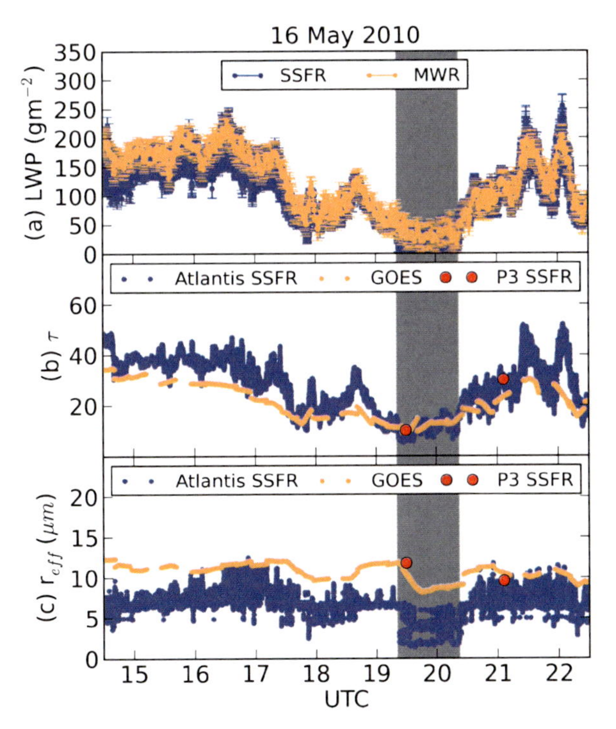

Fig. 6.10. (a) Liquid water path retrieved from SSFR and from microwave radiometer (MWR) onboard the Atlantis research vessel, from McBride et al. (2010). The gray-shaded area marks a region with enhanced cloud heterogeneity. The error bars indicate the uncertainty of both retrievals. The variability is not due to measurement noise, but reflects natural cloud variability. (b) Optical thickness and (c) effective radius from SSFR onboard the Atlantis research vessel, from GOES (courtesy of NASA LaRC), and from reflectance-based retrievals (SSFR onboard NOAA P-3), from McBride et al. (2010).

shows SSFR results from the CalNex cruise on May 16, 2010, in comparison with (a) liquid water path retrievals from a microwave radiometer and (b, c) with cloud retrievals from the GOES satellite (McBride et al., 2010). The liquid water path for Fig. 6.10(a) was derived from effective radius and optical thickness using a formula derived by Wood and Hartmann (2006) for adiabatic liquid water clouds:

$$LWP = \frac{5}{9}\rho\tau r_{\text{eff}},$$

where ρ is the density of liquid water. This results in a slightly smaller liquid water path than the more common formula $2/3\rho\tau r_{\text{eff}}$, which is based on a vertically homogeneous cloud. The shaded regions indicate times of enhanced cloud variability. The red dots show albedo-based cloud retrievals from an airplane that overflew the ship.

Marshak et al. (2009) and Chiu et al. (2009) analyzed the spectral signature of SWS measurements in the transition zone between cloudy and cloud-free air, which carry information about the aerosol loading, and thus, on the aerosol indirect effect. They are developing a method to retrieve cloud properties under heterogeneous conditions (Chiu et al., 2010).

In summary, the studies of Ehrlich et al. (2009) and McBride et al. (2010, 2011) have shown that within the near-infrared wavelength range, the slowly changing cloud optical properties can be exploited using spectroscopic methods. One virtue of spectroscopy is that is can be applied to relative measurements, lessening the dependence of derived parameters on instrument stability and radiometric accuracy. However, in order to fully utilize the potential of these methods for cloud retrievals, a more rigorous assessment of errors is required. In the next section, we introduce a formal approach for the analysis of information content and error propagation.

6.3.3 Spectral information content and error analysis

Cloud retrieval error analysis usually relies on linear or Gaussian propagation of measurement and model uncertainties into retrieved quantities. These simple propagation formulae conceal some implicit assumptions, for example the correlation between individual error sources, or the shape of the probability density function (PDF) of input and output parameters.

In this section, we present an approach by Vukicevic et al. (2010), similar to Bayesian estimation theory, that takes into account the nonlinear nature of errors and can establish the Shannon information content (Shannon and Weaver, 1949) of spectral measurements with respect to cloud parameters. The approach is unique in that it can distinguish between systematic errors (such as radiometric uncertainty and model input bias) and random errors (such as dark current fluctuations). It treats the inverse problem of remote sensing in terms of probability density functions of measured and retrieved quantities. Due to the nonlinear dependence of cloud reflectance on optical thickness and effective radius, a Gaussian PDF of input parameters will generally result in a non-Gaussian PDF of output parameters. Therefore, standard error propagation formulae are inadequate.

Using this approach, Vukicevic et al. (2010) characterized an SSFR cloud retrieval that are based on pre-computed radiative transfer tables. The retrieval is a modification of a method proposed by Twomey and Cocks (1989) where the measured values for reflectance or albedo, x_k, at five different wavelengths $\lambda_\mathrm{k} = \{515, 745, 1015, 1240, 1625 \text{ nm}\}$ are compared to a set of model results, $y_\mathrm{k}(\tau, R_\mathrm{eff})$, and

$$\zeta^2 = \frac{1}{60} \sum_{k=1}^{5} \left((5-k)^2 (x_k - y_k)^2 + (k-1)^2 \left(\frac{x_k}{x_1} - \frac{y_k}{y_1} \right)^2 \right)$$

is minimized with respect to τ and R_eff. The weights are based on physical arguments (Coddington et al., 2010). The framework of Vukicevic et al. (2010) allows to improve retrieval accuracy through determining the number, placement, and weightings of retrieval wavelengths. Successively adding input at other wavelengths can be understood as a narrowing of the parameters' PDF prior to and after the addition of the wavelength to the set of observations. The retrieval error is the width of the posterior PDF of the retrieved parameters. In the case of the discrete wavelengths, Fig. 6.11 shows the narrowing of the PDF for the effective radius and optical thickness, as well as the gain in information after each step, for a cloud with $\tau = 60$ and $R_\mathrm{eff} = 20$ μm. After adding albedo information at 745 nm and 1015 nm, the PDF of R_eff remains nearly flat. Only after adding albedo at 1250 nm and 1625 nm does the information content with respect to effective radius (Fig. 6.11(c)) increase significantly. Perhaps unexpectedly, the accuracy of the optical thickness increases with the addition of each wavelength. Given the fact that the entire reflectance or albedo spectrum can be predicted from just two wavelengths, one might expect that adding more wavelengths is redundant and would not narrow the PDFs of the optical thickness. However, since each of the spectral albedo measurements is associated with an error, accumulating several spectral reflectance

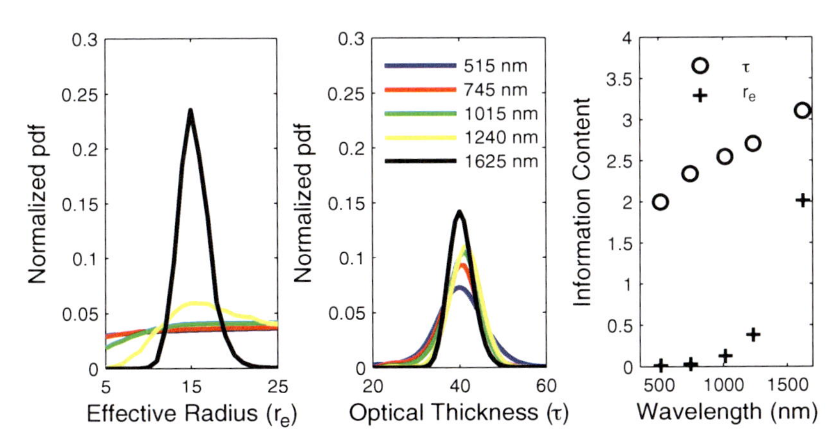

Fig. 6.11. Adapted from Vukicevic et al. (2010): Posterior marginal PDFs of (a) effective radius and (b) optical thickness after adding information from five wavelengths from an SSFR cloud observation. (c) Wavelength-by-wavelength increase in the Shannon information content with respect to effective radius and optical thickness.

values with similar physical dependence on the cloud parameters gradually reduces the random part of the error, and is perceived as a narrowing PDF. The systematic error, on the other hand, can only be reduced by applying spectroscopic methods as presented in section 6.2. Isolating random and systematic errors in a rigorous way is crucial to evaluate the accuracy of new retrieval techniques, and thus to take full advantage of spectral resolution in the future.

6.3.4 Aerosols above clouds

The presence of a homogeneous aerosol layer above a homogeneous cloud layer is the simplest case of clouds and aerosols in close proximity. Haywood et al. (2004) predicted that neglecting overlying absorbing aerosols in MODIS cloud retrievals would lead to an underestimation in the retrieved effective radius. Coddington et al. (2010) examined the impact of overlying aerosol on cloud retrievals using aircraft measurements from INTEX–A (Intercontinental Chemical Transport Experiment). During this experiment, the SSFR was mounted on one aircraft with the Ames Airborne Sunphotometer (AATS-14: Russell et al., 1999) which provided aerosol optical thickness above the flight level of the plane. They used the Twomey and Cocks (1989) technique described above to retrieve cloud optical thickness and effective radius from SSFR spectral albedo, with a flight geometry as shown in Fig. 6.12. On the low-level leg (above the cloud but below the aerosol layer) they obtained the actual cloud properties, as well as aerosol optical thickness from AATS-14. On the high-level leg (above the cloud and aerosol layer) they retrieved cloud properties in a similar way as they would be perceived from an instrument like MODIS.

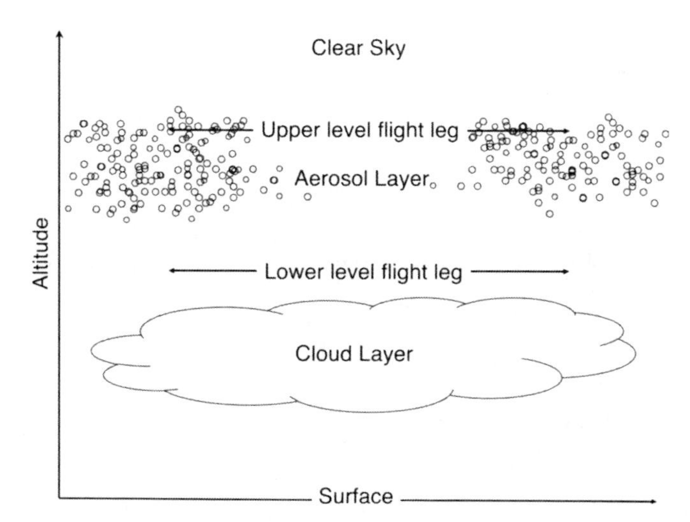

Fig. 6.12. Flight leg geometry from INTEX-NA.

Results showed that a low bias in effective radius was indeed possible when ignoring the absorbing aerosol. The optical thickness was also slightly underestimated. The decrease of spectral albedo due to a typical non-absorbing and absorbing aerosol layer (from INTEX-A) above a cloud is shown in Fig. 6.13. While a scattering aerosol layer does not change the cloud albedo, the absorbing aerosol reduces it by almost 10% at 350 nm. Coddington et al. (2010) generated three different lookup tables, specifically for this experiment. The first one was based on a cloud without overlying aerosols (similar to the MODIS cloud retrieval); the second and third one included an overlying non-absorbing ($\varpi = 1.0$) and partially absorbing ($\varpi = 0.8$) aerosol layer, with an optical thickness that incorporated typical measurements by AATS-14. Since the SSFR retrieval is based on five, rather than two wavelengths, the retrieval residual (the difference between the measured and best-fit modeled spectrum) carries some information about the adequacy of a particular lookup table for the case under study. However, due to cloud heterogeneities, which also affect the spectral shape of the cloud albedo, it is quite difficult to clearly discern the signal from the aerosols from the overwhelmingly large signal of the underlying cloud. We will discuss this further in the context of spectral consistency in section 6.5.

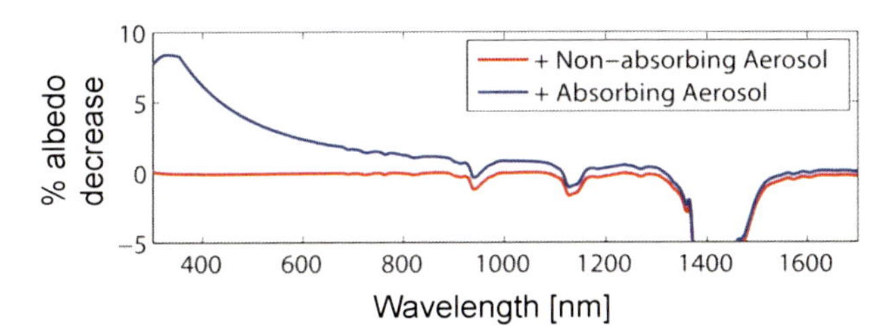

Fig. 6.13. Percentage decrease in spectral cloud albedo due to an overlying aerosol layer (from Coddington et al., 2010).

Vukicevic et al. (2010) represent the percentage change in spectral albedo due to an overlying aerosol layer as a systematic error in the model PDF for four different $\{\tau, R_{\mathrm{eff}}\}$ pairs. The decrease in albedo (Fig. 6.13) caused a shift in the 2D representation of the marginal posterior (i.e., solution) PDF with respect to the expected outcome if only a cloud were present (white dots in Fig. 6.14). For optically thick clouds (Figs. 6.14(c) and 6.14(d)), an absorbing aerosol leads to an underestimation of the optical thickness, but no underestimation of the effective radius. However, the effective radius is overestimated for optically thin clouds. This can be explained by the topology of the lookup tables with and without absorbing aerosols. In the future, it may be possible to use the spectral fingerprint of absorbing aerosols above a cloud layer (as shown in Fig. 6.14) to extract some information about its properties from spectral albedo or reflectance.

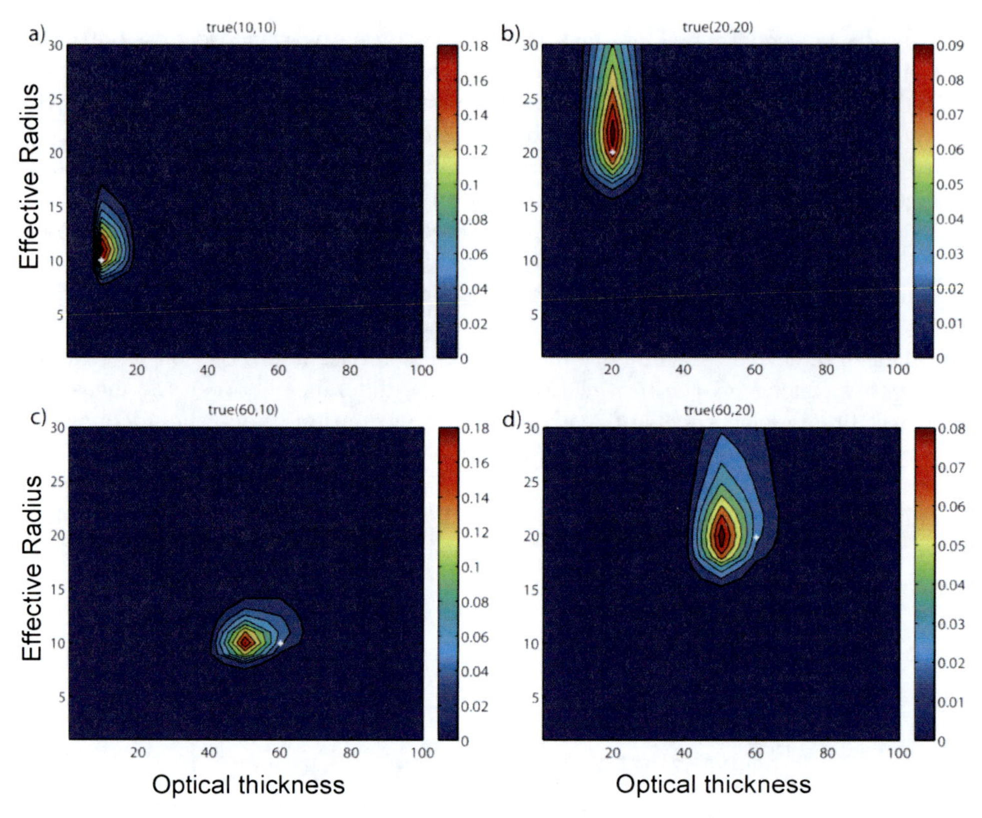

Fig. 6.14. From Vukicevic et al. (2010): Posterior PDFs of optical thickness and effective radius. The true values are marked with a white dot. The shift in the PDF with respect to the truth is due to the overlying absorbing aerosol layer, causing an albedo decrease as shown in Fig. 6.13.

6.4 Constraining spectral aerosol radiative forcing

The direct measurement of radiative forcing is impossible since it would require simultaneous irradiance measurements in the presence and absence of a layer. Most commonly, it is therefore inferred from the optical properties of a layer, using a radiative transfer model. Deriving aerosol radiative forcing in this way is difficult because the input parameters (optical thickness, single scattering albedo and asymmetry parameter) are afflicted with large uncertainties. For example, the instrumentation for measuring aerosol extinction, absorption, or scattering coefficient *in situ* incurs errors due to spectral dependence (often, only three wavelengths are available), inlet characteristics (large particle cut-offs), and desiccation (most inlets remove ambient humidity). Measuring humidity growth factors allows corrections for the latter, but cannot account for lost particles due to inlet aerodynamics. In satellite retrievals, the only available aerosol parameter is often optical thickness (on occasion, single scattering albedo). Assumptions in the radiative transfer models (for example, about the aerosol scattering phase function or surface albedo) cause uncertainties, not only in the aerosol retrieval but also when calculating irradiances.

Closure of *in situ* measured aerosol properties with remote sensing products from the ground (AERONET) or from satellite and aircraft is rarely achieved, because of differing sampling volumes and times.

Although the forcing itself cannot be measured, it is possible to determine forcing *efficiency*, the forcing normalized by the magnitude of the perturbing agent (in this case, aerosol optical thickness), by underflying a gradient in optical thickness. Redemann et al. (2006) define aerosol radiative forcing efficiency as the change in net irradiance per change in optical thickness at a mid-visible wavelength (499 nm): $E_\lambda = dF_\lambda/d\tau_{499}$, and relative forcing efficiency as $e_\lambda = E_\lambda/F_\lambda^\downarrow \times 100\%$, where $F_\lambda^\downarrow$ is the downwelling irradiance at the top of the layer. They used simultaneous optical thickness measurements with the Ames Airborne Tracking Sunphotometer (AATS-14: Russell et al., 1999) and irradiance measurements with SSFR, and derived E_λ as the slope of the regression line between net irradiance and optical thickness. The concept of forcing efficiency is useful because it provides a direct link between aerosol remote sensing and radiative forcing. If the spectral forcing efficiency of a certain aerosol type is known, the forcing can be obtained from satellite-derived optical thickness, circumventing the assumptions in radiative transfer calculations. The gradient method, however, has some limitations: if properties other than optical thickness (for example, surface albedo, single scattering albedo, sun angle) change during the gradient flight leg, those will affect the measured irradiance and bias the forcing efficiency. Therefore, it can only be used over ocean where the surface albedo is very homogeneous. Also, it does not work when gradients in optical thickness are small.

Schmidt et al. (2010a) developed a new approach that can also be used over land surfaces, and for homogeneous aerosol layers. It was adapted from a method that was originally developed for deriving aerosol single scattering albedo from absorption measurements (Bergstrom et al., 2003). It uses upwelling and downwelling irradiance measurements above and below an aerosol layer from SSFR, along with AATS-14 optical thickness measurements. At the fourteen wavelengths of AATS-14 (ranging from 353 nm to 2139 nm), single scattering albedo, asymmetry parame-

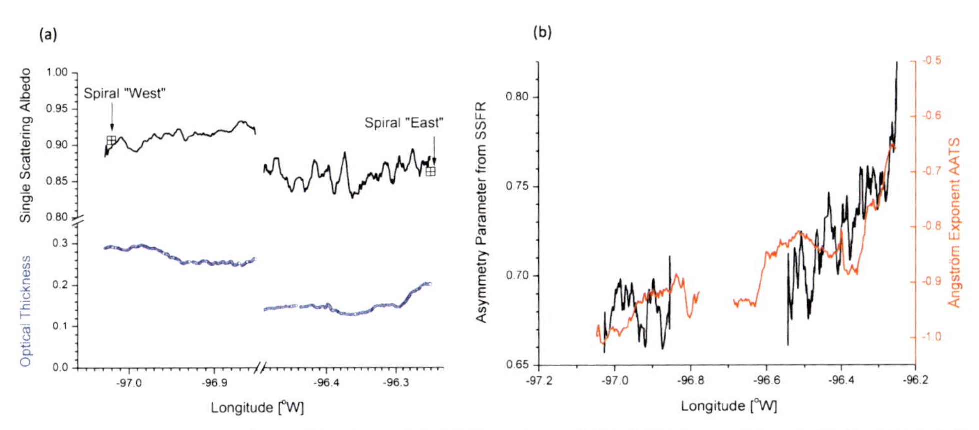

Fig. 6.15. Retrievals from SSFR and AATS-14 from MILAGRO, on March 13th, 2006. (a) Single scattering albedo and optical thickness. (b) Asymmetry parameter and Ångström exponent (from Schmidt et al., 2010a).

ter, and surface albedo are iteratively adjusted in a radiative transfer model until the modeled irradiances match with the measurements. The measurements can be obtained from aircraft spiral ascents and descents, from slant profiles, or from collocated flight legs above and below a layer. Figure 6.15 shows the retrievals of single scattering albedo and asymmetry parameter at 499 nm along collocated flight legs over the Gulf of Mexico on March 13th, 2006, during the MILAGRO (Megacity Initiative – Local and Global Research Observations) experiment (Molina et al., 2010). At both ends of the legs, spirals through the layer were flown, and similar values for ϖ and g as on the collocated legs were obtained. At a longitude of around 96.8°W, a cloud was encountered. These measurements were filtered out. Both the Ångström exponent from AATS-14 and the asymmetry parameter (Fig. 6.15(b)) carry information on particle size, and thus show similar trends along the leg. The values of single scattering albedo, optical thickness, and asymmetry parameter differ on both sides of the cloud, indicating that two different air masses were sampled. The uncertainty of the retrieved parameters increases for the lower optical thickness values encountered to the east of the cloud.

The forcing efficiency is determined at the end of the iterative adjustment of the single scattering albedo, asymmetry parameter, and surface albedo in the radia-

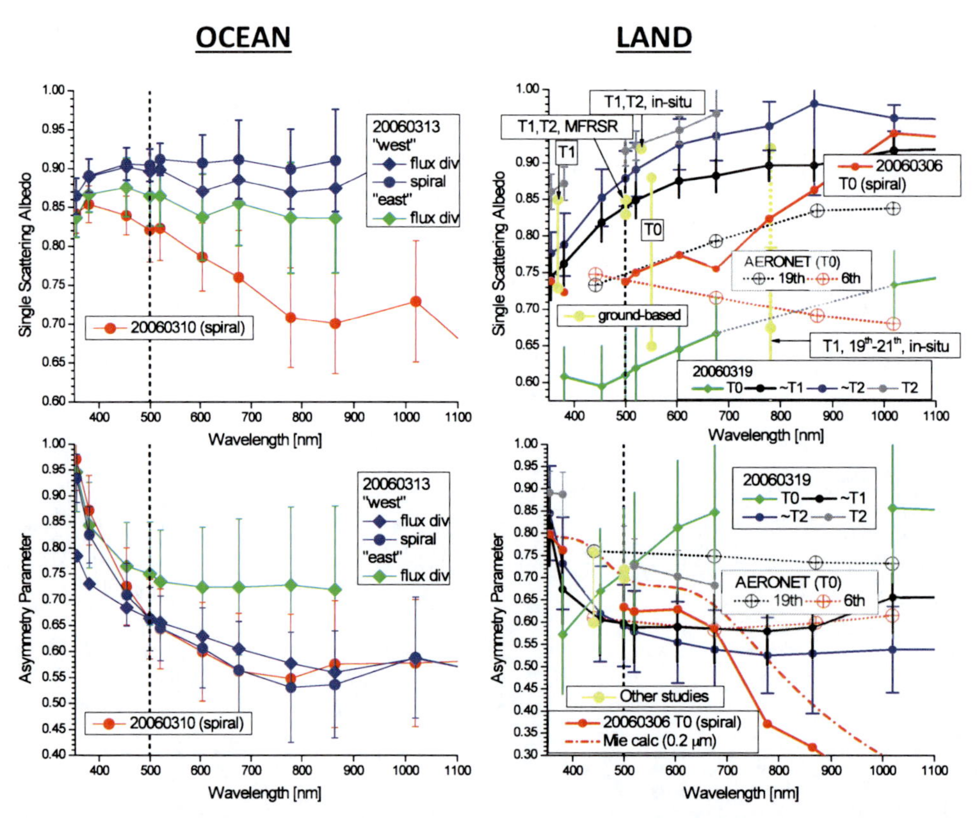

Fig. 6.16. From Schmidt et al. (2010a): Single scattering albedo and asymmetry parameter for MILAGRO cases measured over ocean and land. Over land, results are given in the context of various *in situ* measurements and remote-sensing results.

tive transfer model. By design of the algorithm, it is consistent with the measured irradiances, the local surface albedo, and the aerosol properties. Therefore, radiative closure exercises are not required. In contrast to the gradient method, which generally only provides the forcing efficiency at the bottom of the layer, the new algorithm provides top-of-layer and bottom-of-layer forcing efficiencies, as well as layer absorption.

Fig. 6.16 shows the spectra of single scattering albedo and asymmetry parameter for various cases over ocean (Gulf of Mexico) and land (Mexico City, and its northern outskirts). The blue and green spectra in the left column (representing the same case as shown in Fig. 6.15, at a longitude of 97.2°W – 'West', and 96.2°W – 'East') differ from one another, providing further evidence of two different air masses. Over land, the retrievals were compared with various *in situ* measurements and remote sensing instruments on the ground (details in Schmidt et al., 2010a). The range of single scattering albedo is quite large. Over land, it generally increases from values as low as 0.6 at 500 nm near the sources (Mexico City, site T0) to 0.80–0.90 further downstream of the city (sites T1 and T2). The range of the asymmetry parameter is equally large (0.55–0.75 at 500 nm). It typically decreases with wavelength, but there is one exception (site T0, gray line) where it increases. Mie calculations based on particle size distributions with a 0.2 μm and 0.5 μm mode (red and gray dash-dotted lines) reproduce both the typical and the atypical spectral shape. Although the method by Bergstrom et al. (2003) keeps

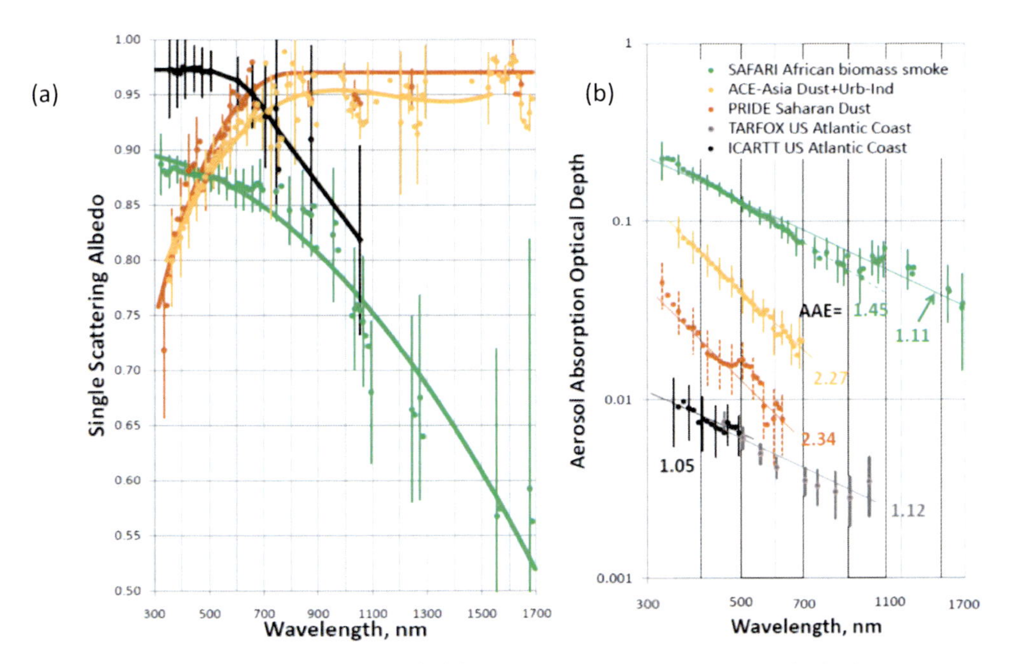

Fig. 6.17. From Russell et al. (2010): (a) Spectra of single scattering albedo from various different experiments, based on Bergstrom et al. (2007). Orange and yellow show dust-type aerosols, green a biomass plume, and black measurements from the US East Coast. (b) The same data are shown as aerosol absorption optical depth. The spectral fits are used to derive absorption Ångström exponents for various aerosol types.

the asymmetry parameter fixed at 0.75, the SSA values retrieved in this way for MILAGRO, shown in Bergstrom et al. (2010), are similar to the ones in Schmidt et al. (2010a). The reason is that absorbed irradiances (the basis for the Bergstrom retrieval) are less sensitive to asymmetry parameter.

The spectral shape of the single scattering albedo can be used to categorize the aerosol by type. Figure 6.17 shows a compilation (Russell et al., 2010) of (a) single scattering albedo spectra and (b) absorption optical depth from various experiments, based on Bergstrom et al. (2007), and provides some context for the single scattering albedo values obtained from MILAGRO. The reddish-brown and yellow spectra represent dust (increasing SSA with wavelength) from the Sahara and from source regions in Asia, respectively. The green spectrum was obtained for a biomass smoke plume in South Africa (SSA decreasing with wavelength). The black spectrum shows measurements from the US East Coast. The SSA spectral shapes measured during MILAGRO suggest dust as primary aerosol type; only the spectrum measured over ocean on 10 March 2006 indicates a biomass plume. Figure 6.17(b) shows that for all of the cases, an absorption Ångström exponent can be assigned to the absorption optical depths by fitting a line to the double logarithmic plot of the absorption optical thickness to wavelength. This exponent can be used to classify plumes by type. A more thorough discussion of classification of various aerosol types by the spectral shape of SSA and Ångström exponent can be found in Bergstrom et al. (2010), and references therein.

Figure 6.18 shows the relative forcing efficiency for a selection of MILAGRO measurements over land and ocean from Schmidt et al. (2010a). Top-of-layer (TOL) and bottom-of-layer (BOL) spectra are marked as dotted and solid lines, respec-

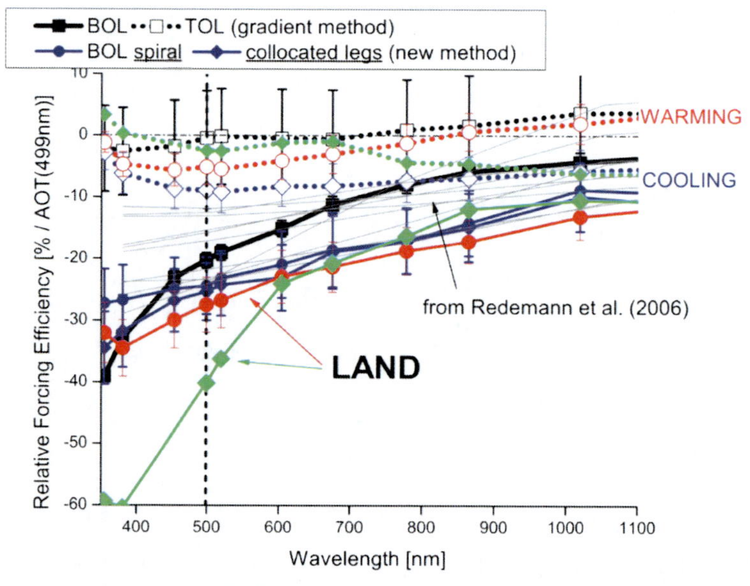

Fig. 6.18. Relative forcing efficiency spectra from INTEX-NA (Redemann et al., 2006, gray lines) and MILAGRO (Schmidt et al., 2010a). All measurements were acquired over ocean, except for the cases labeled "LAND" which indicate the typical range encountered in the Mexico City area (green: strongly absorbing; red: aged aerosol layer).

tively. The difference is related to the layer absorption; for a non-absorbing layer, BOL and TOL forcing efficiencies would be identical. BOL results from Redemann et al. (2006) are overlaid as thin gray lines. They were obtained using the gradient method for INTEX-NA (Intercontinental Chemical Transport Experiment – North America) off the US East Coast in 2004. The spectra from 13 March, 2006, obtained from the new method are shown in blue. The results from the collocated points above and below the layer (diamonds) and from the nearby spiral (circles) agree with one another at all wavelengths. However, the spectrum derived from the gradient method, shown in black, deviates from those results, because the aerosol properties vary considerably throughout the leg.

The red and the green spectra show the forcing efficiency for two cases over land; the green line represents a strongly absorbing aerosol near the sources, the red line an aged airmass. The relative forcing efficiencies acquired over the ocean and over land are quite similar, which is surprising because of the different surface albedo spectra of ocean and land and because of the large range of the retrieved single scattering albedos and asymmetry parameters. The BOL results from INTEX-NA (Redemann et al., 2006) are similar to those found in MILAGRO. In most cases, the BOL relative forcing efficiencies ranged from −40% to −10% (cooling), with the lowest values occurring at the shortest wavelengths. The TOL values ranged from −10% to +5% (weak cooling to warming).

In summary, aerosol radiative forcing can be constrained by airborne measurements of the relative forcing efficiency. Despite the large ranges of surface albedo, single scattering albedo and asymmetry parameter, the relative forcing efficiency are remarkably similar. The spectral shape of single scattering albedo is indicative of the aerosol type, while the asymmetry parameter and Ångström exponent provide information on the particle size distribution. The combination of irradiance and optical thickness measurements on one aircraft is a way to obtain aerosol optical properties, surface albedo, and forcing simultaneously, which are mutually consistent – thus providing a strong measurement constraint on the aerosol direct effect. The forcing efficiency establishes the link between single-wavelength satellite retrievals of aerosol optical thickness and the spectral aerosol radiative effect. In contrast to the gradient method by Redemann et al. (2006), the new algorithm can also be used over land surfaces. However, the measurements are more difficult to make because collocated flight legs above or below the layer, or spiral measurements through the layer are required.

6.5 The spectral signature of heterogeneous clouds

As shown in the previous section, it is not straightforward to derive aerosol radiative forcing from satellite observations, because of the spectral dependence of optical thickness, single scattering albedo, and asymmetry parameter. In contrast, shortwave radiative forcing and absorption of a single-layer homogeneous liquid water cloud can be calculated from satellite-derived optical thickness and effective radius retrievals, at least in principle. Such a cloud, however, is a construct of models and rarely occurs in nature. Real clouds are spatially heterogeneous, which causes numerous biases in remote sensing that have been discussed extensively in the literature during the past two decades (Davis and Marshak, 2010). Ice crystals in clouds introduce another level of complication, and the general case of multi-layer multi-phase heterogeneous clouds is nearly intractable. Deriving irradiances and thus forcing and absorption from satellite observations of radiances is currently accomplished by angular distribution models, which account for various different cloud types in a statistical manner (Loeb et al., 2005). The accuracy of satellite-derived cloud properties can only be assessed with aircraft observations since they cover enough ground and provide enough statistics to directly compare cloud fields from the perspective of an aircraft and an overflying satellite.

To date, mapping *in situ* microphysical measurements *directly* onto satellite observations has been impossible for heterogeneous clouds because the sampling volumes of *in situ* probes cannot acquire sufficient statistics to fully represent the satellite field-of-view. A possible path for connecting detailed cloud microphysics with radiation fields is the statistical, *indirect*, assessment of cloud scenes such as those observed from aircraft using surrogate (Schmidt et al., 2007) or physical 3D-cloud generators (Schmidt et al., 2009; see section 6.6). A *direct* validation of satellite products is possible with airborne remote sensing instruments if their fields-of-view can be collocated in time and space with satellite pixels. NASA designated several aircraft for the validation of their satellite fleet. The high-flying ER-2 is equipped with instrumentation that matches remote sensors onboard NASA's A-Train polar-orbiting satellite constellation. The MODIS Airborne Simulator (MAS) (King et al., 1996), for example, mimics the functionality of MODIS.

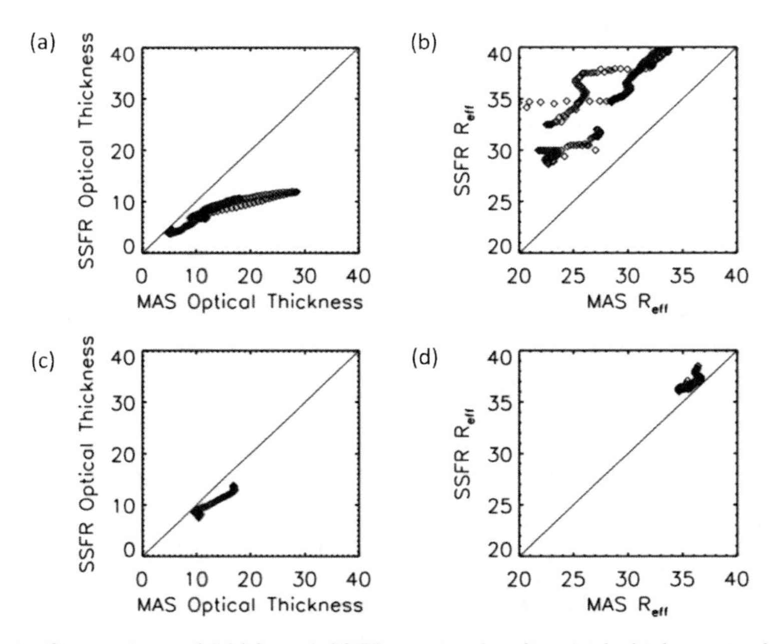

Fig. 6.19. Comparison of MAS and SSFR retrievals of optical thickness and effective radius for two cases from TC^4 : a heterogeneous cloud (a and b), and a homogeneous cloud (c and d). The MAS retrievals are averaged within the half-power footprint of SSFR (from Kindel et al., 2010a).

During the TC[4] (Tropical Composition, Cloud and Climate Coupling) experiment (Costa Rica, July/August 2007), the ER-2 was equipped with MAS and the MODIS/ASTER (MASTER) airborne simulator (Hook et al., 2001), alongside with SSFR (section 6.2.2), Cloud Physics Lidar (CPL: McGill et al., 2004), Cloud Radar System (CRS: Li et al., 2004), and other sensors. MAS and MASTER, for example, were used for a validation of MODIS on the cloud product level (i.e., optical thickness and effective radius) (King et al., 2010), but not on the radiance level because of incompatible viewing geometries of airborne and satellite imagery. Unlike most other instruments onboard the ER-2, the SSFR irradiance measurements have no space-borne counterpart. The original incentive for deploying SSFR was to constrain the radiative energy budget of clouds. But SSFR was also used as a remote-sensing instrument. In contrast to reflectance-based (radiance) retrieval products, albedo-derived (irradiance) cloud optical thickness and effective radius represent the hemispherically averaged cloud properties, which can be translated directly into the cloud forcing and absorption, *without* resorting to angular distribution models. Introducing budget-equivalent cloud properties thus bridges the gap between energy budget and remote sensing. The question is whether retrievals from irradiance and radiance (averaged over the lower hemisphere) are mutually consistent. The case in Fig. 6.19 illustrates how SSFR and MAS retrievals compare in a homogeneous (Figs. 6.19(c) and 6.19(d)) and a heterogeneous case (Figs. 6.19(a) and 6.19(b)) from TC[4] (Kindel et al., 2010). The MAS optical thickness and effective radius were averaged within a circle of 13 km diameter. About 50% of the measured upwelling irradiance originates from within this circle, the SSFR half-power footprint. As expected, MAS and SSFR retrievals are close to one another for homogeneous clouds. For heterogeneous scenes, however, SSFR optical thickness (effective radius) values are always lower (higher) than those obtained from MAS. One explanation is that SSFR incorporates the properties of the entire scene beneath the aircraft, including any clear-sky areas, whereas MAS retrievals pertain to cloud-pixels only. Clear-sky areas within the field of view of SSFR lower the albedo below that of a cloud-only scene, lowering the hemispherical averaged optical thickness. At the same time, the larger absorption at near-infrared wavelengths by the dark ocean leads to an erroneous increase in retrieved effective radius. The difference between the MAS and SSFR perspective of the same cloud underlines the difficulty of deriving the radiative energy budget from space. Considerable biases may arise when deriving the forcing and absorption from radiance-based cloud retrievals. There is a possibility that the differences in MAS and SSFR retrievals are caused by radiometric uncertainty. One way to mitigate this problem could be the use of spectral slopes of normalized reflectance rather than reflectance itself in future retrievals, as demonstrated by McBride et al. (2011) for transmittance-based retrievals (section 6.3.2).

6.5.1 The spectral consistency approach

In order to rule out radiometric issues as cause for discrepancies, Kindel et al. (2010) introduced a spectral self-consistency approach, in loose analogy with Baran and Labonnote (2007). SSFR-retrieved cloud optical thickness and effective radius (based on albedos at only two wavelengths, 870 nm and 1600 nm) were used to re-calculate the albedo spectrum across the entire wavelength range of SSFR. The re-calculated spectrum was compared with the original SSFR spectrum that was the basis for the retrieval. SSFR measurements are self-consistent with respect to the retrieved cloud properties if no residuals occur between original and reproduced spectrum. If residuals do occur anywhere across the spectrum, they indicate either instrument issues, such as wavelength-dependent changes in the response function, or the inadequacy of the underlying radiative transfer model that was used for the retrievals and for recreating the cloud spectral albedo. For homogeneous ice clouds, the only significant residuals between measured and reproduced spectra occur at wavelengths between 1000 nm and 1300 nm. No such discrepancies were observed for homogeneous water clouds (Kindel et al., 2011). This suggests that the library of ice single scattering properties (the same one as used in MODIS collection 5 ice retrievals: Yang et al., 2007) may be spectrally inconsistent. Currently, a new version of the single scattering libraries (Baum et al., 2011) are being analyzed.

Figure 6.20(a) shows the measured albedo above a thick heterogeneous ice cloud (black line), and the reproduced albedo for $\{\tau, R_{\text{eff}}\} = \{53.2, 30\ \mu\text{m}\}$, retrieved at $\{870\ \text{nm},\ 1600\ \text{nm}\}$ (blue circles). Figure 6.20(b) shows the difference between reproduced and measured albedo. The black lines show the range of the spectral residual along an extended flight leg; the red symbols show the mean spectral residual. While the residuals between 1000 nm and 1300 nm can be attributed to the ice crystal scattering library, the effects between 400 nm and 1000 nm (not observed above homogeneous clouds) point to cloud heterogeneities as possible cause. This was corroborated by the spatial variability in the MAS imagery. Since the plane-parallel albedo bias leads to a constant offset across the conservative scattering wavelengths, the albedo at 870 nm should be affected in the same way as 500 nm, and the spectral residual should be identical to zero in this range. Any departure from this behavior indicates that cloud heterogeneities have an effect that is *not* spectrally neutral. When retrieving optical thickness and effective radius of

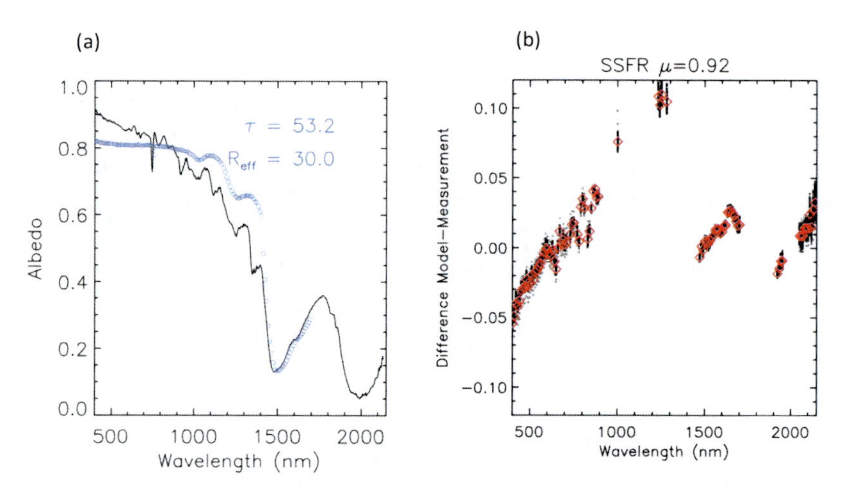

Fig. 6.20. (a) Measured and modeled albedo for a heterogeneous cloud case from Kindel et al. (2010a), (b) spectral residuals (difference between modeled and measured albedo) for individual points along the leg (black) and averaged over the leg (red).

a cloud with large sub-pixel variability, the equivalent homogeneous cloud optical thickness and effective radius does not have the same spectral properties as the heterogeneous clouds, and those two parameters alone are *not* sufficient to fully characterize the radiative effect of the cloud. Although these effects were found in irradiance measurements, a similar effect is expected in satellite observations since the pixel size of the imagers is comparable to the half-power footprint of SSFR. In future experiments, the reflectance from space-borne or airborne cloud observations should be analyzed with respect to their spectral consistency.

The spectral signature of heterogeneous clouds is probably related to a combination of horizontal photon transport, molecular scattering, and surface reflectance, and their different spectral dependencies. This can be explored only by looking at the cloud scenes in detail, using a 3D radiative transfer model (next section).

6.5.2 Observing and modeling 3D cloud effects – apparent absorption

To understand the spectral signature of heterogeneous clouds, one can use fields of optical thickness and effective radius as retrieved from MAS as input to 3D radiative transfer calculations and compare modeled and measured spectral irradiance point-by-point along the flight track. In this way, the combined effects of horizontal photon transport, interactions with the surface, and molecular scattering are taken into account, as long as the model cloud properly represents the true cloud structure.

An advantage of TC^4 was the occurrence of closely collocated flight legs between the ER-2 and the NASA/University of North Dakota DC-8, which was also equipped with an SSFR. The coordination was possible through the NASA Real Time Mission Monitor (RTMM), which allows real-time display of all flight tracks at mission control on the ground, as well as a direct downlink of in-flight data. Using this information, the ground crew could direct pilots to target regions of interest while keeping aircraft in close proximity. During TC^4, it was possible to align DC-8 and ER-2 along the same ground track over hundreds of kilometers with a temporal separation of less than two minutes.

With the ER-2 flying above and the DC-8 flying below a cloud layer, it was possible to simultaneously measure spectral irradiance above and below heterogeneous cloud fields while obtaining the horizontal structure of cloud parameters from MAS onboard the ER-2. In addition, CRS and CPL retrieved the vertical distribution along the nadir track of the ER-2. Figure 6.21 shows the ER-2 measurements of (a) MAS horizontal distribution of optical thickness, (b) CRS radar reflectivity as a cross-section through the cloud, and (c) SSFR spectral albedo along a flight track on July 17, 2007 (Schmidt et al., 2010b). In this case, the flight track was nearly 200 km long. The flight track of the DC-8 underneath the ER-2 is shown as a dotted line in the center of the MAS swath, and at just above 8 km within the CRS reflectivity profile. The cloud field was very complex, with the optical thickness ranging from zero (white areas in Fig. 6.21(a)) to 100 (saturation value for MAS, black areas). In addition to the outflow of a convective cell between 6 and 12 km altitude, scattered boundary layer clouds were present at an altitude of 2 km.

In this extremely heterogeneous cloud scene it is expected that, at least in some locations, the net horizontal photon transport will be non-zero. The combination of

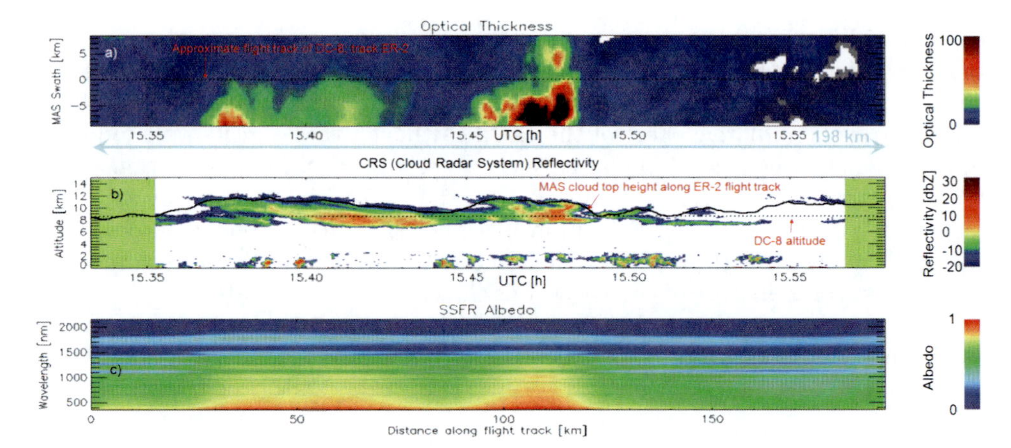

Fig. 6.21. Heterogeneous cloud case from TC[4] (July 17, 2007). (a) MAS optical thickness, (b) CRS radar reflectivity; (c) SSFR spectral cloud albedo (from Schmidt et al., 2010b).

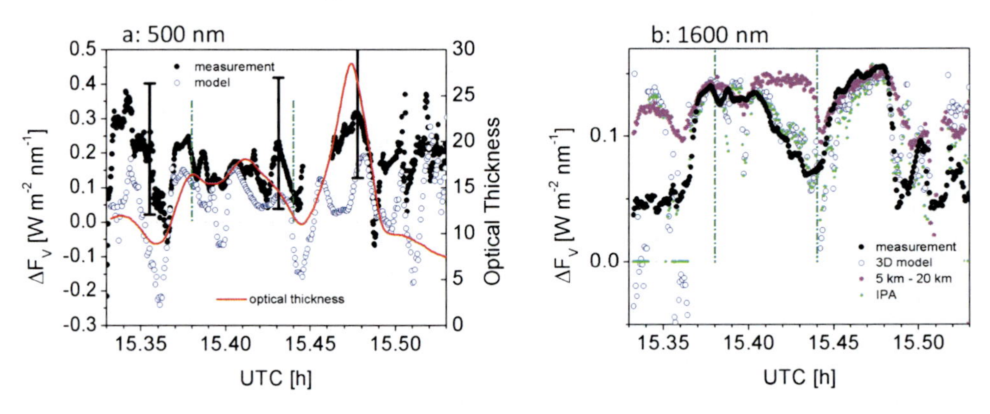

Fig. 6.22. Modeled and measured apparent absorption (or vertical net irradiance difference, ΔF_V) along the flight leg shown in Figure 6.21, for 500 nm (a) and 1600 nm (b). The red line in (a) is the MAS optical thickness, averaged over the half-power footprint of SSFR. Measurement uncertainties are indicated as black bars. In (b), the green symbols show the modeled true absorption, obtained from the independent pixel approximation (IPA).

SSFR onboard the ER-2 and the DC-8 allowed to characterize net horizontal photon transport spectrally from the difference between top-of-cloud and bottom-of-cloud spectral net irradiance. This is called *apparent* absorption because it is equivalent to true absorption only when the horizontal divergence in the sampling volume is zero. Not only can it be non-zero, but of either sign, meaning *apparent emission* is also possible when net horizontal photon transport is neglected. Some studies defined apparent absorption without true absorption. Here, apparent absorption comprises true (physical) absorption of solar radiation within a particular sampling volume, and the net photon transport through the sides of this volume. In the absence of physical absorbers positive (negative) apparent absorption is entirely caused by photon loss (gain) through the sampling volume sides.

Figure 6.22 shows the measured and modeled apparent absorption derived from the vertical net irradiance difference, ΔF_V, along the flight track from Fig. 6.21, at a conservative scattering wavelength (500 nm, Fig. 6.22(a)), and at a wavelength with non-zero physical absorption (1600 nm, Fig. 6.22(b)). Measurements (black symbols) are shown with error bars. For 500 nm, they reach up to 100% of the value of the absorbed irradiance because ΔF_V is the small difference of two large quantities. The modeled apparent absorption (blue symbols) traces the time series along the leg quite reliably, although averaged over the entire leg, the modeled values are biased low with respect to the measurements (about 0.1 W m^{-2} nm^{-1}). It should be pointed out that the vertical structure of the model cloud has a bearing on the agreement between measurements and model, since the DC-8 flew partially within, not below the cloud. Therefore, it was essential to use the CRS profile for vertically distributing the liquid cloud water for each MAS pixel below the airplane. For off-nadir pixels of MAS, where CRS was not available, the same profile was used, but shifted up or down to align the CRS cloud top from nadir observations with the off-nadir cloud top retrieved by MAS. As shown in Fig. 6.21(b), the MAS-retrieved cloud top height along nadir (black lines) reproduces the cloud top variation observed by CRS. At 1600 nm, the model–measurement agreement is much better because true absorption far outweighs horizontal photon transport. Also, the horizontal distances over which photons are transported are shorter in presence of absorption (Platnick, 2001).

Figure 6.22 illustrates the difficulties of cloud absorption measurements and supports the statement by Marshak et al. (1999), that it is nearly impossible to 'harvest' true cloud absorption from aircraft measurements even if two aircraft are perfectly aligned in time and space, let alone from a single aircraft measurements that flies legs above and below a dynamically changing cloud field. When flying broadband radiometers rather than spectrometers, horizontal photon transport can dominate over true absorption in the spectrally integrated signal. This is immediately obvious in Fig. 6.22 where the apparent absorption at 500 nm and 1600 nm are comparable in magnitude. The literature suggests two methods to extract true absorption data from such measurements: (1) grand averaging over legs that are long enough to ensure that net horizontal transport converges to zero over the extended sampling volume; (2) using apparent absorption at a visible wavelength to identify where the apparent absorption vanishes, and correct all wavelengths for net horizontal photon transport under the assumption that it is spectrally neutral (Ackerman and Cox, 1981). Grand averaging may be problematic, given the bias between measurements and model in Fig. 6.22(a): While modeled domain-averaged apparent absorption at 500 nm is identical to zero, due to energy conservation and periodic boundary conditions in the 3D model, this constraint does not apply to the measurements, and convergence to zero horizontal transport cannot be guaranteed. As seen in Fig. 6.22(a), photon loss areas (positive apparent absorption) are loosely correlated with the cloud optical thickness (red line). At visible wavelengths, areas with high optical thickness lose photons to optically thinner areas over scales on the order of tens of kilometers. This net photon transport can even occur across the boundaries set by the model. A net photon export beyond the MAS swath, into optically thin or clear-sky areas, would lead to an overall positive bias in apparent absorption, even when averaging over nearly 200-km flight legs.

The filtering technique by Ackerman and Cox (1981) may be equally problematic, as illustrated in Fig. 6.23 where the apparent absorption is shown as a function of wavelength, at two locations along the flight leg. In this case, the apparent absorption was normalized by the incident irradiance on top of the layer and displayed in percent absorptance. The measured spectra (red and blue lines) have a distinct spectral shape, with increasing values throughout the visible. Even if the apparent absorption equals zero at some visible wavelengths (for example, around 420 nm for the red spectrum), this does not guarantee that the apparent absorption vanishes throughout the visible wavelength range. The unusual spectral shape of apparent absorption is reproduced in part by the 3D model (full circles), within the measurement uncertainty. The domain-averaged true absorption is obtained from the independent pixel approximation (IPA, dashed line) where horizontal photon transport is disabled in the radiative transfer model. The leg-averaged apparent absorption (black line) is close to true absorption in the near-infrared but disagrees up to 10% at conservative scattering wavelengths. Integrated over the SSFR wavelength range, the measured apparent absorption is 85% higher than the true absorption. The spectral shape of the absorptance mirrors that of the measured albedo from Fig. 6.23(a), which indicates that the deviations of the spectral albedo from 1D model calculations can, in part, also be ascribed to net horizontal photon transport. A full explanation for the peculiar spectral shape of albedo and absorptance is still missing. Preliminary calculations indicate that spherical effects cannot explain the spectral shape (Kokhanovsky, private communication). Further measurements in heterogeneous cloud scenes, along with 3D radiative transfer cal-

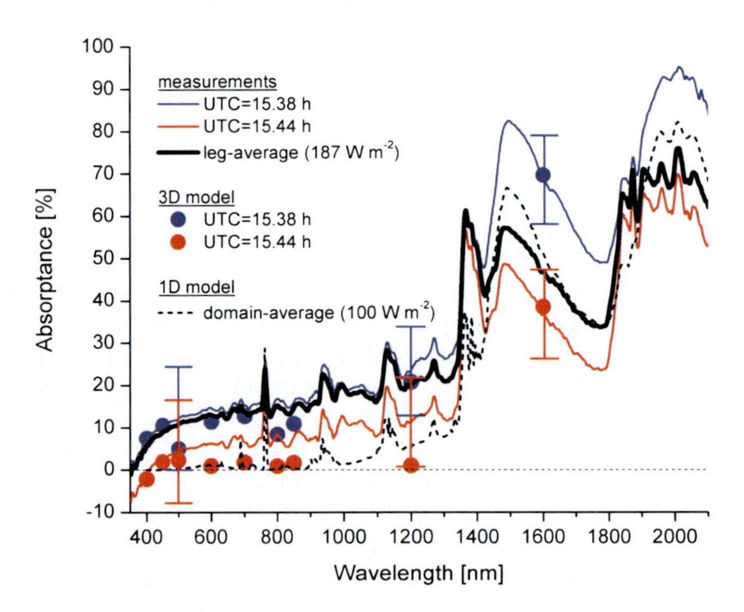

Fig. 6.23. Measured and modeled spectral absorptance for two different times along the flight leg. The thick black line shows the measured apparent absorption, averaged over the entire leg. The dotted black line shows the domain-averaged modeled true absorption (calculated with IPA).

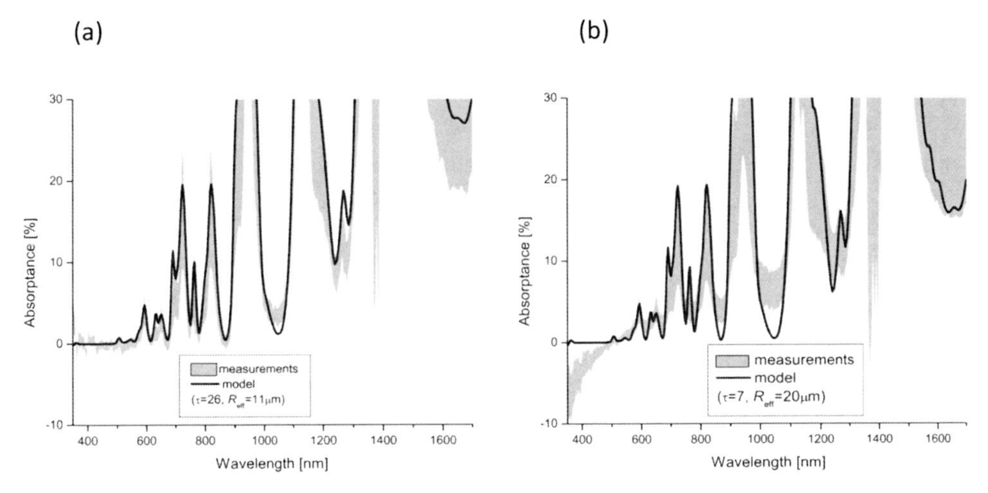

Fig. 6.24. Measured and modeled spectral absorptance of two water cloud layers from TC^4 (a: July 29, 2007; b: August 6, 2007). The gray shaded area represents the mean measured spectrum ± standard deviation, the black line shows the modeled absorptance spectrum. Values above 30% absorptance (mainly water vapor absorption) are cut off.

culations, provide a path towards broader understanding of 3D effects, and their unique spectral signatures can be used to identify the source of anomalies.

6.5.3 Attributing cloud absorption to causes

A goal of cloud irradiance measurements is to constrain forcing and absorption, and to determine the role of various contributors such as water vapor and condensed water to the total forcing or absorption. But given the difficulties outlined above, is it possible to extract this information from the measurements with any reasonable certainty? Judging from Fig. 6.22(a), only a handful of data points along a 200 km leg would satisfy the filter criterion suggested by Ackerman and Cox (1981), while the leg-average would lead to an overestimation of true absorption. If the interest lies in the true absorption of a cloud field *as a whole*, rather than in characterizing the radiative effects *at each point*, the data recovery can be increased with a random sampling technique by Kindel et al. (2011), inspired by Marshak et al. (1999). For homogeneous boundary layer clouds, which are much more amenable to absorption measurements than the highly heterogeneous ice clouds from the previous section, above- and below-cloud measurements were paired on the basis of their 500 nm net irradiance. This technique works for collocated measurements from two aircraft, as well as for one-aircraft sequential measurements above and below a layer. In either case, the spatial structure is re-sorted by imposing a match in the spectrum at 500 nm, rather than by location. True absorption of the cloud layer is estimated from the mean difference of all the matching net irradiance pairs on top and at the bottom of the layer. With this randomized sampling, a considerable amount of spectra can be identified for further analysis.

Fig. 6.24 shows the range of measured cloud absorption (gray shaded area) for two water cloud cases during TC^4, as obtained from the randomized sampling technique. From the collection of the associated top-of-cloud albedo spectra, an optical thickness and effective radius can be derived in the same way as in section 6.5.1, and the expected spectral absorptance can be re-calculated with a 1D model (black line). In the case of Fig. 6.24(a) ($\tau = 26$, $R_{\text{eff}} = 11$ μm), the modeled absorptance reproduced the measurement results across the entire SSFR wavelength range, outside of the water vapor bands. In the case of Fig. 6.24(b) ($\tau = 7$, $R_{\text{eff}} = 20$ μm), the measured absorptance in the visible had a similar shape as observed in Fig. 6.23, whereas the model showed no absorption. The case shown in Fig. 6.24(b) shows a higher degree of spatial heterogeneity (higher optical thickness and absorptance standard deviation), although to a much lesser degree than the ice cloud in section 6.5.2. Regardless of the sampling approach (point-by-point, as in the previous section, or randomized), a spectrally *increasing* absorptance appears to be an indicator of cloud heterogeneity. Absorbing aerosols within or above the cloud, on the other hand, would lead to a spectrally *decreasing* positive absorptance, and can be clearly separated from heterogeneity effects. By restricting analysis to cases with a spectrally flat absorptance, the confounding effects of horizontal photon transport can be minimized, and the total amount of true absorption to liquid water (ice) absorption, gas absorption can be attributed by decomposing the absorption spectrum into its constituents.

This is done in Fig. 6.25. The spectrum of absorbed irradiance is shown in Fig. 6.25(a), identical to Fig. 6.24(a) without normalization by the incident irradiance. The spectrally integrated absorbed irradiance of the cloud is 100.4 W m^{-2}. Figure 6.25(b) shows the gas absorption spectrum obtained by switching off the cloud in the radiative transfer model. Water vapor dominates the gas absorption by far and it varies more than any other absorbing species in this spectral range. Figure 6.25(c) shows the enhancement of gas absorption due to cloud multiple scattering. It is obtained by switching off cloud absorption and switching on cloud scattering in the model, and subtracting the clear-sky gas absorption spectrum from the result. Finally, the cloud particle absorption (Fig. 6.25(d)) is obtained by subtracting gas absorption (Fig. 6.25(b)) and multiple-scattering gas absorption enhancement (Fig. 6.25(c)) from the total absorption spectrum (Fig. 6.25(a)).

Together, the clear-sky gas absorption (58.7 W m^{-2}) and the multiple-scattering enhancement of the gas absorption (16.4 W m^{-2}) account for 75% of the total absorption (Fig. 6.26). The absorption by cloud droplets (25.2 W m^{-2}, 25%) is only 50% larger than the enhanced gas absorption. In ice clouds, the partitioning of absorption would look quite different because the water vapor concentration is lower at ice cloud altitudes, and because the absorption maxima of ice are displaced with respect to those of water vapor – far more than those of liquid water, which have significant regions of overlap with vapor absorption bands. Constraining modeling attribution studies with observations can only be achieved with spectrally resolved measurements as presented in Kindel et al. (2011).

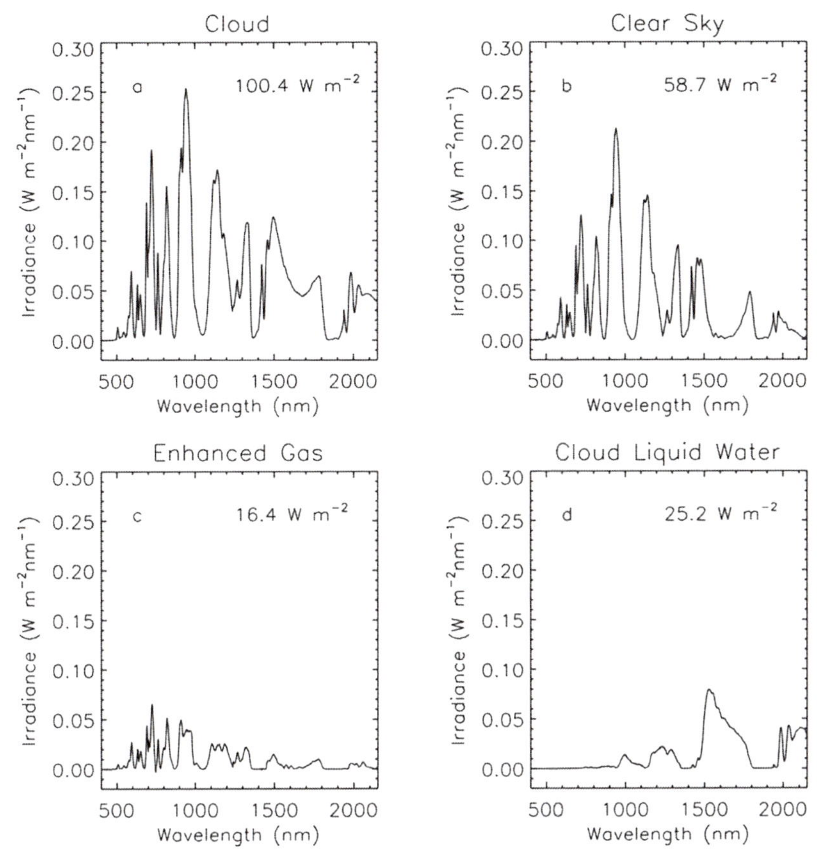

Fig. 6.25. Absorbed irradiance spectrum from July 29, 2007, into its constituents: (a) total, (b) clear-sky, (c) enhanced gas absorption through multiple scattering, and (d) absorption by the cloud drops.

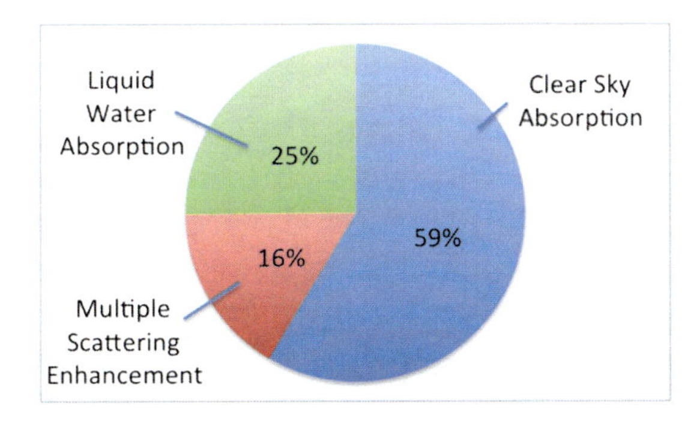

Fig. 6.26. Fractional absorption for the case from Fig. 6.25.

6.6 Heterogeneous clouds and aerosols

Aerosols in close proximity to homogeneous or heterogeneous clouds represent a great challenge to space-borne remote sensing. To date, passive imagery cannot detect aerosols when clouds are present in the same pixel or within a distance of tens of kilometers. Non-absorbing aerosol particles above a homogeneous cloud can often be neglected for the reflectance and the radiative forcing of the combined aerosol–cloud layer. In contrast, an undetected absorbing aerosol may bias the retrieval of an underlying cloud layer (Haywood et al., 2004; Coddington et al., 2010) and alter its radiative forcing. Polarimetry from space improves the retrieval of aerosol properties above homogeneous clouds. However, as demonstrated by the measurement examples in the previous section, heterogeneous clouds introduce various spectral effects that are problematic for passive imagery, with or without polarimetric capabilities. The reason is that the signatures of heterogeneous clouds are not spectrally neutral and change the background signal of the atmosphere in which the aerosol is embedded. Therefore, even distant clouds (tens of kilometers away) can bias aerosol retrievals (Wen et al., 2007; Marshak et al., 2008). In some cases, clouds are too small to be detected with imagery (Koren et al., 2008), and cannot be distinguished from the surrounding aerosols. Therefore, some studies have suggested dropping the separation between clouds and aerosol altogether, and to regard them as an entity (Koren et al., 2007; Charlson et al., 2007). Such a cloud–aerosol continuum can be distinctively diverse from the clouds or the aerosol layer by themselves. For example, the radiative properties of polluted scattered boundary layer clouds as shown in Fig. 6.27a cannot be reproduced by summing up the properties of the heterogeneous cloud field and the aerosol layer in which it is embedded: if $p_{\rm cld} = \{\tau_{\rm cld}, R_{\rm eff}, c\}$ represents the cloud properties with optical thickness, effective drop radius, and cloud fraction; and $p_{\rm aer} = \{\tau_{\rm aer}, a, \varpi_{\rm aer}, g_{\rm aer}\}$ the aerosol properties with optical thickness, Ångström parameter, single scattering albedo, and asymmetry parameter, the spectral forcing $f_\lambda(\{p_{\rm cld}, p_{\rm aer}\})$ is generally *not* equal to $f_\lambda(p_{\rm cld}) + f_\lambda(p_{\rm aer})$. The same applies for reflectance, absorptance, and transmittance.

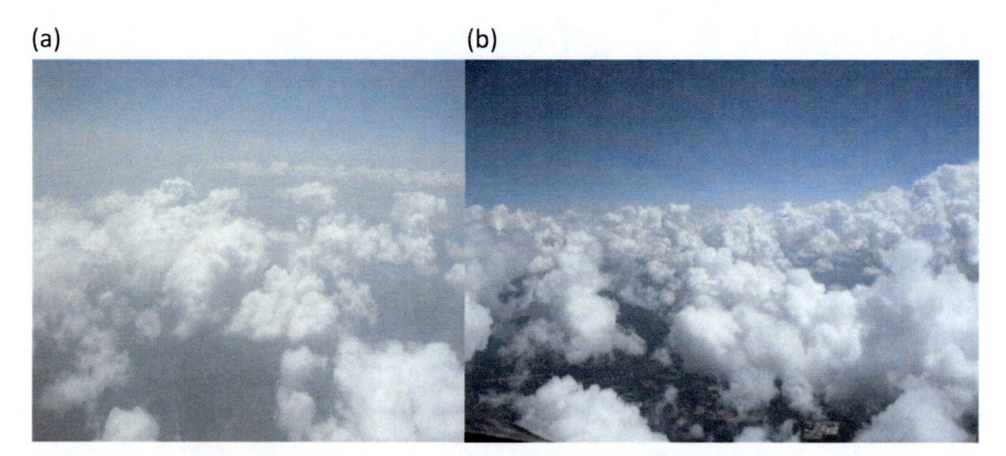

Fig. 6.27. Polluted (a) and clean (b) broken cloud scene during GoMACCS (photographs taken from the CIRPAS Twin Otter, courtesy Armin Sorooshian).

How can the combined forcing of heterogeneous clouds and aerosols be measured? In principle, aircraft observations suffer from the same limitations as encountered for homogeneous aerosol layers (section 6.4) since forcing, unlike irradiances, cannot be measured directly. To solve this problem, a *combination* of measurements and model calculations is used. Spectral irradiance above and below a layer is measured along with cloud and aerosol properties, such as profiles of liquid water content, droplet effective radius, aerosol size distribution, single scattering albedo, extinction, or optical thickness. The forcing above and below the layer can then be obtained from the difference between the measured net irradiances in presence of the cloud-aerosol layer, and the calculated clear-sky net irradiances. Calculating the clear-sky irradiances also requires an independent measurement of surface albedo. The difficulty is to link the measured cloud–aerosol properties with the irradiances and thus the forcing at each location across the domain. For heterogeneous clouds, this causes problems at three levels. (1) In most experiments, cloud–aerosol properties and irradiances above and below the layer cannot be measured at the same time and have to be sampled sequentially with a single aircraft. Since the layer properties change in space and time, it is virtually impossible to map observed irradiances directly to the associated cloud properties. (2) Even for ideal conditions such as during TC^4 (collocated irradiance measurements above and below a layer, combined with cloud imagery), horizontal photon transport can bias the observed irradiances, and thus obfuscate the relationship between cloud and aerosol properties on the one hand, and absorption or forcing on the other. (3) When extrapolating from small-scale aircraft measurements to large-scale satellite observations, it is necessary to sample the cloud structure across the entire satellite pixel. In order to eliminate heterogeneity-related biases, one-point and two-point statistics are required.

Rather than relating cloud properties and irradiance fields directly, Schmidt et al. (2007) employed an indirect statistical approach: microphysical measurements of liquid water content and effective radius from in-cloud horizontal flight legs and ascents or descents through the layer were used as input to cloud generators, which reproduced the 3D distribution of cloud microphysics. Subsequently, a 3D radiative transfer model was used to calculate spectral irradiances from the 3D cloud fields, which were compared with aircraft measurements through histograms (one-point statistics) and power spectra (two-point statistics). If those two matched, the measured irradiance fields were consistent with the microphysical measurements, and the regenerated 3D cloud field could be used to infer quantities that could not be measured directly, such as the forcing, absorption, irradiance or radiance at a specific location, or for the domain as a whole. The thus-derived quantities can be regarded as measurement-validated model results, in a statistical sense.

Fig. 6.28(a) shows a cloud field that was reproduced by the Iterative Amplitude Adapted Fourier Transform Algorithm (IAAFT: Venema et al., 2006), for a case encountered on September 14, 2002, during the INSPECTRO experiment (Influence of Clouds on the Spectral Actinic Flux in the Lower Troposphere: Kylling et al., 2005). IAAFT used microphysical cloud data along three flight legs (marked in blue) and a vertical profile to create surrogate clouds with the same probability density functions and power spectra as the input. On the flight legs the algorithm was nudged to match the measurements exactly. Figure 6.28(b) shows the his-

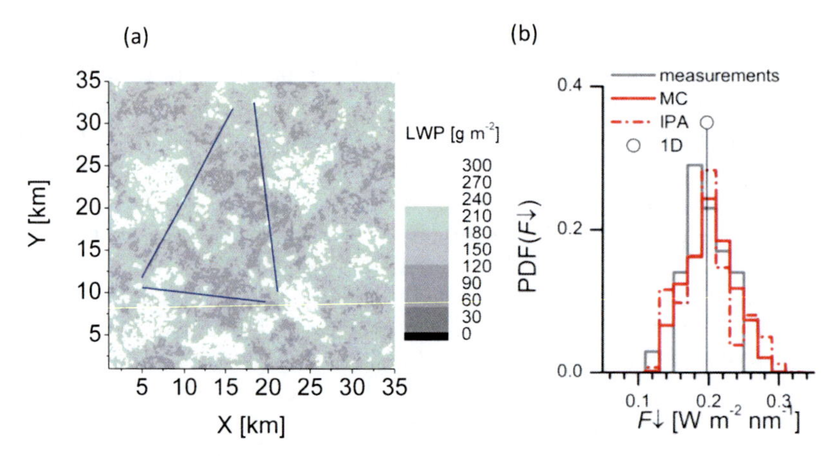

Fig. 6.28. Adapted from Schmidt et al. (2007): (a) Liquid Water Path (LWP) of a stratocumulus scene generated by IAAFT for INSPECTRO flight on September 14, 2002. The blue lines show the in-cloud flight legs. (b) Below-cloud downwelling irradiance measurements (gray) and model results (red) at 500 nm as histograms.

tograms of measured (gray) and modeled (red) downwelling irradiance (at 500 nm) below the cloud field. In this case, it is possible to use the domain-averaged cloud properties in a 1D model to reproduce the mean value of the transmitted irradiance. Both 3D and IPA calculations reproduced mean values and variability. The difference between full 3D calculations and the independent pixel approximation becomes obvious in Fig. 6.29(a) where power spectra of the transmitted irradiance along below-cloud legs are shown. While the IPA calculations (dash-dotted line) trace the $-5/3$ power law, which originates in the underlying LWP cloud field, geometric smoothing leads to a variance reduction in the 3D modeled irradiance fields for scales below about 5 km (solid line). This scale break also occurs in the measurements. At scales below 1.6 km, the limited number of photons used in the Monte Carlo simulations leads to white photon noise which can be suppressed by increasing the number of photons in the model runs. Measurement noise sets in at scales around 0.5 km. Figure 6.29(b) shows the downwelling irradiance below a broken cloud field. In this case, 1D calculations become meaningless, and the independent pixel approximation fails to reproduce the measurements. It predicts the location of the clear-sky mode at 0.8 W m^{-2} nm^{-1}, which is lower than in the measurements because the effects of cloud focusing are neglected. The cloud-transmittance mode at 0.3 W m^{-2} nm^{-1} is not reproduced at all. The areas under the cloud-transmittance and the clear-sky modes are similar in model and measurements, which shows that the cloud cover is correctly reproduced by IAAFT. Even if this were not the case, the matching locations of modeled and measured clear-sky and cloud-transmittance modes give confidence in the statistical approach.

One of the 3D effects in broken clouds demonstrated in Fig. 6.29(b) is the enhancement of downwelling irradiance in cloud gaps. When broken clouds are embedded in an aerosol layer, these 3D effects are superimposed by aerosol scattering and absorption. This was studied by Schmidt et al. (2009) for a case from GoMACCS (Gulf of Mexico Atmospheric Composition and Climate Study), which

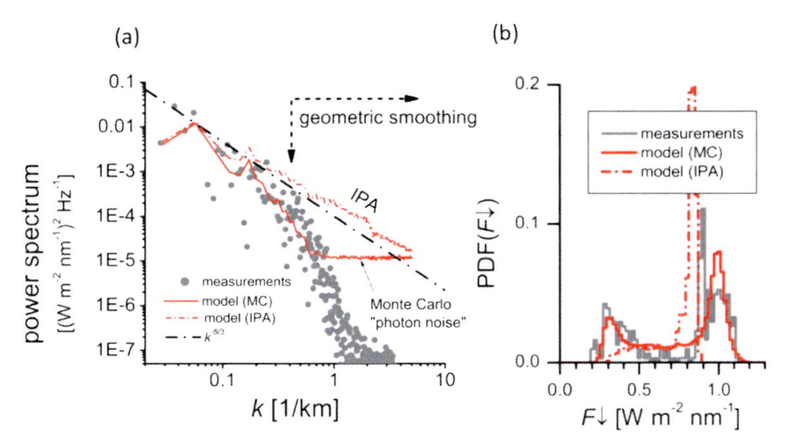

Fig. 6.29. (a) Power spectrum along a below-cloud flight leg (gray) and across the model domain (red), for the overcast case from Fig. 6.28. (b) Histogram of downwelling irradiance below a broken cloud field, measured on September 22, 2002.

targeted convective boundary layer clouds in the polluted, urban-industrial region around Houston (Lu et al., 2008). They used Large Eddy Simulations (LES: Jiang and Feingold, 2006) to create cloud fields, which were validated against airborne cloud microphysical measurements (Jiang et al., 2008). The LES input included profiles of rawinsondes and aircraft measurements of aerosol properties. The LES cloud–aerosol fields were used to calculate irradiance fields with MYSTIC (section 6.2.4) with some additional input such as spectral surface albedo. Figure 6.30 shows a comparison of measured and modeled downwelling irradiance at 500 nm below a cloud–aerosol field from August 15, 2006, which looks quite similar to the case shown in Fig. 6.29(b). In this case, the area under the cloud-transmittance mode (labeled "CLD") and the clear-sky mode (labeled "GAP") is not reproduced by the calculations. This indicates a slight underestimation of the cloud fraction

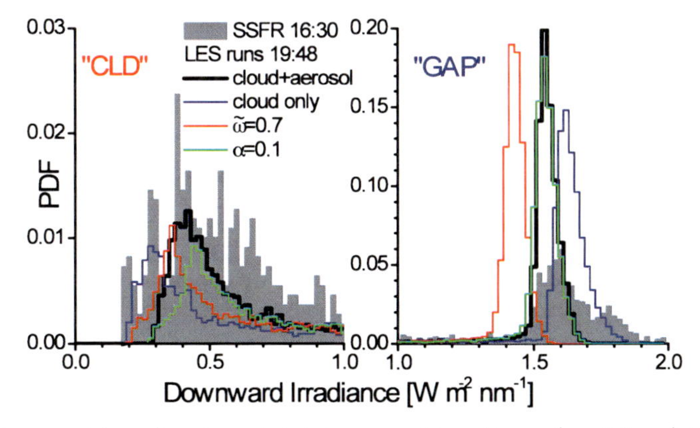

Fig. 6.30. Measured (gray) and modeled downward irradiance (at 500 nm) below polluted broken boundary layer clouds from August 15, 2006, separated into cloud-transmittance ("CLD") and clear-sky (direct beam, "GAP") modes. For the cloud–aerosol runs (black PDF), single scattering albedo was 0.8 (0.7 for the red PDF), and surface albedo 0.1 (0.035 for the green PDF).

by the LES, which did not account for cloud advection into the model domain. Nevertheless, the location of the modes leads to the following conclusions: (1) In the main model run (black PDF), the full aerosol-cloud field was used in the calculations, with a fixed single scattering albedo of 0.8 (based on AERONET during closest clear-sky event), and a surface albedo of 0.035 (obtained from SSFR measurements on a different day). It reproduces the irradiance measurements for both modes; the secondary clear-sky mode at 1.8 W m^{-2} nm^{-1} is due to a large cloud edge effect. (2) For an aerosol-free 'clean' run (blue PDF), the clear-sky mode was biased high with respect to the measurements, and the cloud-transmittance mode was biased low. In cloud gaps, the aerosol scatters radiation out of the direct beam and thus decreases the clear-sky downward irradiance. At the same time, the shadow regions receive additional diffuse radiation, and the cloud-transmittance mode shifts to higher values. The model reproduces the measurements only if the aerosol between the broken clouds is included in the calculations. (3) Increasing the aerosol absorption (red PDF: single scattering albedo decreased to 0.7) leads to decreased irradiance in both modes, and does not reproduce the measurements. (4) A change in surface albedo beyond the measured value slightly increases the downward irradiance below clouds (due to increased reflection from the cloud bottom). Changes in the aerosol properties lead to deviations of the model results from the measurements. This was shown here for single scattering albedo, but there is also sensitivity to optical thickness and asymmetry parameter.

Figure 6.31 shows the location of the two modes from Fig. 6.30 as a function of wavelength. Across the entire visible wavelength range, the measured downward irradiance underneath the clouds (red spectrum) was only reproduced when including aerosols in the model runs (red full circles); otherwise, it was underestimated (red open circles). This was similar for the irradiance under cloud gaps (blue spectra), where cloud-only runs were higher than measured. Above 800 nm, the 3D effects

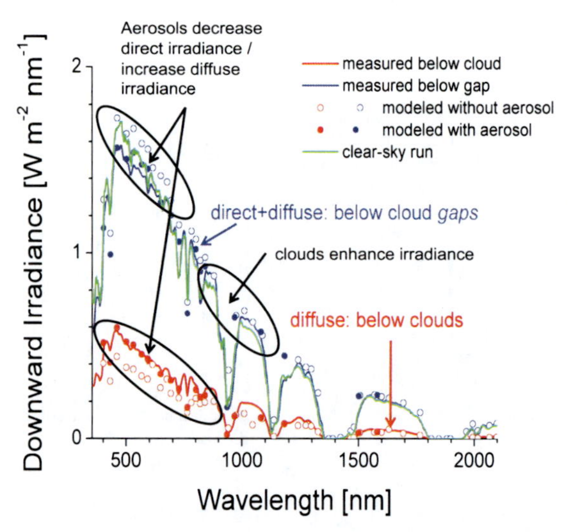

Fig. 6.31. Location of the modes from Fig. 6.30 as a function of wavelength. The blue (red) spectrum shows the measurements in cloud gaps (below clouds); the symbols show the 3D calculations with (full circles) and without aerosols (open circles). The green spectrum shows a 1D calculation under clear-sky conditions (no clouds or aerosols).

lead to an enhancement of downward irradiance in cloud gaps. This can be seen when comparing a clear-sky model run (green spectrum) with the measurements in cloud gaps (blue spectrum). Below 700 nm, the aerosol counteracts the 3D enhancement of downward irradiance and decreases it below its clear-sky value. Both effects combined lead to a unique spectral signature of the cloud–aerosol layer. In the shadow zone (red spectra), the aerosol has an effect in the opposite direction, compared to the sunlit areas (blue spectra). The contrasting effects of the aerosol in shadow and sunlit areas, as well as the spectral shape of combined 3D and aerosol effects in the sunlit areas may be the starting point for a retrieval of aerosol properties *between* clouds.

Although the forcing cannot be measured directly, it is possible to measure the *apparent forcing* in broken cloud fields as defined by Schmidt et al. (2009), by making the following replacements: in the definition of the forcing, the perturbed net irradiance, F_{cld}, is replaced by $cF_{\text{CLD}} + (1-c)F_{\text{GAP}}$, where c is the cloud fraction, F_{CLD} is the net irradiance measured underneath the clouds, and F_{GAP} is the measured net irradiance in cloud gaps:

$$f_{\text{rel}} = \frac{F_{\text{cld}} - F_{\text{clear}}}{F^{\downarrow}} \rightarrow \frac{cF_{\text{CLD}} + (1-c)F_{\text{GAP}} - F_{\text{GAP}}}{F^{\downarrow}_{\text{GAP}}} = f^{\text{app}}_{\text{rel}}$$

In the above formula, the forcing is defined relative to the incident irradiance on top of the cloud layer, which is replaced by $F^{\downarrow}_{\text{GAP}}$. F_{CLD} and F_{GAP} can be derived from the histograms of measured net irradiance; the cloud fraction c can be inferred from the area under the clear-sky mode divided by the total area. The apparent forcing is equal to the true forcing if no 3D effects occur, and in absence of aerosols. The difference between true forcing and apparent forcing is illustrated in Fig. 6.32. The true bottom-of-layer cloud forcing (open red circles)

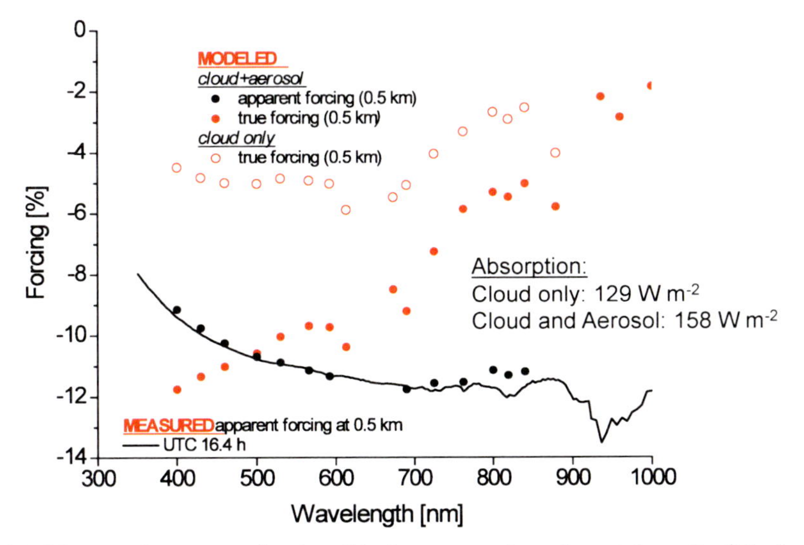

Fig. 6.32. Measured apparent forcing (black spectrum), and model results (black circles). The true forcing of the cloud–aerosol layer (red full circles) or the forcing of the cloud layer (red open circles) can only be modeled.

is only about −2% above 800 nm, and −5% below 600 nm. The contrast between these two different wavelength regimes is due to the shape of the surface albedo, which increases between 600 and 800 nm. The combined cloud–aerosol forcing (red closed circles) deviates considerably from the forcing of the cloud layer alone, especially at short wavelengths where the aerosol is optically thick. In this case, the aerosol increases the absorption by about 20%, from 129 W m^{-2} to 158 W m^{-2}. The spectral shape of the *apparent forcing* is substantially different from the true forcing, due to 3D effects acting at all wavelengths, and aerosol scattering and absorption counteracting the 3D effects at short wavelengths. It is obvious that the apparent absorption cannot be used as a proxy for the true absorption. But since it can directly be derived from measurements, unlike true forcing, it provides the link between model and observations. In this sense, the true absorption of the cloud–aerosol layer can be regarded as a measurement-validated product.

6.7 Summary

Throughout the previous sections, we have illustrated the value of spectrally-resolved observations for cloud–aerosol research. We demonstrated with airborne irradiance measurements and model calculations that cloud variability and spectral cloud properties are closely related. Generally, the signature from heterogeneous clouds is different from their homogeneous counterparts. This means that the two parameters optical thickness and effective radius, retrieved from current dual-channel techniques, do not reproduce the spectral albedo of the cloud field, with potentially important ramifications for the clouds' radiative forcing and absorption.

We introduced the spectral consistency approach, which is capable of detecting cloud heterogeneities by comparing the predicted spectral albedo from retrieved optical thickness and effective radius to the spectral measurements. The spectral consistency approach is not limited to the detection of cloud heterogeneities; it was also used to identify biases in ice crystal single scattering properties, and could be employed to detect radiometric calibration inconsistencies of future spectral imagers. Cloud absorption can be attributed to causes by spectral decomposition. This was tested for boundary layer clouds.

Spectrally resolved measurements help in understanding biases in cloud research, as for example the mismatch between measured and modeled cloud absorption. Broadband measurements in the past allowed only speculations as to what caused the observed discrepancies: gas absorption, undetected large droplets, or horizontal photon transport, to name a few. We have presented first measurements and 3D model results of apparent absorption, the difference of net irradiance on top and at the bottom of a cloud layer, at individual points of an extended coordinated flight leg of two aircraft. Unlike true absorption, apparent absorption can be measured directly. The 3D model generally reproduced the spectral shape of the measured absorption. In contrast to common belief, the bias was not spectrally neutral in the UV and visible wavelength range. The point-by-point observations constitute the foundation for understanding these spectral effects. It is an ongoing effort to identify the mechanism which leads to the spectral bias, and to examine whether a similar effect can be found in radiance observations.

Heterogeneous clouds surrounded by aerosols are another domain of spectral observations. The aerosol scattering and absorption superimpose the 3D effects in the cloud field. Combined, aerosol and 3D effects impose a unique spectral signature upon the cloud–aerosol field, with different effects in sunlit and shadow zones. We introduced apparent forcing, which, in contrast to true forcing, can be derived from measurements, and can be linked to 3D model calculations in a statistical manner. The difference between true and apparent forcing is due to 3D and aerosol effects.

We also discussed various ways to use spectral information under homogeneous conditions. We described a new technique for deriving aerosol forcing efficiency along with single scattering albedo and asymmetry parameter from the layer optical thickness and irradiance measurements above and below the layer, which works not only over water, but also above land surfaces. Aerosols can be classified by means of the spectral shape of the single scattering albedo and asymmetry parameter, as well as the absorption Ångström exponent, which describes the spectral dependency of aerosol absorption optical thickness. We pointed out that the derived relative spectral forcing efficiencies of various types of aerosols are remarkably similar over land and water, despite different surface albedos and aerosol properties. This finding is at odds with model predictions, which exhibit a wide range of the spectral forcing efficiencies. The tight range of the observations could be used to constrain the direct aerosol effect in satellite observations and modeling applications, but needs to be explained in the course of future experiments.

Furthermore, we reviewed ways to characterize the vertical layering of mixed-phase clouds, which capitalize on the fact that the photon penetration depth into a cloud layer is a function of wavelength, most notably in the near-infrared. While polarimetry will be able to separately characterize the ice and liquid phase on the very top of a cloud, multiple scattering does not allow to retrieve such information for deeper layers, which, at least to some extent, *can* be explored with spectrally-resolved unpolarized observations. The best way to characterize mixed-phase clouds would be a combination of polarized and un-polarized passive imagers, combined with active sensors to assess the vertical structure.

Finally, we proposed the term 'cloud spectroscopy'. In ranges where liquid water or ice absorbs, wavelength-dependent bulk physical properties translate into a spectral slope in transmittance or reflectance, which can replace the traditional observation at a single wavelength and reduce the sensitivity of cloud retrievals to the radiometric uncertainty. The power of this new approach was demonstrated with a new retrieval based on cloud transmittance. When replacing the near-infrared transmittance by the spectral slope of the near-infrared transmittance, the error bars can be reduced considerably. We described the theoretical framework, which is currently being laid for the assessment of spectral information content and retrieval uncertainty.

The conclusion from the discussed measurement and modeling examples is that spectral observations advance cloud–aerosol remote sensing and budget applications considerably and may be the key to resolving some of the problems that research is currently confronted with. Many of the findings presented here are related to irradiances. However, some of them may be applicable to radiance as well. Research in this direction is ongoing, and will continue with the deployment of new spectral imagers in orbit.

Acknowledgments

This research was funded through numerous grants from NASA, NOAA, and the German Research Foundation. Odele Coddington, Bruce Kindel, and Patrick McBride (University of Colorado), André Ehrlich (University of Leipzig, Germany), Jens Redemann and Philip Russell (BAERI/NASA Ames) contributed to the manuscript with graphs, data, text, and discussions. We are also grateful to Warren Gore and Antony Trias (NASA Ames) who built and continuously supported the SSFR instrument.

References

Ackerman, S. A., and S. K. Cox, 1981: Aircraft observations of shortwave fractional Albedo of non-homogeneous clouds, *J. Appl. Meteorol.*, **20**, 1510–1515.

Arking, A., and J. D. Childs, 1985: Retrieval of cloud cover parameters from multispectral satellite images, *J. Appl. Meteorol.*, **24**, 323–333.

Bannehr, L., and R. Schwiesow, 1993: A technique to account for the misalignment of pyranometers installed on aircraft, *J. Atmos. Oceanic Technol.*, **10**, 774–777.

Baran, A. J., and L.-C. Labonnote, 2007: A self consistent scattering model for cirrus. I: The solar region, *Q. J. R. Meteorol. Soc.*, **133**, 1899-1912.

Baum, B.A., P. Yang, A. J. Heymsfield, C. G. Schmitt, Y. Xie, A. Bansemer, Y.-X. Hu, and Z. Zhang, 2011: Improvements in Shortwave Bulk Scattering and Absorption Models for the Remote Sensing of Ice Clouds, *J. Appl. Meteorol. and Climatology*, **50**, 1037–1056.

Bergstrom, R. W., P. Pilewskie, B. Schmid, and P. B. Russell, 2003: Estimates of the spectral aerosol single scattering albedo and aerosol radiative effects during SAFARI 2000, *J. Geophys. Res.*, **108**(D13), 8474, doi:10.1029/2002JD002435.

Bergstrom, R. W., P. Pilewskie, P. B. Russell, J. Redemann, T. C. Bond, P. K. Quinn, and B. Sierau, 2007: Spectral absorption properties of atmospheric aerosols, *Atmos. Chem. Phys.*, **7**, 5937–5943, doi:10.5194/acp-7-5937-2007.

Bergstrom, R. W., K. S. Schmidt, O. Coddington, P. Pilewskie, H. Guan, J. M. Livingston, J. Redemann, and P. B. Russell, 2010: Aerosol spectral absorption in the Mexico City area: results from airborne measurements during MILAGRO/INTEX B, *Atmos. Chem. Phys.*, **10**, 6333–6343, doi:10.5194/acp-10-6333-2010.

Cess, R. D., M. H. Zhang, P. Minnis, L. Corsetti, E. G. Dutton, B. W. Forgan, D. P. Garber, W. L. Gates, J. J. Hack, E. F. Harrison, X. Jing, J. T. Kiehl, C. N. Long, J. J. Morcrette, G. L. Potter, V. Ramanathan, B. Subasilar, C. H. Whitlock, D. F. Yound, and Y. Zhou, 1995: Absorption of solar radiation by clouds: Observations versus models, *Science*, **267**, 496–499.

Charlson, R. J., A. S. Ackerman, F. A. M. Bender, T. L. Anderson, and Z. Liu, 2007: On the climate forcing consequences of the albedo continuum between cloudy and clear air, *Tellus, Ser. B*, **59**, 715–727.

Chiu, J. C., A. Marshak, Y. Knyazikhin, P. Pilewskie, and W. J. Wiscombe, 2009: Physical interpretation of the spectral radiative signature in the transition zone between cloud-free and cloudy regions, *Atmos. Chem. Phys.*, **9**, 1419–1430, doi:10.5194/acp-9-1419-2009.

Chiu, J. C., A. Marshak, Y. Knyazikhin, and W. J. Wiscombe, 2010: Spectrally-invariant behavior of zenith radiance around cloud edges simulated by radiative transfer, *Atmos. Chem. Phys.*, **10**, 11295–11303, doi:10.5194/acp-10-11295-2010.

Chylek, P., S. Robinson, M. K. Dubey, M. D. King, Q. Fu, W. B. and Clodius, 2006: Comparison of near-infrared and thermal infrared cloud phase detections, *J. Geophys. Res.*, **111**, D20203, doi:10.1029/2006JD007140.

Clark, R.N., G. A. Swayze, I. Leifer, K. E. Livo, R. Kokaly, T. Hoefen, S. Lundeen, M. Eastwood, R. O. Green, N. Pearson, C. Sarture, I. McCubbin, D. Roberts, E. Bradley, D. Steele, T. Ryan, R. Dominguez et al., 2010: A method for quantitative mapping of thick oil spills using imaging spectroscopy, *U.S. Geological Survey Open-File Report* 2010–1167, 51 pp.

Coddington, O., K. S. Schmidt, P. Pilewskie, W. J. Gore, R. W. Bergstrom, M. Roman, J. Redemann, P. B. Russell, J. Liu, and C. C. Schaaf, 2008: Aircraft measurements of spectral surface albedo and its consistency with ground-based and space-borne observations, *J. Geophys. Res.*, **113**, D17209, doi:10.1029/2008JD010089.

Coddington, O. M., P. Pilewskie, J. Redemann, S. Platnick, P. B. Russell, K. S. Schmidt, W. J. Gore, J. Livingston, G. Wind, and T. Vukicevic, 2010: Examining the impact of overlying aerosols on the retrieval of cloud optical properties from passive remote sensing, *J. Geophys. Res.*, **115**, D10211, doi:10.1029/2009JD012829.

Cracknell, A. P., 1997: *The Advanced Very High Resolution Radiometer*, Taylor & Francis, London.

Davis, A. B., and A. Marshak, 2010: Solar radiation transport in the cloudy atmosphere: a 3D perspective on observations and climate impacts, *Rep. Prog. Phys.*, **73** (1010), 70 pp., doi:10.1088/0034-4885/73/2/026801.

de Graaf, M., and P. Stammes, 2005: SCIAMACHY Absorbing Aerosol Index – calibration issues and global results from 2002–2004, *Atmos. Chem. Phys.*, **5**, 2385–2394, doi:10.5194/acp-5-2385-2005.

Ehrlich, A., E. Bierwirth, M. Wendisch, J.-F. Gayet, G. Mioche, A. Lampert, and J. Heintzenberg, 2008: Cloud phase identification of Arctic boundary-layer clouds from airborne spectral reflection measurements: Test of three approaches, *Atmos. Chem. Phys.*, **8**, 7493–7505.

Ehrlich, A., M. Wendisch, E. Bierwirth, J.-F. Gayet, G. Mioche, A. Lampert, and B. Mayer, 2009: Evidence of ice crystals at cloud top of Arctic boundary-layer mixed-phase clouds derived from airborne remote sensing, *Atmos. Chem. Phys.*, **9**, 9401–9416, doi:10.5194/acp-9-9401-2009.

Gayet, J.-F., I. S. Stachlewska, O. Jourdan, V. Shcherbakov, A. Schwarzenboeck, and R. Neuber, 2007: Microphysical and optical properties of precipitating drizzle and ice particles obtained from alternated Lidar and in situ measurements, *Ann. Geophys.*, **25**, 1487–1497.

Green, R. O., and B. Pavri, 2000: AVIRIS in-flight calibration experiment, sensitivity analysis, and intraflight stability, *Proceedings of the Ninth JPL Airborne Earth Science Workshop*, R. Green (ed.), Pasadena, CA.

Green, R. O., M. L. Eastwood, and C. M. Sarture, 1998: Imaging spectroscopy and the Airborne Visible Infrared Imaging Spectrometer (AVIRIS), *Remote Sens. Environ.*, **65**: (3) 227–248.

Grzegorski, M., M. Wenig, U. Platt, P. Stammes, N. Fournier, and T. Wagner, 2006: The Heidelberg iterative cloud retrieval utilities (HICRU) and its application to GOME data, *Atmos. Chem. Phys.*, **6**, 4461–4476, doi:10.5194/acp-6-4461-2006.

Harrison, L., J. Michalsky, and J. Berndt, 1994: Automated multifilter rotating shadow-band radiometer: an instrument for optical depth and radiation measurements, *Appl. Optics*, **33**, 5118–5125.

Haywood, J. M., S. R. Osborne, and S. J. Abel, 2004: The effect of overlying absorbing aerosol layers on remote sensing retrievals of cloud effective radius and cloud optical depth, *Q. J. R. Meteorol. Soc.*, **130**, 779–800.

Holben B.N., T.F. Eck, I. Slutsker, D. Tanré, J.P. Buis, A. Setzer, E. Vermote, J.A. Reagan, Y. Kaufman, T. Nakajima, F. Lavenu, I. Jankowiak, and A. Smirnov, 1998: AERONET – A federated instrument network and data archive for aerosol characterization, *Rem. Sens. Environ.*, **66**, 1–16.

Hook, S. J., K. J. Thome, M. Fitzgerald, and A. B. Kahle, 2001: The MODIS/ASTER airborne simulator (MASTER)—A new instrument for earth science studies, *Remote Sens. Environ.*, **76**, 93–102, doi:10.1016/ S0034-4257(00)00195-4.

Jiang, H., and G. Feingold, 2006: Effect of aerosol on warm convective clouds: Aerosol-cloud-surface flux feedbacks in a new coupled large eddy model, *J. Geophys. Res.*, **111**, D01202, doi:10.1029/2005JD006138.

Jiang, H., G. Feingold, H. H. Jonsson, M.-L. Lu, P. Y. Chuang, R. C. Flagan, and J. H. Seinfeld, 2008: Statistical comparison of properties of simulated and observed cumulus clouds in the vicinity of Houston during the Gulf of Mexico Atmospheric Composition and Climate Study (GoMACCS), *J. Geophys. Res.*, **113**, D13205, doi:10.1029/2007JD009304.

Joiner, J. and P. K. Bhartia, 1995: The determination of cloud pressures from rotational Raman scattering in satellite backscatter ultraviolet measurements, *J. Geophys. Res.*, **100**, 23019–23026.

Kalesse, H., K. S. Schmidt, R. Buras, M. Wendisch, B. Mayer, P. Pilewskie, M. King, L. Tian, G. Heymsfield, S. Platnick, 2011: The impact of crystal shape and spatial variability on the remote sensing of ice cloud optical thickness and effective radius – a TC4 case study, submitted to *J. Geophys. Res.*

Kindel, B. C., K. S. Schmidt, P. Pilewskie, B. A. Baum, P. Yang, and S. Platnick, 2010: Observations and modeling of ice cloud shortwave spectral albedo during the Tropical Composition, Cloud and Climate Coupling Experiment (TC4), *J. Geophys. Res.*, **115**, D00J18, doi:10.1029/2009JD013127.

Kindel, B.C., 2010: Cloud shortwave spectral radiative properties: Airborne hyperspectral measurements and modeling of irradiance, *Ph.D. thesis*, University of Colorado.

Kindel, B.C., P. Pilewskie, K. S. Schmidt, O. Coddington, 2011: Spectral absorption of marine stratus clouds: Measurements and modeling, under review, *J. Geophys. Res.*

King, M. D., Y. J. Kaufman, W. P. Menzel, and D. Tanre, 1992: Remote- sensing of cloud, aerosol, and water-vapor properties from the Moderate Resolution Imaging Spectrometer (MODIS), *IEEE Trans. Geosci. Remote Sens.*, **30**, 2–27.

King, M. D, W. P. Menzel, P. S. Grant, J. S. Myers, G. T. Arnold, S. E. Platnick, L. E. Gumley, S. C. Tsay, C. C. Moeller, M. Fitzgerald, K. S. Brown, and F. G. Osterwisch, 1996: Airborne scanning spectrometer for remote sensing of cloud, aerosol, water vapor and surface properties, *J. Atmos. Oceanic Technol.*, **13**, 777–794.

King, M. D., S. Platnick, G. Wind, G. T. Arnold, and R. T. Dominguez, 2010: Remote sensing of radiative and microphysical properties of clouds during TC4: Results from MAS, MASTER, MODIS, and MISR, *J. Geophys. Res.*, **115**, D00J07, doi:10.1029/2009JD013277.

Knap, H. W., P. Stammes, and R. B. A. Koelemeijer, 2002: Cloud thermodynamic-phase determination from near-infrared spectra of reflected sunlight, *J. Atmos. Sci.*, **59**, 83–96.

Koelemeijer, R. B. A., P. Stammes, J. W. Hovenier, and J. F. de Haan, 2002: Global distributions of effective cloud fraction and cloud top pressure derived from oxygen A band spectra measured by the Global Ozone Monitoring Experiment: Comparison to ISCCP data, *J. Geophys. Res.*, **10** (D12), 4151, 10.1029/2001JD000840.

Kokhanovsky, A. A., O. Jourdan, and J. P. Burrows, 2006: The cloud phase discrimination from a satellite, *IEEE Geosci. Rem. Sens. Lett.*, **3**, 103–106.

Kokhanovsky, A. A., S. Platnick, and M.D. King, 2011: Remote sensing of terrestrial clouds from space using backscattering and thermal emission techniques, in *The Remote Sensing of Tropospheric Composition from Space*, J. P. Burrows et al. (eds.), Physics of Earth and Space Environments.

Koren, I., L. A. Remer, Y. J. Kaufman, Y. Rudich, and J. V. Martins, 2007: On the twilight zone between clouds and aerosols, *Geophys. Res. Lett.*, **34**, L08805, doi:10.1029/2007GL029253.

Koren, I., L. Oreopoulos, G. Feingold, L. A. Remer, and O. Altaratz, 2008: How small is a small cloud?, *Atmos. Chem. Phys.*, **8**, 3855–3864, doi:10.5194/acp-8-3855-2008.

Kylling, A., A. R. Webb, R. Kift, G. P. Gobbi, L. Ammannato, F. Barnaba, A. Bais, S. Kazadzis, M. Wendisch, E. Jäkel, S. Schmidt, A. Kniffka, S. Thiel, W. Junkermann, M. Blumthaler, R. Silbernagl, B. Schallart, R. Schmitt, B. Kjeldstad, T. M. Thorseth, R. Scheirer, and B. Mayer, 2005: Spectral actinic flux in the lower troposphere: measurement and 1D simulations for cloudless, broken cloud and overcast situations. *Atmos. Chem. Phys.*, **5**, 1975–1997.

Li, L., G. M. Heymsfield, P. E. Racette, L. Tian, and E. Zenker, 2004: A 94 GHz cloud radar system on a NASA high-altitude ER-2 aircraft, *J. Atmos. Ocean. Technol.*, **21**, 1378–1388.

Liew, S.C., and L. K. Kwoh, L. K., 2003: Mapping optical parameters of coastal sea waters using the Hyperion Imaging Spectrometer: intercomparison with MODIS ocean color products, *Geoscience and Remote Sensing Symposium Proceedings, IEEE International*, vol. 1, 549–551, doi: 10.1109/IGARSS.2003.1293838

Loeb, N. G., S. Kato, K. Loukachine, and N. Manalo-Smith, 2005: Angular distribution models for top-of-atmosphere radiative flux estimation from the clouds and the Earth's radiant energy system instrument on the Terra satellite. part I: Methodology, *J. Atmos. Oceanic Technol.*, **22**, 338–351.

Lu, M.-L., G. Feingold, H. H. Jonsson, P. Y. Chuang, H. Gates, R. C. Flagan, and J. H. Seinfeld, 2008: Aerosol-cloud relationships in continental shallow cumulus, *J. Geophys. Res.*, **113**, D15201, doi:10.1029/2007JD009354.

Marshak, A., W. Wiscombe, A. Davis, L. Oreopoulos, and R. Cahalan, 1999: On the removal of the effect of horizontal fluxes in two-aircraft measurements of cloud absorption, *Q. J. R. Meteorol. Soc.*, **125**, 2153–2170, doi:10.1002/qj.49712555811.

Marshak, A., Y. Knyazikhin, K. D. Evans, and W. J. Wiscombe, 2004: The RED versus NIR plane to retrieve broken-cloud optical depth from ground-based measurements, *J. Atmos. Sci.*, **61** (15), 1911–1925.

Marshak, A., G. Wen, J. A. Coakley Jr., L. A. Remer, N. G. Loeb, and R. F. Cahalan, 2008: A simple model for the cloud adjacency effect and the apparent bluing of aerosols near clouds, *J. Geophys. Res.*, **113**, D14S17, doi:10.1029/2007JD009196.

Marshak, A., Y. Knyazikhin, J. C. Chiu, and W. J. Wiscombe, 2009: Spectral invariant behavior of zenith radiance around cloud edges observed by ARM SWS, *Geophys. Res. Lett.*, **36**, L16802, doi:10.1029/2009GL039366.

Mayer, B., 2009: Radiative transfer in the cloudy atmosphere, *Euro. Phys. J. Conf.*, **1**, 75–99, doi:10.1140/epjconf/e2009-00912-1.

Mayer, B., and A. Kylling, 2005: Technical note: The libRadtran software package for radiative transfer calculations—Description and examples of use, *Atmos. Chem. Phys.*, **5**, 1855–1877, doi:10.5194/acp-5-1855-2005.

McBride, P. J., K. S. Schmidt, P. Pilewskie, S. Lance, P. Minnis, K. M. Bedka, D. E. Wolfe, 2010: Cloud property retrievals from surface spectral transmittance and airborne spectral reflectance: Comparisons with satellite, microwave, and *in situ* observations during CalNex, *Presentation during fall meeting of the American Geophysical Union*; San Francisco, December 2010.

McBride, P. J., P. Pilewskie, K. S. Schmidt, S. Kittelman, and D. Wolfe, 2011: A spectral method for retrieving cloud optical thickness and effective radius from surface-based transmittance measurements, *Atmos. Chem. Phys. Discuss.*, **11**, 1053–1104, doi:10.5194/acpd-11-1053.

McFarlane, S. A., R. T. Marchand, and T. P. Ackerman, 2005: Retrieval of cloud phase and crystal habit from Multiangle Imaging Spectroradiometer (MISR) and Moderate Resolution Imaging Spectroradiometer (MODIS) data, *J. Geophys. Res.*, **110**, D14201, doi:10.1029/2004JD004831.

McGill, M. J., L. Li, W. D. Hart, G. M. Heymsfield, D. L. Hlavka, P. E. Racette, L. Tian, M. A. Vaughan, and D. M. Winker, 2004: Combined lidar-radar remote sensing: Initial results from CRYSTAL-FACE, *J. Geophys. Res.*, **109**, D07203, doi:10.1029/2003JD004030.

Mlawer, E. J., S. J. Taubman, P. D. Brown, M. J. Iacono, and S. A. Clough, 1997: Radiative transfer for inhomogeneous atmospheres: RRTM, a validated correlated-k model for the longwave, *J. Geophys. Res.*, **102**(D14), 16,630–16,682.

Molina, L. T., S. Madronich, J. S. Gaffney, E. Apel, B. de Foy, J. Fast, R. Ferrare, S. Herndon, J. L. Jimenez, B. Lamb, A. R. Osornio-Vargas, P. Russell, J. J. Schauer, P. S. Stevens, R. Volkamer, and M. Zavala, 2010: An overview of the MILAGRO 2006 Campaign: Mexico City emissions and their transport and transformation, *Atmos. Chem. Phys.*, **10**, 8697–8760, doi:10.5194/acp-10-8697-2010.

Nakajima, T., and M. King, 1990: Determination of the optical thickness and effective particle radius of clouds from reflected solar radiation measurements. Part I: Theory, *J. Atmos. Sci.*, **47**, 1878–1893.

Penning de Vries, M. J. M., S. Beirle, and T. Wagner, T., 2009: UV Aerosol Indices from SCIAMACHY: introducing the SCattering Index (SCI), *Atmos. Chem. Phys.*, **9**, 9555–9567, doi:10.5194/acp-9-9555-2009.

Pignatti, S., R. M. Cavalli, V. Cuomo, L. Fusilli, S. Pascucci, and M. Poscolieri, 2009: Evaluating Hyperion capability for land cover mapping in a fragmented ecosystem: Pollino National Park, Italy, *Remote Sens. Env.*, **113**, 622–634.

Pilewskie, P., and S. Twomey, 1987: Discrimination of ice from water in clouds by optical remote sensing, *Atmos. Res.*, **21**, 113–122.

Pilewskie, P., J. Pommier, R. Bergstrom, W. Gore, S. Howard, M. Rabbette, B. Schmid, P. V. Hobbs, and S. C. Tsay, 2003: Solar spectral radiative forcing during the Southern African Regional Science Initiative, *J. Geophys. Res.*, **108**(D13), 8486, doi:10.1029/2002JD002411.

Platnick, S., 2000: Vertical photon transport in cloud remote sensing problems, *J. Geophys. Res.*, **105**, 22,919–22,935, doi:10.1029/ 2000JD900333.

Platnick, S., 2001: Approximations for horizontal photon transport in cloud remote sensing problems, *J. Quant. Spectrosc. Radiat. Transfer*, **68**, 75–99, doi:10.1016/S0022-4073(00)00016-9.

Platt, U., and J. Stutz, 2008: *Differential Optical Absorption Spectroscopy, Principles and Applications*, Physics of Earth and Space Environments, Springer, Berlin.

Rabbette, M., and P. Pilewskie, 2002: Principal component analysis of Arctic solar irradiance spectra, *J. Geophys. Res.*, **107**(C10), 8049, doi:10.1029/2000JC000566.

Redemann, J., P. Pilewskie, P. B. Russell, J. M. Livingston, S. Howard, B. Schmid, J. Pommier, W. Gore, J. Eilers, and M. Wendisch, 2006: Airborne measurements of spectral direct aerosol radiative forcing in the Intercontinental Chemical Transport Experiment/Intercontinental Transport and Chemical Transformation of anthropogenic pollution, 2004, *J. Geophys. Res.*, **111**, D14210, doi:10.1029/2005JD006812.

Rothman, L., et al., 2005: The HITRAN 2004 molecular spectroscopic database, *J. Quant. Spectrosc. Radiat. Transfer*, **96**, 139–204, doi:10.1016/ j.jqsrt.2004.10.008.

Russell, P. B., J. M. Livingston, P. Hignett, S. Kinne, J. Wong, A. Chien, R. Bergstrom, and P. V. Hobbs, 1999: Aerosol-induced radiative flux changes off the United States mid-Atlantic coast: Comparison of values calculated from Sunphotometer and in situ data with those measured by airborne pyranometer, *J. Geophys. Res.*, **104**(D2), 2289–2307.

Russell, P. B., R. W. Bergstrom, Y. Shinozuka, A. D. Clarke, P. F. DeCarlo, J. L. Jimenez, J. M. Livingston, J. Redemann, O. Dubovik, and A. Strawa, 2010: Absorption Angstrom Exponent in AERONET and related data as an indicator of aerosol composition, *Atmos. Chem. Phys.*, **10**, 1155–1169, doi:10.5194/acp-10-1155-2010.

Schmidt, K. S., V. Venema, F. Di Giuseppe, R. Scheirer, M. Wendisch, and P. Pilewskie, 2007: Reproducing cloud microphysical and irradiance measurements using three 3D cloud generators, *Q. J. R. Meteorol. Soc.*, **133**, 765–780.

Schmidt, K. S., G. Feingold, P. Pilewskie, H. Jiang, O. Coddington, and M. Wendisch, 2009: Irradiance in polluted cumulus fields: Measured and modeled cloud–aerosol effects, *Geophys. Res. Lett.*, **36**, L07804, doi:10.1029/2008GL036848.

Schmidt, K. S., P. Pilewskie, R. Bergstrom, O. Coddington, J. Redemann, J. Livingston, P. Russell, E. Bierwirth, M. Wendisch, W. Gore, M. K. Dubey, and C. Mazzoleni, 2010a: A new method for deriving aerosol solar radiative forcing and its first application within MILAGRO/INTEX-B, *Atmos. Chem. Phys.*, **10**, 7829–7843, doi:10.5194/acp-10-7829-2010.

Schmidt, K. S., P. Pilewskie, B. Mayer, M. Wendisch, B. Kindel., S. Platnick, M. D. King, G. Wind, G. T. Arnold, L. Tian, G. Heymsfield, and H. Kalesse, 2010b: Apparent absorption of solar spectral irradiance in heterogeneous ice clouds, *J. Geophys. Res.*, **115**, D00J22, doi:10.1029/2009JD013124.

Shannon, C., and W. Weaver, 1949: The mathematical theory of communication, University of Illinois, Urbana.

Twomey, S., and T. Cocks, 1989: Remote sensing of cloud parameters from spectral reflectance in the near-infrared, *Beitr. Phys. Atmos.*, **62**, 172–179.

Venema, V., S. Meyer, S. G. Garcia, A. Kniffka, C. Simmer, S. Crewell, U. Löhnert, T. Trautmann, and A. Macke, 2006: Surrogate cloud fields generated with the iterative amplitude adapted Fourier transform algorithm. *Tellus*, **58B**, 104–120.

Vukicevic, T., O. Coddington, and P. Pilewskie, 2010: Characterizing the retrieval of cloud properties from optical remote sensing, *J. Geophys. Res.*, **115**, D20211, doi:10.1029/2009JD012830.

Wagner, T., S. Beirle, T. Deutschmann, E. Eigemeier, C. Frankenberg, M. Grzegorski, C. Liu, T. Marbach, U. Platt, and M. Penning de Vries, 2008: Monitoring of atmospheric trace gases, clouds, aerosols and surface properties from UV/vis/NIR satellite instruments, *J. Opt. A: Pure Appl. Opt.*, **10**, 104019 doi: 10.1088/1464-4258/10/10/104019.

Wen, G., A. Marshak, R. F. Cahalan, L. A. Remer, and R. G. Kleidman, 2007: 3-D aerosol-cloud radiative interaction observed in collocated MODIS and ASTER images of cumulus cloud fields, *J. Geophys. Res.*, **112**, D13204, doi:10.1029/2006JD008267.
Wendisch, M., D. Müller, D. Schell, and J. Heintzenberg, 2001: An airborne spectral albedometer with active horizontal stabilization, *J. Atmos. Oceanic Technol.*, **18**, 1856–1866.
Wood, R., and D. L. Hartmann, 2006: Spatial variability of liquid water path in marine low cloud: the importance of mesoscale cellular convection, *J. Climate*, **19**(9), 1748–1764.
Yang, P., L. Zhang, S. L. Nasiri, B. A. Baum, H.-L., Huang, M. D. King, and S. Platnick, 2007: Differences between collection 4 and 5 MODIS ice cloud optical/microphysical products and their impact on radiative forcing simulations, *IEEE Transactions on Geoscience and Remote Sensing*, **45**, 2886–2899.
Yoshida, Y., and Asano, S., 2005: Effects of the vertical profiles of cloud droplets and ice particles on the visible and near-infrared radiative properties of mixed-phase stratocumulus clouds, *J. Meteor. Soc. Japan*, **83**, 471–480.

7 The retrieval of snow characteristics from optical measurements

Alexander A. Kokhanovsky and Vladimir V. Rozanov

7.1 Introduction

Retrievals of snow grain size using ground/satellite optical measurements have been performed by several research groups (Bourdelles and Fily, 1993; Fily et al., 1997; Zege et al., 1998; Polonsky et al., 1999; Nolin and Dozier, 1993, 2000; Nolin and Liang, 2000; Painter et al., 1998, 2003; Stamnes et al., 2007; Hori et al., 2007; Zege et al., 2008; Lyapustin et al., 2009). The aim of this chapter is to discuss the information content of corresponding measurements and also to present the fast semi-analytical snow grain size retrieval algorithm.

Retrievals are based on the fact that the snow reflectance in the near-IR (say, above 800 nm) decreases for larger particles because larger grains are relatively more absorptive as compared to smaller grains. Actually the same physical mechanism is used for the optical sizing of droplets/grains in cloudy media. The retrievals for snow fields are complicated by the fact that corresponding pixels can be contaminated by forest, soil, vegetation, or slash. Also snow horizontal (e.g., sastrugi) and vertical (e.g., buried ice/dust layers) inhomogeneity can bias retrievals.

In addition, the determination of the concentration c of pollutants in snow is discussed. The value of c can be assessed from measurements in the UV and visible, where pure snow reflects almost 100% of incident radiation and polluted snow absorbs a significant portion of light. The effect of absorption leads to snow darkening, which can be accurately measured and used for the determination of concentration of pollutants (at least, for heavy pollution events).

7.2 Forward model

In most snow retrieval algorithms it is assumed that snow can be modeled as an ice cloud layer positioned on the ground. Therefore, snow reflectance can be derived from the solution of the corresponding radiative transfer equation. In particular, if the assumption of a semi-infinite snow layer is used, then the Ambartsumian nonlinear integral equation can be used for studies of radiative transfer in snow (Mishchenko et al., 1999). The parameters of this equation are the single scattering albedo ω_0 and phase function $p(\theta)$ (θ is the scattering angle). Alternatively, other radiative transfer solvers (e.g., SCIATRAN (www.iup.physik.uni-bremen.de/sciatran)) can be used assuming a large value of snow optical thickness (say, 5000). The single scattering albedo can be calculated as (Kokhanovsky and Nauss, 2005):

$$\omega_0 = 1 - \beta, \tag{1}$$

$$\beta = \beta_\infty(1 - \exp(-\alpha\ell)). \tag{2}$$

Here β is the probability of photon absorption (PPA), $\alpha = 4\pi\chi/\lambda$ is the temperature-dependent bulk ice absorption coefficient, χ is the imaginary part of the ice refractive index at the wavelength λ. This formula was obtained fitting geometrical optics results derived with the Monte Carlo code described by Macke et al. (1996). The value of β_∞ corresponds to the limiting case of an ice crystal which absorbs all radiation penetrating inside the particle ($\alpha\ell \to \infty$). It can be calculated using the model of spherical particles because total reflection from an impenetrable sphere and a randomly oriented non-spherical impenetrable convex particle coincides (van de Hulst, 1957). It follows that $\beta_\infty = 0.47$ at the refractive index $n = 1.31$ (for ice in the visible). The particle absorption length (PAL) ℓ is proportional (Kokhanovsky and Nauss, 2005) to the effective grain size (EGS) $a_{ef} = 3V/\Sigma$ (V is the average volume of grains and Σ is their average surface area):

$$\ell = Ka_{ef} \tag{3}$$

with the parameter K *depending on the shape of particles.* For weakly absorbing particles, it follows from Eqs. (2), (3) that:

$$\beta = \beta_\infty K\alpha a_{ef}. \tag{4}$$

We found using geometrical optics Monte Carlo simulations and fitting procedure implemented in ORIGIN that $K = 2.63$ for fractals and, therefore, $\beta = D\alpha a_{ef}$ with $D \approx 1.24$.

The phase function was modeled using the assumption of fractal grains (Macke et al., 1996) (see Fig. 7.1, where other possible snow phase functions are given as well). It is assumed that the phase function does not depend on the size of grains. This is a correct assumption in the geometrical optics domain, if particles are not absorbing (outside diffraction peak). The assumption of spherical particles is not very realistic and must be discarded (Tanikawa et al., 2006; Xie et al., 2006).

7.3 The information content of snow spectral reflectance

7.3.1 Theory

The radiance I over a snow field as detected on a satellite depends on the snow properties and also on atmospheric parameters in the propagation channel. The snow parameter of interest in this work is the effective grain size. The retrievals of EGS can be effected by the concentration of pollutants (CP) c. Therefore, it is of importance to derive both parameters simultaneously. So here we will study the sensitivity of the reflection function to the determination of both EGS and CP.

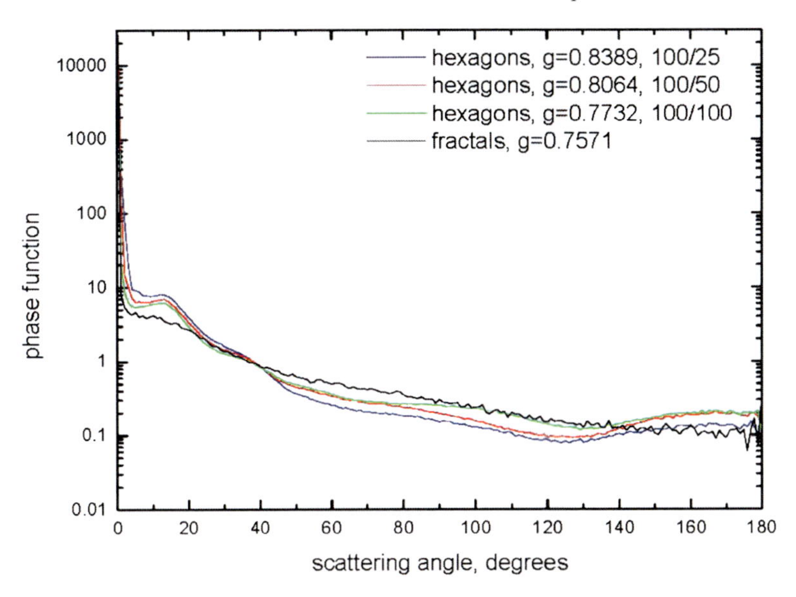

Fig. 7.1. The dependence of the phase function on the shape of particles. The following shapes were considered: hexagons with the length to the diameter ratios equal to 100/25, 100/50, 100/100 micrometers and fractal particles (Macke et al., 1996) at the wavelength 550 nm (Kokhanovsky et al., 2011). The diameter is defined as the distance between opposite sides of the hexagon. For all particles the surface was assumed to be rough in the calculations. The asymmetry parameter g is given for each curve. It follows that these diverse shapes do not produce very different phase functions as is the case for spheres, where the reduced scattering in the side-scattering region (around 90°) takes place. Also glories and rainbow characteristics of spheres are not observed for irregularly shaped particles.

The derivatives of the snow reflectance defined as

$$R = \frac{\pi I}{\mu_0 E_0} \tag{5}$$

($\mu_0 = \cos\vartheta_0, \vartheta_0$ is the solar zenith angle (SZA), E_0 is the incident light irradiance) with respect to these parameters are given by:

$$D_a = \frac{\partial R}{\partial a_{ef}}, \quad D_c = \frac{\partial R}{\partial c}. \tag{6}$$

They help to understand if given measurements can be used to retrieve the pair (a_{ef}, c). Clearly, derivatives depend on the viewing and illumination geometry (solar zenith angle (SZA) ϑ_0, viewing zenith angle (VZA) ϑ, and the relative azimuthal angle (RAA) φ), the spectral channel, values of (a_{ef}, c), and also on the atmospheric conditions (primarily through the aerosol optical thickness (AOT) τ). So, quite generally, we can write:

$$D = f\big(\vartheta_0, \vartheta, \varphi, \lambda, a_{ef}, c, \tau\big). \tag{7}$$

The task of this section is to understand how the derivatives D_a and D_c are influenced by various parameters given in Eq. (7). For this we use the software code

SCIATRAN (http://www.iup.uni-bremen.de/sciatran/). The derivatives are calculated through the following chain of equations.

First of all the weighting function (WF) W is introduced. We define it as (e.g., in the case of WF, W_a for a_{ef} of a homogeneous snow layer):

$$W_a = \frac{\partial R}{\partial \ln a_{ef}} = a_{ef} D_a. \tag{8}$$

Clearly, this is a dimensionless quantity. Then, it follows, e.g., for the reflectance function at the effective radius a_{ef}:

$$R(a_{ef}) = R(\bar{a}_{ef}) + [a_{ef} - \bar{a}_{ef}] W_a / \bar{a}_{ef}, \tag{9}$$

if *a priori* assumed radius $\bar{a}_{ef}$ is close to a_{ef} (so the linear approximation is valid). Clearly, if $W_a = 0$, then the reflectance is not sensitive to a_{ef}. Similar equations can be written for WFs with respect to the concentration of impurities (W_c) and also AOT (W_τ). There are different ways to calculate weighting functions. One possibility is the numerical calculation of ratios $M = \Delta R / \Delta(\ln x)$, where x is equal to a_{ef}, c, or τ depending on the case considered. In SCIATRAN yet another approach for the calculation of derivatives is followed. It is faster as compared to the calculation of ratios M and also more accurate.

In particular, it is assumed that the variation of the reflectance δR due to the variation of the effective radius profile $\delta a_{ef}(z)$ inside the snow layer of the thickness H can be presented in the following form:

$$\delta R(\lambda) = \int_0^1 w_a(\lambda, z) \delta a_{ef}(z)\, dz. \tag{10}$$

Here z is the vertical coordinate divided by the thickness of the layer$H$. It follows that the information on the function $w_a(\lambda, z)$ is of a great importance for understanding how changes in the profile $a_{ef}(z)$ influence the variation in reflectance. The WF W_a is related to $w(\lambda, z)$ via the following equation:

$$W_a(\lambda, z) = w_a(\lambda, z) a_{ef}(z). \tag{11}$$

Then it follows that:

$$\delta R(\lambda) = \int_0^1 W_a(\lambda, z)\, [\delta a_{ef}(z) / a_{ef}(z)]\, dz \tag{12}$$

or

$$\delta R(\lambda) = \sum_{k=1}^{N_k} J_a(\lambda, z_k) \frac{a_{ef}(z_k) - \bar{a}_{ef}(z_k)}{\bar{a}_{ef}(z_k)}, \tag{13}$$

where the summation is performed for the number of layers N_k inside of snow layer specified in the input of SCIATRAN and

$$J_a(\lambda, z_k) = W_a(\lambda, z_k) \Delta z_k \tag{14}$$

are corresponding Jacobians related to the sub-layer of thickness Δz_k. For a homogeneous layer it follows that:

$$\delta R(\lambda) = [\delta a_{ef}/a_{ef}]\, W_a(\lambda) \tag{15}$$

and we return to the same expression as written above:

$$R(a_{ef}) = R(\bar{a}_{ef}) + \Delta R = R(\bar{a}_{ef}) + [a_{ef} - \bar{a}_{ef}]\, W_a/\bar{a}_{ef}. \tag{16}$$

The WF $W_a(\lambda, z_k)$ contains information not only on the dependence of R on a_{ef} but also on the sensitivity of the reflectance to the changes in the radii of grains at different layers inside the snow.

The derivatives

$$W_a(\lambda) = \sum_{k=1}^{N_k} J_a(\lambda, z_k) \tag{17}$$

and also Jacobians $J_a(\lambda, z_k)$ are the main parameters discussed in the next section. The corresponding derivatives and Jacobians with respect to the concentration of pollutants and AOT are also considered.

As follows from Eq. (13), W_a in Eq. (17) gives the change in the reflectance (δR) if the change in the radius is equal to 100%. The technique to derive $w_a(\lambda, z)$ using the solution of direct and adjoint radiative transfer equations is described by Rozanov et al. (2007).

7.3.2 Results

The results of numerical experiments on the sensitivity studies are shown in Figs. 7.2–7.9. Let us analyze them now. All results were obtained using SCIATRAN and assuming that snow can be modeled as an ice cloud with the optical thickness 5000 at the ground level (with fractal phase function shown in Fig. 7.1). It was assumed that snow impurities (soot) are present in the form of Rayleigh scatterers and they influence only absorption and not scattering processes in a snow layer. The LOWTRAN aerosol maritime model implemented in SCIATRAN with aerosol optical thickness $\tau(550\text{ nm}) = 0.05$ was used. Also molecular scattering (but not absorption) has been taken into account. SCIATRAN is able to simulate satellite signals to account for the gaseous absorption. However, this was not needed for this work because only channels almost free of gaseous absorption have been selected.

We show the dependence of Jacobians $J_c(\lambda, z)$ for soot concentration on the distance from the snow bottom for several wavelengths in Fig. 7.2. The top of snow layer is located at 1 m height. It follows that Jacobians are different from zero only in the upper snow layers. They are about zero at depths 20 cm (and larger) from the top in the visible. Therefore, the concentration of pollutants at very deep layers cannot be retrieved from satellite observations. The maximum of the sensitivity is at some distance from the top of the layer and then the sensitivity decreases with the distance from the top. Most sensitivity comes from the upper 5 cm of snow, if the visible channels are used. The penetration depth depends on the grain size, being greater for smaller grains. The sensitivity to soot concentration decreases with the wavelength and it happens more rapidly for larger snow grains (see, e.g., the blue

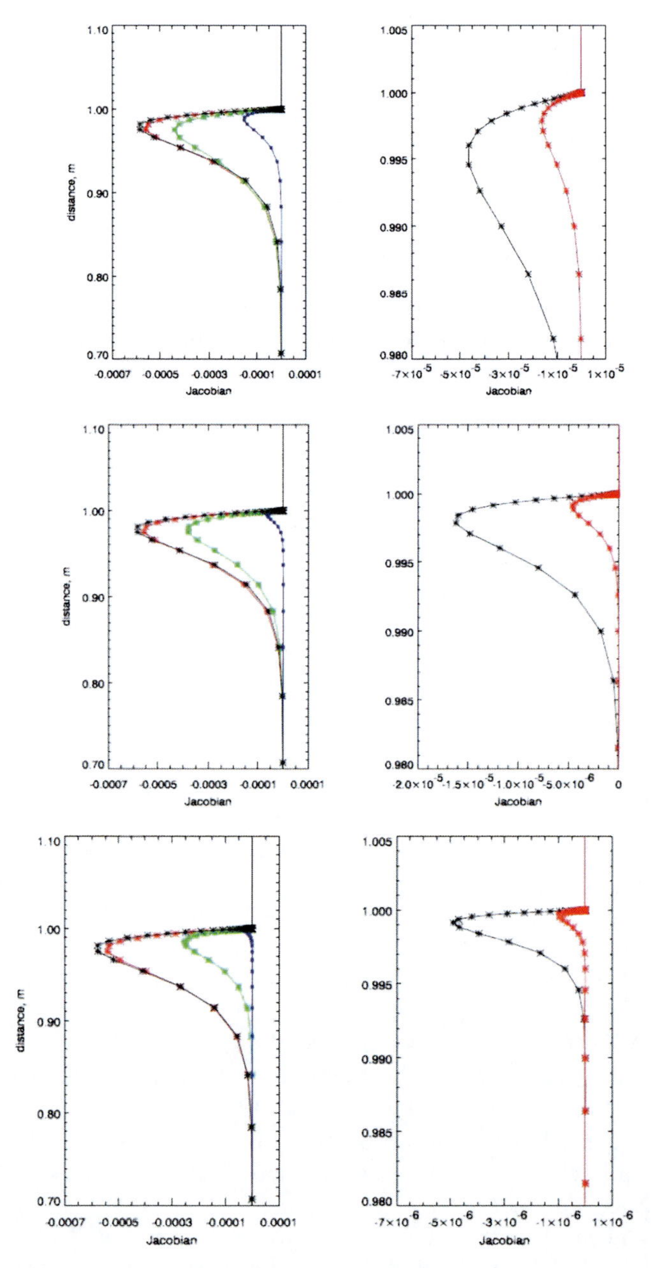

Fig. 7.2. Dependence of the Jacobian for soot concentration $J_c(\lambda, z)$ on the distance from the snow bottom at the wavelengths 400, 443, 555, 670 nm (left, larger Jacobians at maximum correspond to a smaller wavelength) and at wavelengths 865 and 1029 nm (right, larger Jacobians at maximum correspond to a smaller wavelength). The LOWTRAN aerosol model with the aerosol optical thickness (AOT) equal to 0.05 was used. The snow geometrical thickness is equal to 1 m and the length of side of fractal particles is equal to 50 micrometers (upper panel), 300 micrometers (middle panel), 750 micrometers (lower panel). The concentration of soot is equal to 10^{-8}. The solar zenith angle (SZA) is equal to 60 degrees and the observation is at the nadir direction.

line in Fig. 7.2). The main conclusion is that the shortest possible wavelength must be used for the soot concentration retrievals (e.g., at 400 nm). Then the dependence of retrievals on the grain size can be largely ignored (see Fig. 7.2).

Figure 7.3 is similar to Fig. 7.2 but now the Jacobians for the effective radius of snow grains are given. The green curves correspond to the measurements at 1240 nm. The upper panel of the figure corresponds to the case of $a_{ef} = 50\ \mu$m. It follows that the use of 1240 nm channel is preferable (at least for the considered geometry). The channel at 1020 nm is the second choice (red line on the left plots). Channels located at 0.865, 1.61 and 2.2 μm can be used as well. The channel at 2.2 μm has slightly better performance (especially for large particles) as compared to the channel at 1.61 μm. The radiance in the 865–1240 nm range is sensitive just to the snow properties at the top (e.g., 5–10 cm from the top depending on the wavelength and also the size of particles; see Fig. 7.3, left panels). The radiance at 1610 nm and 2190 nm is sensitive to changes of the crystals' sizes only at a very top of the layer. As a matter of fact, the corresponding penetration depth is smaller than the EGS at these wavelengths (see right panels in Fig. 7.3). Therefore, retrievals in the spectral range 865–1240 nm range are most preferable with somewhat larger penetration depths (but smaller sensitivity) at 865 nm.

The dependence of the derivatives W_a, W_c, and W_τ on the wavelength is given in Fig. 7.4. The value of W_a is given by Eq. (17). The derivatives with respect to the concentration of a pollutant c and the aerosol optical thickness τ are defined in a similar way as for the value of a_{ef}. It follows from the analysis of this figure that the sensitivity to the soot concentration in snow disappears for larger wavelengths. This is generally the case for the aerosol optical thickness as well. The behavior of the derivative with respect to the grain size (at a fixed wavelength) is non-monotonous. At $a_{ef} = 50\ \mu$m, the greatest sensitivity comes from the longer wavelengths. For the sizes of 300 and 750 μm, the greatest sensitivity comes from the wavelengths in the middle of the spectral interval studied (1.02 and 1.24 μm). Therefore, there is no such simple linear relationship to the wavelength in the sensitivity as in the case of the soot concentration. This is due to the fact that the influence of the size of grains on the reflectivity is small, both at low values of the absorption parameter $b = \alpha a_{ef}$ and also at high values of this parameter. Therefore, the optimum is located at some middle value of absorption, which occurs at 1240 nm for most practical situations. This also follows from the asymptotic radiative transfer theory. As it is shown by Kokhanovsky and Zege (2004) (see also the next section): $R \sim \exp(-\nu\sqrt{b})$ (ν is the parameter not depending on b) in the limit of weak light absorption by snow ($\beta \to 0$). This means that $W_a \sim \sqrt{b}\exp(-\nu\sqrt{b})$ and $W_a \to 0$ at high and low values of b as it was discussed above.

It follows that $W_\tau(\lambda) \ll W_a(\lambda)$ and, therefore, generally the information on AOT is of no importance for the grain size retrievals. This conclusion is valid only at AOT $= 0.05$ usual for clear polar conditions (Tomasi et al., 2007). The case of larger aerosol load (Arctic haze, for example) will be considered later on. Also we find that $W_c(\lambda) > W_\tau(\lambda)$ at channels 400 and 443 nm. Therefore, these channels can be used for the soot concentration determination. However, the influence of uncertainty in the value of AOT on the retrieval of soot concentration is much larger as compared to the case of a_{ef} retrieval. We conclude that without accurate retrieval of AOT, the retrieval of soot concentration in snow is not possible. The

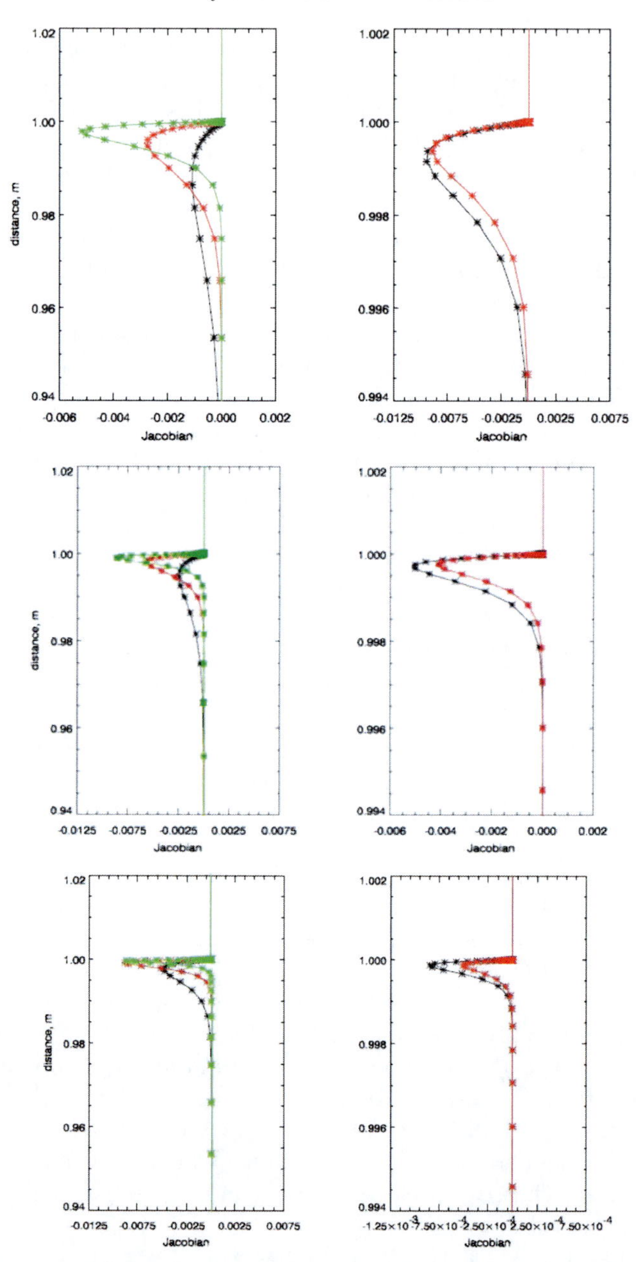

Fig. 7.3. Dependence of the Jacobian $J_a(\lambda, z)$ for effective radius of ice crystals in snow on the distance from the snow bottom at the wavelengths 865 nm (black curve), 1020 nm (red curve), and 1240 nm (green curve) (left panel). The same except at the wavelengths 1610 nm (red) and 2190 nm (black) (right panel). The LOWTRAN aerosol model with the aerosol optical thickness (AOT) equal to 0.05 was used. The snow geometrical thickness is equal to 1 m and the length of side of fractal particles is equal to 50 micrometers (upper panel), 300 micrometers (middle panel), 750 micrometers (lower panel). The concentration of soot is equal to 10^{-8}. The solar zenith angle (SZA) is equal to $60°$ and the observation is at the nadir direction.

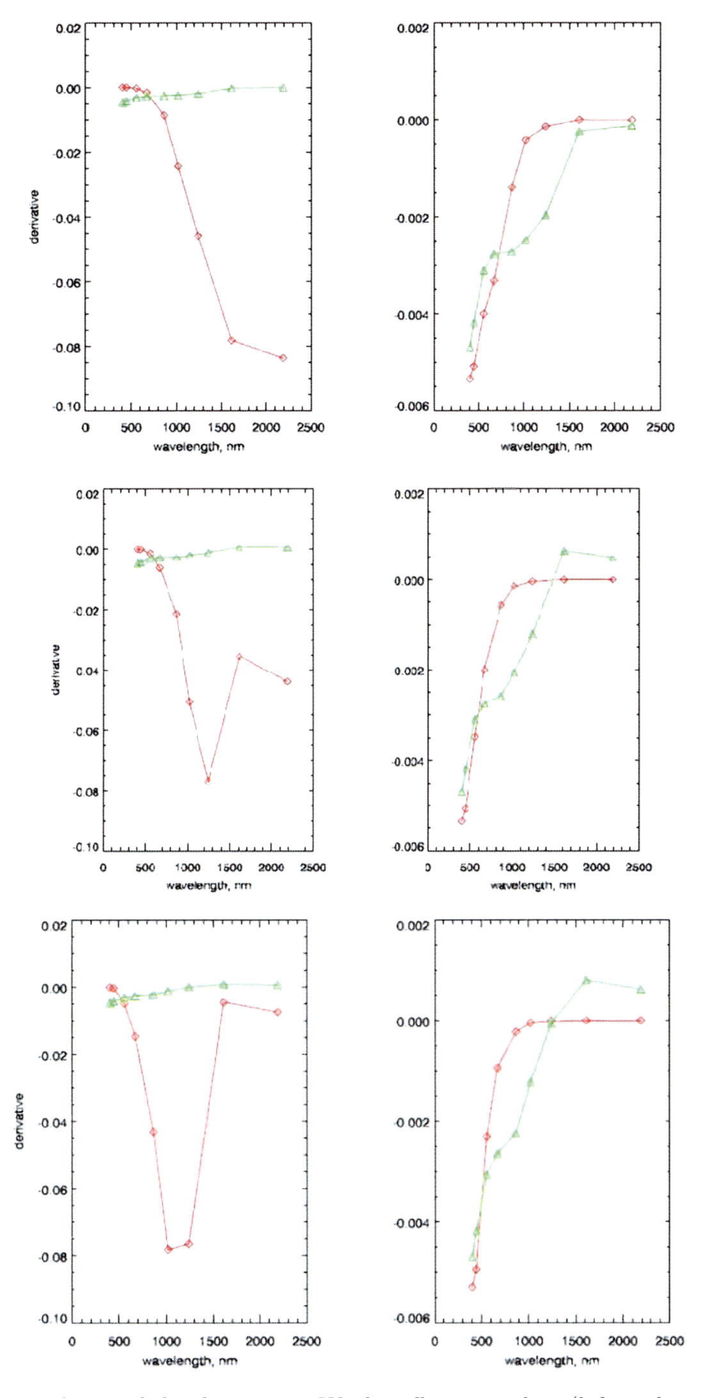

Fig. 7.4. Dependence of the derivatives W_a for effective radius (left, red curve) and W_c for the soot concentration (right, red curve) on the wavelength (AOT(555 nm) = 0.05). Other parameters are the same as in Fig. 7.3. Green lines give derivatives W_τ for the aerosol optical thickness at 555 nm.

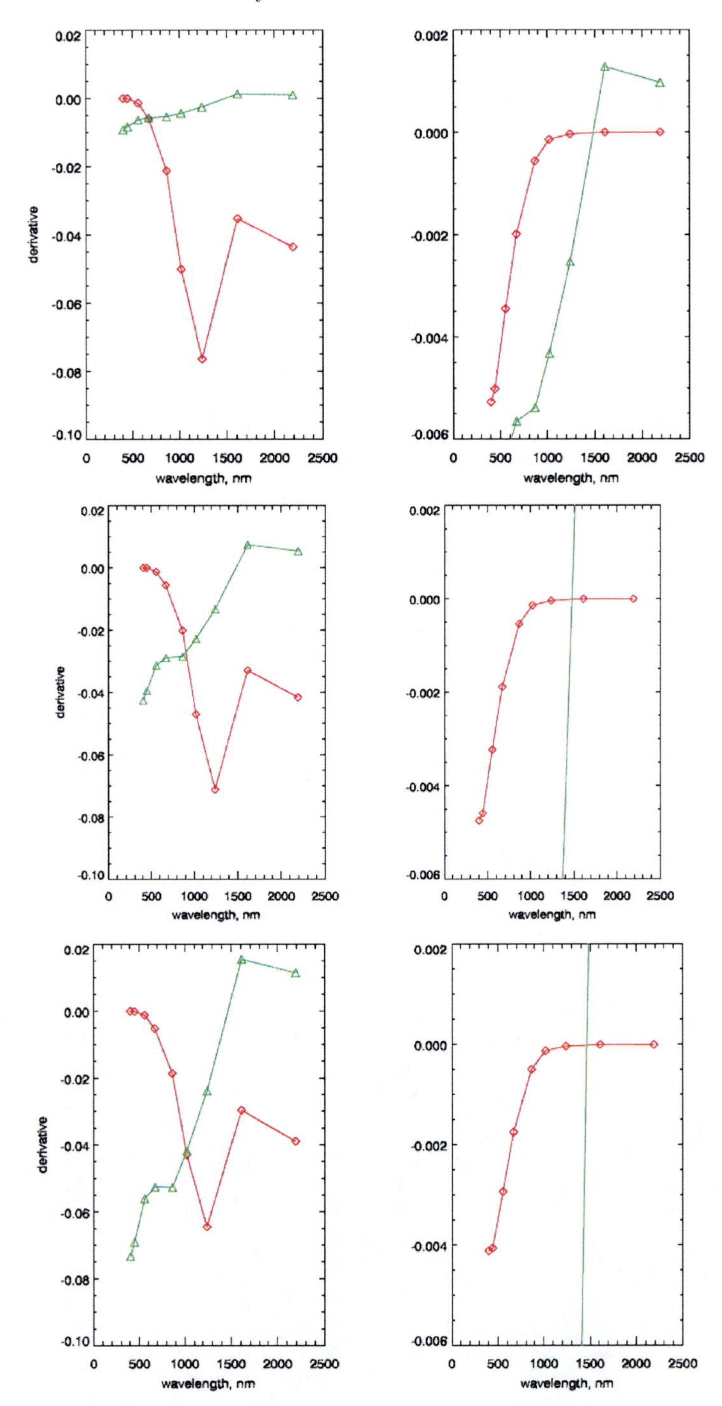

Fig. 7.5. The same as in Fig. 7.4 except at AOT = 0.1 (upper panel), 0.5 (middle panel), 1.0 (lower panel). The length side of fractal particles is 300 micrometers.

derivatives in Fig. 7.4 give the change in reflectance due to 100% change of a_{ef}, c, τ, respectively. We conclude that the signal at AOT = 0.05 is influenced in the similar way by the soot concentration in snow and the aerosol optical thickness. So accurate retrieval of AOT is of paramount importance for soot concentration determination. This is also confirmed by data shown in Fig. 7.5, which is similar to Fig. 7.4 except AOT = 0.1 (upper figures), AOT = 0.5 (middle figures), AOT = 1.0 (lower figures) are used in calculations. In all these cases (see figures on the right side) $W_c < W_\tau$ and, therefore, retrieval of soot concentration (at least at the level studied, $c = 10^{-8}$) is hardly possible without accurate information on aerosol optical thickness, which is difficult to retrieve over snow. As a matter of fact, both soot in snow and suspended aerosol particles can lead to the decrease of the registered signal and there is no technique in place to separate these two equally important contributions.

In Figs. 7.6–7.8 the dependence of derivatives on the viewing zenith angle, the solar zenith angle, and the relative azimuth, respectively, is presented. The main conclusion is that the viewing geometry influences grain size retrieval to a lesser extent as compared to the retrieval of soot concentration. Generally, retrieval of the

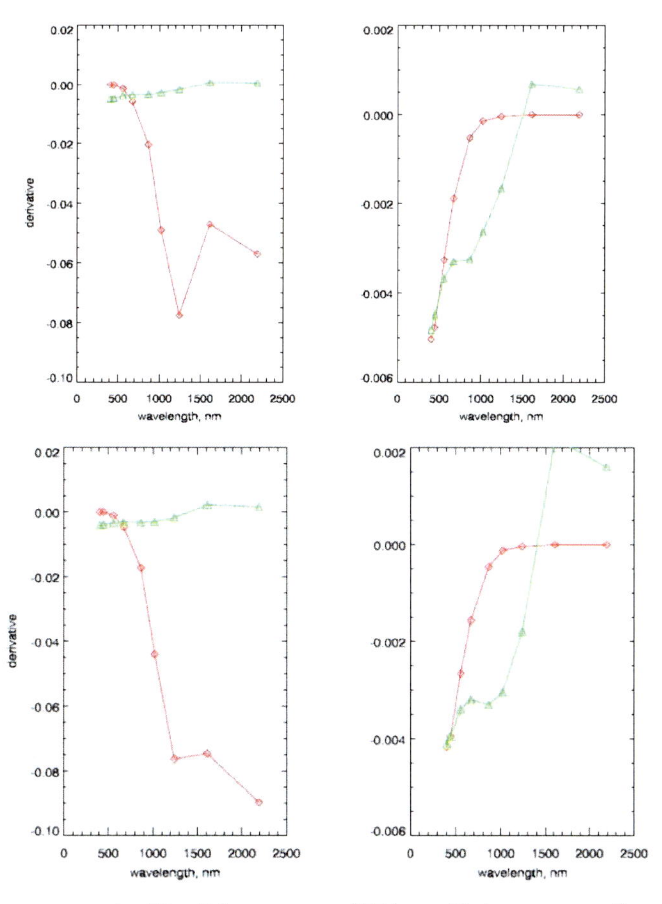

Fig. 7.6. The same as in Fig. 7.5 except at VZA = 25 (upper panel) and 50° (lower panel). The solar zenith angle is 60° and the relative azimuth is 0°.

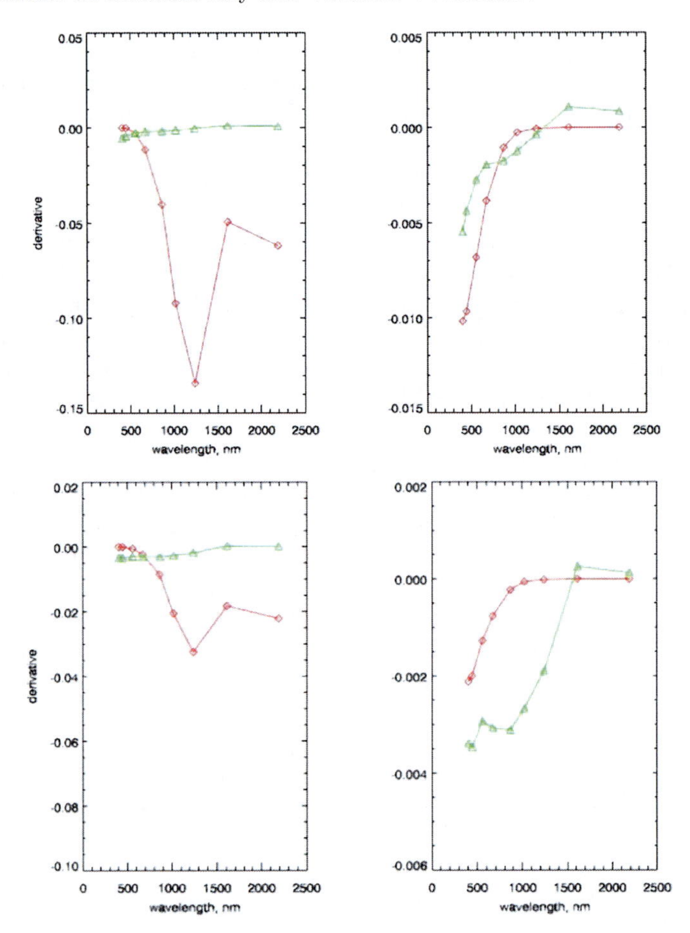

Fig. 7.7. The same as in Fig. 7.6 except at SZA = 45 (upper panel) and 75° (lower panel).

soot concentration is more problematic at larger VZAs due to long paths of light in the atmosphere. One can notice (see Fig. 7.6) that the increase of VZA leads to the increase in the sensitivity to the EGS at wavelengths 1.6 and 2.1 μm. This is due to the fact that the snow brightness increases with VZA at these channels. Figure 7.7 shows the sensitivity of derivatives to the solar zenith angle. It follows that the sensitivity to the grain size increases for the high Sun. However, in the regions, where there is permanent snow cover and generally, where snowfalls occur, the Sun is low and the sensitivity to the grain size (and also the soot concentration) decreases considerably. For the solar zenith angle 45° (and also for smaller SZAs), there is an enhanced sensitivity of measurements to the soot concentration. However, such high solar zenith angles usually do not occur in the snow-covered regions. The sensitivity to the soot concentration drops significantly at SZA = 75°. As follows from Fig. 7.8, the value of azimuth is also of importance as far as the sensitivity is of concern. The sensitivity is more pronounced at the relative azimuthal angle equal to zero degrees at the wavelengths 1.6 and 2.1 μm. For shorter wavelengths, the influence of the relative azimuthal angle on the sensitivity of retrievals is quite low.

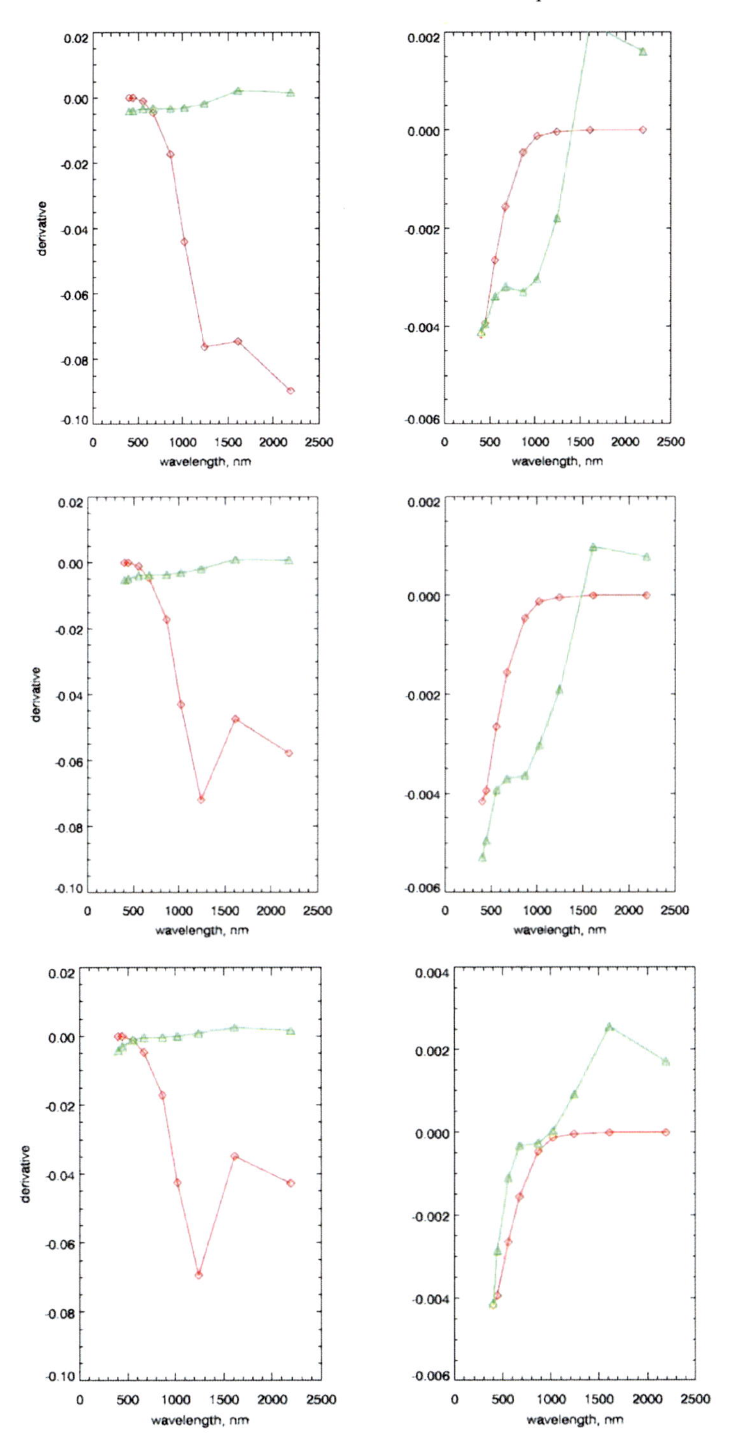

Fig. 7.8. The same as in Fig. 7.7 except at VZA = 50°, SZA = 60°, azimuths 0° (upper panel), 90° (middle panel) and 180° (lower panel).

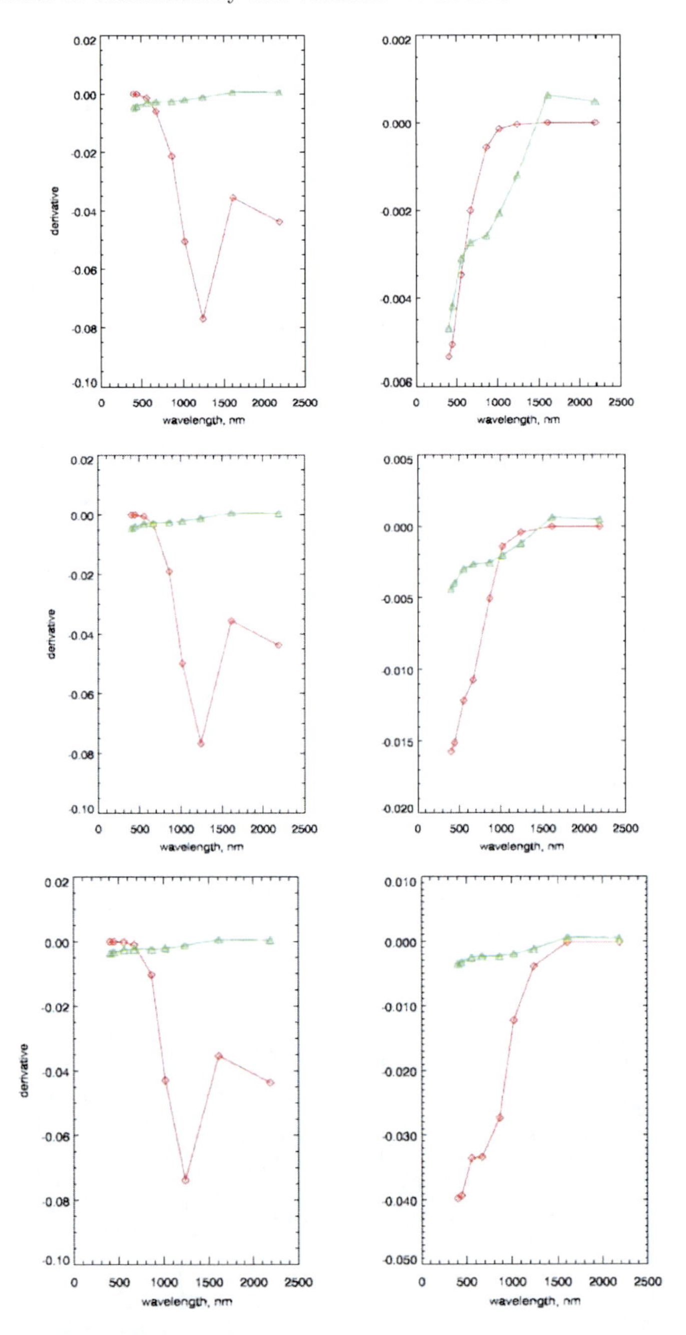

Fig. 7.9. Dependence of the derivatives W_a for effective radius (left, red curve) and W_c for the soot volumetric concentration (right, red curve) on the wavelength. Other parameters are as in Fig. 7.3. Green lines give derivatives W_τ for the aerosol optical thickness at 555 nm. The concentration of soot is equal to 10^{-8} (upper panel), 10^{-7} (middle panel), 10^{-6} (lower panel). The solar zenith angle (SZA) is equal to 60° and the observation is at the nadir direction. The length of side of fractal particles is equal to 300 micrometers.

All results shown above correspond to the background level of soot concentration ($c = 10^{-8}$). Clearly, for higher soot concentrations, the sensitivity of reflectance to the soot concentration is greater (see Fig. 7.9) and, therefore, c can be retrieved even if atmospheric correction is performed with considerable errors.

7.4 Retrieval algorithm: FORCE

7.4.1 Theory

The developed retrieval algorithm for the EGS determination is based on the look-up table (LUT) approach. In particular, the Fourier components of the reflection function in the visible (for a non-absorbing snow) are tabulated using the code developed by Mishchenko et al. (1999). The code solves the Ambartsumian nonlinear integral equation for the harmonics $R^m(\mu, \mu_0)$ of the reflection function. These harmonics are stored in LUTs. Then the reflection function at any relative azimuth angle is found as

$$R(\mu, \mu_0, \varphi) = R^0(\mu, \mu_0) + 2 \sum_{m=1}^{M_{\max}} R^m(\mu, \mu_0) \cos(m\varphi). \quad (18)$$

Here $\mu = \cos\vartheta$ and the value of $M_{\max}$ is chosen from the condition that the next term does not contribute more than 0.01% in the sum (18). In principle one more dimension (for a given phase function) in this LUT is needed and this is the dimension of the single scattering albedo.

We use the following representation valid as $\omega_0 \to 1$ (Zege et al., 1991; Kokhanovsky, 2006):

$$R(\mu, \mu_0, \varphi) = R_0(\mu, \mu_0, \varphi) A^{f(\mu, \mu_0, \varphi)}, \quad (19)$$

where

$$A = \exp\left\{-4s/\sqrt{3}\right\}, \; s = \sqrt{\frac{1-\omega_0}{1-g\omega_0}}, \; f = \frac{u(\mu_0)u(\mu)}{R_0(\mu, \mu_0, \varphi)}, \; u(\mu) = \frac{3}{7}(1+2\mu). \quad (20)$$

Here R_0 is the reflection function of a semi-infinite snow layer under the assumption that the single scattering albedo is equal to one. It is calculated using Eq. (18). For pure snow, the experimentally measured value of R_0 (say, at 443 nm) can be used. This speeds up the retrievals.

The only approximation as compared to the exact RT calculations involved is the use of the term A^f in Eq. (19) to characterize light absorption by snow. The accuracy of this approximation is studied in Figs. 7.10–7.13. It follows that errors are below 6% as compared to SCIATRAN calculations at the wavelengths 0.52–1.24 μm and SZA $= 54°$ for all azimuthal angles. So these short wavelengths will be used here for the inverse problem solution. In the case of MERIS onboard ENVISAT instrument (http://envisat.esa.int/instruments/meris/) channels 443 and 865 nm, the errors are smaller than 2% at the VZA $< 40°$ typical for MERIS observations. This is well inside the calibration error of MERIS.

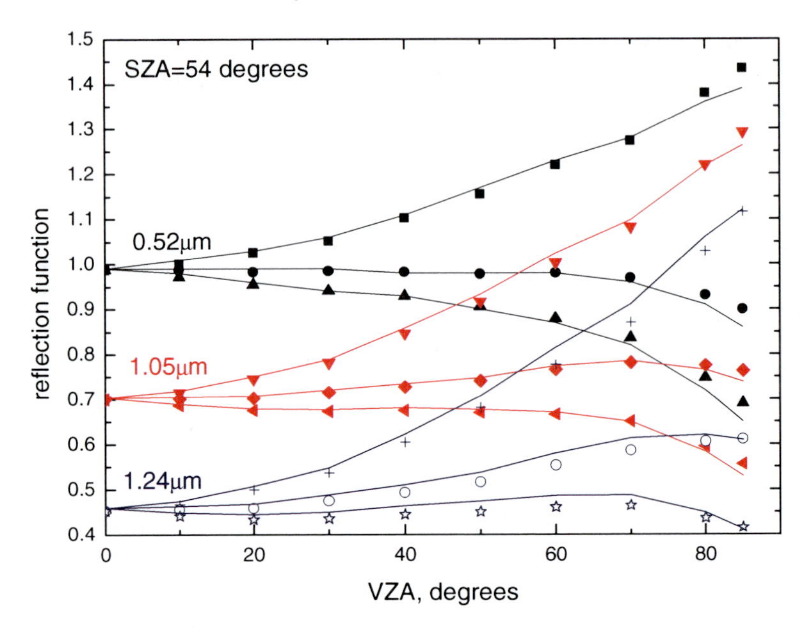

Fig. 7.10. The dependence of the reflection function on the viewing zenith angle (lines: approximate theory with the use of LUTs for the reflection function of a non-absorbing semi-infinite snow; points: SCIATRAN calculations) for selected wavelengths and the SZA equal to 54°. Lower lines for each wavelength correspond to the larger relative azimuthal angle equal to 0°, 90°, and 180°. The results are obtained using LUT for the function $R_0(\mu, \mu_0, \varphi)$ through the application of the nonlinear integral equation of Ambartsumian (Mishchenko et al., 1999).

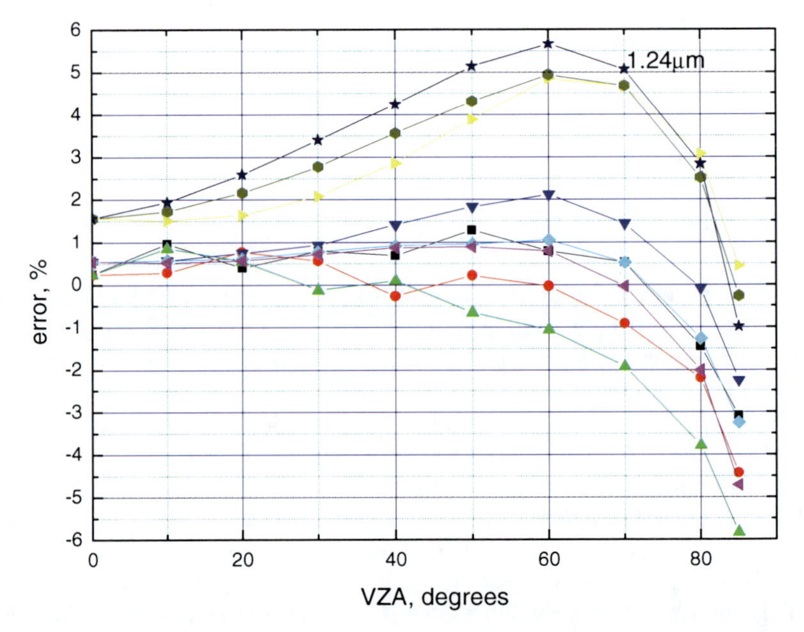

Fig. 7.11. Error of the approximation shown in Fig. 7.10. The upper curves correspond to the calculations at 1.24 μm. Lower curves correspond to calculations at 0.52 and 1.05 μm.

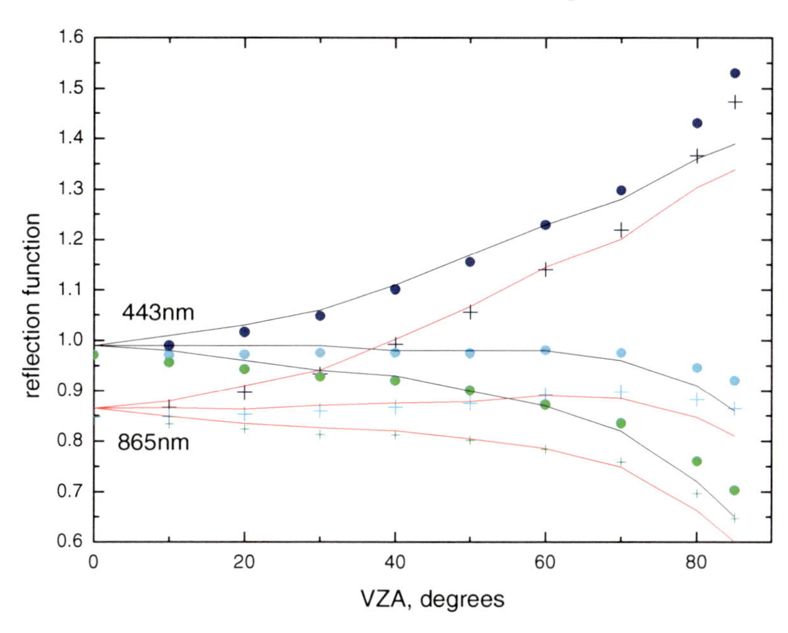

Fig. 7.12. The difference between reflectances at three azimuthal angles (0°, 90°, 180° from top down) at wavelengths 443 and 865 nm according to SCIATRAN (points) and the approximation (lines). The SZA is equal to 54°. SSA = 1 at the wavelength 443 nm. SSA = 0.9994 at the wavelength 865 nm. LUTs are used.

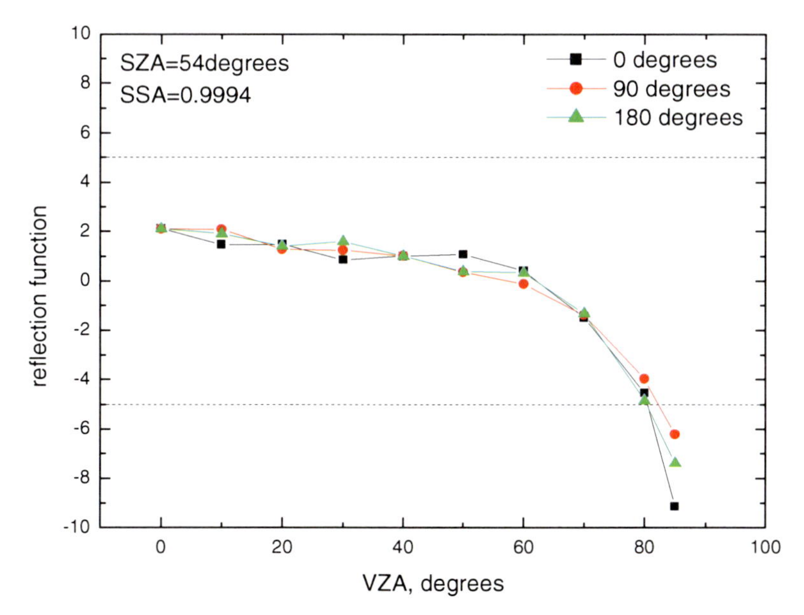

Fig. 7.13. Errors corresponding to the results shown in Fig. 7.12 at SSA = 0.9994 ($\lambda = 865$ nm).

MERIS does not have channels above 0.9 μm and, therefore, the approximation proposed here is very relevant to the interpretation of MERIS observations over snow fields. This is due to the fact that the snow albedo (and the accuracy of the approximation) increases for shorter wavelengths. The forward model itself (e.g., the flat snow surface assumption) and also errors of atmospheric correction introduce much larger uncertaintis as compared to differences between approximate and exact theories.

Eq. (19) can be used for the analytical determination of ω_0 and, therefore, a_{ef} from the snow reflection function measurements. As a matter of fact in the case of small grains and the MERIS wavelengths, even simpler approximation can be used. This approximation follows from Eq. (19) as $\omega_0 \to 1$:

$$R(\mu, \mu_0, \varphi) = R_0(\mu, \mu_0, \varphi) - \frac{4s}{\sqrt{3}} u(\mu) u(\mu_0). \tag{21}$$

Eq. (19) also enables the determination of the snow spectral albedo:

$$A(\lambda) = (R_{mes}(\lambda)/R_0)^{1/f} \tag{22}$$

from measurements of the spectral reflection function just at one observation geometry. It is assumed that the atmospheric correction has already been performed and the influence of atmosphere is removed from the value of $R_{mes}(\lambda)$.

One can write for channels 1(0.443 μm) and 2(0.865 μm) in the approximation under study:

$$R_1 = R_0 \exp(-\gamma\sqrt{\beta_1}), \tag{23}$$

$$R_2 = R_0 \exp(-\gamma\sqrt{\beta_2}), \tag{24}$$

where indices 1 and 2 signify the channel,

$$\gamma = \frac{4f}{\sqrt{3(1 - g\omega_0)}}. \tag{25}$$

We will neglect the difference of ω_0 from 1.0 in the denominator of Eq. (25).

Here we assume that there is some light absorption by snow even in the visible (e.g., due to soot). The value of probability of photon absorption can be written as

$$\beta = \frac{N_i C_{abs,i} + N_s C_{abs,s}}{N_i C_{ext,i} + N_s C_{ext,s}}. \tag{26}$$

Here

$$N_s = \frac{c_s}{\bar{V}_s} \tag{27}$$

is the number concentration of soot particles, $\bar{V}_s$ is their average volume, c_s is the volumetric concentration of soot (the fraction of volume filled by soot), $C_{abs,\alpha}$ is the average absorption cross-section of soot particles, $C_{ext,s}$ is the average extinction cross-section of soot particles. Parameters with the index 'i' have the same meaning as described above except for ice.

We will neglect the contribution of soot to the general light extinction in snow. Then it follows:

$$\beta = \beta_i + \beta_s, \tag{28}$$

where $\beta_i = C_{abs,i}/C_{ext,i}$ is given by Eq. (2) and

$$\beta_s = \frac{\bar{V}_i c_s C_{abs,s}}{\bar{V}_s c_i C_{ext,i}} \quad . \tag{29}$$

The average extinction cross-section of the ice grains $C_{ext,i}$ can be estimated as follows (Kohkanovsky, 2006):

$$C_{ext,i} = \frac{\bar{\Sigma}_i}{2}. \tag{30}$$

Here $\bar{\Sigma}_i$ is the average surface area of grains. Taking into account that $C_{ext,i}/\bar{V}_i = 1.5a_{ef}^{-1}$ in this approximation and also assuming that $C_{abs,s}/\bar{V}_s = B\alpha_s$, which is true in the Rayleigh domain for small soot particles ($B = 0.84$ at the soot refractive index $n = 1.75$ (van de Hulst, 1957), $\alpha_s = 4\pi\chi_s/\lambda$, $\chi_s = 0.46$), we derive:

$$\beta_s = \frac{2}{3} B c \alpha_s a_{ef} \tag{31}$$

where

$$c = c_s/c_i \tag{32}$$

is the relative soot concentration.

The mass absorption coefficient of soot $\sigma_{abs} = C_{abs}/\rho_s \bar{V}_s$ is equal to $B\alpha_s/\rho_s$ in the considered approximation. Here ρ_s is the soot density. Assuming that $B = 0.84$, $\chi_s = 0.46$, $\lambda = 443$ nm, $\rho_s = 1$ g/cm^3, one derives: $\sigma_{abs} = 8.4$ g/m^2, which is close to the modern estimates of this parameter (7.5 ± 1.2 m^2/g (Bond and Bergstrom, 2006; Flanner et al., 2007)).

Therefore, we can write:

$$R_1 = R_0 \exp\left[-\gamma\sqrt{\frac{2}{3} B \alpha_{s,1} c a_{ef}}\right], \tag{33}$$

$$R_2 = R_0 \exp(-\gamma\sqrt{\beta_{i,2} + \frac{2}{3} B \alpha_{s,2} c a_{ef}}). \tag{34}$$

Here we neglected light absorption by ice at the first wavelength. These two equations can be used to find both the size of ice crystals and the concentration of pollutants. It follows from the first equation for $X = ca_{ef}$:

$$X = \frac{3}{2B\gamma^2\alpha_{s,1}} \ln^2 r_1 \tag{35}$$

and, therefore,

$$\beta_2 = \frac{\ln^2 r_2}{\gamma^2} - \frac{2}{3} B X \alpha_{s,2}, \tag{36}$$

where X is determined from Eq. (33). Here we introduced the normalized reflectance: $r_i \equiv R_i/R_0$. The EGS can be found from Eqs. (2), (3), (36):

$$a_{ef} = K\alpha_{i,2}^{-1} \ln \left[\frac{\beta_\infty}{\beta_\infty - \beta_2} \right]. \tag{37}$$

Then the concentration of soot is determined as $c = X/a_{ef}$. In practice, one measures the concentration of soot as the fraction of soot mass in a given mass of snow $c_f = c_s\rho_s/c_i\rho_i$, where ρ_s is the density of soot and ρ_i is the density of ice. Therefore, for the transformation of the satellite-derived c to the ground measured values of c_f, one must use the multiplier $\eta = \rho_s/\rho_i$:

$$c_f = \eta c. \tag{38}$$

We will assume that $\eta \equiv 1$ in this study. It is known that $\rho_i = 0.917$ g/cm^3. The density of soot depends on its structure. It varies in the range 1–2 g/cm^3. The assumption of $\eta \equiv 1$ is consistent with the lower limit of this variability.

For MODIS instrument (http://modis.gsfc.nasa.gov/about/), the channel at 1.24 μm is available in addition to 0.865 μm channel. The Global Imager (GLI, JAXA, currently not in operation) had several channels relevant to snow remote sensing (e.g., located at 0.865, 1.05 and 1.24 μm). The applications of the asymptotic radiative transfer theory for these sensors are given by Zege et al. (1998, 2008), Polonsky et al. (1999), Tedesco and Kokhanovsky (2007), and Lyapustin et al. (2009). The algorithms listed above differ from the algorithm presented here as follows:

- Zege et al. (2008) retrieval is based on the analysis of multiple channels (3 for MODIS (0.645,0.859,1.24) and 4 for GLI (0.68, 0.865, 1.05, 1.24). The obvious superiority of their technique is in the fact that uncertainty related to the modeling of R_0 can be substantially reduced due to the use of ratios of reflectances given by Eqs. (23), (24). The shortcomings are due to the fact that the algorithm is valid only for vertically homogeneous snow. Otherwise, the grain size is not constant along the vertical, and therefore, it differs at different channels due to different penetration depths. This effect is not taken into account by Zege et al. (2008).
- Lyapustin et al. (2009) propose two methods: one based on ratios of reflectivities at two close wavelengths, where light absorption by ice is not the same and another method actually very similar to the method described here. It is based on ratio of reflectivities at absorbing and non-absorbing bands. The first method can be applied only to the vertically homogeneous snow as correctly stated by Lyapustin et al. (2009). Therefore, they concentrate on the second method in the corresponding retrievals.
- The method of Tedesco and Kokhanovsky (2007) is based on a single wavelength measurement in the infrared. Therefore, the concentration of pollutants cannot be assessed and evaluated. The correspondence between different retrieval methods based on asymptotic theory is discussed in Appendix.

We note that in addition there are methods based on look-up tables of reflectivities. They produce similar results as methods discussed above if applied for similar setups and channels.

Generally, the wavelength 1.24 μm is the best for retrievals in the case of a homogeneous snow because then even heavy pollution does not influence the results of the grain size retrieval (therefore, one can put $X = 0$ in the expression for β_2 and derive the following simplified equation: $\beta_2 = \gamma^{-2} \ln^2 r_2$, which can be used in conjunction with Eq. (37) for the retrievals of a_{ef}). For vertically inhomogeneous snow, this wavelength brings information only from the top of the layer and may not be consistent with grains at deeper layers seen by the 443-nm wavelength used for the snow pollution retrieval. Even if measurements at 865 nm are used, there is quite large mismatch in the volume of snow sensed using 443 nm and 865 nm wavelengths. As follows from Figs. 7.2 and 7.3, the Jacobians for the soot concentration (at 443 nm) approach zero at the distance of 20 cm from the top layer and the values of Jacobians for the EGS vanish already at 2–5 cm from the snow top depending on the wavelength. Therefore, possible soot layer deposited at, say, 5 cm from the snow top will influence the signal in the visible but not at 865 nm. This makes application of the dual-wavelength algorithm not possible in this case and one should use the single channel algorithm outlined above.

7.4.2 Synthetic retrievals

To understand the sensitivity of the reflected radiation to the probability of photon absorption and the effective grain radius, we have performed a number of numerical experiments. In particular, we have implemented the retrieval algorithm in the numerical code and studied the influence of possible errors of forward model on the retrieval of PPA and a_{ef}. In particular, we have assumed that the measured reflectance differs by $\varsigma = \pm 5\%$ or $\pm 10\%$ from the forward model due to inherent calibration errors, errors of the forward model, cloud screening, atmospheric correction, etc. The resulting retrievals at the solar zenith angle equal to 54° and nadir observations are given in Fig. 7.14(a). It follows that the positive bias in the measured reflection function leads to underestimation of PPA (and otherwise for the negative bias). At $\varsigma = 0$, the algorithm retrieves input parameters with errors below several fraction of a percent (see the green line in Fig. 7.14(a)), which is the prove of the algorithm with the synthetic data. It follows from Figs. 7.14(b) that the error of PPA retrieval increases considerably as $\beta \to 0(\alpha a_{ef} \to 0)$. In particular, errors smaller than 20% (at a reasonable estimate of $\varsigma \sim 5\%$) are possible only if $\beta \geq 0.01$ ($\omega_0 \geq 0.99$). Therefore, it is of importance to use the spectral interval, where PPA varies in the range 0.01–0.02. The upper limit is needed to ensure small errors of the assumed asymptotic theory. It follows from Fig. 7.15 that the use of the wavelengths 1020 and 1240 nm is superior for the usually occurring grain sizes (0.05–1 mm). The value of ω_0 at $\lambda = 865$ nm is always smaller than 0.01 producing a reduced sensitivity to the effective grain radius. This is confirmed by Figs. 7.16(a) and (b), where we see that the value of $\varsigma = \pm 5\%$ makes retrievals at the wavelength 865nm possible with the accuracy better than 50% only at radii larger than 0.4 mm. Otherwise, retrievals are characterized by quite large errors. This is an unfortunate situation because the wavelength of 865 nm is the largest which can be used for retrievals using MERIS. The wavelengths of 885 and 900 nm are contaminated by the uncertainty in the water vapor vertical column and also they are not much different with respect to the sensitivity to the grain size retrievals

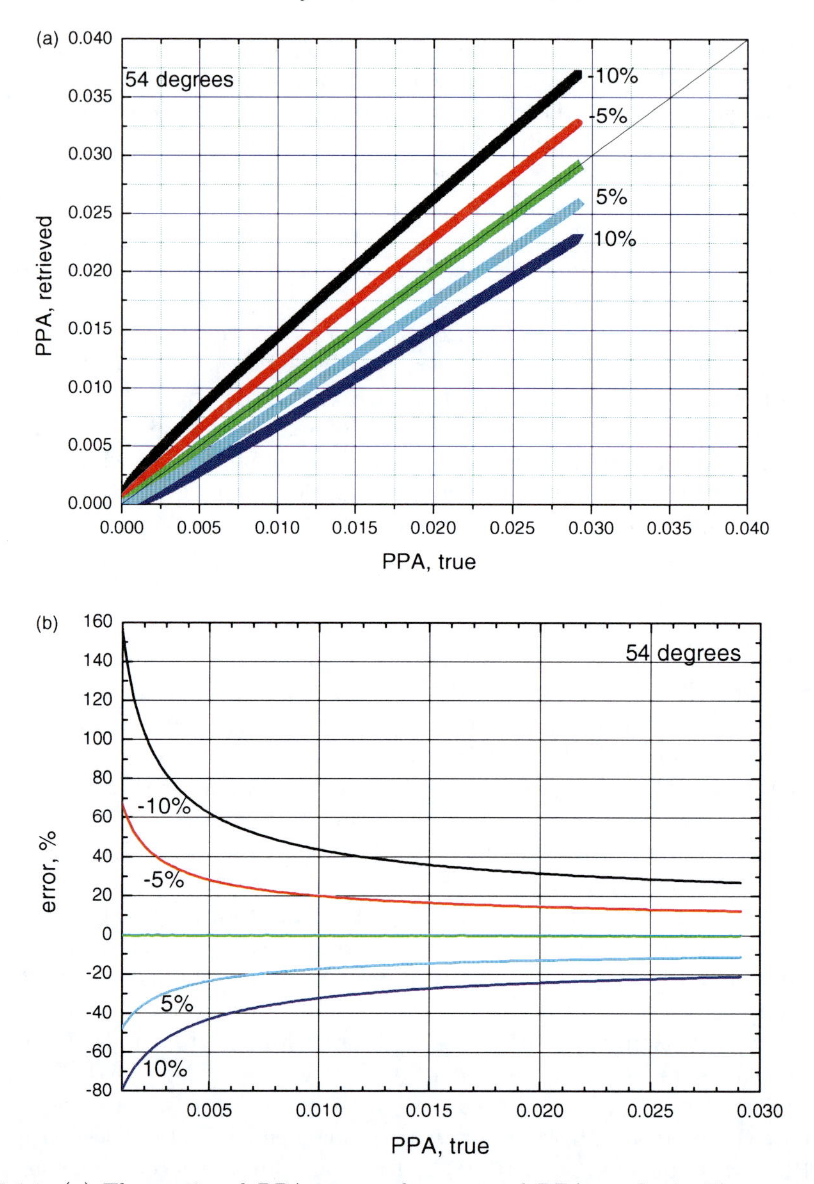

Fig. 7.14. (a) The retrieved PPA versus the assumed PPA at the incidence angle 54° and the nadir observations. (b) The error of the retrieved PPA versus the assumed PPA at the incidence angle 54° and the nadir observations.

as compared to the wavelength 865nm. At the wavelength 1020 nm, the errors of retrievals are smaller than 20%, if $\varsigma = \pm 5\%$at $a_{ef} \geq 0.2$ mm (see Fig. 7.17(a)). The errors reach maximum of 50% at $a_{ef} = 0.1$mm. The errors are still lower at 1240 nm (see Fig. 7.17(b)) but then the use of asymptotic theory is in question (at least for large sizes of grains).

The measurements at the wavelength 443 nm, where absorption of light by ice grains is small, can be used to retrieve the soot concentration in snow as described

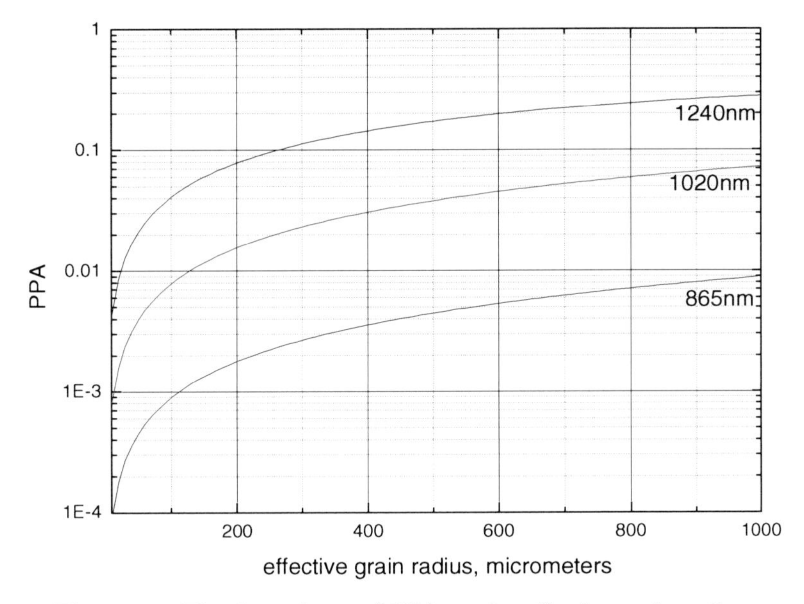

Fig. 7.15. The dependence of PPA on the effective grain radius.

above. To get the concentration of pollutants, one must determine grain size (at least for large grains). Errors in the determination of the grain size influence the retrieval of soot concentration. In particular, errors are large, if the concentration of pollutants is small. Then the retrieved values of β_s are not reliable. One can estimate β_s assuming that $c = 300$ ng/g, which is quite a high concentration of pollutants. Then $\beta = 0.0002$ at $a_{ef} = 0.1$ mm and 0.002 at $a_{ef} = 1$ mm. Clearly, these values of PPA are so small that the reliable determination of β and also $c \leq 300$ ng/g is hardly possible. And the values of the concentration above 300 ng/g are extremely rare (Hansen and Nazarenko, 2004; Flanner et al., 2007). Therefore, it is proposed to make retrievals of β (at 443 nm) only if the change of reflectance at this wavelength is considerable as compared to the case of pure snow. We show the difference of the calculated reflection functions of snow at PPA = 0.0002 and 0.002 in Figs. 7.18 at incident angles 54° and 75° as compared to the pure snow case. It follows that indeed retrievals at such levels of pollution are hardly possible due to errors of the forward model. The concentration of soot in the Arctic is just 10–30 ng/g (Flanner et al., 2007) and PPA is always smaller than 0.0002. Then the retrievals of soot concentration are doubtful. The comparison of retrievals with ground measurements also confirms this conclusion (Aoki et al., 2007).

Because the analytical dependence of the reflectance on relevant parameters is provided in the framework of our approach, it is also easy to calculate corresponding errors analytically. To simplify calculations, we use Eq. (24). Then after differentiation and some algebraic calculations, it follows for the uncertainty in the value of $b = \alpha a_{ef}$:

$$\frac{db}{b} = K_{amp}\frac{dR}{R}, \tag{39}$$

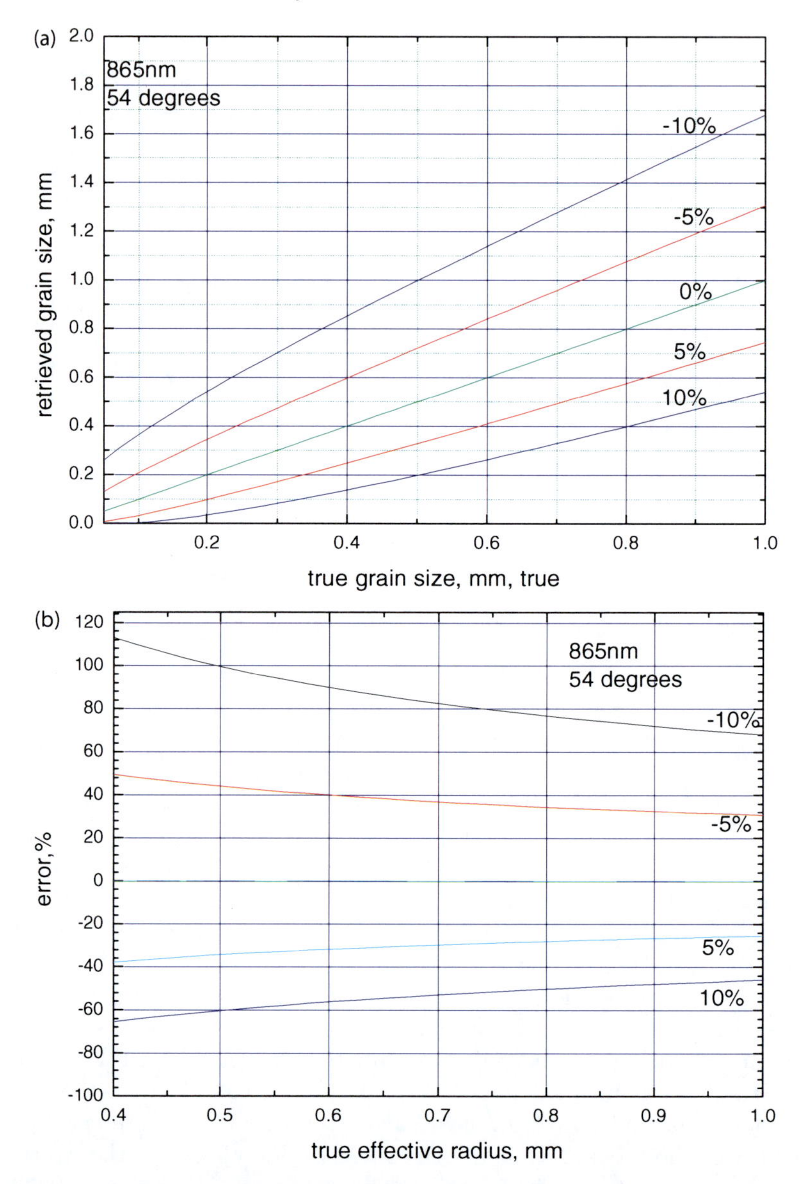

Fig. 7.16. (a) The retrieved effective grain radius versus the assumed grain radius at $\lambda = 865$ nm and SZA $= 54°$. (b) The error of the retrieved EGS as the function of the assumed effective radius of grains retrieved at $\lambda = 865$ nm.

where the error amplification coefficient is given by the following equation:

$$K_{amp} = \frac{2}{\gamma\sqrt{\beta}} \tag{40}$$

and

$$\beta = K\beta_{\infty}\alpha a_{ef} \tag{41}$$

under assumption of weak absorption ($b \to 0$).

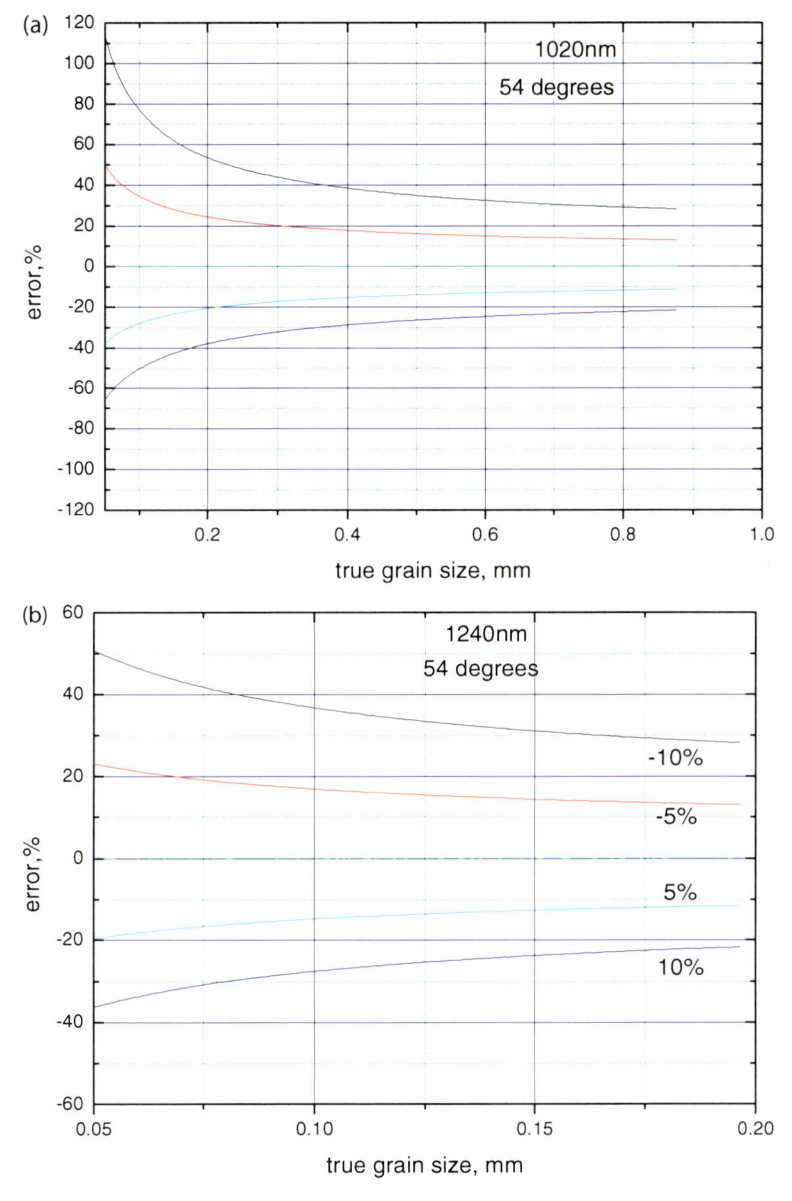

Fig. 7.17. (a) The error of retrieved EGS as the function of the assumed effective radius of grains at $\lambda = 1.02\ \mu$m. (b) The error of the retrieved EGS as the function of the assumed effective radius of grains at $\lambda = 1.24\ \mu$m.

It follows that $K_{amp} \to \infty$ as $b \to 0$, the finding which follows from the numerical experiment given above and also from the sensitivity study presented in the previous section. Because the function $u(\mu)$ increases with the cosine of the angle, we conclude that K_{amp} is somehow reduced for oblique observation and illumina-

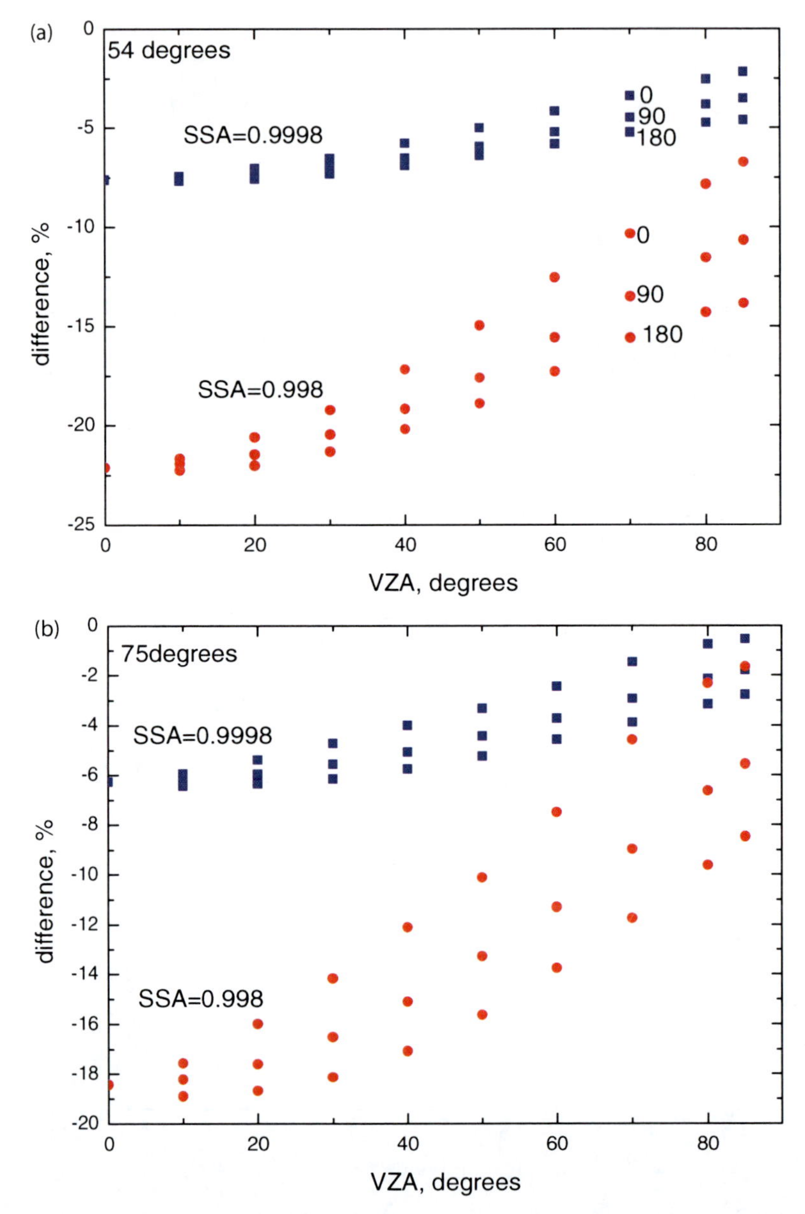

Fig. 7.18. (a) The difference of the reflection function from that for the case of a nonabsorbing snow as the function of the viewing zenith angle at azimuths 0°, 90°, and 180°, SZA = 54° (SSA = 0.998 and SSA = 0.9998). (b) The difference of the reflection function from that for the case of a nonabsorbing snow as the function of the viewing zenith angle at azimuths 0°, 90°, and 180°, SZA = 75° (SSA = 0.998 and SSA = 0.9998).

tion conditions (at least for angular regions, where R_0 is constant or decreases, e.g., see cases with the RAA in the range 90–180°). The error amplification coefficient is proportional to $\sqrt{1-g}$ and, therefore, it is smaller for more extended in the forward direction phase functions.

7.4.3 Application of the algorithm to MERIS data

7.4.3.1 Atmospheric correction

As far as atmospheric correction over snow is concerned, we used the LUT of atmospheric reflectances calculated with SCIATRAN for a prescribed aerosol model. Anyway, the influence of atmosphere on grain size retrievals is small, as was proved in the sensitivity studies. Therefore, such an approach is well justified (at least for a clear atmosphere). In the case of polluted atmospheres, the atmospheric contribution must be assessed using measurements at the neighboring pixels containing open water or polynyas. We demonstrate the importance of atmospheric correction in Figs. 7.19 and 7.20, where results of SCIATRAN calculations are given for the case of surface albedo equal to 0.8 and overlying aerosol layer. It follows that the atmosphere substantially changes the values of reflectance as compared to the case $A = 0.8$ especially at large solar and viewing zenith angles and for shorter wavelengths. An interesting point to note is that the atmospheric contribution can increase or decrease the satellite signal over the snow field depending on the geometry and, therefore, relative contribution of the atmosphere to the signal.

The following simplified atmospheric correction algorithm is proposed. It is supposed that the correction for the gaseous absorption can be performed as:

$$R_{cor} = T_{gas} R_M, \tag{42}$$

where T_{gas} is the gaseous transmittance calculated as

$$T_{gas} = \exp(\upsilon \mathrm{M}), \tag{43}$$

M is the trace gas vertical column (O_3, H_2O as obtained, e.g., from MERIS observations), $\upsilon = 1/\mu + 1/\mu_0$, R_M is the measured reflectance. The obtained reflectances R_{cor} (except at 760 and 900 nm channels) are corrected for Rayleigh and aerosol scattering using the pre-calculated LUTs. It is assumed that the aerosol optical thickness is equal to 0.05 (see, e.g., Tomasi et al., 2007) in the calculation of LUTs. The aerosol model is WMO coarse maritime aerosol model with no absorption assumed. The angle grid was one degree for SZA(0(1)89°), VZA(0(1)80°), and relative azimuth angle (0(1)180°). Therefore, the largest possible mismatch of MERIS data and those in LUTs is 0.5°. The error due to this mismatch is smaller than 0.01 in the snow albedo, which is acceptable due to other complications inherent to snow properties retrievals. This is illustrated in Fig. 7.21, where we show the retrieved albedo (retrieved using the pre-calculated LUTs described above) under the assumption that true snow albedo is equal to 0.8. The synthetic data were generated not for the same grid as in LUT but for the 0.5-degree shifted LUT (for angles, the same shift was assumed). The algorithm performance is very good

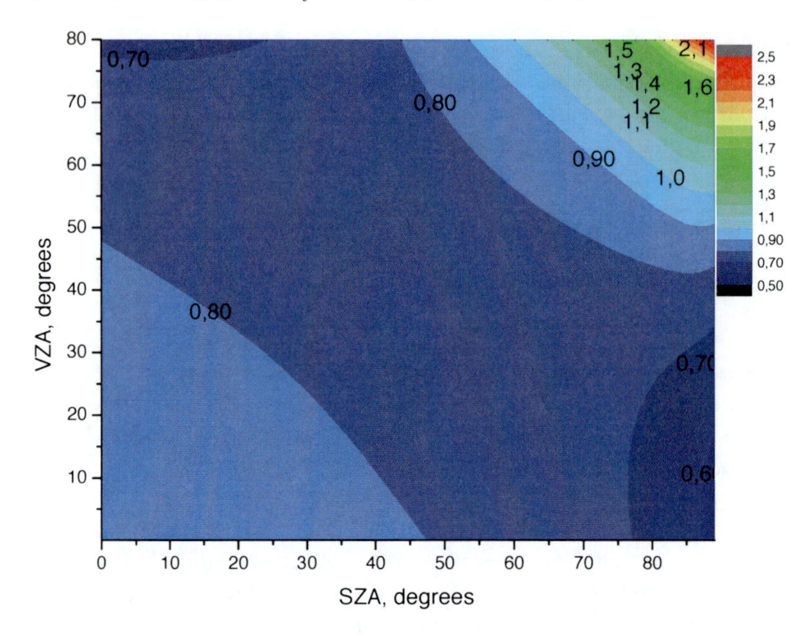

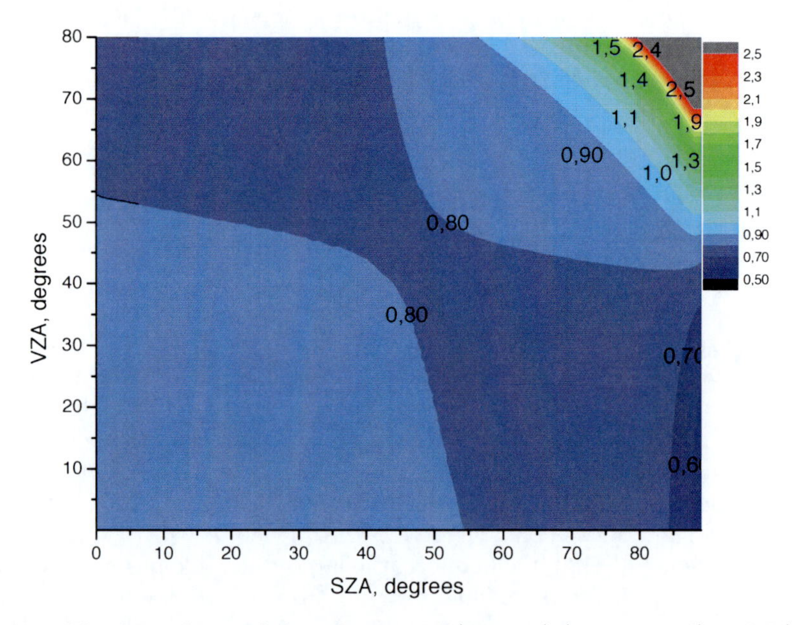

Fig. 7.19a. The dependence of the reflectance $R(443\text{ nm})$ (upper panel) and $R(865\text{ nm})$ (lower panel) on the SZA and VZA at the relative azimuth angle equal to 0.0. The case of the underlying Lambertian surface with albedo 0.8 is presented for the maritime coarse aerosol model with aerosol optical thickness equal to 0.05. The values of reflectance are above 0.8 in the lower left corner and well below 0.8 in the upper lower corner. This means that the Lambertian surface looks brighter from space, if observed from the nadir direction at SZA = 0.0. It looks darker for the nadir observation and the oblique solar light incidence (say, SZA = 80°).

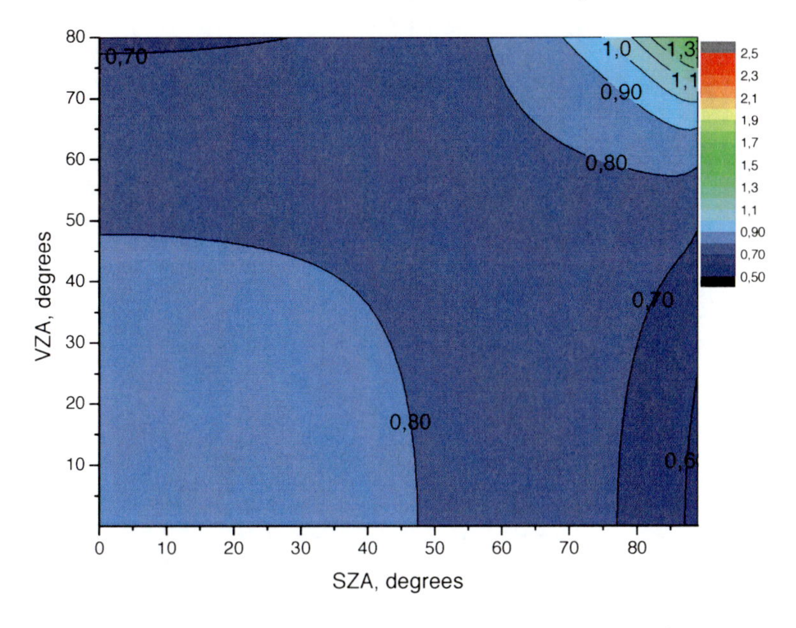

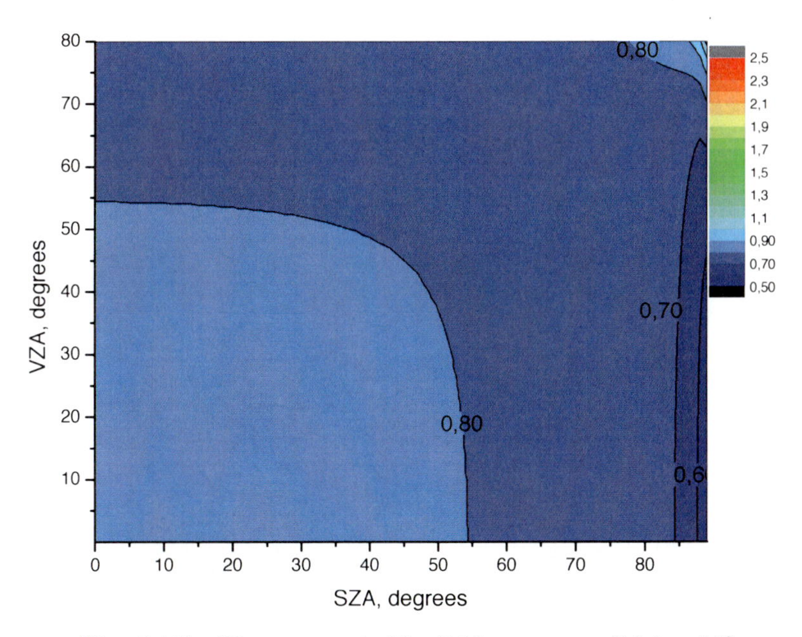

Fig. 7.19b. The same as in Fig. 7.19a except at RAA = 90°.

till VZA is equal to 55°, which is of importance for AATSR measurements. The largest MERIS VZA is 42° and then the performance of the algorithm is even better. The true albedo is retrieved if no shift in the grid is applied and the same snow and atmosphere models are used in the solution of direct and inverse problems. The change of the aerosol model in the retrieval process (from maritime coarse to maritime fine) does not bring complications, if the aerosol optical thickness was

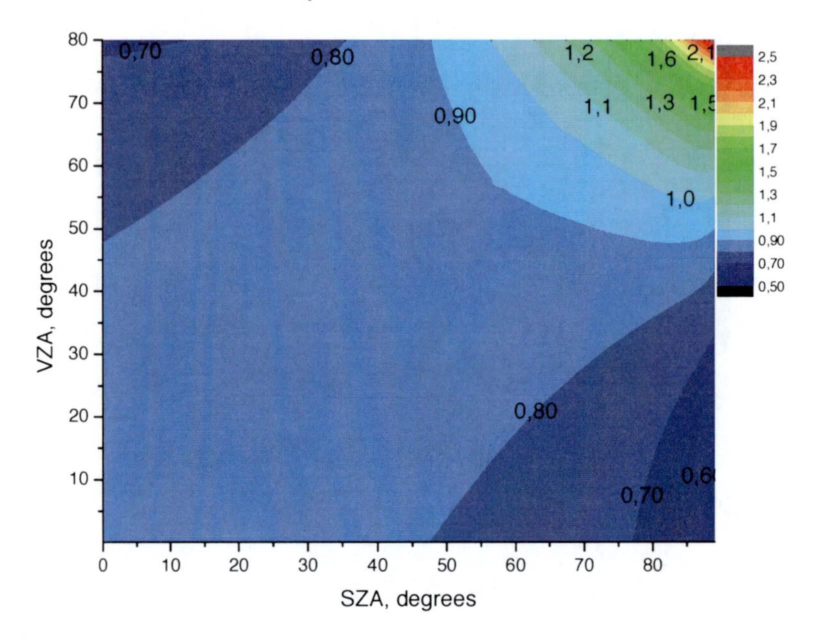

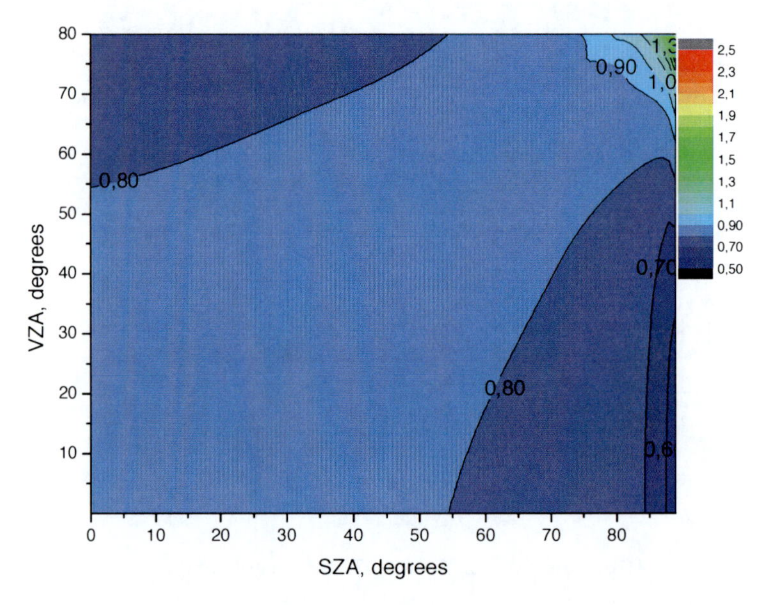

Fig. 7.19c. The same as in Fig. 7.19a except at RAA = 180°.

assumed to be the same (0.05). It follows that errors of retrievals are negligible even in the case where the aerosol model is not correctly selected (the errors are less than 0.01 in the snow albedo) as far as MERIS observations are of concern (VZA < 42°).

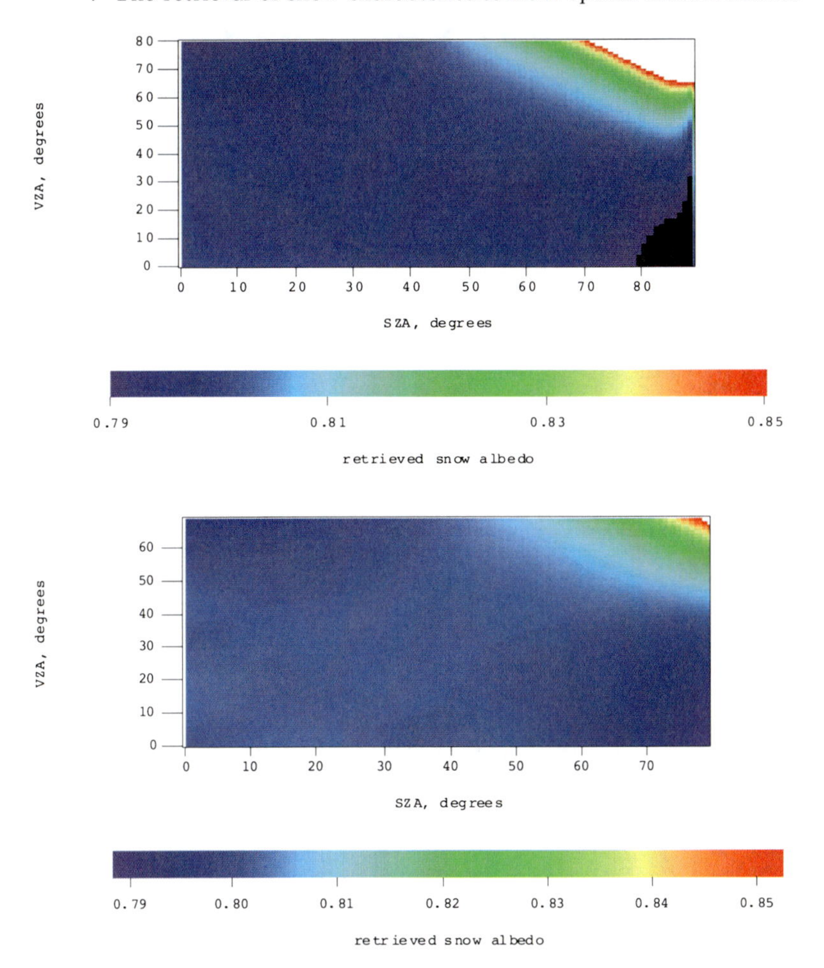

Fig. 7.20. The retrieved snow albedo as the function of the satellite and solar zenith angles. The aerosol optical thickness is equal to 0.05 and the fine-mode maritime aerosol phase function was used in synthetic data preparation. The relative azimuth was varied in the range 0(1)180°. The true snow albedo is equal to 0.8. The upper panel differs from the lower panel due to the different regions of change of SZA and VZA. The range of SZA and VZA change is lower in the figure shown at the bottom.

For the atmospheric correction, it is assumed that MERIS reflectance can be presented as

$$R = R_b + \frac{TR_s}{1 - Ar}. \tag{44}$$

Here R_b is the reflectance for a black underlying surface (stored in LUTs for molecular-aerosol atmosphere with the assumed aerosol optical thickness of 0.05 and coarse maritime aerosol model with no absorption), R_s is the snow reflectance, T is the atmospheric transmittance stored in LUTs, r is the spherical albedo stored in LUTs, $A = 0.8$ is the assumed snow albedo. The term Ar is quite small. Therefore, the assumption on A does not have a large impact on the retrieved snow reflectance:

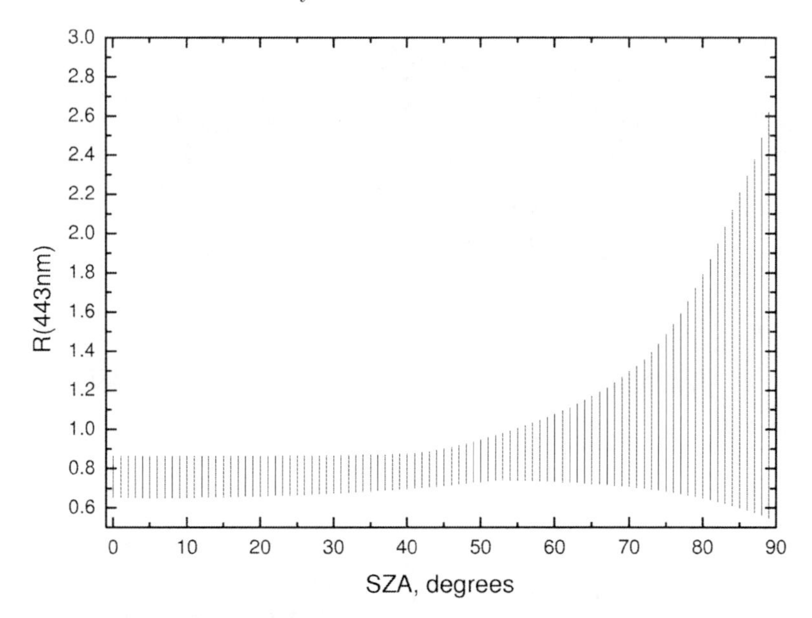

Fig. 7.21. The dependence of the reflectance of the snow field with the assumed Lambertian albedo 0.8 on the SZA. The vertical lines give the variability due the variation of the viewing zenith angle (0(1)80°) and relative azimuth (0(1)180°). The atmosphere can increase or decrease the albedo of the snow field depending on the geometry. The aerosol optical thickness is 0.05 and the coarse maritime aerosol model is assumed.

$$R_s = (R - R_b)(1 - Ar)/T. \tag{45}$$

The algorithm described above is of importance only for MERIS channels below 800 nm, where the soot concentration and snow albedo is retrieved. The atmospheric correction is less important at 865 nm and also at 1020 and 1240 nm, where the snow grain size is usually retrieved.

7.4.3.2 The retrieval of snow grain size using MERIS observations

The MERIS browse image of the snow field under clear sky in Greenland is shown in Fig. 7.22. The corresponding maps of reflectances at 443 nm, and 865 nm, and also in the oxygen A-band (762 nm) are given in Fig. 7.23a and 7.23b. The line across the image in the case of measurements at 762 nm shows the border between two cameras of MERIS (the so-called instrumental smile effect).

A lot of clouds are present in the region. The retrieved grain size is shown in Fig. 7.24a and 7.24b (after atmospheric correction and cloud screening procedures have been applied). The average EGS is around 0.2 mm for the whole scene and 0.15 mm for the left part of the scene, Unfortunately, *in situ* data for EGS at this location during the satellite measurements are not available to us.

We show the results of the retrieved concentration of pollutants in Fig. 7.25 (in ng/g). The concentrations are very low as one might expect for Arctic. Generally, as follows from the sensitivity studies given above, the accurate determination of soot concentration from a satellite is difficult in the Arctic due to the low concentration

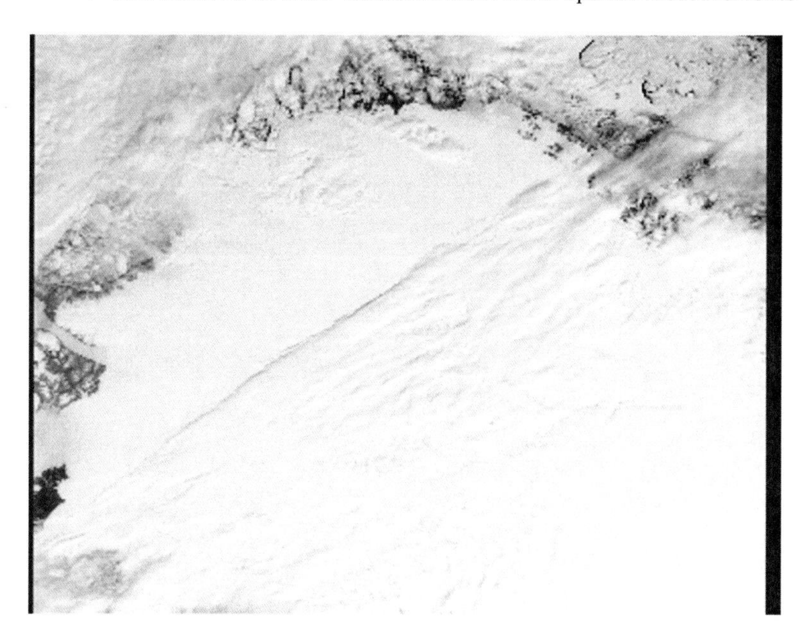

Fig. 7.22. Browse image of the scene analyzed. The retrievals have been performed for the clear sky portion of this image (Kokhanovsky et al., 2011).

of pollutants there. Although, as is seen from Fig. 7.19b, the magnitude of c is determined in a correct way.

7.5 Conclusions

In this work we discussed the snow grain size retrieval algorithm FORCE. The work is also relevant to the area of the determination of the specific area of snow using optical measurements (Schneebeli and Sokratov, 2004; Matzl, 2006; Matzl and Schneebeli, 2006; Gallet et al., 2009; Kokhanovsky and Schreier, 2009). The correlation coefficient between satellite and ground measurements of EGS is in the range 0.6–0.7 (Kohkanovsky et al., 2011). The small values of the correlation coefficient could be due to the different definitions of sizes in the ground and satellite measurements. Also we proposed techniques for the cloud screening and atmospheric correction of satellite images over snow. The algorithm must be improved in future. The current version of the algorithm was implemented in the ESA software package BEAM and free for use by the remote sensing community.

Several simplifications have been used in the algorithm development. In particular, it was assumed that snow is vertically homogeneous. In reality snow has a layered structure, as discussed by Colbeck (1991). The layering arises from a sequence of storms, reworking of the snow surface into a distinctive horizon which is subsequent buried, or the generation of certain types of horizons within the snow profile. Not only is the sequence of these buried layers unique from year to year and highly variable with location, but each layer also evolves as the snowy season progresses (Colbeck, 1982, 1983). Dust and soot can be deposited in such layers and

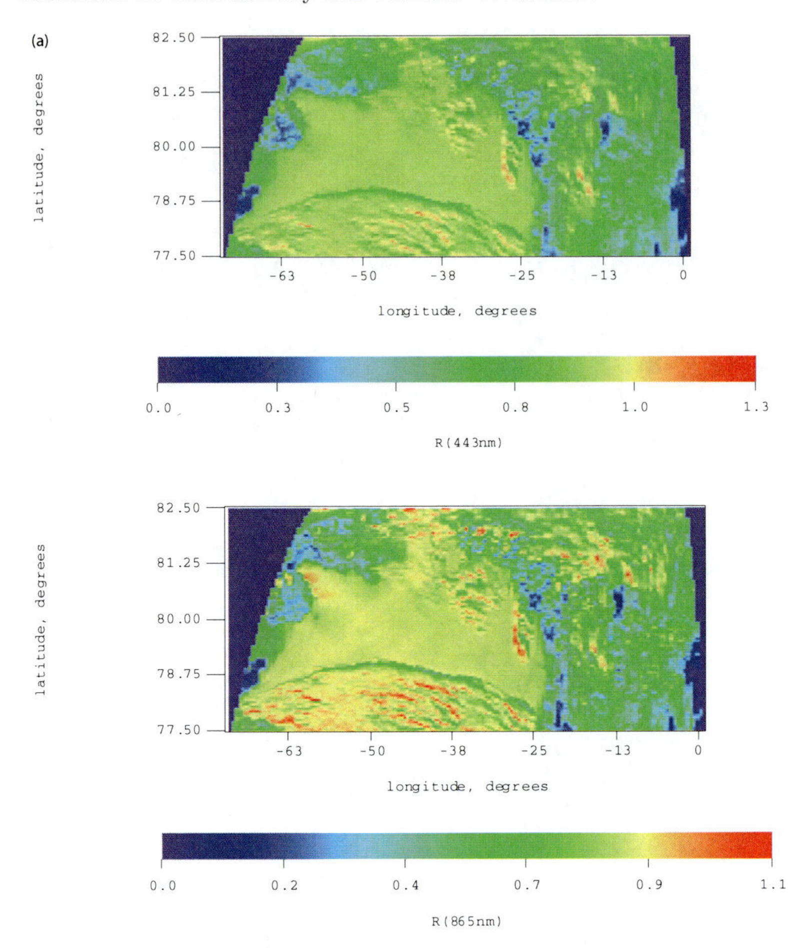

Fig. 7.23a. Maps of reflectances at 443 nm (upper panel) and 865 nm (lower panel) for the browse image shown in Fig. 7.22 (Kokhanovsky et al., 2011).

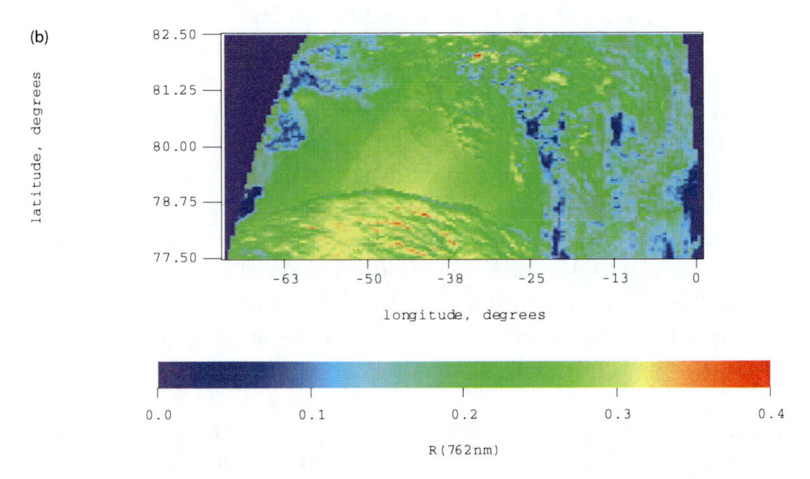

Fig. 7.23b. The map of reflectance at 762 nm (Kokhanovsky et al., 2011).

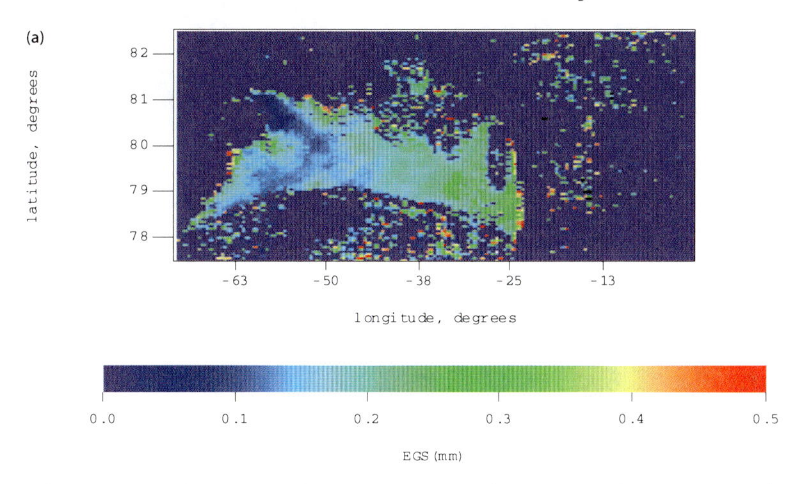

Fig. 7.24a. The retrieved snow grain size (Kokhanovsky et al., 2011).

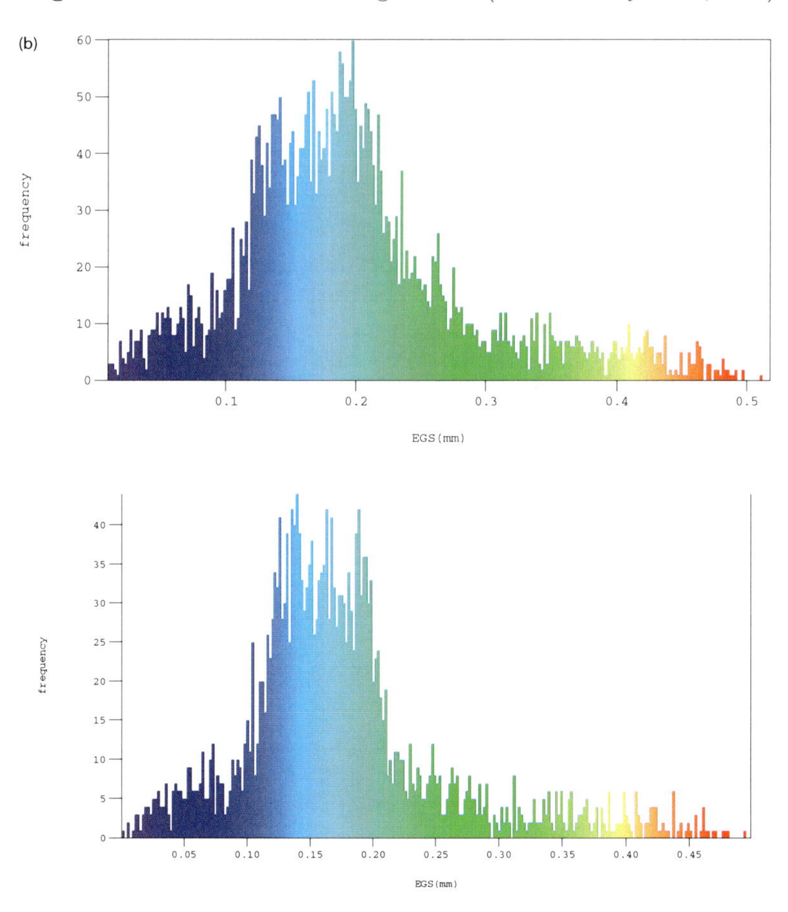

Fig. 7.24b. The retrieved snow grain size histogram (the lower panel corresponds to retrievals in the left part of the image) (Kokhanovsky et al., 2011).

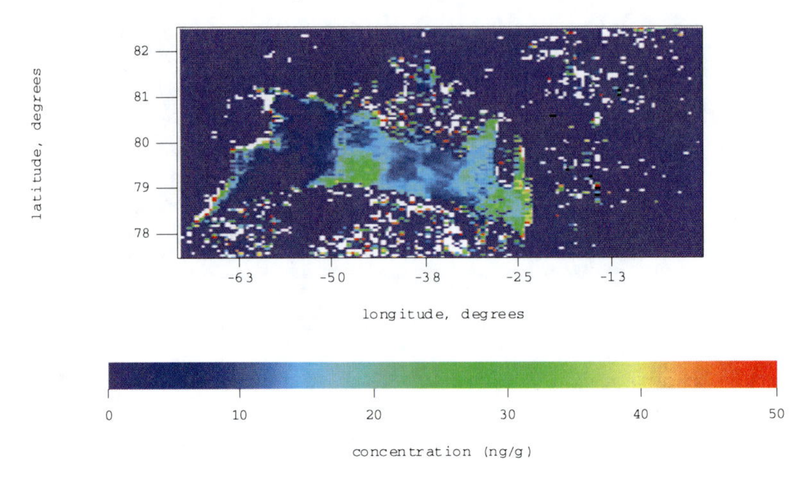

Fig. 7.25. The retrieved concentration of pollutants (Kokhanovsky et al., 2011).

then covered by the fresh snow. Because in standard retrieval algorithms vertically homogeneous snow is assumed, the pollutant content derived will be that of an entire snow column, which does not correspond to reality. Moreover, in retrievals one needs to assume the refractive index of the pollutants. The refractive index is considerably different for dust and soot. Also the absorption and scattering cross-sections of soot and dust particles are considerably different. Therefore, wrong *a priori* assumptions on the type of pollutants (soot, dust, red algae on the surface of snow) prevent correct retrievals of the concentration of pollutants. In principle, the type of pollutants can be distinguished from spectral measurements of the snow reflectance because, e.g., dust and soot have different spectral bulk absorption coefficients (e.g., red and grey colours). However, this is possible only in the case of thin layers of fresh snow over dirty snow or in the case of freshly polluted snow.

Also it follows that the structure of snow and also shapes/sizes of crystals are very different from the top to the bottom of the snow layer. This peculiarity is also not accounted for in the forward model. The snow grain size is retrieved using infrared measurements. But it is a well known fact that the imaginary part of the refractive index of ice changes with wavelength (and temperature). This means that light with different wavelengths will penetrate to different depths. Therefore, the use of multiple wavelengths, in principle, can reveal the vertical distribution of the snow grains (Li et al., 2001; Zhou et al., 2003). Using one-wavelength retrieval, only one grain size for a given depth is retrieved. Importantly, the snow penetration depth is not fixed for a given wavelength: it also depends on the grain size itself. Generally, it is lower for larger wavelengths. Therefore, it is of importance to report at which wavelength the retrievals have been performed. If pollution is not uniformly distributed in snow but rather contained in distinct layers (e.g., dust), then one cannot ignore light scattering by pollutants. Then both absorption and scattering effects by pollutants must be considered. Usually, in retrievals of EGS, the pollution is assessed assuming the homogeneous distribution of snow layer. If pollution is in a layer well below the snow surface, it plays no role in the EGS retrieval, but it can play some role if pollutants are close to the surface and retrievals are made at short wavelength (say 865 nm) and grains are small.

The radiative transfer models used extensively in the snow optics assume that snow has no structures on the surfaces. For satellite ground scenes (e.g., 1km), the horizontal inhomogeneity of snow (e.g., sastrugi) may influence the snow reflectance and, therefore, the retrieved snow grain size considerably (Warren et al., 1998). We found that generally, the reflectance decreases, if sastrugi present and the decrease could be on the order of 5-30% depending on PPA (Zhuravleva and Kokhanovsky, 2011). It is smaller for smaller PPA. The patches of vegetation penetrating through snow or trees make retrievals not possible or difficult. Therefore, it is of importance to make not only cloud screening but also only 100% snow covered ground scenes (without forest and vegetation) must be used in retrievals of the grain size, snow albedo, and the concentration of snow pollutants. This is due to the fact that there is a limitation with respect to the complexity of the forward model, which can be used in the retrieval process. Although there are some reports on the retrieval of subpixel snow properties (see, e.g., Painter et al., 1998, 2003, 2009). Retrieval of snow properties in the mountainous regions is also of problem. Then effects of shadowing are evident and 3-D radiative transfer models are needed with the known topography and illumination conditions at a given location.

The retrieval of grain sizes for the polluted case (both for polluted snow and for polluted atmosphere) can cause problems if the channel at 865 nm is used for retrievals. This is due to the fact that the signal at 865 nm can be influenced by pollution (Painter et al., 2007) and this influence is difficult to assess *a priori*. For instance, there is the problem of the possible presence of soot in the atmosphere and in the snow. For longer wavelengths, the influence of pollution is reduced considerably. Although Dozier et al. (2009) report that there are cases, where the pollution (e.g., dust) influences snow reflectance at all wavelengths in the visible and near-infrared (up to $\lambda = 1.4\ \mu$m). The retrieval of the pollution level depends on the type of pollution (Warren and Wiscombe, 1980; Warren, 1982; Painter and Dozier, 2004; Painter et al., 2007). The uncertainty in the imaginary part of the ice refractive index in the visible (Warren and Brandt, 2008) can also play a role.

As noted by Peltoniemi (2007), snow becomes less reflective at larger densities of snow. The radiative transfer theory can be applied at very low densities (actually not possible for snow on ground) only and, therefore, this darkening will be interpreted as the presence of pollutants – although, in fact, the snow is fresh and clean. In particular, this could be the reason behind the observed reduction of fresh snow reflectance in the visible as compared to radiative transfer simulations (see, e.g., Fig. 7.5 Dozier et al. (2009)).

The radiative transfer model described above is valid for dry snow only. During the melting season, water can accumulate in snow. Then the model must be changed, taking into account the darkening of the snow due to the presence of liquid water in it. Modifications both of snow absorption and scattering are important. The snow grains become more spherical and grow in size. Clusters are formed (Colbeck, 1979). The presence of liquid films between grains reduces the scattering and also it leads to more extended in the forward direction phase functions. This for sure will bias retrievals if it is not accounted for in the retrieval procedure. Another possibility is the dependence of the snow bulk absorption coefficient on temperature (e.g., for extremely low temperatures such as those possible, e.g., in the Antarctic (Grundy and Schmitt, 1998)), which is not accounted for in the current retrieval algorithm. The issues highlighted above are a subject of our ongoing research on forward and inverse models in snow optics.

Acknowledgments

This work was supported by the ESA Project Snow_Radiance and BMBF Project CLIMSLIP (Germany) and JAXA (Japan). We thank Andreas Macke and Michael Mishchenko for providing their codes used in this study. We are grateful to ESA for providing MERIS reduced resolution data.

Appendix. The relationship between different retrieval approaches based on the asymptotic radiative transfer theory

Snow reflectance can be presented in the following way using asymptotic radiative transfer theory (Zege et al., 1991):

$$R(\mu, \mu_0, \varphi) = R_0(\mu, \mu_0, \varphi) \exp\{-pf(\mu, \mu_0, \varphi)\}, \tag{7.A1}$$

where

$$p = 4s/\sqrt{3},\ s=\sqrt{\frac{1-\omega_0}{1-g\omega_0}},\ f = \frac{u(\mu_0)u(\mu)}{R_0(\mu, \mu_0, \varphi)},\ u(\mu) = \frac{3}{7}(1+2\mu). \tag{7.A2}$$

Here R_0 is the reflection function of a semi-infinite snow layer under the assumption that the single scattering albedo is equal to one, the values (μ, μ_0, φ) give the cosines of observation and incidence angles, and also the relative azimuthal angle, respectively, ω_0 is the single scattering albedo, and g is the asymmetry parameter.

Various snow retrieval algorithms differ with respect to the actual implementation of Eq. (7.A1) in the retrieval code and also due to different approximations of the relationship between the similarity parameter s(or p) and a_{ef}.

Tedesco and Kokhanovsky (2007) proposed the relationship between p and the effective grain size a_{ef} in the following form:

$$p = A\sqrt{\gamma_i a_{ef}}, \tag{7.A3}$$

where $A = 5.09$, $\gamma_i = 4\pi\chi_i/\lambda$ and χ_i is the ice refractive index. The value of p can be found from Eq. (7.A1):

$$p = \frac{1}{f}\ln\left\{\frac{R_0}{R}\right\} \tag{7.A4}$$

and then the approximate solution for the function $R_0(\mu, \mu_0, \varphi)$ is used to find p, and, therefore, EGS (see Eq. (7.A3)) from the measured reflectance R at a given wavelength. The wavelength 1.24 μm is used because of the following factors:

- this reflectance at the wavelength is the most sensitive to the grain size for most situations occurring in practice;

- the influence of pollutants in snow and in the atmosphere is minimal at this wavelength;
- the retrievals have a direct meaning and relevance to the EGS at the subsurface snow layer.

A very similar method was used by Langlois et al. (2010) for the interpretation of near-infrared photography of vertical snow walls. However, they assumed that $A = 5.66$. The shortcoming of this algorithm (called SARA) is in fact that the concentration of pollutants cannot be assessed. Also the influence of pollutants on the reflectance at 1.24 μm (possible for heavy pollution events) cannot be estimated. The algorithm FORCE described in this chapter is the extension of SARA to solve this problem. Then it is assumed that p is related to a_{ef} and the concentration of soot c using the following equation, which approximately holds for the vertically homogeneous snow:

$$p = \frac{4}{\sqrt{3(1-g)}}\sqrt{\frac{2}{3}Bc\gamma_s a_{ef} + \beta_\infty(1 - \exp(-\gamma_i \kappa a_{ef}))}, \tag{7.A5}$$

where $g = 0.76$, $\kappa = 2.63$, $\beta_\infty = 0.47$, $B = 0.84$, $\gamma_s = 4\pi\chi_s/\lambda$, $\chi_s = 0.46$ is the soot refractive index in the visible (spectrally neutral) . Eq. (7.A3) follows from Eq. (7.A5) assuming that $c \equiv 0$ and $\gamma_i a_{ef} \to 0$ (but with slightly different $A = 5.24$). In this case there are two unknowns (c, a_{ef}) and they can be found, if measurements at two wavelengths are performed. In the visible, absorption by ice can be ignored and Eq. (7.A4) is used to find p, and, therefore, ca_{ef} assuming that the second term under the square root in Eq. (7.A5) can be neglected. Then p and, therefore, EGS can be found from Eqs. (7.A4), (7.A5) applied at the second wavelength (for known product $x = ca_{ef}$ determined from measurements in the visible). Finally, the soot concentration is derived as $c = x/a_{ef}$. The value of R_0 is obtained from look-up tables calculated using the Ambartsumian nonlinear integral equation. The shortcoming of this method is in the assumption that the snow grain size does not vary along the vertical.

Lyapustin et al. (2009) used the technique similar to SARA (Tedesco and Kokhanovsky, 2007). However, instead of calculations of R_0, the corresponding measured value (in the visible) was used. Such an assumption is valid in the case of pure snow (say, as in Greenland as studied by Lyapustin et al. (2009)). This makes retrievals faster and also makes it possible to account for possible errors in modeled R_0 due to possible close-packed media and 3D effects.

Finally, Zege et al. (2008) proposed the multi-channel retrieval technique based on the assumption of a vertically and horizontally homogeneous semi-infinite snow layer. In particular, they proposed to use three channels of MODIS to derive not only c and a_{ef}, but also R_0 from the measurements itself. They used Eq. (7.A1) for three wavelengths with p defined as follows:

$$p = A\sqrt{b\gamma_s c a_{ef} + \gamma_i a_{ef}}. \tag{7.A6}$$

Zege et al. (2008) used the value of A=6 in their retrievals but state that it can vary from 3.5 to 6.5 depending on snow type. It is assumed that the value of $b = 0.43$. Eq. (7.A3) follows from Eq. (7.A6) if $c = 0$. Eq. (7.A5) can be presented in the form given by Eq. (7.A6) if it is assumed that $\gamma_i a_{ef} \to 0$. The resulting constants

A and b slightly differ from those used by Zege et al. (2008) ($A = 5.24$, $b = 0.45$). More work must be done to assess the values of (A, b) from *in situ* measurements of natural snow covers.

References

Ackerman, S., Strabala, K., Menzel, P., Frey, R., Moeller, C., Gumley, L., Baum, B., Wetzel, S., Seemann, S. and Zhang, H., 2006: Discriminating clear sky from cloud with MODIS algorithm theoretical basis document (MOD35), at http://modis.gsfc.nasa.gov/data/atbd/atbd_mod06.pdf.

Aoki, T., Hori, M., Motoyoshi, H., Tanikawa, T., Hachikubo, A., Sugiura, K., Yasunari, T. J., Storvold, R., Eide, H. A., Stamnes, K., Li, W., Nieke, J., Nakajima, Y., and Takahashi, F., ADEOS-II/GLI snow/ice products – Part II, 2007: Validation results using GLI and MODIS data. *Remote Sensing of Environment*, **111**, 274–290.

Bond, T. C., and Bergstrom, R. W., 2006: Light absorption by carbonaceous particles: an investigative review, *Aerosol Science and Technology*, **40**, 27–67.

Bourdelles, B., and Fily, M., 1993: Snow grain-size determination from Landsat imagery over Terre Adelie, Antarctica, *Annals Glaciology*, **17**, 86–92.

Chandrasekhar, S., 1960: *Radiative Transfer*, New York: Dover.

Colbeck, S. C., 1979: Grain cluster in wet snow, *Journal of Colloid and Interface Sciences*, **72**, 371–384.

Colbeck, S. C., 1982: An overview of seasonal snow metamorphism, *Review of Geophysics and Space Physics*, **20**, 45–61.

Colbeck, S. C., 1983: Theory of metamorphism of dry snow, *Journal of Geophysical Research*, **88**, 5475–5482.

Colbeck, S. C., 1991: The layered character of snow covers, *Reviews of Geophysics*, **29**, 81–96.

Domine F., Albert, M., Huthwelker, T., Jacobi, H.-W., Kokhanovsky, A., Lehning, M., Picard, G., and Simpson, W. R., 2008: Snow physics as relevant to snow photochemistry, *Atmospheric Chemistry and Physics*, **8**, 171–208.

Dozier, J., 1987: Recent research in snow hydrology, *Reviews of Geophysics*, **25**(2), 153–161.

Dozier, J., and Painter, T. H., 2004: Multispectral and hyperspectral remote sensing of alpine snow properties, *Annual Review of Earth and Planetary Sciences*, **32**, 465–494.

Dozier, J., Green, R.O., Nolin, A. W., and Painter, T. H., 2009: Interpretation of snow properties from imaging spectrometry, *Remote Sensing of Environment*, **113**, S25–S37.

Fily, M., Bourdelles, B., Dedieu J. P., and Sergent, C., 1997: Comparison of in situ and Landsat Thematic Mapper derived snow grain characteristics in the Alps, *Remote Sensing of Environment*, **59**, 452–460.

Flanner, M. G., Zender, C. S., Randerson, J. T., Rasch, P. J., 2007: Present-day climate forcing and response from black carbon in snow, *Journal of Geophysical Research*, **112**, D11202, doi:10.1029/2006JD008003.

Gallet, J.-C., Domine, F., Zender, C. S., and Picard G., 2009: Measurements of the specific surface area of snow using infrared reflectance in an integrating sphere at 1310 and 1550 nm, *Cryosphere*, **3**, 167–182,

Garret, T. J., Hobbs, P.V., and Gerber, H., 2001: Shortwave, single scattering properties of arctic ice clouds, *Journal of Geophysical Research*, **106**, D14, 15,555–15,172.

Grundy W. M., and Schmitt B., 1998: The temperature-dependent near-infrared absorption spectrum of hexagonal H_2O ice, *Journal Geophysical Research*, **103**, E11, 25809–25822.

Hansen J., and Nazarenko L., 2004: Soot climate forcing via snow and ice albedos, *Proceedings of National Academy of Sciences of USA*, **101**, 423–428.

Hori, M., Aoki, T., Stamnes, K., and Li, W., 2007: ADEOS-II/GLI snow/ice products – Part III: Retrieved results, *Remote Sensing of Environment*, **111**, 291–336.

Kokhanovsky, A. A., 1998: On light scattering in random media with large densely packed particles, *Journal of Geophysical Research*, **103**(D6), 6089–6096.

Kokhanovsky, A. A., and Nauss, T., 2005: Satellite based retrieval of ice cloud properties using semianalytical algorithm, *Journal of Geophysical Research*, **110**, D19206, doi: 10.1029/2004JD005744, 2005.

Kokhanovsky, A. A., and Schreier, M., 2008: The determination of snow albedo using combined AATSR and MERIS observations, in Proc. of the 2nd MERIS/(A)AATSR Workshop, Frascati, Italy, 22–26 September 2008 (ESA SP-666, November 2008).

Kokhanovsky, A. A., and Schreier M., 2009: The determination of snow specific area, albedo and effective grain size using AATSR spaceborne observations, *International Jounal Remote Sensing*, **30**, 4, 919–933.

Kokhanovsky, A. A., and Zege, E. P., 2004: Scattering optics of snow, *Applied Optics*, **43**, 1589–1602.

Kokhanovsky, A. A., Aoki, T., Hachikubo, A., Hori, M., and Zege, E. P., 2007: Reflective properties of natural snow: approximate asymptotic theory versus in situ measurements, *IEEE Transactions on Geosciences and Remote Sensing*, **43**, 1529–1535.

Kokhanovsky, A. A., Rozanov V., Aoki T., et al., 2011: Sizing snow grains using backscattered solar light, *International Journal Remote Sensing*, in press.

Langlois, A., Royer, A., Motpetit, B., et al., 2010: On the relationship between snow grain morphology and in situ near infrared calibrated reflectance photographs, *Cold regions Science and Technology*, **61**, 34–42.

Legagneux, L., Cabanes, A., and Domine, F., 2002: Measurement of the specific surface area of 176 snow samples using methane adsorption at 77 K, *Journal of Geophysical Research*, **107** (D17), 4335, doi: 10.1029/2001JD001016.

Lenoble, J., 1985: *Radiative Transfer in Scattering and Absorbing Atmospheres: Standard Computational Procedures*, Hampton: Deepak.

Li, W., Stamnes, K., Chen, B., and Xiong, X., 2001: Snow grain size retrieved from near-infrared radiances at multiple wavelengths, *Geophysical Research Letters*, **28**, 1699–1702, doi: 10.1029/2000GL011641.

Lyapustin. A., Tedesco, M., Wang, Y., Aoki, T., Hori, M., and Kokhanovsky, A., 2009: Retrieval of snow grain size over Greenland from MODIS, *Remote Sensing of Environment*, **113**, 1976–1987.

Macke, A., J. Mueller, E. Raschke, 1996: Single scattering properties of atmospheric ice crystals, *Journal Atmospheric Science*, **53**, 2813–2825.

Matzl, M., 2006: Quantifying the stratigraphy of snow profiles, PhD Thesis, Swiss Federal Institute of Technology, Zürich.

Matzl, M., and Schneebeli,M., 2006: Measuring specific surface area of snow by near-infrared photography, *Journal of Glaciology*, **52**, 558–564.

Mishchenko, M. I., Dlugach, J. M., Yanovitskij, E. G., and Zakharova, N. T., 1999: Bidirectional reflectance of flat, optically thick particulate layers: an efficient radiative transfer solution and applications to snow and soil surfaces, *Journal of Quantitative Spectroscopy and Radiative Transfer*, **63**, 409–432.

Muinonen, K., Nusiainen, T., Fast, P., Lumme, K., and Peltoniemi, J. I., 1996: Light scattering by Gaussian random particles: ray optics approximation, *Journal of Quantitative Spectroscopy and Radiative Transfer*, **55**, 577–601.

Nolin, A.W., and Dozier, J., 1993: Estimating snow grain size using AVIRIS data, *Remote Sensing of Environment*, **44**, 231–238.

Nolin, A. W., and Dozier, J., 2000: A hyperspectral method for remotely sensing the grain size of snow, *Remote Sensing of Environment*, **74**, 207– 216.

Nolin, A. W., and Liang, S., 2000: Progress in bidirectional reflectance modeling and applications for surface particulate media: snow and soils, *Remote Sensing Reviews*, **18**, 307–342.

Odermatt D., Schläpfer, D., Lehning, M., Schwikowski, M., Kneubühler, M., and Itten, I. K., 2005: Seasonal study of directional reflectance properties of snow, EARSeL eProceedings, 4, 203–214.

Painter, T. H., Dozier, J., Roberts, D. A., Davis, R. E., and Greene, R. O., 2003: Retrieval of subpixel snow-covered area and grain size from imaging spectrometer data, *Remote Sensing of Environment*, **85**, 64–77.

Painter, T. H., Roberts, D. A., Green, R. O., and Dozier, J., 1998: The effect of grain size on spectral mixture analysis of snow-covered area from AVIRIS data, *Remote Sensing of Environment*, **65**, 320–332.

Painter, T. H., and Dozier, J., 2004: Measurements of the hemispherical–directional reflectance of snow at fine spectral and angular resolutions, *Journal of Geophysical Research*, **109**, doi: 10.1029/2003JD004458.

Painter T. H., Barrett, A. P., Landry, C. C., Neff, J. C., Cassidy, M. P., Lawrence, C. R., McBride, K. E., and Farmer, G. L., 2007: Impact of disturbed desert soils on duration of mountain snow cover, *Geophysical Research Letters*, **34**, L12502, doi:10.1029/2007GL030284.

Painter, T. H., Rittger, K., McKenzie, C., Slaughter, P., Davis, R. E., and Dozier, J., 2009: Retrieval of subpixel snow covered area, grain size, and albedo from MODIS, *Remote Sensing of Environment*, **113**, 868–879.

Polonsky, I. N., Zege, E. P., Kokhanovsky, A. A., Katsev, I. L., and Prikhach, A. S., 1999: The retrieval of the effective radius of snow grains and control of snow pollution with GLI data, Geoscience and Remote Sensing Symposium, IGARSS '99 Proceedings, IEEE 1999 International, 2, 28 June–2 July 1999, pp. 1071–1073 vol. 2, DOI: 10.1109/IGARSS.1999.774536.

Rozanov, A. A., Rozanov, V. V., Buchwitz, M., Kokhanovsky, A. A., and Burrows, J. P., 2005: SCIATRAN 2.0-a new radiative transfer model for geophysical applications in the 175–2400 nm spectral range, *Advances in Space Research*, **36**, 1015–1019.

Rozanov, V. V., Rozanov, A. V., and Kokhanovsky, A. A., 2007: Derivatives of the radiation field and their application to the solution of inverse problems, in A. A. Kokhanovsky (ed.), *Light Scattering Reviews*, vol. 2, pp. 205–265, Chichester, UK: Praxis.

Sandmeier, S. R., and Itten, K. I., 1999: A field goniometer system (FIGOS) for acquisition of hyperspectral BRDF data, *IEEE Transactions on Geoscience and Remote Sensing*, **37**, 978–986.

Schneebeli, M., and Sokratov, S. A., 2004: Tomography of temperature gradient metamorphism of snow and associated changes in heat conductivity, *Hydrological Processes*, **18**, 3655–3665.

Stamnes, K., Li, W., Eide, H., Aoki, T., Hori, M., and Storvold, R., 2007: ADEOS-II/GLI snow/ice products – Part I: Scientific basis, *Remote Sensing of Environment*, **111**, 258–273.

Tanikawa, T., Aoki, T., Hori, M., Hachikubo, A., and Aniya, M., 2006: Snow bidirectional reflectance model using nonspherical snow particles and its validation with field measurements, *EARSeL Proceedings*, **5**, 137–145.

Tedesco, M., and Kokhanovsky A. A., 2007: The semi-analytical snow retrieval algorithm and its application to MODIS data, *Remote Sensing of Environment*, **110**, 317–331.

Tomasi, C., Vitale, V., and Lupi, A., 2007: Aerosols in polar regions: A historical overview based on optical depth and in situ observations, *Journal of Geophysical Research*, **112**, D16205, doi: 10.1029/2007JD008432.

van de Hulst, H. C., 1957: *Light Scattering by Small Particles*, New York: John Wiley.

Warren, S. G., 1982: Optical properties of snow, *Reviews of Geophysics*, **20**, 67–89.

Warren, S. G., 1984: Optical constants of ice from the ultraviolet to the microwave. *Applied Optics*, **23**, 1206–1225.

Warren S. G., and Brandt, R. E., 2008: Optical constants of ice from the ultraviolet to the microwave: A revised compilation, *Journal of Geophysical Research*, **113**, D14220, doi:10.1029/2007JD009744.

Warren, S. G., and Wiscombe, W. J., 1980: A model for the spectral albedo of snow. II: Snow containing atmospheric aerosols. *Journal of Atmospheric Sciences*, **37**, 2734–2733.

Warren, S. G., Brandt, R. E., and Hinton, P., 1998: Effects of surface roughness on bidirectional reflectance of Antarctic snow, *Journal of Geophysical Research*, **103**, 25789–25807.

Xie, Y., Yang, P., Gao, B.-C., Kattawar, G. W., and Mishchenko, M. I., 2006: Effect of ice crystal shape and effective size on snow bidirectional reflectance, *Journal of Quantitative Spectroscopy and Radiative Transfer*, **100**, 457–469.

Zege, E. P., Ivanov A. P., and Katsev, I. L., 1991: Image transfer through a scattering medium, Berlin: Springer.

Zege, E. P., Kokhanovsky, A. A., Katsev, I. L., Polonsky, I. N., and Prikhach, A. S., 1998: The retrieval of the effective radius of snow grains and control of snow pollution with GLI data, in Proceedings of Conference on Light Scattering by Nonspherical Particles: Theory, Measurements, and Applications, Mishchenko, M. I., Travis, L. D., and Hovenier, J. W., eds. American Meteorological Society, Boston, MA, 288–290.

Zege, E. P., Katsev, I. L., Malinka, A., Prikhach, A. S., and Polonsky, I. N., 2008: New algorithm to retrieve the effective snow grain size and pollution amount from satellite data, *Annals of Glaciology*, **49**, 139–144.

Zhou, X., Li, S., and Stamnes, K., 2003: Effects of vertical inhomogeneity on snow spectral albedo and its implications for remote sensing of snow, *Journal of Geophysical Research*, **108**, 4738, 2003. doi: 10.1029/2003JD003859.

Zhuravleva, T., and Kokhanovsky, A., 2011: Influence of surface roughness on the reflective properties of snow, *J. Quant. Spectr. Rad. Transfer*, **112**, 1353–1368.

Index